ACTIVE™ PHYSICAL SCIENCE

Dr. Arthur Eisenkraft
Gary Freebury

IT's ABOUT TIME®
HERFF JONES EDUCATION DIVISION

84 Business Park Drive, Armonk, NY 10504 Phone (914) 273-2233
Fax (914) 273-2227 Toll Free (888) 698-TIME (8463) www.its-about-time.com

It's About Time, Founder
Laurie Kreindler

Creative/Art Director
John Nordland

Design/Production
Burmar Technical Corporation
Murray Fuchs
Robin Hoffmann
Kadi Sarv
Jennifer Von Holstein

Technical Art
Burmar Technical Corporation
Jennifer Von Holstein
Dennis Falcon

Illustrations
Tomas Bunk

Safety Reviewers
Jack Breazale
Gregory Puskar
Ed Robeck

Project Managers
Barbara Zahm
John Nordland

Project Editor
Ruta Demery

Associate Editors
Carl Heltzel
Al Mari

Physics Reviewers
John Hubrisz
Dwight Neuenschwander
John Roeder
Patty Rourke

Photo Research
Jennifer Von Holstein

Active Physical Science™

All student activities in this textbook have been designed to be as safe as possible, and have been reviewed by professionals specifically for that purpose. As well, appropriate warnings concerning potential safety hazards are included where applicable to particular activities. However, responsibility for safety remains with the student, the classroom teacher, the school principals, and the school board.

Printed and bound in the United States of America

ISBN #1-58591-319-7

1 2 3 4 5 VH 08 07 06 05 04

This project was supported, in part, by the
National Science Foundation

Opinions expressed are those of the authors and not necessarily those of the
National Science Foundation

Project Director, Active Physics

Arthur Eisenkraft has taught high school physics for over 28 years and is currently the Distinguished Professor of Science Education and a Senior Research Fellow at the University of Massachusetts, Boston. Dr. Eisenkraft is the author of numerous science and educational publications. He holds U.S. Patent #4447141 for a Laser Vision Testing System (which tests visual acuity for spatial frequency).

Dr. Eisenkraft has been recognized with numerous awards including: Presidential Award for Excellence in Science Teaching, 1986 from President Reagan; American Association of Physics Teachers (AAPT) Excellence in Pre-College Teaching Award, 1999; AAPT Distinguished Service Citation for "excellent contributions to the teaching of physics", 1989; Science Teacher of the Year, Disney American Teacher Awards in their American Teacher Awards program, 1991; Honorary Doctor of Science degree from Rensselaer Polytechnic Institute, 1993. Tandy Technology Scholar Award 2000.

In 1999 Dr. Eisenkraft was elected to a 3-year cycle as the President-Elect, President and Retiring President of the National Science Teachers Association (NSTA), the largest science teacher organization in the world. In 2003, he was elected a fellow of the American Association for the Advancement of Science (AAAS).

Dr. Eisenkraft has been involved with a number of projects and chaired many competition programs, including: the Toshiba/NSTA ExploraVisions Awards (1991 to the present); the Toyota TAPESTRY Grants (1990 to the present); the Duracell/NSTA Scholarship Competitions (1984 to 2000). He was a columnist and on the Advisory Board of *Quantum* (a science and math student magazine that was published by NSTA as a joint venture between the United States and Russia; 1989 to 2001). In 1993, he served as Executive Director for the XXIV International Physics Olympiad after being Academic Director for the United States Team for six years. He has served on a number of committees of the National Academy of Sciences including the content committee that helped write the National Science Education Standards.

Dr. Eisenkraft has appeared on *The Today Show*, *National Public Radio*, *Public Television*, *The Disney Channel* and numerous radio shows. He serves as an advisor to the ESPN Sports Figures Video Productions.

He is a frequent presenter and keynote speaker at National Conventions. He has published over 100 articles and presented over 200 papers and workshops. He has been featured in articles in *The New York Times*, *Education Week*, *Physics Today*, *Scientific American*, *The American Journal of Physics* and *The Physics Teacher*.

Content Specialist, Active Chemistry

Gary Freebury, a noted chemistry teacher, educator, and writer worked as the Project Manager and Editor for the 3 Prototype chapters of *Active Chemistry* which are currently being field tested and he will continue to serve as the Project Manager and Editor to the project.
He will be responsible for the writing of any introductory materials, producing the table of contents, indices, glossary, and reference materials. He will have a critical role in maintaining the integrity of safety standards across all units. He will also coordinate any modifications, changes, and additions to the materials based upon pilot and field-testing results.

Mr. Freebury has been teaching chemistry for more than 35 years. He has been the Safety Advisor for Montana Schools, past director of the Chemistry Olympiad, past chairman of the Montana Section of the American Chemical Society (ACS), member of the Executive Committee of the Montana Section of the ACS, and a past member of the Montana Science Advisory Council.
Mr. Freebury has been the regional director and author of Scope, Sequence and Coordination (SS&C) – Integrated Science Curriculum and Co-director of the National Science Foundation supported Chemistry Concepts four-year program. He earned a B.S. degree at Eastern Montana College in mathematics and physical science, and an M.S. degree in chemistry at the University of Northern Iowa.

UNIT 1: ACTIVE PHYSICS
PRIMARY AND CONTRIBUTING AUTHORS

Active Physics
was developed by a team of leading physicists, university educators and classroom teachers with financial support from the National Science Foundation.

NSF Program Officer
Gerhard Salinger
Instructional Materials Development (IMD)

Principal Investigators
Bernard V. Khoury
American Association of Physics Teachers

Dwight Edward Neuenschwander
American Institute of Physics

Project Director
Arthur Eisenkraft
University of Massachusetts

Primary and Contributing Authors
Richard Berg
University of Maryland
College Park, MD

Howard Brody
University of Pennsylvania
Philadelphia, PA

Chris Chiaverina
New Trier Township High School
Crystal Lake, IL

Ron DeFronzo
Eastbay Ed. Collaborative
Attleboro, MA

Ruta Demery
Blue Ink Editing
Stayner, ON

Carl Duzen
Lower Merion High School
Havertown, PA

Jon L. Harkness
Active Physics Regional Coordinator
Wausau, WI

Ruth Howes
Ball State University
Muncie, IN

Douglas A. Johnson
Madison West High School
Madison, WI

Ernest Kuehl
Lawrence High School
Cedarhurst, NY

Robert L. Lehrman
Bayside, NY

Salvatore Levy
Roslyn High School
Roslyn, NY

Tom Liao
SUNY Stony Brook
Stony Brook, NY

Charles Payne
Ball State University
Muncie, IN

Mary Quinlan
Radnor High School
Radnor, PA

Harry Rheam
Eastern Senior High School
Atco, NJ

Bob Ritter
University of Alberta
Edmonton, AB, CA

John Roeder
The Calhoun School
New York, NY

John J. Rusch
University of Wisconsin, Superior
Superior, WI

Patty Rourke
Potomac School
McLean, VA

Ceanne Tzimopoulos
Omega Publishing
Medford, MA

Larry Weathers
The Bromfield School
Harvard, MA

David Wright
Tidewater Comm. College
Virginia Beach, VA

Consultants
Peter Brancazio
Brooklyn College of CUNY
Brooklyn, NY

Robert Capen
Canyon del Oro High School
Tucson, AZ

Carole Escobar

Earl Graf
SUNY Stony Brook
Stony Brook, NY

Jack Hehn
American Association of Physics Teachers
College Park, MD

Donald F. Kirwan
Louisiana State University
Baton Rouge, LA

Gayle Kirwan
Louisiana State University
Baton Rouge, LA

James La Porte
Virginia Tech
Blacksburg, VA

Charles Misner
University of Maryland
College Park, MD

Robert F. Neff
Suffern, NY

Ingrid Novodvorsky
Mountain View High School
Tucson, AZ

John Robson
University of Arizona
Tucson, AZ

Mark Sanders
Virginia Tech
Blacksburg, VA

Brian Schwartz
Brooklyn College of CUNY
New York, NY

Bruce Seiger
Wellesley High School
Newburyport, MA

Clifford Swartz
SUNY Stony Brook
Setauket, NY

Barbara Tinker
The Concord Consortium
Concord, MA

Robert E. Tinker
The Concord Consortium
Concord, MA

Joyce Weiskopf
Herndon, VA

Donna Willis
American Association of Physics Teachers
College Park, MD

Safety Reviewer
Gregory Puskar
University of West Virginia
Morgantown, WV

Equity Reviewer
Leo Edwards
Fayetteville State University
Fayetteville, NC

Physics at Work

Alex Straus
writer
New York, NY

Mekea Hurwitz
photographer

Physics InfoMall

Brian Adrian
Bethany College
Lindsborg, KS

First Printing Reviewer

John L. Hubisz
North Carolina State University
Raleigh, NC

Unit Reviewers

Robert Adams
Polytech High School
Woodside, DE

George A. Amann
F.D. Roosevelt High School
Rhinebeck, NY

Patrick Callahan
Catasaugua High School
Center Valley, PA

Beverly Cannon
Science and Engineering
Magnet High School
Dallas, TX

Barbara Chauvin

Elizabeth Chesick
The Baldwin School
Haverford, PA

Chris Chiaverina
New Trier Township High School
Crystal Lake, IL

Andria Erzberger
Palo Alto Senior High School
Los Altos Hills, CA

Elizabeth Farrell Ramseyer
Niles West High School
Skokie, IL

Mary Gromko
President of Council of State Science
Supervisors
Denver, CO

Thomas Guetzloff

Jon L. Harkness
Active Physics Regional Coordinator
Wausau, WI

Dawn Harman
Moon Valley High School
Phoenix, AZ

James Hill
Piner High School
Sonoma, CA

Bob Kearney

Claudia Khourey-Bowers
McKinley Senior High School

Steve Kliewer
Bullard High School
Fresno, CA

Ernest Kuehl
Roslyn High School
Cedarhurst, NY

Jane Nelson
University High School
Orlando, FL

Mary Quinlan
Radnor High School
Radnor, PA

John Roeder
The Calhoun School
New York, NY

Patty Rourke
Potomac School
McLean, VA

Gerhard Salinger
Fairfax, VA

Irene Slater
La Pietra School for Girls

Pilot Test Teachers

John Agosta

Donald Campbell
Portage Central High School
Portage, MI

John Carlson
Norwalk Community
Technical College
Norwalk, CT

Veanna Crawford
Alamo Heights High School
New Braunfels, TX

Janie Edmonds
West Milford High School
Randolph, NJ

Eddie Edwards
Amarillo Area Center for Advanced
Learning
Amarillo, TX

Arthur Eisenkraft
Fox Lane High School
Bedford, NY

Tom Ford

Bill Franklin

Roger Goerke
St. Paul, MN

Tom Gordon
Greenwich High School
Greenwich, CT

Ariel Hepp

John Herrman
College of Steubenville
Steubenville, OH

Linda Hodges

Ernest Kuehl
Lawrence High School
Cedarhurst, NY

Fran Leary
Troy High School
Schenectady, NY

Harold Lefcourt

Cherie Lehman
West Lafayette High School
West Lafayette, IN

Kathy Malone
Shady Side Academy
Pittsburgh, PA

Bill Metzler
Westlake High School
Thornwood, NY

Elizabeth Farrell Ramseyer
Niles West High School
Skokie, IL

Daniel Repogle
Central Noble High School
Albion, IN

Evelyn Restivo
Maypearl High School
Maypearl, TX

Doug Rich
Fox Lane High School
Bedford, NY

John Roeder
The Calhoun School
New York, NY

Tom Senior
New Trier Township High School
Highland Park, IL

John Thayer
District of Columbia Public Schools
Silver Spring, MD

Carol-Ann Tripp
Providence Country Day
East Providence, RI

Yvette Van Hise
High Tech High School
Freehold, NJ

Jan Waarvick

Sandra Walton
Dubuque Senior High School
Dubuque, IA

Larry Wood
Fox Lane High School
Bedford, NY

Field Test Coordinator

Marilyn Decker
Northeastern University
Acton, MA

Field Test Workshop Staff

John Carlson

Marilyn Decker

Arthur Eisenkraft

Douglas Johnson

John Koser

Ernest Kuehl

Mary Quinlan

Elizabeth Farrell Ramseyer

John Roeder

Field Test Evaluators

Susan Baker-Cohen

Susan Cloutier

George Hein

Judith Kelley

all from Lesley College,
Cambridge, MA

Field Test Teachers and Schools

Rob Adams
Polytech High School
Woodside, DE

Benjamin Allen
Falls Church High School
Falls Church, VA

Robert Applebaum
New Trier High School
Winnetka, IL

Joe Arnett
Plano Sr. High School
Plano, TX

Bix Baker
GFW High School
Winthrop, MN

Debra Beightol
Fremont High School
Fremont, NE

Patrick Callahan
Catasaugua High School
Catasaugua, PA

George Coker
Bowling Green High School
Bowling Green, KY

Janice Costabile
South Brunswick High School
Monmouth Junction, NJ

Stanley Crum
Homestead High School
Fort Wayne, IN

Russel Davison
Brandon High School
Brandon, FL

Christine K. Deyo
Rochester Adams High School
Rochester Hills, MI

Jim Doller
Fox Lane High School
Bedford, NY

Jessica Downing
Esparto High School
Esparto, CA

Douglas Fackelman
Brighton High School
Brighton, CO

Rick Forrest
Rochester High School
Rochester Hills, MI

Mark Freeman
Blacksburg High School
Blacksburg, VA

Jonathan Gillis
Enloe High School
Raleigh, NC

Karen Gruner
Holton Arms School
Bethesda, MD

Larry Harrison
DuPont Manual High School
Louisville, KY

Alan Haught
Weaver High School
Hartford, CT

Steven Iona
Horizon High School
Thornton, CO

Phil Jowell
Oak Ridge High School
Conroe, TX

Deborah Knight
Windsor Forest High School
Savannah, GA

Thomas Kobilarcik
Marist High School
Chicago, IL

Sheila Kolb
Plano Senior High School
Plano, TX

Todd Lindsay
Park Hill High School
Kansas City, MO

Malinda Mann
South Putnam High School
Greencastle, IN

Steve Martin
Maricopa High School
Maricopa, AZ

Nancy McGrory
North Quincy High School
N. Quincy, MA

David Morton
Mountain Valley High School
Rumford, ME

Charles Muller
Highland Park High School
Highland Park, NJ

Fred Muller
Mercy High School
Burlingame, CA

Vivian O'Brien
Plymouth Regional High School
Plymouth, NH

Robin Parkinson
Northridge High School
Layton, UT

Donald Perry
Newport High School
Bellevue, WA

Francis Poodry
Lincoln High School
Philadelphia, PA

John Potts
Custer County District
High School
Miles City, MT

Doug Rich
Fox Lane High School
Bedford, NY

John Roeder
The Calhoun School
New York, NY

Consuelo Rogers
Maryknoll Schools
Honolulu, HI

Lee Rossmaessler
Mott Middle College High School
Flint, MI

John Rowe
Hughes Alternative Center
Cincinnati, OH

Rebecca Bonner Sanders
South Brunswick High School
Monmouth Junction, NJ

David Schilpp
Narbonne High School
Harbor City, CA

Eric Shackelford
Notre Dame High School
Sherman Oaks, CA

Robert Sorensen
Springville-Griffith Institute
and Central School
Springville, NY

Teresa Stalions
Crittenden County High School
Marion, KY

Roberta Tanner
Loveland High School
Loveland, CO

Anthony Umelo
Anacostia Sr. High School
Washington, D.C.

Judy Vondruska
Mitchell High School
Mitchell, SD

Deborah Waldron
Yorktown High School
Arlington, VA

Ken Wester
The Mississippi School for Mathematics and Science
Columbus, MS

Susan Willis
Conroe High School
Conroe, TX

UNIT 2: ACTIVE CHEMISTRY
PRIMARY AND CONTRIBUTING AUTHORS

Active Chemistry Writers

Gary Freebury
Kalispell, MT

Mary Gromko
Colorado Springs, CO

Carl Heltzel
Transylvania University
Lexington, KY

John Roeder
The Calhoun School
New York, NY

Hannah Sevian
University of Massachusetts – Boston Campus, Boston, MA

Sandra Smith
Colorado Springs, CO

Michael Tinnesand
American Chemical Society
Washington, D.C.

Active Chemistry Consultants

James Davis
Chemistry Professor Emeritus Harvard University

George Miller
University of California at Irvine
Irvine, CA

Carlo Parravano
Merck Institute for Science Education
Rahway, NJ

Maren Reeder
Merck Institute for Science Education
Rahway, NJ

Active Chemistry Pilot Testers

Rob Adams
Wyoming, DE

Ina Ahern
Plymouth Regional High School
Plymouth, NH

John Bibb
Georgetown, DE

Amy Biddle
Pinkerton Academy
Derry, NH

Robert Dayton
Rush-Henrietta High School
Henrietta, NY

Barbara Duch
Education Resource Ctr.
Newark, DE

Gabriel Duque
North Miami Senior High School
Miami, FL

Liz Garcia
Carson High School
Carson, CA

Laura Hajdukiewicz
The Bromfield School
Harvard, MA

Jonathon Haraty
SAGE School
Springfield, MA

Carl Heltzel
Transylvania University
Lexington, KY

Natalie Hiller
Philadelphia Public Schools
Philadelphia, PA

Penny Hood
Stanwood High School
Stanwood, WA

Tamilyn Ingram
Menifee County High School
Frenchburg, KY

Barbara Jeffries
Casey Co. High School
Liberty, KY

Diane Johnson
Lewis County High School
Vanceburg, KY

Gerry LaFontaine
Toll Gate High School
Warwick, RI

Jo Larmore
Laurel, DE

Jeffrey Little
Pikeville High School
Pikeville, KY

Kathy Lucas
Casey Co. High School
Liberty, KY

Barbara Malkas
Taconic High School
Pittsfield, MA

Barbara Martin
Reading High School
Reading, MA

Robert Mayton
Allen Central High School
Eastern, KY

Vicki Mockbee
Caesar Rodney High School
Camden, DE

Brenda Mullins
Knott County Central High School
Hindman, KY

Jim Nash
Stanwood High School
Stanwood, WA

Barry North
Rivendell Interstate Regional High School
Orford, NH

Angela Pence
Wolfe County High School
Campton, KY

Roy Penix
Prestonburg High School
Prestonburg, KY

Robin Ringland
Stanwood High School
Stanwood, WA

George Robertson
Stanwood High School
Stanwood, WA

Lance Rudiger
Potsdam High School
Potsdam, NY

John Scali
Newark, DE

Noreen Scarpitto
Reading High School
Reading, MA

Fred Schiess
Stanwood High School
Stanwood, WA

Hannah Sevian
Chelsea High School
Chelsea, MA

Angela Skaggs
Magoffin County High School
Salyersville, KY

Mary Stoukides
Toll Gate High School
Warwick, RI

Jim Swanson
Saugus High School
Saugus, MA

Kathy Swingle
Rehoboth, DE

Jeffrey Scott Townsend
Powell Co. High School
Stanton, KY

Josh Underwood
Deming High School
Mount Olivet, KY

Elaine Weil
Berlin, MD

Jen Wilson
Wilmington, DE

You can do any science. Here are the reasons why. . .

The following features make it that much easier to understand the science principles you will be studying. You will find that regardless of which science discipline you approach, whether it be physics, or chemistry, the way you go about learning each of them is the same. Every step along the way will prepare you for the following step. Using all these features together will help you actually learn about science and see how it works for you every day, everywhere. Look for these features in each chapter of *Active Physical Science.*

4

Thrills and Chills

Scenario

You are excited and scared as you sit back into the seat. You pull the safety restraints into place. The next thing you know, you are beginning a slow but steady ascent into the sky. Then, with a sudden jolt, you reach the top. This is where the thrill or nightmare begins. You hurtle down the track at ever-increasing speeds. You are flung against one side of your seat as you scream around a curve. You shriek as you hang upside down, fortunately, firmly secured to your seat. All the time, your stomach has no idea where it is or you are. Finally, you come to rest where you began. What a ride! Want to go again?

Roller coasters have been enjoyed for many years. However, the roller coaster that may appeal to you, may not appeal to your parents or other friends and relatives.

206

Challenge

A roller coaster, called the Terminator Express, has been designed for an amusement park. Your challenge is to take the roller coaster design and modify it for a select group of riders. For instance, you may decide that you will modify the roller coaster so that young children can experience the thrill of a roller coaster in a safe and non-threatening way. You may prefer to design the roller coaster for adults that are a bit squeamish about the big hills and sharp turns. They want to experience the thrill of a roller coaster ride, but are ready to pass on the death-defying action. On the other hand, you may choose to design a roller coaster for daredevils that are ready to handle any thrill you can provide. You may also wish to design a roller coaster for people who are physically challenged or visually impaired. However, in each situation, you must ensure that the roller coaster is safe and nobody will be in danger.

Criteria

You will make the design based on whatever group you choose. Your design will have to meet certain criteria. All designs must have certain components including:

- **at least two hills**
- **one horizontal curve**

In addition, you will have to provide evidence that the ride is safe. Safety data will include the height, speed, and acceleration of the roller coaster at five designated locations. Finally, you will have to calculate the work required to get the roller coaster rolling.

You will present your design as both a model and a written poster or report.

Make a list of criteria for each of the required elements of the roller coaster. What should a roller coaster design that deserves an "A" look like? Are there other elements that you think should be included in a roller coaster grade? How much should each element be worth? Discuss these questions with your small group and the class.

Active Physical Science

1 Scenario

Each chapter begins with a realistic event or situation. You might actually have experienced the event, or you can imagine yourself participating in a similar situation at home, in school, or in your community. Chances are you probably never thought about the science involved in each case.

2 Challenge

This feature presents you with a challenge that you can expect to complete by the end of the chapter. As you progress through the chapter you will accumulate all the knowledge you need to successfully complete the challenge.

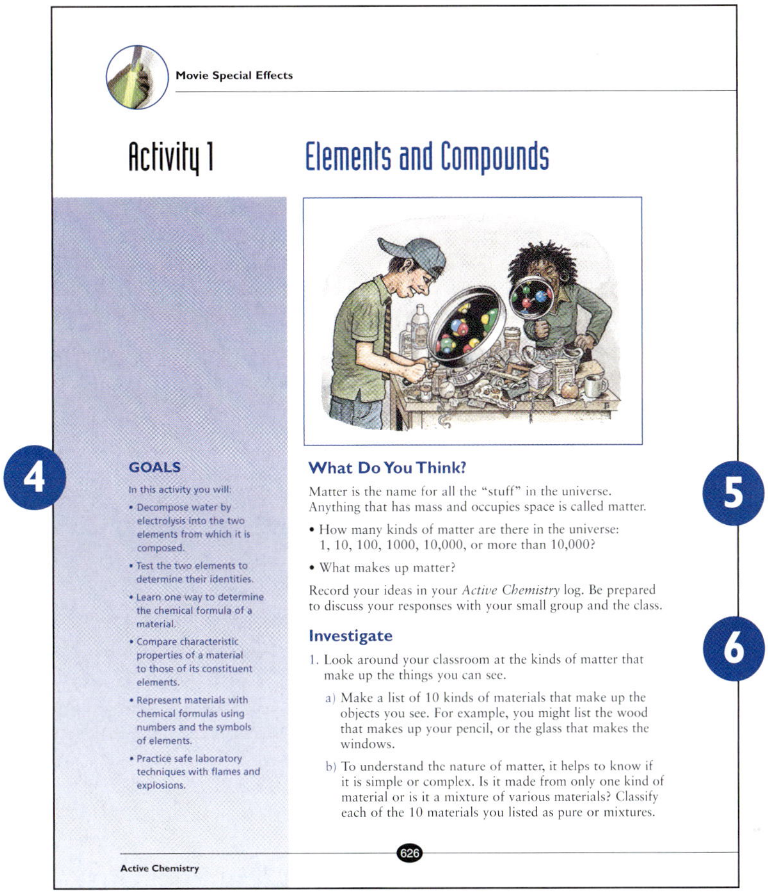
Movie Special Effects

Activity 1 Elements and Compounds

GOALS

In this activity you will:

- Decompose water by electrolysis into the two elements from which it is composed.
- Test the two elements to determine their identities.
- Learn one way to determine the chemical formula of a material.
- Compare characteristic properties of a material to those of its constituent elements.
- Represent materials with chemical formulas using numbers and the symbols of elements.
- Practice safe laboratory techniques with flames and explosions.

What Do You Think?

Matter is the name for all the "stuff" in the universe. Anything that has mass and occupies space is called matter.

- How many kinds of matter are there in the universe: 1, 10, 100, 1000, 10,000, or more than 10,000?
- What makes up matter?

Record your ideas in your *Active Chemistry* log. Be prepared to discuss your responses with your small group and the class.

Investigate

1. Look around your classroom at the kinds of matter that make up the things you can see.
   a) Make a list of 10 kinds of materials that make up the objects you see. For example, you might list the wood that makes up your pencil, or the glass that makes the windows.
   b) To understand the nature of matter, it helps to know if it is simple or complex. Is it made from only one kind of material or is it a mixture of various materials? Classify each of the 10 materials you listed as pure or mixtures.

Active Chemistry 626

3 Criteria

Before you begin the chapter and the challenge, you and your classmates, along with your teacher, will explore exactly how you will be graded. You will review the criteria and expectations for solving the challenge, and make decisions about how your work should be evaluated.

4 Goals

At the beginning of each activity you are provided with a list of goals that you should be able to achieve by completing your science inquiry.

5 What Do You Already Know?

Before you start each activity you will be asked one or two questions to consider. You will have a chance to discuss your ideas with your group and your class. You are not expected to come up with the "right" answer, but to share your current understanding and reasoning.

6 Investigate

In *Active Physical Science* you learn by doing science. In your small groups, or as a class, you will take part in scientific inquiry by doing hands-on experiments, participating in fieldwork, or searching for answers using the Internet and reference materials.

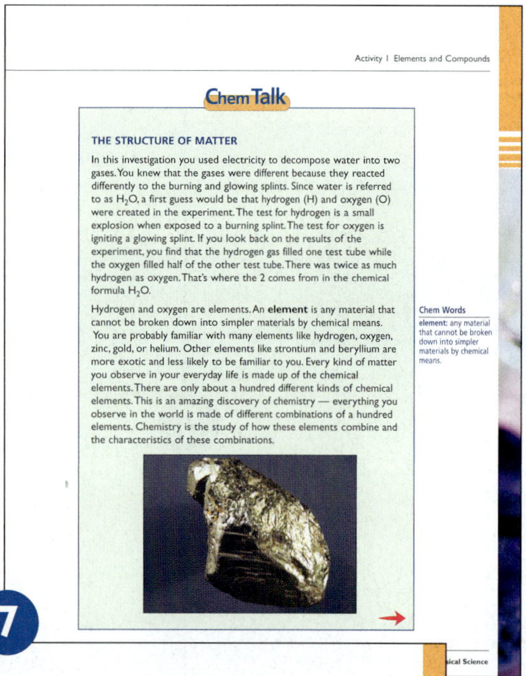

Activity 1 Elements and Compounds

Chem Talk

THE STRUCTURE OF MATTER

In this investigation you used electricity to decompose water into two gases. You knew that the gases were different because they reacted differently to the burning and glowing splints. Since water is referred to as H_2O, a first guess would be that hydrogen (H) and oxygen (O) were created in the experiment. The test for hydrogen is a small explosion when exposed to a burning splint. The test for oxygen is igniting a glowing splint. If you look back on the results of the experiment, you find that the hydrogen gas filled one test tube while the oxygen filled half of the other test tube. There was twice as much hydrogen as oxygen. That's where the 2 comes from in the chemical formula H_2O.

Hydrogen and oxygen are elements. An **element** is any material that cannot be broken down into simpler materials by chemical means. You are probably familiar with many elements like hydrogen, oxygen, zinc, gold, or helium. Other elements like strontium and beryllium are more exotic and less likely to be familiar to you. Every kind of matter you observe in your everyday life is made up of the chemical elements. There are only about a hundred different kinds of chemical elements. This is an amazing discovery of chemistry — everything you observe in the world is made of different combinations of a hundred elements. Chemistry is the study of how these elements combine and the characteristics of these combinations.

Chem Words

element: any material that cannot be broken down into simpler materials by chemical means.

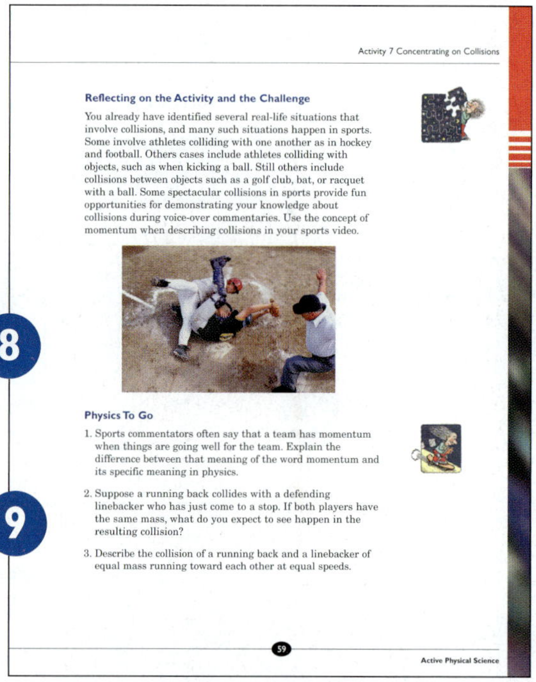

Activity 7 Concentrating on Collisions

Reflecting on the Activity and the Challenge

You already have identified several real-life situations that involve collisions, and many such situations happen in sports. Some involve athletes colliding with one another as in hockey and football. Others cases include athletes colliding with objects, such as when kicking a ball. Still others include collisions between objects such as a golf club, bat, or racquet with a ball. Some spectacular collisions in sports provide fun opportunities for demonstrating your knowledge about collisions during voice-over commentaries. Use the concept of momentum when describing collisions in your sports video.

Physics To Go

1. Sports commentators often say that a team has momentum when things are going well for the team. Explain the difference between that meaning of the word momentum and its specific meaning in physics.
2. Suppose a running back collides with a defending linebacker who has just come to a stop. If both players have the same mass, what do you expect to see happen in the resulting collision?
3. Describe the collision of a running back and a linebacker of equal mass running toward each other at equal speeds.

59 Active Physical Science

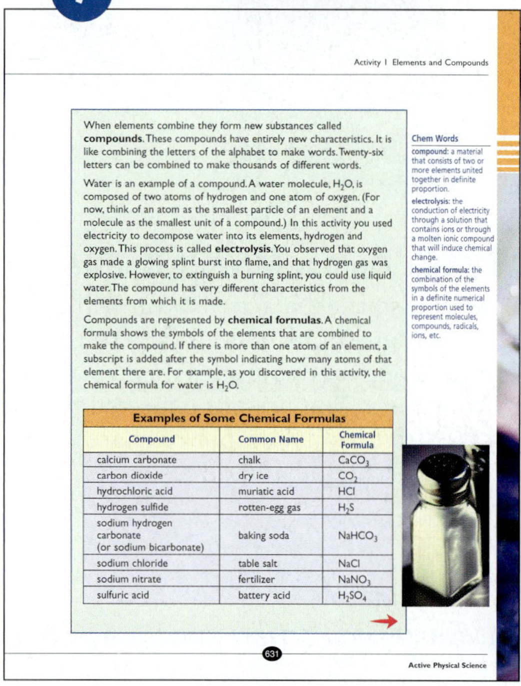

Activity 1 Elements and Compounds

When elements combine they form new substances called **compounds**. These compounds have entirely new characteristics. It is like combining the letters of the alphabet to make words. Twenty-six letters can be combined to make thousands of different words.

Water is an example of a compound. A water molecule, H_2O, is composed of two atoms of hydrogen and one atom of oxygen. (For now, think of an atom as the smallest particle of an element and a molecule as the smallest unit of a compound.) In this activity you used electricity to decompose water into its elements, hydrogen and oxygen. This process is called **electrolysis**. You observed that oxygen gas made a glowing splint burst into flame, and that hydrogen gas was explosive. However, to extinguish a burning splint, you could use liquid water. The compound has very different characteristics from the elements from which it is made.

Compounds are represented by **chemical formulas**. A chemical formula shows the symbols of the elements that are combined to make the compound. If there is more than one atom of an element, a subscript is added after the symbol indicating how many atoms of that element there are. For example, as you discovered in this activity, the chemical formula for water is H_2O.

Examples of Some Chemical Formulas

Compound	Common Name	Chemical Formula
calcium carbonate	chalk	$CaCO_3$
carbon dioxide	dry ice	CO_2
hydrochloric acid	muriatic acid	HCl
hydrogen sulfide	rotten-egg gas	H_2S
sodium hydrogen carbonate (or sodium bicarbonate)	baking soda	$NaHCO_3$
sodium chloride	table salt	NaCl
sodium nitrate	fertilizer	$NaNO_3$
sulfuric acid	battery acid	H_2SO_4

Chem Words

compound: a material that consists of two or more elements united together in definite proportion.

electrolysis: the conduction of electricity through a solution that contains ions or through a molten ionic compound that will induce chemical change.

chemical formula: the combination of the symbols of the elements in a definite numerical proportion used to represent molecules, compounds, radicals, ions, etc.

631 Active Physical Science

7 Reading Sections

These sections provide text, illustrations, and photographs that will give you a greater insight into the concepts you explored in the activity. Equations and formulas are provided with easy-to-understand explanations. Science Words that may be new or unfamiliar to you are defined and explained. In some chapters Checking Up questions are included to guide you in your reading.

8 Reflecting on the Activity and the Challenge

Each activity helps to prepare you to be successful in the challenge. This feature gives you a brief summary of the activity. It will help you relate the activity that you just completed to the challenge. It's another piece of the chapter jigsaw puzzle.

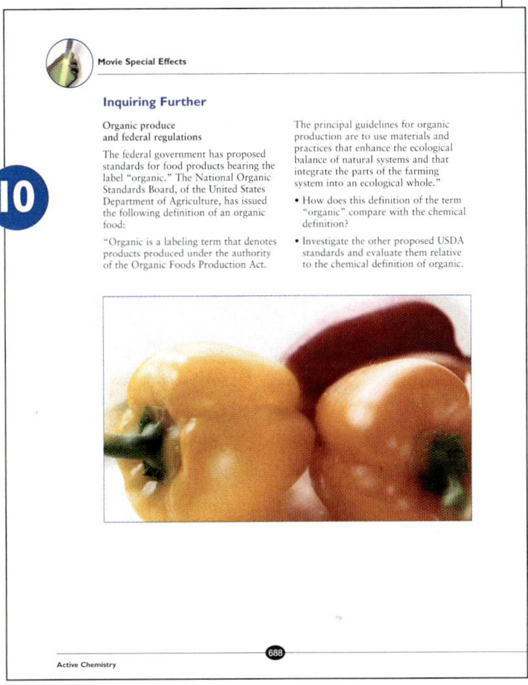

10

Movie Special Effects

Inquiring Further

Organic produce and federal regulations

The federal government has proposed standards for food products bearing the label "organic." The National Organic Standards Board, of the United States Department of Agriculture, has issued the following definition of an organic food:

"Organic is a labeling term that denotes products produced under the authority of the Organic Foods Production Act. The principal guidelines for organic production are to use materials and practices that enhance the ecological balance of natural systems and that integrate the parts of the farming system into an ecological whole."

- How does this definition of the term "organic" compare with the chemical definition?
- Investigate the other proposed USDA standards and evaluate them relative to the chemical definition of organic.

Active Chemistry 688

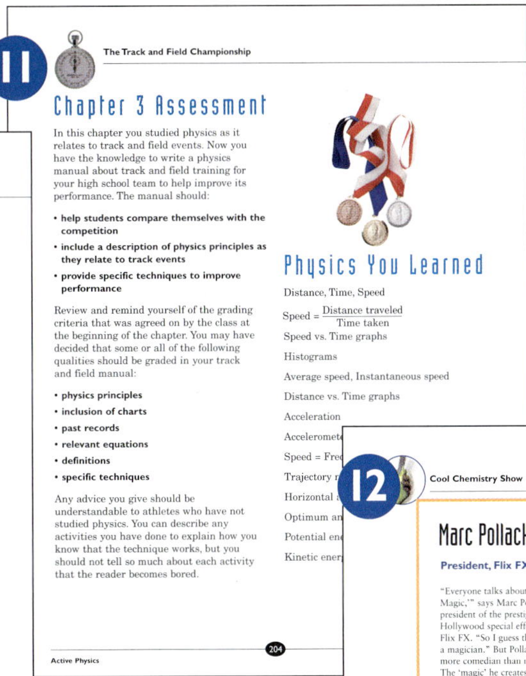

11

The Track and Field Championship

Chapter 3 Assessment

In this chapter you studied physics as it relates to track and field events. Now you have the knowledge to write a physics manual about track and field training for your high school team to help improve its performance. The manual should:

- help students compare themselves with the competition
- include a description of physics principles as they relate to track events
- provide specific techniques to improve performance

Review and remind yourself of the grading criteria that was agreed on by the class at the beginning of the chapter. You may have decided that some or all of the following qualities should be graded in your track and field manual:

- physics principles
- inclusion of charts
- past records
- relevant equations
- definitions
- specific techniques

Any advice you give should be understandable to athletes who have not studied physics. You can describe any activities you have done to explain how you know that the technique works, but you should not tell so much about each activity that the reader becomes bored.

Physics You Learned

Distance, Time, Speed

$$\text{Speed} = \frac{\text{Distance traveled}}{\text{Time taken}}$$

Speed vs. Time graphs

Histograms

Average speed, Instantaneous speed

Distance vs. Time graphs

Acceleration

Acceleromet

Speed = Fre

Trajectory r

Horizontal

Optimum an

Potential en

Kinetic ener

Active Physics 204

12

Cool Chemistry Show

Chemistry at Work

Marc Pollack

President, Flix FX

"Everyone talks about 'Movie Magic,'" says Marc Pollack, president of the prestigious Hollywood special effects company Flix FX. "So I guess that makes me a magician." But Pollack is clearly more comedian than magician. The 'magic' he creates for movies like *Blackhawk Down*, *Men In Black* and *Cast Away*, in addition to scores of television commercials, museum installments and Las Vegas casinos, is the product not of mysterious hocus-pocus but rather fundamental principles of science. "One of the most important aspects of our work," he continues, "is to push the limits of how chemicals are designed to be used." Among other things, Pollack and his crew at Flix FX use vacuum-forming thermo-plastics to make tin-based silicon molds for everything from prehistoric creatures to futuristic robots. Through a combination of trial and error experimentation and traditional research science, they've perfected the process. "Silicon is what we call an R.T.V.," Pollack explains. "That stands for room temperature vulcanization. So depending on the type and amount of catalyst we use, the mold will cure at different rates and with slightly differing properties." By manipulating the ratio of silicon to catalysts they can make strong, realistic molds in the most efficient way possible. "Increasing the amount of catalyst will speed up the curing process but too much catalyst will shorten the life of the mold," he says. "Every job is different so determining that balance is one of our many challenges."

Pollack, who is now a master in the art of using chemicals like silicon, polypropylene, urethane and urethane elastimers, is not a chemist by trade. He actually graduated from film school at SUNY Purchase in the hopes of becoming "the next Steven Spielberg." Then, through a twist of fate, he became a special effects nut and eventually founded Flix FX in 1990. "Now," Pollack says, "Spielberg may one day come to me."

Special effects — Pollack creates both mechanical and physical — is an industry in a constant state of transformation. "The industry is always trying new stuff and that's exciting," Pollack says. "For instance, someone just developed a great water-based breakaway glass for stunts called Smash Glass. It's similar to fiberglass without the dangerous elements associated with that material and can be made to break into either large chunks or tiny little pieces. I can't wait to get my hands on it and break it over someone's head. That's part of my job these days and I love it."

Active Chemistry 832

9 Science to Go

Questions in this feature ask you to use the key principles and concepts introduced in the activity. You may also be presented with new situations in which you will be asked to apply what you have learned. They are excellent as a study guide, helping you review and understand what is most important from the activity. You will also be provided with suggestions for ways in which you can organize your work and get ready for the challenge.

10 Inquiring Further

This feature stretches your thinking. It provides lots of suggestions for deepening your understanding of the concepts and skills developed in the activity. Also, if you're looking for more challenging or in-depth problems, questions, and exercises, you'll find them right here.

11 Chapter Assessment

How do you measure up? Here is your opportunity to share what you have actually learned. Using the activities as a guide, you can now complete the challenge you were presented at the beginning of the chapter.

12 Science at Work

Science is an integral part of many fascinating careers. This feature introduces some people working in different fields that involve the principles of science.

Unit 1

ACTIVE PHYSICS

Chapter 1: Physics in Action

Chapter 2: Safety

Chapter 3: The Track and Field Championship

Chapter 4: Thrills and Chills

Chapter 5: Let Us Entertain You

Chapter 6: Designing the Universal Dwelling

Chapter 7: Electricity For Everyone

Chapter 8: Toys for Understanding

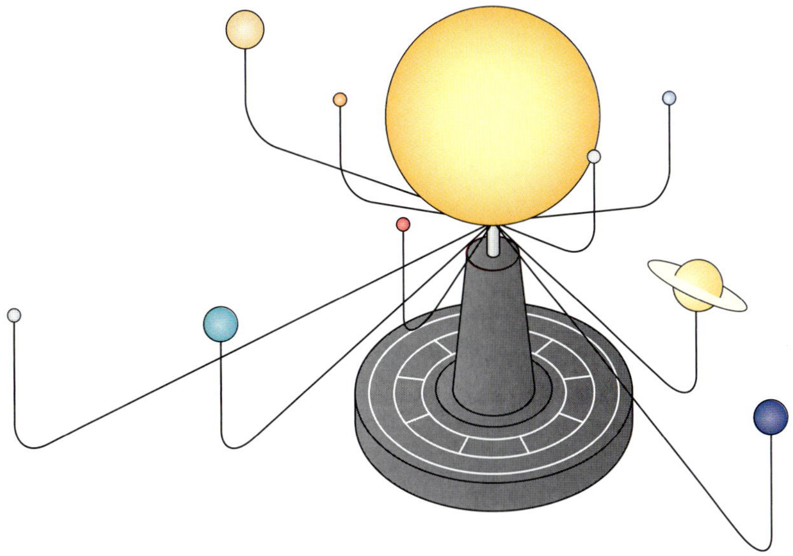

Chapter 9: Atoms on Display

Unit 2

ACTIVE CHEMISTRY

Chapter 10: Movie Special Effects

Chapter 11: The Periodic Table

Chapter 12: Cool Chemistry Show

Glossaries

INQUIRY

Would you like present a Cool Chemistry Show to wow a group of elementary students? Look for and communicate with extraterrestrials? Or develop a booklet to help people to decide how to vote? This year science will be filled with fascinating challenges. This is *Active Physics*, and *Active Chemistry*. This is *Active Physical Science*. It's a science course that introduces you to the fundamental laws and theories of two large disciplines of science and asks that you use your creativity, interest, and expertise to complete challenges in the same way that professionals do.

When scientists, artists, musicians, writers, and engineers are hired to solve a problem, their first responsibility is to understand the nature of the assignment. Often, they must do research to better acquaint themselves with the topic. The person assigning the job has a sense of what is needed, but does not know exactly what the final product will be. For instance, a television producer may need an

announcer for a major upcoming sports event. The producer would like the sports event to have some educational as well as entertainment value. He or she must look for someone who not only has knowledge about the sport, but also have knowledge about the physics of the sport. In the first chapter of *Active Physics—Unit 1*, you will be required to do a "science commentary" on a short sports video. To complete this project, you will be helped along the way with activities that will give you the knowledge and expertise to be successful.

As you complete Chapter Challenges, you will begin to combine your science knowledge with your other interests and create something that will represent the best efforts of you and your team. You will be learning about the disciplines of physics, and chemistry. You will investigate how waves produce music. You will discover how the behavior of atoms and electrons can make some foods delicious and others poisonous.

Hundreds of teachers, professors, artists, editors, and science education professionals who have worked on these materials have taken into account the best information on how students can achieve and how students can be engaged in their own learning. We have taken the philosophy and recommendations of the National Science Education Standards and created a course that will engage you in asking the types of questions that scientists ask; will have you explore the world in the way that scientists explore the world; and will require you to show us that you really understand the meaning and implications of the scientific principles.

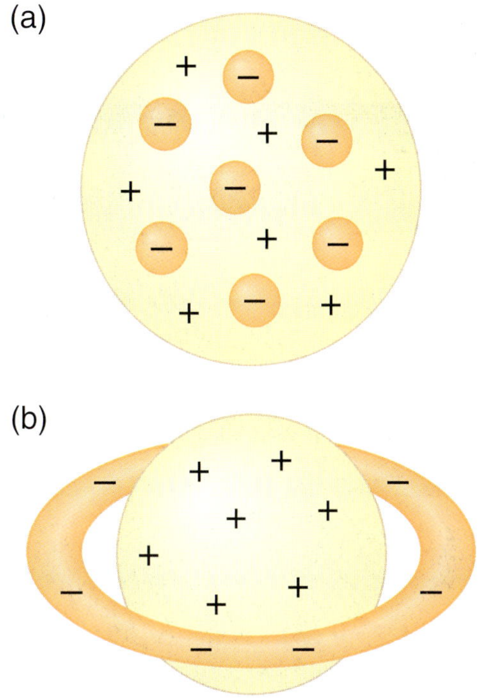

The development of the chapters in this book has been funded by the National Science Foundation. They have gone through a process that included feedback from students much like you in classrooms across the country. Based on that feedback, the materials have been improved so that you can learn science, enjoy what you are learning, and be successful in the course.

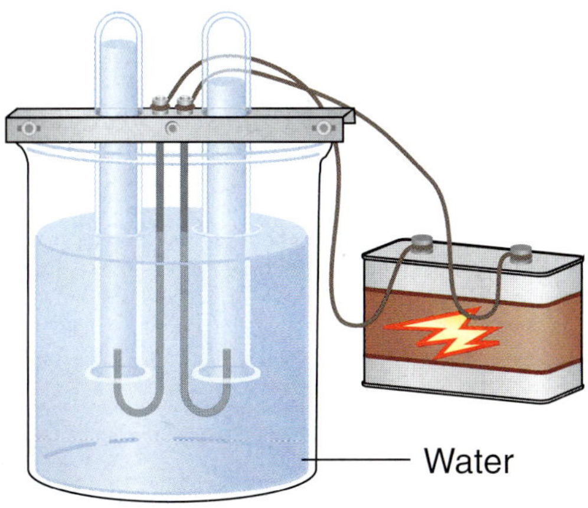

This year will be an introduction to two major sciences. Scientists in each discipline have ways of learning that are unique to their disciplines and yet the sciences are also similar. The energy

released in an earthquake travels in waves, so does sound, as does the light that is given off when electrons move from one orbit to another. Measuring the waves involved in each of these examples requires different tools and different apparatus. You will be introduced to these sciences and will become acquainted with the similarities and differences among the disciplines.You will begin to think like a scientist.

Unit 1

Active Physics®

Chapter 1: Physics in Action

Chapter 2: Safety

Chapter 3: The Track and Field Championship

Chapter 4: Thrills and Chills

Chapter 5: Let Us Entertain You

Chapter 6: Designing the Universal Dwelling

Chapter 7: Electricity for Everyone

Chapter 8: Toys for Understanding

Chapter 9: Atoms on Display

Making Connections

You are about to begin a very active year continuing your exploration of science. This year you will get chances to explore the world around you in a very meaningful way. Active Physical Science is a science course that introduces you to two large disciplines of science—physics and chemistry.

Science is an attempt to find out the secrets of nature. How can air bags help to save lives? How does a roller coaster car stay on the track when it is upside down? How do flour, milk, butter, and sugar become a cake? Physical scientists investigate everything from the structure of an atom to the formation of the universe. Enjoy your active investigation of physical science.

What Do You Think?

- What is science?
- How are physics and chemistry similar? How are they different?

Record your ideas about this in your log. Be prepared to discuss your response with your small group and the class.

For You To Try

- Bertrand Russell (1872-1970) an English philosopher and mathematician said that, "Science is what you know. Philosophy is what you don't know." Research to find quotes from other famous people about what they think that science is. Which one do you think comes closest to what you think that science is?
- Rarely does a week go by without a new scientific study or discovery being reported in the media. Review the news for the last few months to find what new studies or discoveries have been make recently. Research and report on one of these.

Chapter 1

PHYSICS IN ACTION

Scenario

Have you ever dreamed of being a sports analyst for Monday Night Football with millions of people listening to every word you say? What about the sports commentator for the Summer Olympics? Imagine interviewing the Most Valuable Player (MVP) after an NBA championship game or interviewing the Olympic Gold Medalist in women's figure skating. What type of credentials are needed to have such a glamorous career in sportscasting? Should you major in journalism in college or be a retired professional athlete if you desire to land such a lucrative and exciting job? Could the study of physics be a key to becoming a sports analyst? Could a student with a knowledge of physics bring to the TV viewer a different perspective that might provide a new outlook on sporting events?

Challenge

PBS has decided that it wants to televise certain sporting events and that they would like these programs to have some educational as well as entertainment value. As a test of this idea, you are to provide the voice-over on a sports video and to explain the physics of the action appearing on the screen. Here is your chance to audition for a job in sportscasting. Each student (or group of students) will do a "science commentary" on a short (2–3 minute) sports video.

To assess how well you understand this material, you (or your group) are to do one of these:

- **submit a written script**
- **narrate live**
- **dub onto the video soundtrack**
- **record on an audiocassette**

Your task is not to give a play-by-play description of the sporting event or give the rules of the game but rather to go a step beyond and educate the audience by describing to them the rules of nature that govern the event. This approach will give the viewer (and you) a different perspective on both sports and physics. The laws of physics cover not only obscure phenomena in the lab, but everyday events in the real world as well.

Criteria

What criteria should be used to evaluate a voice-over dialogue or script of a sporting event? Since the intention is to provide an analysis of and interest in the physics of sports, the voice-over should include the use of physics terms and physics principles. All of these terms and principles should be used correctly. How many of these terms and principles would constitute an excellent job? Would it be enough to use one physics term correctly and explain how one physics principle is illustrated in the sport? Should use of one physics term and one physics principle be a minimum standard to get minimal credit for this assessment? Discuss in your small groups and your class and decide on reasonable expectations for the physics criteria for the assessment.

Since the assessment requires a product that will be a part of television, another aspect of the criteria for success would be the entertainment quality of the voice-over. Does a commentator who adds humor or drama receive a higher rating than someone who has similar physics content but has added no excitement or interest to the broadcast? How does one weigh the value of the entertainment quality and the value of relevant physics? What are reasonable expectations for the entertainment aspect of the voice-over? Discuss and decide as a class.

Although many people may be in the broadcast booth, a voice-over becomes the product of one person—the commentator or the scriptwriter. Although you will be working in cooperative groups during the chapter, each person will be responsible for a voice-over or script for a sporting event. As a team of two or three, you may wish to work together and share different aspects of the job, but the output of work per person should be the same. That is why one voice-over will be required of each person irrespective of whether individuals prefer to work independently or in groups.

Activity 1 A Running Start and Frames of Reference

GOALS

In this activity you will:

- Understand and apply Galileo's Principle of Inertia.
- Understand and apply Newton's First Law of Motion.
- Recognize inertial mass as a physical property of matter.

What Do You Think?

Many things that happen in athletics are affected by the amount of "running start" speed an athlete can produce.

- **What determines the amount of horizontal distance a basketball player travels while "hanging" to do a "slam dunk" during a fast break?**
- **How do figure skaters keep moving across the ice at high speeds for long times while seldom "pumping" their skates?**

Record your ideas about these questions in your *Active Physics* log. Be prepared to discuss your responses with your small group and the class.

For You To Do

1. Use a salad bowl and a ball to explore the question, "When a ball is released to roll down the inside surface of a salad bowl, is the motion of the ball up the far side of the bowl the 'mirror image' of the ball's downward motion?" Use a nonpermanent pen to mark a starting position for the ball near the top edge of the bowl. Use a

flexible ruler to measure, in centimeters, the distance along the bowl's curved surface from the bottom-center of the bowl to the mark.

a) Make a table similar to the one below in your log.

Start	Trial	Starting Distance (cm)	Recovered Distance (cm)	$\frac{\text{Recovered Distance}}{\text{Starting Distance}}$
High	1			
	2			
	3			
Medium	1			
	2			
	3			
Low	1			
	2			
	3			

b) Record the measured distance in your table as the High Starting Distance.

Do not use a glass bowl, if possible.

2. Prepare to observe and mark the position on the far side of the bowl where the ball stops when it is released from the starting position. Release the ball from the starting position and mark the position where it stops. Measure the distance from the bottom-center to the stop mark.

a) Record the distance in your table as the High Recovered Distance.

3. Repeat **Step 2** above two more times to see if the results are consistent.

a) Record the data for all three trials.

4. Mark two more starting positions on the surface of the bowl, one for Medium Starting Distance and another for Low Starting Distance.

a) Measure and record each of the new distances.

b) Observe, mark, measure and record the recovered distances for three trials at each of the medium and low starting positions.

c) Complete the table by calculating and recording the value of the ratio of the Recovered Distance to the Starting Distance for each trial. (The ratio is the Recovered Distance divided by the Starting Distance.)

Example:

If the Recovered Distance is 6.0 cm for a Starting Distance of 10.0 cm, the value of the ratio is $\frac{6.0 \text{ cm}}{10.0 \text{ cm}} = 0.6$.

d) For each of the three starting distances, to what extent is the motion of the ball up the far side of the bowl the "mirror image" of the downward motion? Use data as evidence for your answer.

e) Does the fraction of the starting distance "recovered" when going up the far side of the bowl depend on the amount of starting distance? Describe any pattern of data that supports your answer.

5. Repeat the activity but roll the ball along varying slopes during its upward motion. Make a track that has the same slope on both sides, as shown below. Your teacher will suggest how high the ends of the track sections should be elevated. This time, concentrate on comparing the vertical height of the ball's release position to the vertical height of the position where the ball stops.

a) Measure and record the vertical height (not the distance along the track) from which the ball will be released at the top end of the left-hand section of track.

b) Prepare to observe and mark the position on the right-hand section of track where the ball stops when it is released from the starting position. Release the ball from the top end of the left-hand section of track and mark the position where it stops. Measure and record the vertical height of the position where the ball stops.

c) Calculate the ratio of the recovered height to the starting height. How is this case, and the result, similar to what you did when using the salad bowl? How is it different?

6. Leave the left-hand starting section of track unchanged, but change the right-hand section of track so that it has less slope and is at least long enough to allow the ball to recover the starting height. The track should be arranged approximately as shown below.

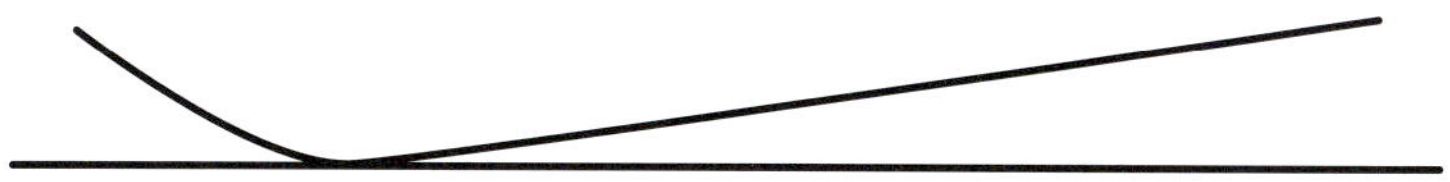

a) Predict the position where the ball will stop on the right-hand track if it is released from the same height as before on the left-hand track. Mark the position of your guess on the right-hand track and explain the basis for your prediction in your log.

7. Release the ball from the same height on the left-hand section of track as before and mark the position where the ball stops on the right-hand section of track.

a) How well did you guess the position? Why do you think your guess was "on" or "off"?

b) Measure the vertical height of the position where the ball stopped and again calculate the ratio of the recovered height to the starting height. Did the ratio change? Why, do you think, did the ratio change or not change?

8. Imagine what would happen if you again did not change the left-hand starting section of track, but changed the right-hand section of track so that it would be horizontal, as shown below.

a) How far along the horizontal track would the ball need to roll to recover its starting height (or most of it)? How far do you think the ball would roll?

b) When rolling on the horizontal track, what would "keep the ball going"?

FOR YOU TO READ

Inertia

Italian philosopher Galileo Galilei (1564–1642), who can be said to have introduced science to the world, noticed that a ball rolled down one ramp seems to seek the same height when it rolls up another ramp. He also did a "thought experiment" in which he imagined a ball made of extremely hard material set into motion on a horizontal, smooth surface, similar to the final track in **For You To Do**. He concluded that the ball would continue its motion on the horizontal surface with constant speed along a straight line "to the horizon" (forever). From this, and from his observation that an object at rest remains at rest unless something causes it to move, Galileo formed the Principle of **Inertia**:

Inertia is the natural tendency of an object to remain at rest or to remain moving with constant speed in a straight line.

Isaac Newton, born in England on Christmas day in 1642 (within a year of Galileo's death), used Galileo's Principle of Inertia as the basis for developing his First Law of Motion, presented in **Physics Talk**. Crediting Galileo and others for their contributions to his thinking, Newton said, "If I have seen farther than others, it is because I have stood on the shoulders of giants."

Running Starts

Running starts take place in many sporting activities. Since there seems to be this prior motion in many sports, there must be some advantage to it.

In sports where the objective is to maximize the speed of an object or the distance traveled in air, the prior motion may be essential. When a javelin is thrown, at the instant of release it has the same speed as the hand that is propelling it.

- The hand has a forward speed relative to the elbow, the elbow has a forward speed relative to the shoulder (because the arm is rotating around the elbow and shoulder joints), and the shoulder has a forward speed relative to the ground because the body is rotating and the body is also moving forward.
- The javelin speed then is the sum of each of the above speeds. If the thrower is not running forward, that speed does not add into the equation.

You can write a **velocity** equation to show the speeds involved.

$$v_{javelin} = v_{hand} + v_{elbow} + v_{shoulder} + v_{ground}$$

Motion captures everyone's attention in sports. Starting, stopping, and changing direction (**accelerations**) are part of the motion story, and they are exciting components of many sports. Ordinary, straight-line motion is just as important but is easily overlooked.

PHYSICS TALK

Newton's First Law of Motion

Isaac Newton included Galileo's Principle of Inertia as part of his **First Law of Motion**:

In the absence of an unbalanced force, an object at rest remains at rest, and an object already in motion remains in motion with constant speed in a straight-line path.

Newton also explained that an object's mass is a measure of its inertia, or tendency to resist a change in motion.

Here is an example of how Newton's First Law of Motion works:

Inertia is expressed in kilograms of mass. If an empty grocery cart has a mass of 10 kg and a cart full of groceries has a mass of 100 kg, which cart would be more difficult to move (have a greater tendency to remain at rest)? If both carts already were moving at equal speeds, which cart would be more difficult to stop (would have a greater tendency to keep moving)? Obviously in both cases, the answer is the cart with more mass.

Physics Words

inertia: the natural tendency of an object to remain at rest or to remain moving with constant speed in a straight line.

acceleration: the change in velocity per unit time.

frame of reference: a vantage point with respect to which position and motion may be described.

FOR YOU TO READ

Frames of Reference

In this activity, you investigated Newton's First Law. In the absence of external forces, an object at rest remains at rest and an object in motion remains in motion. If you were challenged to throw a ball as far as possible, you would probably now be sure to ask if you could have a running start. If you run with the ball prior to throwing it, the ball gets your speed before you even try to release it. If you can run at 5 m/s, then the ball will get the additional speed of 5 m/s when you throw it. When you do throw the ball, the ball's speed is the sum of your speed before releasing the ball, 5 m/s, and the speed of the release.

→

It may be easier to understand this if you think of a toy cannon that could be placed on a skateboard. The toy cannon always shoots a small ball forward at 7 m/s. This can be checked with multiple trials. The toy cannon is then attached to the skateboard. A release mechanism is set up so that the cannon continues to shoot the ball forward at 7 m/s when the skateboard is at rest. When the skateboard is given an initial push, the skateboard is able to travel at 3 m/s. If the cannon releases the ball while the skateboard is moving, the ball's speed is now measured to be 10 m/s. From where did the additional speed come? The ball's speed is the sum of the ball's speed from the cannon plus the speed of the skateboard. 7 m/s + 3 m/s = 10 m/s.

You may be wondering if the ball is moving at 7 m/s or 10 m/s. Both values are correct — it depends on your **frame of reference**. The ball is moving at 7 m/s relative *to* the skateboard. The ball is moving at 10 m/s relative *to* the Earth.

Imagine that you are on a train that is stopped at the platform. You begin to walk toward the front of the train at 3 m/s. Everybody in the train will agree that you are moving at 3 m/s toward the front of the train. This is your speed *relative to* the train. Everybody looking into the train from the platform will also agree that you are moving at 3 m/s toward the front of the train. This is your speed *relative to* the platform.

Imagine that you are on the same train, but now the train is moving past the platform at 9 m/s. You begin to walk toward the front of the train at 3 m/s. Everybody in the train will agree that you are moving at 3 m/s toward the

front of the train. This is your speed *relative to* the train. Everybody looking into the train from the platform will say that you are moving at 12 m/s (3 m/s + 9 m/s) toward the front of the train. This is your speed relative *to* the platform.

Whenever you describe speed, you must always ask, "*Relative to what?*" Often, when the speed is relative to the Earth, this is assumed in the problem. If your frame of reference is the Earth, then it all seems quite obvious. If your frame of reference is the moving train, then different speeds are observed.

In sports where you want to provide the greatest speed to a baseball, a javelin, a football, or a tennis ball, that speed could be increased if you were able to get on a moving platform. That being against the rules and inappropriate for many reasons, an athlete will try to get the body moving with a running start, if allowed. If the running start is not permitted, the athlete tries to move every part of his or her body to get the greatest speed.

Sample Problem 1

A sailboat has a constant velocity of 22 m/s east. Someone on the boat prepares to toss a rock into the water.

a) Before being tossed, what is the speed of the rock with respect to the boat?

b) Before being tossed, what is the speed of the rock with respect to the shore?

c) If the rock is tossed with a velocity of 16 m/s east, what is the rock's velocity with respect to shore?

d) If the rock is tossed with a velocity of 16 m/s west, what is the rock's velocity with respect to shore?

Strategy: Before determining a velocity, it is important to check the frame of reference. The rock's velocity with respect to the boat is different from the velocity with respect to the shore. The direction of the rock also impacts the final answer.

Givens:

v_b = 22 m/s east

v_r = 16 m/s (direction varies)

Solution:

a) With respect to the boat, the rock's velocity is 0 m/s.
The rock is moving at the same speed as the boat, but you wouldn't notice this velocity if you were in the boat's frame of reference.

b) With respect to shore, the rock's velocity is 22 m/s east.
The rock is on the boat, which is traveling at 22 m/s east. Relative to the shore, the boat and everything on it act as a system traveling at the same velocity.

c) With respect to the shore, the rock's velocity is now 38 m/s east.
It is the sum of the velocity values. Since each is directed east, the relative velocity is the sum of the two.

$$v = v_b + v_r$$
$$= 22 \text{ m/s east} + 16 \text{ m/s east}$$
$$= 38 \text{ m/s east}$$

d) With respect to shore, the rock's velocity is now 6 m/s east.
Since the directions are opposite, the relative velocity is the difference between the two.

$$v = v_b - v_r$$
$$= 22 \text{ m/s east} - 16 \text{ m/s west}$$
$$= 6 \text{ m/s east}$$

Sample Problem 2

A quarterback on a football team is getting ready to throw a pass. If he is moving backward at 1.5 m/s and he throws the ball forward at 10.0 m/s, what is the velocity of the ball relative to the ground?

Strategy: Use a negative sign to indicate the backward direction. Add the two velocities to find the velocity relative to the ground.

Givens:

Solution:

Add the velocities.

10.0 m/s + (–1.5 m/s) = 8.5 m/s

The ball is moving forward at 8.5 m/s relative to the ground.

Reflecting on the Activity and the Challenge

Running starts can be observed in many sports. Many observers may not realize the important role that inertia plays in preserving the speed already established when an athlete engages in activities such as jumping, throwing, or skating from a running start. "Immovable objects," such as football linemen, illustrate the tendency of highly massive objects to remain at rest and can be observed in many sports. You should have no problem finding a great variety of video segments that illustrate Newton's First Law.

Physics To Go

1. Provide three illustrations of Newton's First Law in sporting events. Describe the sporting event and which object when at rest stays at rest, or when in motion stays in motion. Describe these same three illustrations in the manner of an entertaining sportscaster.

2. Find out about a sport called curling (it is an Olympic competition that involves some of the oldest Olympians) and how this sport could be used to illustrate Newton's First Law of Motion.

3. When a skater glides across the ice on only one skate, what kind of motion does the skater have? Use principles of physics as evidence for your answer.

4. Use what you have learned in **Activity 1** to describe the motion of a hockey puck between the instant the puck leaves a player's stick and the instant it hits something. (No "slap shot" allowed; the puck must remain in contact with the ice.)

5. Why do baseball players often slide into second base and third base, but almost never slide into first base after hitting the ball? (The answer depends on both the rules of baseball and the laws of physics.)

6. Do you think it is possible to arrange conditions in the "real world" to have an object move, unassisted, in a straight line at constant speed forever? Explain why or why not.

7. You are pulling your little brother in his red wagon. He has a ball, and he throws it straight up into the air while you are pulling him forward at a constant speed.
   a) What will the path of the ball look like to your little brother in the wagon?
   b) What will the path of the ball look like to a little girl who is standing on the sidewalk watching you?
   c) If your brother throws the ball forward at a velocity of 2.5 m/s while you are pulling the wagon at a velocity of 4.5 m/s, at what speed does the girl see the ball go by?

8. A track and field athlete is running forward with a javelin at a velocity of 4.2 m/s. If he throws the javelin at a velocity relative to him of 10.3 m/s, what is the velocity of the javelin relative to the ground?

9. You are riding the train to school. Since the train car is almost empty, you and your friend are throwing a ball back and forth. The train is moving at a velocity of 5.6 m/s. Suppose you throw the ball to each other at the same speed, 2.4 m/s.
   a) What is the velocity of the ball relative to the tracks when the ball is moving toward the front of the car?
   b) What is the velocity of the ball relative to the tracks when it is moving toward the back of the car?
   c) What if you and your friend throw the ball perpendicular to the aisle of the train? What is the ball's velocity then?

10. Two athletes are running toward the pole vault. One is running at 3.8 m/s and the other is running at 4.3 m/s.
    a) What is their velocity relative to each other?
    b) If they leave the ground at their respective velocities, which one has the energy to go higher in the vault? Explain.

11. While riding a horse, a competitor shoots an arrow toward a target. The speed of the arrow as it reaches the target is 85 m/s. If the horse was traveling at 18 m/s, at what speed did the arrow leave the bow? (Assume the horse and arrow are traveling in the same direction.)

Activity 2 Push or Pull—Adding Vectors

GOALS

In this activity you will:

- Recognize that a force is a push or a pull.
- Identify the forces acting on an object.
- Determine when the forces on an object are either balanced or unbalanced.
- Calibrate a force meter in arbitrary units.
- Use a force meter to apply measured amounts of force to objects.
- Compare amounts of acceleration semiquantitatively.
- Understand and apply Newton's Second Law of Motion.
- Understand and apply the definition of the newton as a unit of force.
- Understand weight as a special application of Newton's Second Law.

What Do You Think?

Moving a football one yard to score a touchdown requires strategy, timing, and many forces.

- **What is a force?**
- **Can the same force move a bowling ball and a ping-pong ball?**

Record your ideas about these questions in your *Active Physics* log. Be prepared to discuss your responses with your small group and the class.

For You To Do

1. Use a flexible ruler as a "force meter". Use washers to make a scale of measurement for (that is, to calibrate) the meter in pennyweights. The force you are using to calibrate the meter is gravity, the force with which Earth pulls downward on every

object near its surface. *Carefully* clamp the plastic strip into position as shown in the diagram on page 15.

2. Draw a line on a piece of paper. Hold the paper next to the plastic strip so that the line is even with the edge of the strip. Mark the position of the end of the strip on the reference line and label the position as the "zero" mark.

3. Place one washer on the top surface of the strip near the strip's outside end. Notice that the strip bends downward and then stops. Hold the paper in the original position and mark the new position of the end of the strip. Label the mark as "1 pennyweight."

4. Repeat **Step 3** for two, three, and four coins placed on the strip. In each case mark and label the new position of the end of the strip.

   a) Copy the reference line and the calibration marks from the piece of paper into your log.

5. Practice holding one end of the "force meter" (plastic strip) in your hand and pushing the free end against an object until you can bend the strip by forces of 1, 2, 3, and 4-pennyweight amounts. To become good at this, you will need to check the amount of bend in the strip against your calibration marks as you practice.

6. Use the force meter to push an object such as a tennis ball with a continuous 1-pennyweight force. You will need to keep up with the object as it moves and to keep the proper bend in the force meter. You may need to practice a few times to be able to do this.

   a) In your log, record the amount of force used, a description of the object, and the kind of motion the object seemed to have.

7. Repeat **Step 6** three more times, pushing on the same object with steady (constant) 2, 3, and 4-pennyweight amounts of force.

   a) Record the results in your log for each amount of force.

8. Based on your observations, complete the statement: "The greater the constant, unbalanced force pushing on an object,..."
   a) Write the completed statement in your log.
9. Select an object that has a small mass. Use the force meter to push on the object with a rather large, steady force such as 3- or 4-pennyweight amounts.
   a) Record the amount of force used, a description of the object pushed (especially including its mass, compared to the other objects to be pushed) and the kind of motion the object seemed to have.
10. Repeat **Step 9** using the same amount of force to push objects of greater and greater mass.
    a) Record the results in your log for each object.
11. Based on your observations, complete the statement: "When equal amounts of constant, unbalanced force are used to push objects having different masses, the more massive object..."
    a) Write the completed statement in your log.

Physics Words

Newton's Second Law of Motion: if a body is acted upon by an external force, it will accelerate in the direction of the unbalanced force with an acceleration proportional to the force and inversely proportional to the mass.

weight: the vertical, downward force exerted on a mass as a result of gravity.

PHYSICS TALK

Newton's Second Law of Motion

Based on observations from experiments similar to yours, Isaac Newton wrote his **Second Law of Motion**:

The acceleration of an object is directly proportional to the unbalanced force acting on it and is inversely proportional to the object's mass. The direction of the acceleration is the same as the direction of the unbalanced force.

If 1 N (newton) is defined as the amount of unbalanced force that will cause a 1-kg mass to accelerate at 1 m/s^2 (meter per second every second), the law can be written as an equation:

$$F = ma$$

where F is expressed in newtons (symbol N), mass is expressed in kilograms (kg), and acceleration is expressed in meters per second every second (m/s^2).

By definition, the unit "newton" can be written in its equivalent form: $(kg)m/s^2$.

Newton's Second Law can be arranged in three possible forms:

$$F = ma \qquad a = \frac{F}{m} \qquad m = \frac{F}{a}$$

FOR YOU TO READ

Weight and Newton's Second Law

Newton's Second Law explains what "weight" means, and how to measure it. If an object having a mass of 1 kg is dropped, its free fall acceleration is roughly 10 m/s^2.

Using Newton's Second Law,

$$F = ma$$

the force acting on the falling mass can be calculated as

$$F = ma$$
$$= 1 \text{ kg} \times 10 \text{ m/s}^2 \text{ or } 10 \text{ N}$$

The 10-N force causing the acceleration is known to be the gravitational pull of Earth on the 1-kg object. This gravitational force is given the special name **weight**. Therefore, it is correct to say, "The weight of a 1-kg mass is ten newtons."

What is the weight of a 2-kg mass? If dropped, a 2-kg mass also would accelerate due to gravity (as do all objects in free fall) at about 10 m/s^2. Therefore, according to Newton's Second Law, the weight of a 2-kg mass is equal to

$$2 \text{ kg} \times 10 \text{ m/s}^2 \text{ or } 20 \text{ N}$$

In general, to calculate the numerical value of an object's weight in newtons, it is necessary only to multiply the numerical value of its mass by the numerical value of the g (acceleration due to gravity), which is about 10m/s^2.

$$\text{Weight} = mg$$

The preceding equation is the "special case" of Newton's Second Law that must be applied to any situation in which the force causing an object to accelerate is Earth's gravitational pull.

Where There's Acceleration, There Must Be an Unbalanced Force

There are lots of different everyday forces. You just read about the force of gravity. There is also the force of a spring, the force of a rubber band, the force of a magnet, the force of your hand, the force of a bat hitting a ball, the force of friction, the buoyant force of water, and many more. Newton's Second Law tells you that accelerations are caused by unbalanced external forces. It doesn't matter what kind of force it is or how it originated. If you observe an acceleration (a change in velocity), then there must be an unbalanced force causing it.

When you apply a force, if the object has a small mass, the acceleration may be quite large for a given force. If the object has a large mass, the acceleration will be smaller for the same applied force. Occasionally, the mass is so large that you are not even able to measure the acceleration because it is so small.

If you push on a go-cart with the largest force possible, the cart will accelerate a great deal. If you push on a car with that same force, you

→

will measure a much smaller acceleration. If you were to push on the Earth, the acceleration would be too small to measure. Can you convince someone that a push on the Earth moves the Earth? Why should you believe something that you can't measure? If you were to assume that the Earth does not accelerate when you push on it, then you would have to believe that Newton's Second Law stops working when the mass gets too big. If that were so, you would want to determine how big is "too big." When you conduct such experiments, you find that the acceleration gets less and less as the mass gets larger and larger. Eventually, the acceleration gets so small that it is difficult to measure. Your inability to measure it doesn't mean that it is zero. It just means that it is smaller than your best measurement. In this way, you can assume that Newton's Second Law is always valid.

All of these statements are summarized in Newton's Second Law as you read in **Physics Talk**:

$$F = ma$$

or in forms that emphasize the acceleration and the mass

$$a = \frac{F}{m} \text{ and } m = \frac{F}{a}$$

Sample Problem 1

A tennis racket hits a ball with a force of 150 N. While the 275-g ball is in contact with the racket, what is its acceleration?

Strategy: Newton's Second Law relates the force acting on an object, the mass of the object, and the acceleration given to it by the force. Use the form of the equation that emphasizes acceleration to find the acceleration. The force unit, the newton, is defined as the amount of force needed to give a mass of 1.0 kg an acceleration of 1.0 m/s^2. Therefore, you will need to change the grams to kilograms.

Givens:

$F = 150.0$ N
$m = 275$ g

Solution:

$275 \text{ g} = 0.275 \text{ kg}$

$$a = \frac{F}{m}$$
$$= \frac{150 \text{ N}}{0.275 \text{ kg}}$$
$$= 545 \text{ m/s}^2$$

Sample Problem 2

As the result of a serve, a tennis ball ($m_t = 58$ g) accelerates at 43 m/s^2.

a) What force is responsible for this acceleration?

b) Could an identical force accelerate a 5.0-kg bowling ball at the same rate?

Strategy: Newton's Second Law states that the acceleration of an object is directly proportional to the applied force and indirectly proportional to the mass ($F = ma$).

Givens:

$a = 43 \text{ m/s}^2$
$m_t = 58 \text{ g} = 0.058 \text{ kg}$
$m_b = 5.0 \text{ kg}$

Solution:

a)

$$F = m_t a$$
$$= 0.058 \text{ kg} \times 43 \text{ m/s}^2$$
$$= 2.494 \text{ N or } 2.5 \text{ N}$$

b) Since the mass of the bowling ball is much greater than that of the tennis ball, an identical force will result in a smaller acceleration.
(You can calculate the acceleration.)

$$a = \frac{F}{m_b}$$

$$= \frac{2.5\ \text{N}}{5.0\ \text{kg}}$$

$$= 0.50\ \text{m/s}^2$$

Adding Vectors

A vector is a quantity that has both magnitude and direction. Velocity is a vector. In the previous activity you found that the direction in which an object was traveling and the speed at which it was moving are equally important.

Force is also a vector because you can measure how big it is (its magnitude) and its direction. Acceleration is also a vector. The equation for acceleration reminds you that the force and the acceleration must be in the same direction.

Often, more than one force acts on an object. If the two forces are in the same direction, the sum of the forces is simply the addition of the two forces. A 30-N force by one person and a force of 40 N by a second person (pushing in the same direction) on the same desk provides a 70-N force on the desk. If the two forces are in opposite directions, then you give one of the forces a negative value and add them again. If one student pushes on a desk to the right with a force of 30 N and a second student pushes on the same desk to the left with a force of 40 N, the net force on the desk will be 10 N to the left. Mathematically, you would state that 30 N + (–40 N) = – 10 N where the negative sign denotes "to the left."

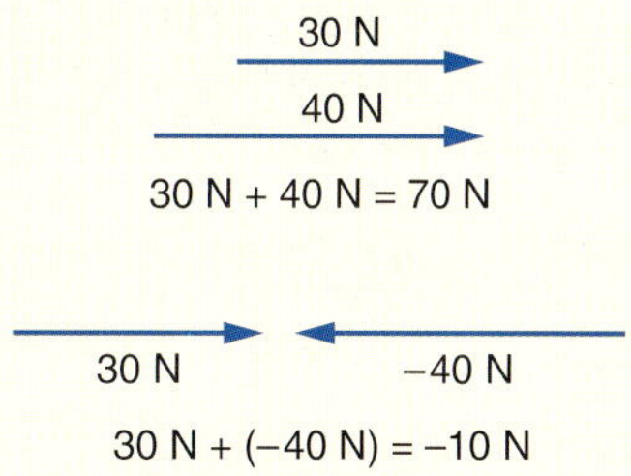

Occasionally, the two forces acting on an object are at right angles. For instance, one student may be kicking a soccer ball with a force of 30 N ahead toward the goal, while the second student kicks the same soccer ball with a force of 40 N toward the sideline. To find the net force on the ball and the direction the ball will travel, you must use vector addition. You can do this by using a vector diagram or the Pythagorean Theorem.

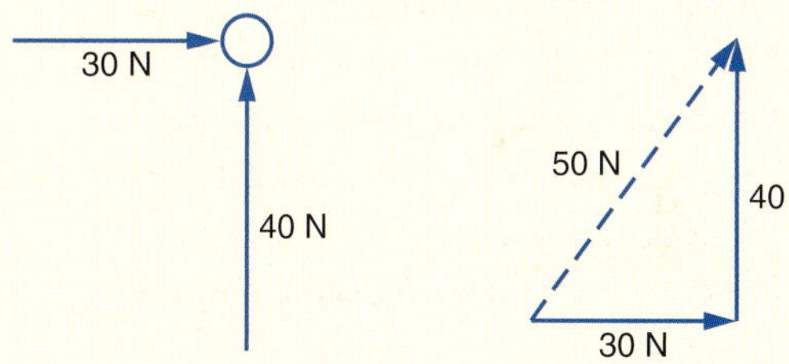

In the vector diagram shown above, the two force vectors are shown as arrows acting on the soccer ball. The magnitudes of the vectors are drawn to scale. The 30-N force may be drawn as 3.0 cm and the 40-N force may be drawn as 4.0 cm, if the scale is 10 N = 1 cm. To add the vectors, slide them so that the tip of the 30-N vector can be placed next to the tail of the 40-N vector (tip to tail method). The sum of the two vectors is then drawn from the tail of the 30-N vector to the tip of the 40-N vector. This *resultant* vector is measured and is found to be 5.0 cm, which is equivalent to 50 N. The angle is measured with a protractor and is found to be 53°.

A second method of finding the resultant vector is to recognize that the 30-N and 40-N force vectors form a right triangle. The resultant is the hypotenuse of this triangle. Its length can be found by the Pythagorean Theorem.

$$a^2 + b^2 = c^2$$
$$30\ \mathrm{N}^2 + 40\ \mathrm{N}^2 = c^2$$
$$900\ \mathrm{N}^2 + 1600\ \mathrm{N}^2 = c^2$$
$$c = \sqrt{2500}\ \mathrm{N}^2$$
$$c = 50\ \mathrm{N}$$

The angle can be found by using the tangent function.

$$\tan\theta = \frac{\text{opposite}}{\text{adjacent}} = \frac{40\ \mathrm{N}}{30\ \mathrm{N}} = 1.33$$
$$\theta = 53°$$

Adding vector forces that are not perpendicular is a bit more difficult mathematically, but no more difficult using scale drawings and vector diagrams. Two other players are kicking a soccer ball in the direction shown in the diagram. The resultant vector force can be determined using the tip to tail approach.

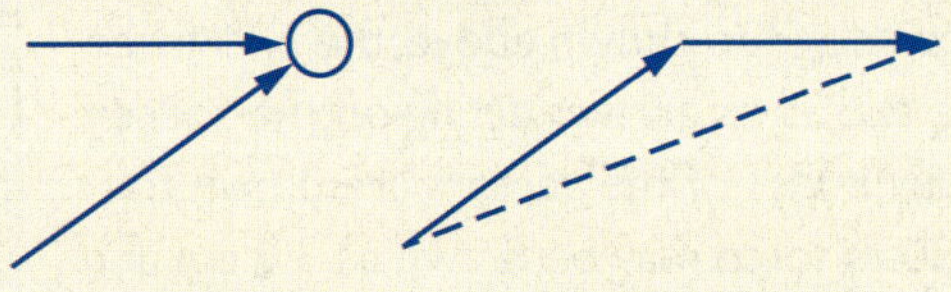

The two arrows in the left diagram correspond to the two players kicking the ball at different angles. The diagram at the right shows the two vectors being added "tip to tail." The resultant vector (shown as a dotted line) represents the net force and is the direction of the acceleration of the soccer ball.

Sample Problem 3

One player applies a force of 125 N in a north direction. Another player pushes with a force of 125 N west. What is the magnitude and direction of the resultant force?

Strategy: Since the forces are acting at right angles, you can use the Pythagorean Theorem to find the resultant force. The direction of the force can be found using the tangent function.

Givens:

$$F_1 = 125\ \mathrm{N}$$
$$F_2 = 125\ \mathrm{N}$$

Solution:

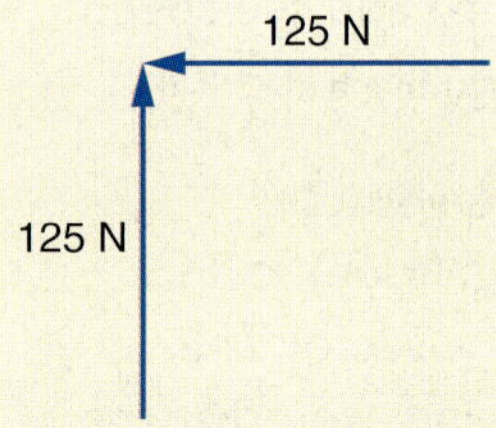

$$F_R^2 = F_1^2 + F_2^2$$
$$F_R = \sqrt{125\ \mathrm{N}^2 + 125\ \mathrm{N}^2}$$
$$= \sqrt{31{,}250\ \mathrm{N}^2}$$
$$= 177\ \mathrm{N}$$
$$\tan\theta = \frac{\text{opposite}}{\text{adjacent}} = \frac{125\,\mathrm{N}}{125\,\mathrm{N}} = 1$$
$$\theta = 45°$$

The resultant force is 177 N, 45° west of north.

Reflecting on the Activity and the Challenge

What you learned in this activity really increases the possibilities for interpreting sports events in terms of physics. Now you can explain why accelerations occur in terms of the masses and forces involved. You know that forces produce accelerations. Therefore, if you see an acceleration occur, you know to look for the forces involved. You can apply this to the sport that you will describe.

Also, you can explain, in terms of mass and weight, why gravity has no "favorite" athletes; in every case of free fall in sports, g has the same value, about 10 m/s^2.

Physics To Go

1. Copy and complete the following table using Newton's Second Law of Motion. Be sure to include the unit of measurement for each missing item.

Newton's Second Law:	***F*** =	***m*** ×	***a***
Sprinter beginning 100-meter dash	?	70 kg	5 m/s^2
Long jumper in flight	800 N	?	10 m/s^2
Shot put ball in flight	70 N	7 kg	?
Ski jumper going down hill before jumping	400 N	?	5 m/s^2
Hockey player "shaving ice" while stopping	−1500 N	100 kg	?
Running back being tackled	?	100 kg	−30 m/s^2

2. The following items refer to the table in **Question 1**:
   a) In which cases in the table does the acceleration match "g," the acceleration due to gravity 10 m/s^2? Are the matches to g coincidences or not? Explain.
   b) The force on the hockey player stopping is given in the table as a negative value. Should the player's acceleration also be negative? What do you think it means for a force or an acceleration to be negative?
   c) The acceleration of the running back being tackled also is given as negative. Should the unbalanced force acting on him also be negative? Explain.

d) In your mind, "play" an imagined video clip that illustrates the event represented by each horizontal row of the preceding table. Write a brief voice-over script for each video clip that explains how Newton's Second Law of Motion is operating in the event. Use appropriate physics terms, equations, numbers, and units of measurement in the scripts.

3. What is the acceleration of a 0.30-kg volleyball when a player uses a force of 42 N to spike the ball?

4. What force would be needed to accelerate a 0.040-kg golf ball at 20.0 m/s^2?

5. Most people can throw a baseball farther than a bowling ball, and most people would find it less painful to catch a flying baseball than a bowling ball flying at the same speed as the baseball. Explain these two apparent facts in terms of:
   a) Newton's First Law of Motion.
   b) Newton's Second Law of Motion.

6. Calculate the weight of a new fast-food sandwich that has a mass of 0.1 kg. Think of a clever name for the sandwich that would incorporate its weight.

7. In the United States, people measure body weight in pounds. Write down the weight, in pounds, of a person who is known to you. (This could be your weight or someone else's.)
   a) Convert the person's weight in pounds to the international unit of force, newtons. To do so, use the following conversion equation:
   Weight in newtons = Weight in pounds × 4.38 newtons/pound
   b) Use the person's body weight, in newtons, and the equation

   $$\text{Weight} = mg$$

   to calculate the person's body mass, in kilograms.

8. Imagine a sled (such as a bobsled or luge used in Olympic competitions) sliding down a 45° slope of extremely slippery ice. Assume there is no friction or air resistance (not really possible). Even under such ideal conditions, it is a fact that gravity could cause the sled to accelerate at a maximum of only 7.1 m/s^2. Why would the "ideal" acceleration of the sled not be g, 10 m/s^2? Your answer is expected only to suggest reasons why, on a 45° hill, the ideal free fall acceleration is "diluted" from 10 m/s^2 to about 7 m/s^2; you are not expected to give a complete explanation of why the "dilution" occurs.

9. If you were doing the voice-over for a tug-of-war, how would you explain what was happening? Write a few sentences as if you were the science narrator of that athletic event.

10. You throw a ball. When the ball is many meters away from you, is the force of your hand still acting on the ball?

11. Carlo and Sara push on a desk in the same direction. Carlo pushes with a force of 50 N, and Sara pushes with a force of 40 N. What is the total resultant force acting on the desk?

12. A car is stuck in the mud. Four adults each push on the back of the car with a force of 200 N. What is the total force on the car?

13. During a football game, two players try to tackle another player. One player applies a force of 50.0 N to the east. A second player applies a force of 120.0 N to the north. What is the total applied force? (Since force is a vector, you must give both the magnitude and direction of the force.)

14. In auto racing, a crash occurs. A red car hits a blue car from the front with a force of 4000 N. A yellow car also hits the blue car from the side with a force of 5000 N. What is the total force on the blue car? (Since force is a vector, you must give both the magnitude and direction of the force.)

15. A baseball player throws a ball. While the 700.0-g ball is in the pitcher's hand, there is a force of 125 N on it. What is the acceleration of the ball?

16. If the acceleration due to gravity at the surface of the Earth is approximately 9.8 m/s^2, what force does the gravitational attraction of the Earth exert on a 12.8-kg object?

17. A force of 30.0 N acts on an object. At right angles to this force, another force of 40.0 N acts on the same object.

    a) What is the net force on the object?
    b) What acceleration would this give a 5.6-kg wagon?

18. Bob exerts a 30.0 N force to the left on a box (m = 100.0 kg). Carol exerts a 20.0 N force on the same box, perpendicular to Bob's.

    a) What is the net force on the box?
    b) Determine the acceleration of the box.
    c) At what rate would the box accelerate if both forces were to the left?

Activity 3 Center of Mass

GOALS

In this activity you will:

- Locate the center of mass of oddly shaped two-dimensional objects.
- Infer the location of the center of mass of symmetrical three-dimensional objects.
- Measure the approximate location of the center of mass of body.
- Understand that the entire mass of an object may be thought of as being located at the object's center of mass.

What Do You Think?

The center of mass of a high jumper using the "Fosbury Flop" (arched back) technique passes below the bar as the jumper's body successfully passes over the bar.

- **What is "center of mass"? What does it mean?**
- **Where is your body's center of mass?**

Record your ideas about these questions in your *Active Physics* log. Be prepared to discuss your responses with your small group and the class.

For You To Do

1. You will be provided four objects made from thin sheets of material in the shapes shown below.

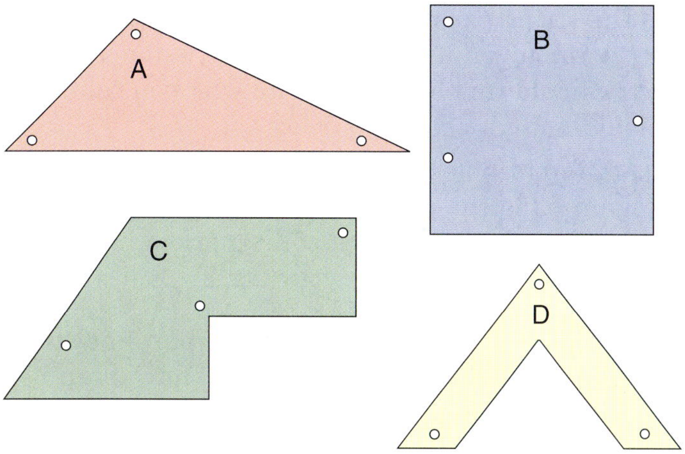

a) Make a sketch in your log to show each shape in a reduced scale of size.

2. Use your intuition and trial and error to locate a point on the flat surfaces of objects *A*, *B*, and *C* where the object will balance on your fingertip. Mark the "balance point" of each object using a nonpermanent method.

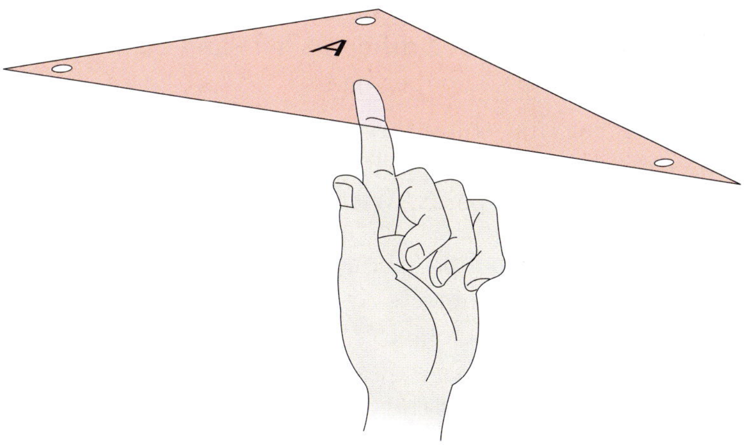

a) Mark the balance points on the sketches in your log.

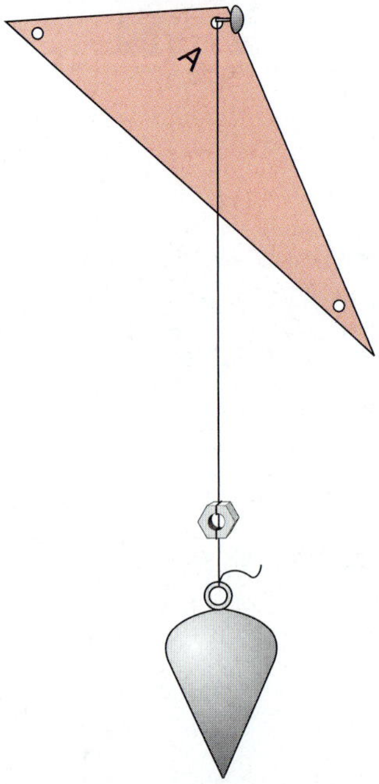

3. To check on the balance points found by the above method for objects A, B and C, use one of the small holes in object A to hang it from a pin as shown, and, also as shown, hang a "plumb bob" (a weight on a string) from the same pin.

   a) Does the string pass over the balance point you marked for object A when you used your finger to balance the object? Should this happen? Write why you think it should or should not happen in your log.

   b) Use a different hole in object A to suspend it from the pin and again hang the plumb bob from the pin. Does the string pass over the balance point marked before? Should it? Write your responses in your log.

4. The intersection of the two lines made where the string passed over the surface of object A could have been used to predict the balance point without first trying to balance the object on your finger. Use the suspension and plumb bob method to check the correspondence of the two methods of finding the balance point for objects B and C.

   a) Record your findings about how well the two methods agree in your log.

5. Locate an "imaginary" balance point for object D. Tape a lightweight piece of paper between the "open arms" of object D and suspend the object and the plumb bob from the pin. Trace the path of the string across the piece of paper. Suspend the object from a different hole and trace the path of the string across the paper again.

   a) Do you agree that the intersection of the two lines on the paper mark the balance point of object D? What is special about this balance point? Write your answers in your log.

6. The above "balance points" that you found for two-dimensional, or "flat," objects A, B, C, and D were, in each case, the location of the object's "center of mass." Do a "thought experiment" (an experiment in your mind) to determine the location of the center of mass of each of the following objects:

   Shot put ball (solid steel) — Basketball
   Banana — Planet Earth
   Baseball bat — Hockey stick

a) For each object, describe in your log how you decided upon the location of the center of mass.

7. The technique that was used to find the center of mass (C of M) relied on the fact that the C of M always lies beneath the point of support when an object is hanging. Similarly, when an object is balanced, the C of M is always above the point of support. To find your C of M, carefully balance on one foot and then the other. Try to keep your arms and legs in roughly identical positions as you shift your weight. Your C of M is located where a vertical meter stick from one foot and the other intersect. Locate this point. The actual C of M is inside your body, since nobody has zero thickness.

a) Record the location of your C of M.

8. Your teacher will balance a hammer on a finger to locate the hammer's C of M and make an obvious mark on the hammer at the C of M. As your teacher drops the hammer into a catch box on the floor, and it twists and turns, notice the movement of the C of M.

a) How does the movement of the C of M compare to the motion of the entire hammer?

Physics Words

center of mass: the point at which all the mass of an object is considered to be concentrated for calculations concerning motion of the object.

Reflecting on the Activity and the Challenge

The **center of mass** is an important concept in any sports activity. The motion of the center of mass of a diver or gymnast is much easier to observe than the movements of the entire body. The sure-fire way of having a football player fall is to move his center of mass away from his support.

Think about the possibilities for using a transparent plastic cover on a TV monitor and using a pen to trace the motion of the center of mass of an athlete executing a free fall jump or dive. This could be used to simulate the light-pen technique used by TV commentators when they comment on football replays. This would seem a good way to add an interesting feature to your TV sports commentary.

Physics To Go

1. When applying a force to make an object move, why is it most effective to have the applied force "aimed" directly at the object's center of mass?

2. "Center of gravity" means essentially the same thing as "center of mass." Why is it often said to be desirable for football players to have a low center of gravity?

3. Stand next to a wall facing parallel to the wall. With your right arm at your side pushing against the wall and with the right edge of your right foot against the wall at floor level, try to remain standing as you lift your left foot. Why is this impossible to do?

4. Think of positions for the human body for which the center of mass might be located outside the body. Describe each position and where you think the center of mass would be located relative to the body for each position.

5. An object tends to rotate (spin) if it is pushed on by a force that is not aimed at the center of mass. How do athletes use this fact to initiate spins before they fly through the air, as in gymnastics, skating, and diving events?

6. Could the suspension technique for finding the center of mass used in **For You To Do** be adapted to locate the center of mass of a three-dimensional object? If you had a crane that you could use to suspend an automobile from various points of attachment, how could you locate the auto's center of mass?

7. Find the center of mass of a baseball bat using the technique that you learned in class.

8. Carefully balance a light object (not too massive) over a table or catch box. Notice that the C of M is directly over the point of support. Move the support a little bit. Explain how this technique can be adapted to tackling in football.

9. Cut out a piece of cardboard in the shape of your state or a country. Find the geographic center of mass of your shape.

Activity 4 Defy Gravity

GOALS

In this activity you will:

- Measure changes in height of the body's center of mass during a vertical jump.
- Calculate changes in the gravitational potential energy of the body's center of mass during a vertical jump.
- Understand and apply the definition of work.
- Recognize that work is equivalent to energy.
- Understand and apply the joule as a unit of work and energy using equivalent forms of the joule.
- Apply conservation of work and energy to the analysis of a vertical jump, including weight, force, height, and time of flight.

What Do You Think?

No athlete can escape the pull of gravity.

- **Does the "hang time" of some athletes defy the above fact?**
- **Does a world-class skater defy gravity to remain in the air long enough to do a triple axel?**

Record your ideas about these questions in your *Active Physics* log. Be prepared to discuss your responses with your small group and the class.

For You To Do

1. Your teacher will show you a slow-motion video of a world-class figure skater doing a triple axel jump. The image of the skater will appear to "jerk," because a video camera completes one "frame," or one complete picture, every $\frac{1}{30}$ s. When the video is played at normal speed, you perceive the action as continuous; played at slow motion, the individual frames can be detected and counted. The duration of each frame is $\frac{1}{30}$ s.

a) Count and record in your log the number of frames during which the skater is in the air.

b) Calculate the skater's "hang time." (Show your calculation in your log.)

$$\text{Time in air (s)} = \text{Number of frames} \times \frac{1}{30}\text{ s}$$

c) Did the skater "hang" in the air during any part of the jump, appearing to "defy gravity"? If necessary, view the slow-motion sequence again to make the observations necessary to answer this question in your log. If your observations indicate that hanging did occur, be sure to indicate the exact frames during which it happened.

2. Your teacher will show you a similar slow-motion video of a basketball player whose hang time is believed by many fans to clearly defy gravity.

a) Using the same method as above for the skater, show in your log the data and calculations used to determine the player's hang time during the "slam dunk."

b) Did the player hang? Cite evidence from the video in your answer.

3. How much force and energy does a person use to do a vertical jump? A person uses body muscles to "launch" the body into the air, and, primarily, it is leg muscles that provide the force. First, analyze only the part of jumping that happens before the feet leave the ground. Find your body mass, in kilograms, and your body weight, in newtons, for later calculations. If you wish not to use data for your own body, you may use the data for another person who is willing to share the information with you. (See **Activity 2**, **Physics To Go**, **Question 7**, for how to convert your body weight in pounds to weight in newtons and mass in kilograms.)

a) Record your weight, in newtons, and mass, in kilograms, in your log.

4. Recall the location of your body's center of mass from **Activity 3**. Place a patch of tape on either the right or left side of your clothing (above one hip) at the same level as your body's center of mass. Crouch as if you are ready to make a vertical jump. While crouched, have an assistant measure the vertical distance, in meters, from the floor to the level of your body's center of mass (C of M).

   a) In your log, record the distance, in meters, from the floor to your C of M in the ready position.

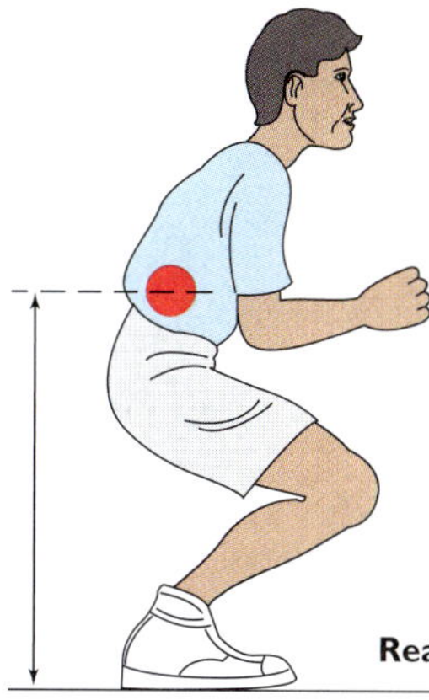

Ready position

5. Straighten your body and rise to your tiptoes as if you are ready to leave the floor to launch your body into a vertical jump, but don't jump. Hold this launch position while an assistant measures the vertical distance from the floor to the level of your center of mass.

   a) In your log, record the distance, in meters, from the floor to your C of M in the launch position.

   b) By subtraction, calculate and record the vertical height through which you used your leg muscles to provide the force to lift your center of mass from the "ready" position to the "launch" position. Record this in your log as legwork height.

   Legwork height = Launch position – Ready position

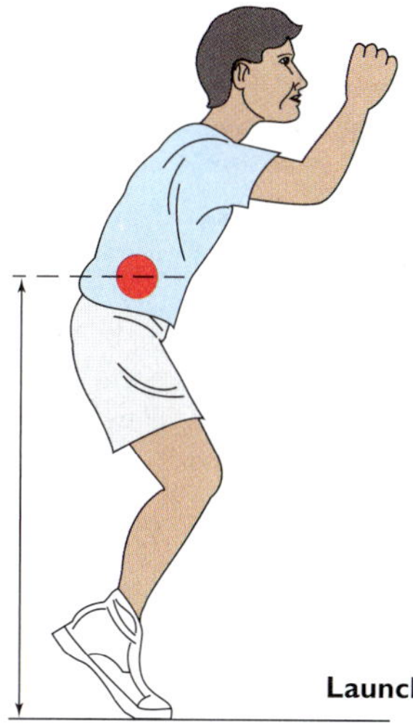

Launch position

6. Now it's time to jump! Have an assistant ready to observe and measure the vertical height from the floor to the level of your center of mass at the peak of your jump. When your assistant is ready to observe, jump straight up as high as you can. (Can you hang at the peak of your jump for a while to make it easier for your assistant to observe the position of your center of mass? Try it, and see if your assistant thinks you are successful.)

Make sure the floor is dry and the area in which you are jumping is clear of obstructions.

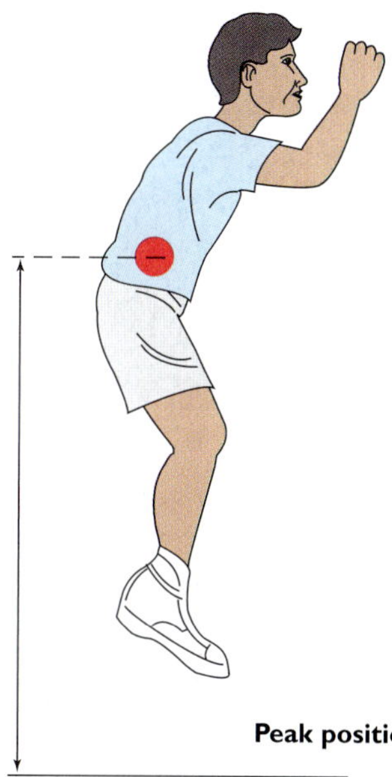

Peak position

Physics Words

work: the product of the displacement and the force in the direction of the displacement; work is a scalar quantity.

potential energy: energy that is dependent on the position of the object.

kinetic energy: the energy an object possesses because of its motion.

a) In your log, record the distance from the floor to C of M at peak position.

b) By subtraction, calculate and record the vertical height through which your center of mass moved during the jump.

Jump height = Peak position – Launch position

7. The information needed to analyze the muscular force and energy used to accomplish your jump—and an example of how to use sample data from a student's jump to perform the analysis—is presented in **Physics Talk** and the Example Analysis.

a) Use the information presented in the **Physics Talk** and Example Analysis sections and the data collected during above **Steps 4** through **6** to calculate the hang time and the total force provided by *your* leg muscles during your vertical jump. Show as much detail in your log as is shown in the Example Analysis.

8. An ultrasonic ranging device coupled to a computer or graphing calculator, which can be used to monitor position, speed, acceleration, and time for moving objects, may be available at your school. If so, it could be used to monitor a person doing a vertical jump. This would provide interesting information to compare to the data and analysis that you already have for the vertical jump. Check with your teacher to see if this would be possible.

FOR YOU TO READ

Conservation of Energy

In this activity you jumped and measured your vertical leap. You went through a chain of energy conversions where the total energy remained the same, in the absence of air resistance. You began by lifting your body from the crouched "ready" position to the "launch" position. The **work** that you did was equal to the product of the applied force and the distance. The work done must have lifted you from the ready position to the launch position (an increase in **potential energy**) and also provided you with the speed to continue moving up (the **kinetic energy**). After you left the ground, your body's potential energy continued to increase, and the kinetic energy decreased. Finally, you reached the peak of your jump, where all of the energy became potential energy. On the way down, that potential energy began to decrease and the kinetic energy began to increase.

When you are in the ready position, you have elastic potential energy. As you move toward the launch position, you have exchanged your elastic potential energy for an increase in gravitational potential energy and an increase in kinetic energy. As you rise in the air, you lose the kinetic energy and gain more gravitational potential energy. You can show this in a table.

Energy → / Position ↓	Elastic potential energy	Gravitational potential energy = mgh	Kinetic energy $= \frac{1}{2} mv^2$
ready position	maximum	0	0
launch position	0	some	maximum
peak position	0	maximum	0

The energy of the three positions must be equal. In this first table, the sum of the energies in each row must be equal. The launch position has both gravitational potential energy and kinetic energy. Using the values in the activity, the total energy at each position is 410 J.

Energy → / Position ↓	Elastic potential energy	Gravitational potential energy = mgh	Kinetic energy $= \frac{1}{2} mv^2$
ready position	410 J	0	0
launch position	0	150 J	260 J
peak position	0	410 J	0

→

In the ready position, all 410 J is elastic potential energy. In the peak position, all 410 J is gravitational potential energy. In the launch position, the total energy is still 410 J but 150 J is gravitational potential energy and 260 J is kinetic energy.

Consider someone the same size, who can jump much higher. Since that person can jump much higher, the peak position is greater, and therefore the gravitational potential energy of the jumper is greater. In the example shown below, the gravitational potential energy is 600 J. Notice that this means the elastic potential energy of the jumper's legs must be 600 J. And when the jumper is in the launch position, the total energy (potential plus kinetic) is also 600 J.

Energy→ / Position↓	Elastic potential energy	Gravitational potential energy = mgh	Kinetic energy $= \frac{1}{2}mv^2$
ready position	600 J	0	0
launch position	0	150 J	450 J
peak position	0	600 J	0

A third person of the same size is not able to jump as high. What numbers should be placed in blank areas to preserve the principle of conservation of energy?

Total energy must be conserved. Therefore, in the launch position the kinetic energy of the jumper must be 50 J. In the peak position, all the energy is in potential energy and must be 200 J.

The conservation of energy is a unifying principle in all science. It is worthwhile to practice solving problems that will help you to see the variety of ways in which energy conservation appears.

Energy→ / Position↓	Elastic potential energy	Gravitational potential energy = mgh	Kinetic energy $= \frac{1}{2}mv^2$
ready position	200 J	0	0
launch position	0	150 J	50 J
peak position	0	200 J	0

A similar example to jumping from a hard floor into the air is jumping on a trampoline (or your bed, when you were younger). If you were to jump on the trampoline, the potential energy from the height you are jumping would provide kinetic energy when you landed on the trampoline. As you continued down, you would continue to gain speed because you would still be losing gravitational potential energy. The trampoline bends and/or the springs holding the trampoline stretch. Either way, the trampoline or springs gain elastic potential energy at the expense of the kinetic energy and the changes in potential energy.

Energy → / Position ↓	Elastic potential energy	Gravitational potential energy = *mgh*	Kinetic energy = $\frac{1}{2}mv^2$
High in the air position	0	2300 J	0
Landing on the trampoline position	0	500 J	1800 J
Lowest point on the trampoline position	2300 J	0	0

A pole-vaulter runs with the pole. The pole bends. The pole straightens and pushes the vaulter into the air. The vaulter gets to his highest point, goes over the bar, and then falls back to the ground, where he lands on a soft mattress. You can analyze the pole-vaulter's motion in terms of energy conservation. (Ignore air resistance.)

A pole-vaulter runs with the pole. (*The vaulter has kinetic energy.*) The pole bends. (*The vaulter loses kinetic energy, and the pole gains elastic potential energy as it bends.*) The pole unbends and pushes the vaulter into the air. (*The pole loses the elastic potential energy, and the vaulter gains kinetic energy and gravitational potential energy.*) The vaulter gets to his highest point (*the vaulter has almost all gravitational potential energy*) goes over the bar, and then falls back to the ground (*the gravitational potential energy becomes kinetic energy*), where he lands on a soft mattress (*the kinetic energy becomes the elastic potential energy of the mattress, which then turns to heat energy*). The height the pole-vaulter can reach is dependent on the total energy that he starts with. The faster he runs, the higher he can go.

The conservation of energy is one of the great discoveries of science. You can describe the energies in words (elastic potential energy, gravitational potential energy, kinetic energy, and heat energy). There is also sound energy, →

light energy, chemical energy, electrical energy, and nuclear energy. The words do not give the complete picture. Each type of energy can be measured and calculated. In a closed system, the total of all the energies at any one time must equal the total of all the energies at any other time. That is what is meant by the conservation of energy.

If you choose to look at one object in the system, that one object can gain energy. For example, in the collision between a player's foot and a soccer ball, the soccer ball can gain kinetic energy and move faster. Whatever energy the ball gained, you can be sure that the foot lost an equal amount of energy. The ball gained energy, the foot lost energy, and the "ball and foot" total energy remained the same. The ball gained energy because work (force x distance) was done on it. The foot lost energy because work (force x distance) was done on it. The total system of "ball and foot" neither gained nor lost energy.

Physics provides you with the means to calculate energies. You may wish to practice some of these calculations now. Never lose sight of the fact that you can calculate the energies because the sum of all of the energies remains the same.

The equations for work, gravitational potential energy, and kinetic energy are given below.

The equation for work is:

$$W = F \cdot d$$

Work is done only when the force and displacement are (at least partially) in the same (or opposite) directions.

The equation for gravitational potential energy is:

$$PE_{gravitational} = mgh = wh$$

The w represents the weight of the object in newtons, where $w = mg$. On Earth's surface, when dealing with g in this course, consider it to be equal to 9.8 m/s^2.(Sometimes we use 10 m/s^2 for ease of calculations.)

The equation for kinetic energy is:

$$KE = \frac{1}{2}mv^2$$

Sample Problem

A trainer lifts a 5.0-kg equipment bag from the floor to the shelf of a locker. The locker is 1.6 m off the floor.

a) How much force will be required to lift the bag off the floor?

b) How much work will be done in lifting the bag to the shelf?

c) How much potential energy does the bag have as it sits on the shelf?

d) If the bag falls off the shelf, how fast will it be going when it hits the floor?

Strategy: This problem has several parts. It may look complicated, but if you follow it step by step, it should not be difficult to solve.

Part (a): Why does it take a force to lift the bag? It takes a force because the trainer must act against the pull of the gravitational field of the Earth. This force is called weight, and you can solve for it using Newton's Second Law.

Part (b): The information you need to find the work done on an object is the force exerted on it and the distance it travels. The distance was given and you calculated the force needed. Use the equation for work.

Part (c): The amount of potential energy depends on the mass of the object, the acceleration due to gravity, and the height of the object above what is designated as zero height (in this case, the floor). You have all the needed pieces of information, so you can apply the equation for potential energy.

Part (d): The bag has some potential energy. When it falls off the shelf, the potential energy becomes kinetic energy as it falls. When it strikes the ground in its fall, it has zero potential energy and all kinetic energy. You calculated the potential energy. Conservation of energy tells you that the kinetic energy will be equal to the potential energy. You know the mass of the bag so you can calculate the velocity with the kinetic energy formula.

Givens:

$$m = 5.0 \text{ kg}$$
$$h = 1.6 \text{ m}$$
$$a = 9.8 \text{ m/s}^2$$

Solution:

a)

$$\begin{aligned} F &= ma \\ &= (5.0 \text{ kg})(9.8 \text{ m/s}^2) \\ &= 49 \text{ kg} \cdot \text{m/s}^2 \text{ or } 49 \text{ N} \end{aligned}$$

b)

$$\begin{aligned} W &= F \cdot d \\ &= (49 \text{ N})(1.6 \text{ m}) \\ &= 78.4 \text{ Nm or } 78 \text{ J (Nm = J)} \end{aligned}$$

c)

$$\begin{aligned} PE_{gravitational} &= mgh \\ &= (5.0 \text{ kg})(9.8 \text{ m/s}^2)(1.6 \text{ m}) \\ &= 78 \text{ J} \end{aligned}$$

Should you be surprised that this is the same answer as **Part (b)**? No, because you are familiar with energy conservation. You know that the work is what gave the bag the potential energy it has. So, in the absence of work that may be converted to heat because of friction, which you did not have in this case, the work equals the potential energy.

d)

$$\begin{aligned} KE &= \tfrac{1}{2}mv^2 \\ v^2 &= \frac{KE}{\tfrac{1}{2}m} \\ &= \frac{78 \text{ J}}{\tfrac{1}{2}(5.0 \text{ kg})} \\ &= 31 \text{ m}^2/\text{s}^2 \\ v &= 5.6 \text{ m/s} \end{aligned}$$

PHYSICS TALK

Work

When you lifted your body from the ready (crouched) position to the launch (standing on tiptoes) position before takeoff during the vertical jump activity, you performed what physicists call work. In the context of physics, the word *work* is defined as:

The work done when a constant force is applied to move an object is equal to the amount of applied force multiplied by the distance through which the object moves in the direction of the force.

You used symbols to write the definition of work as:

$$W = F \cdot d$$

where F is the applied force in newtons, d is the distance the object moves in meters, and work is expressed in joules (symbol, J). At any time it is desired, the unit "joule" can be written in its equivalent form as force times distance, "(N)(m)."

The unit "newton" can be written in the equivalent form "(kg)m/s^2." Therefore, the unit joule also can be written in the equivalent form (kg)m^2/s^2. In summary, the units for expressing work are:

$$1 \text{ J} = 1 \text{ (N)(m)} = 1 \text{ (kg)m}^2\text{/s}^2$$

As you read, it is very common in sports that work is transformed into kinetic energy, and then, in turn, the kinetic energy is transformed into gravitational potential energy. This chain of transformations can be written as:

$$\text{Work} = KE = PE$$

$$Fd = \frac{1}{2}mv^2 = mgh$$

These transformations are used in the analysis of data for a vertical jump.

Example:
Calculation of Hang Time and Force During Vertical Jump

DATA: Body Weight = 100 pounds = 440 N
Body Mass = 44 kg
Legwork Height = 0.35 m
Jump Height = 0.60 m

Analysis:
Work done to lift the center of mass from ready position to launch position without jumping ($W_{R\ to\ L}$):

$$W_{R\ to\ L} = Fd = (\text{Body Weight}) \times (\text{Legwork Height}) = 440\ \text{N} \times 0.35\ \text{m} = 150\ \text{J}$$

Gravitational Potential Energy gained from jumping from launch position to peak position (PE_J):

$$PE_J = mgh = (\text{Body Mass}) \times (g) \times (\text{Jump Height}) = 44\ \text{kg} \times 10\ \text{m/s}^2 \times 0.60\ \text{m} = 260\ (\text{kg})\text{m}^2/\text{s}^2 = 260\ (\text{N})(\text{m}) = 260\ \text{J}$$

The jumper's kinetic energy at takeoff was transformed to increase the potential energy of the jumper's center of mass by 260 J from launch position to peak position. Conservation of energy demands that the kinetic energy at launch be 260 J:

$$KE = \frac{1}{2}mv^2 = 260\ \text{J}$$

This allows calculation of the jumper's launch speed:

$$v = \sqrt{2(KE)/m} = \sqrt{2(260\ \text{J})/(44\ \text{kg})} = 3.4\ \text{m/s}$$

From the definition of acceleration, $a = \Delta v/\Delta t$, the jumper's time of flight "one way" during the jump was:

$$\Delta t = \Delta v/a = (3.4\ \text{m/s}) / (10\ \text{m/s}^2) = 0.34\ \text{s}$$

Therefore, the total time in the air (hang time) was $2 \times 0.34\ \text{s} = 0.68\ \text{s}$.

→

The total work done by the jumper's leg muscles before launch, W_T, was the work done to lift the center of mass from ready position to launch position without jumping, $W_{R\ to\ L}$ = 150 J, plus the amount of work done to provide the center of mass with 260 J of kinetic energy at launch, a total of 150 J + 260 J = 410 J. Rearranging the equation $W = F \cdot d$ into the form $F = W/d$, the total force provided by the jumper's leg muscles, F_T was:

$$F_T = \frac{W_T}{\text{(Legwork Height)}}$$

$$= 410\text{ J} / 0.35\text{ m}$$

$$= 1200\text{ N}$$

Approximately one-third of the total force exerted by the jumper's leg muscles was used to lift the jumper's center of mass to the launch position, and approximately two-thirds of the force was used to accelerate the jumper's center of mass to the launch speed.

Reflecting on the Activity and the Challenge

Work, the force applied by an athlete to cause an object to move (including the athlete's own body as the object in some cases), multiplied by the distance the object moves while the athlete is applying the force explains many things in sports. For example, the vertical speed of any jumper's takeoff (which determines height and "hang time") is determined by the amount of work done against gravity by the jumper's muscles before takeoff. You will be able to find many other examples of work in action in sports videos, and now you will be able to explain them.

Physics To Go

1. How much work does a male figure skater do when lifting a 50-kg female skating partner's body a vertical distance of 1 m in a pairs competition?

2. Describe the energy transformations during a bobsled run, beginning with team members pushing to start the sled and ending when the brake is applied to stop the sled after crossing the finish line. Include work as one form of energy in your answer.

3. Suppose that a person who saw the video of the basketball player used in **For You To Do** said, "He really can hang in the air. I've seen him do it. Maybe he was just having a 'bad hang day' when the video was taken, or maybe the speed of the camera or VCR was 'off.' How do I know that the player in the video wasn't a 'look-alike' who can't hang?" Do you think these are legitimate statements and questions? Why or why not?

4. If someone claims that a law of physics can be defied or violated, should the person making the claim need to provide observable evidence that the claim is true, or should someone else need to prove that the claim is not true? Who do you think should have the burden of proof? Discuss this issue within your group and write your own personal opinion in your log.

5. Identify and discuss two ways in which an athlete can increase his or her maximum vertical jump height.

6. Calculate the amount of work, in joules, done when:

   a) a 1.0-N weight is lifted a vertical distance of 1.0 m.
   b) a 1.0-N weight is lifted a vertical distance of 10 m.
   c) a 10-N weight is lifted a vertical distance of 1.0 m.
   d) a 0.10-N weight is lifted a vertical distance of 100 m.
   e) a 100-N weight is lifted a distance of 0.10 m.

7. List how much gravitational potential energy, in joules, each of the weights in **Question 6** above would have when lifted to the height listed for it.

8. List how much kinetic energy, in joules, each of the weights in **Questions 6** and **7** would have at the instant before striking the ground if each weight were dropped from the height listed for it.

9. How much work is done on a go-cart if you push it with a force of 50.0 N and move it a distance of 43 m?

10. What is the kinetic energy of a 62-kg cyclist if she is moving on her bicycle at 8.2 m/s?

11. A net force of 30.00 N acts on a 5.00-kg wagon that is initially at rest.
    a) What is the acceleration of the wagon?
    b) If the wagon travels 18.75 m, what is the work done on the wagon?

12. Assume you do 40,000 J of work by applying a force of 3200 N to a 1200-kg car.
    a) How far will the car move?
    b) What is the acceleration of the car?

13. A baseball (m = 150.0 g) is traveling at 40.0 m/s. How much work must be done to stop the ball?

14. A boat exerts a force of 417 N pulling a water-skier (m = 64.0 kg) from rest. The skier's speed is now 15.0 m/s. Over what distance was this force exerted?

Activity 5 Run and Jump

GOALS

In this activity you will:

- Understand the definition of acceleration.
- Understand meters per second per second as the unit of acceleration.
- Use an accelerometer to detect acceleration.
- Use an accelerometer to make semiquantitative comparisons of accelerations.
- Distinguish between acceleration and deceleration.

What Do You Think?

The men's high jump record is over 8 feet.

• Pretend that you have just met somebody who has never jumped before. What instructions could you provide to get the person to jump up (that is, which way do you apply the force)?

Record your ideas about this question in your *Active Physics* log. Be prepared to discuss your responses with your small group and the class.

For You To Do

1. Carefully stand on a skateboard or sit on a wheeled chair near a wall. By touching only the wall, not the floor, cause yourself to move away from the wall to "coast" across the floor. Use words and diagrams to record answers to the following questions in your log:

   a) When is your motion accelerated? For what distance does the accelerated motion last? In what direction do you accelerate?

Physics Words

Newton's Third Law of Motion: forces come in pairs; the force of object A on object B is equal and opposite to the force of object B on object A.

b) When is your motion at constant speed? Neglecting the effects of friction, how far should you travel? (Remember Galileo's Principle of Inertia when answering this question.)

c) Newton's Second Law, $F = ma$, says that a force must be active when acceleration occurs. What is the source of the force, the push or pull, that causes you to accelerate in this case? Identify the object that does the pushing on your mass (body plus skateboard) to cause the acceleration. Also identify the direction of the push that causes you to accelerate.

d) Obviously, you do some pushing, too. On what object do you push? In what direction?

e) How do you think, on the basis of both amount and direction, the following two forces compare?
- The force exerted by you on the wall
- The force exerted by the wall on you

2. Do a "thought experiment" about the forces involved when you are running or walking on a horizontal surface. Use words and sketches to answer the following questions in your log:

a) With each step, you push the bottom surface of your shoe, the sole, horizontally backward. The force acts parallel to the surface of the ground, trying to scrape the ground in the direction opposite your motion. Usually, friction is enough to prevent your shoe from sliding across the ground surface.

b) Since you move forward, not rearward, there must be a force in the forward direction that causes you to accelerate. Identify where the forward force comes from, and compare its amount and direction to the rearward force exerted by your shoe with each step.

c) Would it be possible to walk or run on an extremely slippery skating rink when wearing ordinary shoes? Discuss why or why not in terms of forces.

3. Think about the vertical forces acting on you while you are standing on the floor.

a) Copy the diagram of a person at left in your log.

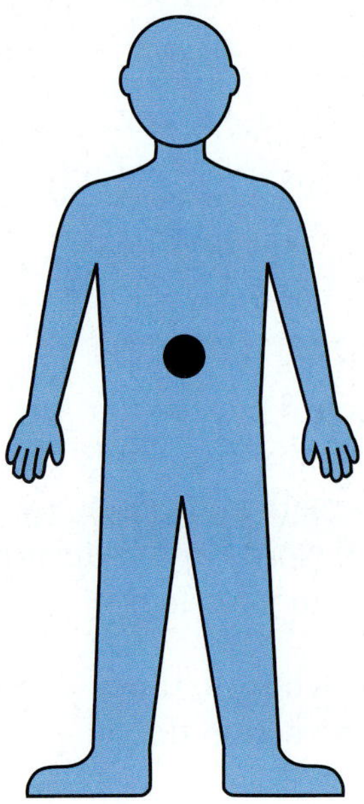

b) Identify all the vertical forces. Use an arrow to designate the size and direction of the force. Draw the forces from the dot.

c) How can you be sure that the force with which you push on the floor and the floor pushes on you are equal?

4. Set up a meter stick with a few books for support as shown.

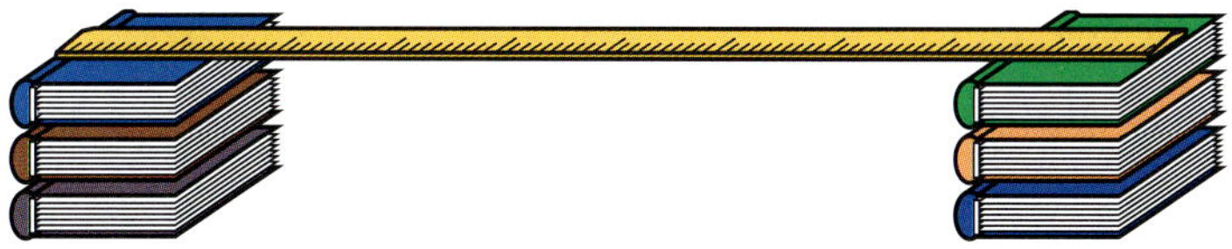

5. Place a washer in the center of the meter stick.

a) In your log, record what happens.

6. Remove the washer and replace it with 100 g (weight of 100 g = 1.0 N). Continue to place 1.0 N weights on the center of the meter stick. Note what happens as you place each weight on the stick.

Do not exceed 10 N of weight.

a) Measure the deflection of the meter stick for each 1.0 N of weight and record the values for these deflections.

b) How does the deflection of the meter stick compare to the weight it is supporting? In your log, sketch a graph to show this relationship.

c) Write a concluding statement concerning the washer and the deflection of the meter stick.

PHYSICS TALK

Newton's Third Law of Motion

Newton's **Third Law of Motion** can be stated as:

For every applied force, there is an equal and opposite force.

If you push or pull on something, that something pushes or pulls back on you with an equal amount of force in the opposite direction. This is an inescapable fact; it happens every time.

Reflecting on the Activity and the Challenge

According to Newton's Third Law, each time an athlete acts to exert a force on something, an equal and opposite force happens in return. Countless examples of this exist as possibilities to include in your video production. When you kick a soccer ball, the soccer ball exerts a force on your foot. When you push backward on the ground, the ground pushes forward on you (and you move). When a boxer's fist exerts a force on another boxer's body, the body exerts an equal force on the fist. Indeed, it should be rather easy to find a video sequence of a sport that illustrates all three of Newton's Laws of Motion.

Physics To Go

1. When an athlete is preparing to throw a shot put ball, does the ball exert a force on the athlete's hand equal and opposite to the force the hand exerts on the ball?

2. When you sit on a chair, the seat of the chair pushes up on your body with a force equal and opposite to your weight. How does the chair know exactly how hard to push up on you—are chairs intelligent?

3. For a hit in baseball, compare the force exerted by the bat on the ball to the force exerted by the ball on the bat. Why do bats sometimes break?

4. Compare the amount of force experienced by each football player when a big linebacker tackles a small running back.

5. Identify the forces active when a hockey player "hits the boards" at the side of the rink at high speed.

6. Newton's Second Law, $F = ma$, suggests that when catching a baseball in your hand, a great amount of force is required to stop a high-speed baseball in a very short time interval. The great amount of force is needed to provide the great amount of deceleration required. Use Newton's Third Law to explain why baseball players prefer to wear gloves for catching high-speed baseballs. Use a pair of forces in your explanation.

7. Write a sentence or two explaining the physics of an imaginary sports clip using Newton's Third Law. How can you make this description more exciting so that it can be used as part of your sports voice-over?

8. Write a sentence or two explaining the concept that a deflection of the ground can produce a force. How can you make this description more exciting so that it can be used as part of your sports voice-over?

Stretching Exercises

Ask the manager of a building that has an elevator for permission to use the elevator for a physics experiment. Your teacher may be able to help you make the necessary arrangements.

1. Stand on a bathroom scale in the elevator and record the force indicated by the scale while the elevator is:
   a) At rest.
   b) Beginning to move (accelerating) upward.
   c) Seeming to move upward at constant speed.
   d) Beginning to stop (decelerating) while moving upward.
   e) Beginning to move (accelerating) downward.
   f) Seeming to move downward at constant speed.
   g) Beginning to stop (decelerating) while moving downward.

2. For each of the above conditions of the elevator's motion, the Earth's downward force of gravity is the same. If you are accelerating up, the floor must be pushing up with a force larger than the acceleration due to gravity.
   a) Make a sketch that shows the vertical forces acting on your body.
   b) Use Newton's Laws of Motion to explain how the forces acting on your body are responsible for the kind of motion—at rest, constant speed, acceleration, or deceleration—that your body has.

Activity 6 The Mu of the Shoe

GOALS

In this activity you will:

- Understand and apply the definition of the coefficient of sliding friction, μ.
- Measure the coefficient of sliding friction between the soles of athletic shoes and a variety of floor surface materials.
- Calculate the effects of frictional forces on the motion of objects.

What Do You Think?

A shoe store may sell as many as 100 different kinds of sport shoes.

- **Why do some sports require special shoes?**

Record your ideas about this question in your *Active Physics* log. Be prepared to discuss your responses with your small group and the class.

For You To Do

1. Take an athletic shoe. Use a spring scale to measure the weight of the shoe, in newtons.

   a) Record a description of the shoe (such as its brand) and the shoe's weight, in your log.

2. Place the shoe on one of two horizontal surfaces (either rough or smooth) designated by your teacher to be used for testing. Attach the spring scale to the shoe as shown below so that the spring scale can be used to slide the shoe across the surface while, at the same time, the amount of force indicated by the scale can be read.

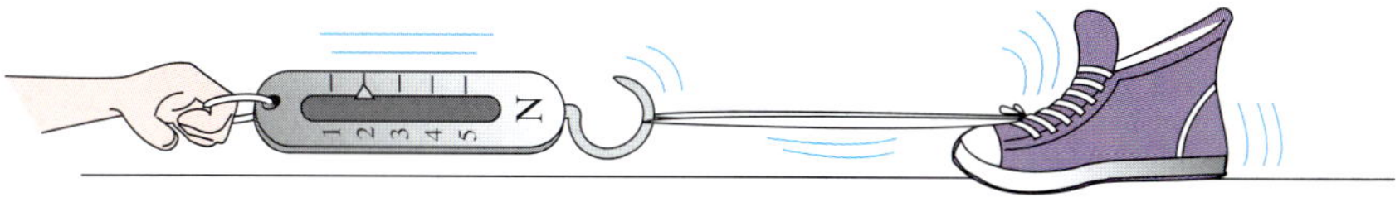

a) Record in your log a description of the surface on which the shoe is to slide.

b) Measure and record the amount of force, in newtons, needed to cause the shoe to slide on the surface at constant speed. Do not measure the force needed to start, or "tear the shoe loose," from rest. Measure the force needed, after the shoe has started moving, to keep it sliding at low, constant speed. Also, be careful to pull horizontally so that the applied force neither tends to lift the shoe nor pull downward on the shoe.

c) Use the data you have gathered to calculate μ, the coefficient of sliding friction for this particular kind of shoe on the particular kind of surface used. Show your calculations in your log.

The coefficient of sliding friction, symbolized by μ, is calculated using the following equation:

$$\mu = \frac{\text{force required to slide object on surface at constant speed}}{\text{perpendicular force exerted by the surface on the object}}$$

Example:

Brand X athletic shoe has a weight of 5 N. If 1.5 N of applied horizontal force is required to cause the shoe to slide with constant speed on a smooth concrete floor, what is the coefficient of sliding friction?

$$\mu_{x \text{ on concrete}} = \frac{1.5\text{ N}}{5.0\text{ N}} = 0.30$$

3. Add "filler" to the shoe to approximately double its weight and repeat the above procedure for measuring the μ of the shoe.

a) Calculate μ for this surface, showing your work in your log.

b) Taking into account possible errors of measurement, does the weight of the shoe seem to affect μ? Use data to answer the question in your log.

c) How do you think the weight of an athlete wearing the shoe would affect μ? Why?

4. Place the shoe on the second surface designated by your teacher and repeat the procedure.

a) Make another sketch to show the forces acting on the shoe.

b) Calculate μ.

c) How does the value of μ for this surface compare to μ for the first surface used? Try to explain any difference in μ.

d) Would it make any difference if you used the empty shoe or the shoe with the filler to calculate μ in this activity? Explain your answer.

Reflecting on the Activity and the Challenge

Many athletes seem more concerned about their shoes than most other items of equipment, and for good reason. Small differences in the shoes (or skates or skis) athletes wear can affect performance. As everyone knows, athletic shoes have become a major industry because people in all "walks" of life have discovered that athletic shoes are great to wear, not only on a track but, as well, just about anywhere. Now that you have studied friction, a major aspect of what makes shoes function well when need exists to be "sure-footed," you are prepared to do "physics commentary" on athletic footgear and other effects of friction in sports. Your sports commentary may discuss the μ of the shoe, the change in friction when a playing field gets wet, and the need for friction when running.

PHYSICS TALK

Coefficient of Sliding Friction, μ

There are not enough letters in the English alphabet to provide the number of symbols needed in physics, so letters from another alphabet, the Greek alphabet, also are used as symbols. The letter μ, pronounced "mu," traditionally is used in physics as the symbol for the "coefficient of sliding friction."

The coefficient of sliding friction, symbolized by μ, is defined as the ratio of two forces:

$$\mu = \frac{\text{force required to slide object on surface at constant speed}}{\text{perpendicular force exerted by the surface on the object}}$$

Facts about the coefficient of sliding friction:

- **μ does not have any units because it is a force divided by a force; it has no unit of measurement.**
- **μ usually is expressed in decimal form, such as 0.85 for rubber on dry concrete (0.60 on wet concrete).**
- **μ is valid only for the pair of surfaces in contact when the value is measured; any significant change in either of the surfaces (such as the kind of material, surface texture, moisture, or lubrication on a surface, etc.) may cause the value of μ to change.**
- **Only when sliding occurs on a horizontal surface, and the pulling force is horizontal, is the perpendicular force that the sliding object exerts on the surface equal to the weight of the object.**

Physics To Go

1. Identify a sport and changing weather conditions that probably would cause an athlete to want to increase friction to have better footing. Name the sport, describe the change in conditions, and explain what the athlete might do to increase friction between the shoes and ground surface.

2. Identify a sport in which athletes desire to have frictional forces as small as possible and describe what the athletes do to reduce friction.

3. If a basketball player's shoes provide an amount of friction that is "just right" when she plays on her home court, can she be sure the same shoes will provide the same amount of friction when playing on another court? Explain why or why not.

4. A cross-country skier who weighs 600 N has chosen ski wax that provides $\mu = 0.03$. What is the minimum amount of horizontal force that would keep the skier moving at constant speed across level snow?

5. A racecar having a mass of 1000 kg was traveling at high speed on a wet concrete road under foggy conditions. The tires on the vehicle later were measured to have $\mu = 0.55$ on that road surface. Before colliding with the guardrail, the driver locked the brakes and skidded 100 m, leaving visible marks on the road. The driver claimed not to have been exceeding 65 miles per hour (29 m/s). Use the equation:

$$\text{Work} = \text{Kinetic Energy}$$

to estimate the driver's speed upon hitting the brakes. (Hint: In this case, the force that did the work to stop the car was the frictional force; calculate the frictional force using the weight of the vehicle, in newtons, and use the frictional force as the force for calculating work.)

6. Identify at least three examples of sports in which air or water have limiting effects on motion similar to sliding friction. Do you think forces of "air resistance" and "water resistance" remain constant or do they change as the speeds of objects (such as athletes, bobsleds, or rowing sculls) moving through them change? Use examples from your own experience with these forms of resistance as a basis for your answer.
7. If there is a maximum frictional force between your shoe and the track, does that set a limit on how fast you can start (accelerate) in a sprint? Does that mean you cannot have more than a certain acceleration even if you have incredibly strong leg muscles? What is done to solve this problem?
8. How might an athletic shoe company use the results of your experiment to "sell" a shoe? Write copy for such an advertisement.
9. Explain why friction is important to running. Why are cleats used in football, soccer, and other sports?
10. Choose a sport and describe an event in which friction with the ground or the air plays a significant part. Create a voice-over or script that uses physics to explain the action.

Activity 7 Concentrating on Collisions

GOALS

In this activity you will:

- Understand and apply the definition of momentum.
- Conduct semiquantitative analyses of the momentum of pairs of objects involved in one-dimensional collisions.
- Infer the relative masses of two objects by staging and observing collisions between the objects.

Moving bowling balls can cause injury to people and property. Be careful!

What Do You Think?

In contact sports, very large forces happen during short time intervals.

- **A football player runs toward the goal line, and a defensive player tries to stop him with a head-on collision. What factors determine whether the offensive player scores?**

Record your ideas about this question in your *Active Physics* log. Be prepared to discuss your responses with your small group and the class.

For You To Do

1. You will stage a head-on collision between two matched bocci balls. Set up a launch ramp for one ball, and find a level area clear of obstructions nearby where the other bocci ball can be at rest.

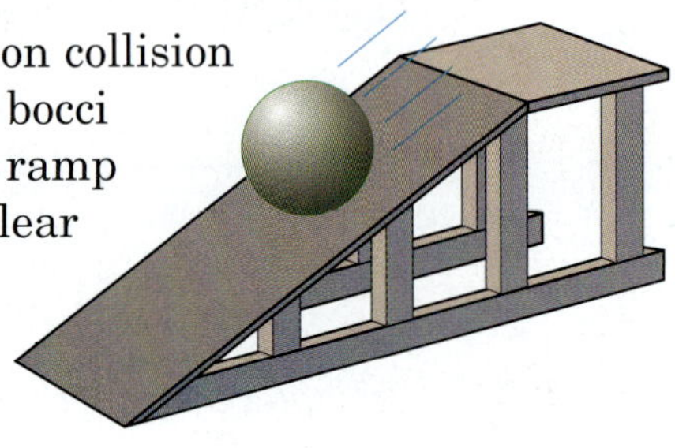

2. Temporarily remove the "target" bocci ball. Find a point of release within the first one-fourth of the ramp's total length that gives the ball a slow, steady speed across the floor. Mark the point of release on the ramp with a piece of tape.

3. Replace the target ball. Adjust the aim until a good approximation of a head-on collision is obtained. Stage the collision.

   a) Record the results in your log. Use a diagram and words to describe what happened to each ball.

4. Repeat the above type of collision, but this time move the release point up the ramp to at least double the ramp distance.

   a) Describe the results in your log.

   b) How did the results of the collision change from the first time?

   c) Identify a real-life situation that this collision could represent.

5. Arrange another head-on collision between the balls, but this time have both balls moving at equal speeds before the collision. Using a second, identical ramp, aim the second ramp so that the second ball's path is aligned with the first ball's path. Mark a release point on the second ramp at a height equal to the mark already made on the first ramp. This should ensure that the balls will have low, approximately equal speeds. On a signal, two persons should release the balls simultaneously from equal ramp heights.

   a) Describe the results in your log.

   b) Identify a real-life situation that this collision could represent.

Physics Words

momentum: the product of the mass and the velocity of an object; momentum is a vector quantity.

6. Repeat **Steps 1, 2, and 3**, but replace the stationary bocci ball with a nerf ball.
   a) Be sure to write all responses, including identification of a similar situation in real life, in your log.
7. Repeat **Steps 1, 2,** and **3**, but in this case have the nerf ball roll down the ramp to strike a stationary bocci ball.
   a) Be sure to write all responses, including identification of a similar situation in real life, in your log.
8. Using your observations, determine the relative mass of a golf ball compared to a wiffle ball by staging collisions between them.
   a) Which ball has the greater mass? How many times more massive is it than the other ball? Describe what you did to decide upon your answer.
   b) Use a scale or balance to check your result. Comment on how well observing collisions between the balls worked as a method of comparing their masses.

FOR YOU TO READ

Momentum

Taken alone, neither the masses nor the velocities of the objects were important in determining the collisions you observed in this activity. The crucial quantity is **momentum** (mass × velocity). A soccer ball has less mass than a bocci ball, but a soccer ball can have the same momentum as a bocci ball if the soccer ball is moving fast. A soccer ball moving very fast can affect a stationary bocci ball more than a soccer ball moving very slowly. This is similar to the damage small pieces of sand moving at very high speeds can cause (such as when a sand blaster is used to clean various surfaces).

Sportscasters often use the term *momentum* in a different way. When a team is doing well, or "on a roll," that team has momentum.

A team can gain or lose momentum, depending on how things are going. This momentum clearly does not refer to the mass of the entire team multiplied by the team's velocity.

Other times, sportscasters use the term momentum to mean exactly how it is defined in the activity (mass × velocity), when they say things such as, "Her momentum carried her out of bounds."

Reflecting on the Activity and the Challenge

You already have identified several real-life situations that involve collisions, and many such situations happen in sports. Some involve athletes colliding with one another as in hockey and football. Others cases include athletes colliding with objects, such as when kicking a ball. Still others include collisions between objects such as a golf club, bat, or racquet with a ball. Some spectacular collisions in sports provide fun opportunities for demonstrating your knowledge about collisions during voice-over commentaries. Use the concept of momentum when describing collisions in your sports video.

Physics To Go

1. Sports commentators often say that a team has momentum when things are going well for the team. Explain the difference between that meaning of the word momentum and its specific meaning in physics.

2. Suppose a running back collides with a defending linebacker who has just come to a stop. If both players have the same mass, what do you expect to see happen in the resulting collision?

3. Describe the collision of a running back and a linebacker of equal mass running toward each other at equal speeds.

4. Suppose that you have two baseball bats, a heavy (38-ounce) bat and a light (30-ounce) bat.
   a) If you were able to swing both bats at the same speed, which bat would allow you to hit the ball the farthest distance? Explain your answer.
   b) How fast would you need to swing the light bat to produce the same hitting effect as the heavy bat? Explain your answer.

5. Why do football teams prefer offensive and defensive linemen who weigh about 300 pounds?

6. What determines who will get knocked backward when a big hockey player checks a small player in a head-on collision?

7. A 100.0-kg athlete is running at 10.0 m/s. At what speed would a 0.10-kg ball need to travel in the same direction so that the momentum of the athlete and the momentum of the ball would be equal?

8. Use the words *mass*, *velocity*, and *momentum* to write a paragraph that gives a detailed "before and after" description of what happens when a moving shuffleboard puck hits a stationary puck of equal mass in a head-on collision.

9. Describe a collision in some sport by using the term *momentum*. Adapt this description to a 15-s dialogue that could be used as part of the voice-over for a video.

Activity 8 Conservation of Momentum

GOALS

In this activity you will:

- Understand and apply the Law of Conservation of Momentum.
- Measure the momentum before and after a moving mass strikes a stationary mass in a head-on, inelastic collision.

What Do You Think?

The outcome of a collision between two objects is predictable.

- **What determines the momentum of an object?**
- **What does it mean to "conserve" something?**

Record your ideas about these questions in your *Active Physics* log. Be prepared to discuss your responses with your small group and the class.

For You To Do

1. From the objects provided arrange to have a head-on collision between two objects of equal mass. Before the collision, have one object moving and the other object at rest. Arrange for the objects to stick together to move as a single object after the collision. Stage a head-on, sticky collision between equal masses. Measure the velocity, in meters per second, of the moving mass before the collision and the velocity of the combined masses after the collision.

a) Prepare a data table in your log similar to the one shown below. Provide enough horizontal rows in the table to enter data for at least four collisions.

Sticky Head-on Collisions: One Object Moving before Collision

Mass of Object 1 (kg)	Mass of Object 2 (kg)	Velocity of Object 1 before Collision (m/s)	Velocity of Object 2 before Collision (m/s)	Mass of Combined Objects after Collision (kg)	Velocity of Combined Objects after Collision (m/s)
1.0	1.0		0.0	2.0	
2.0	1.0		0.0	3.0	
1.0	2.0		0.0	3.0	
			0.0		

b) Record the measured values of the velocities in the first row of the data table.

2. Stage other sticky head-on collisions using the masses listed in the second and third rows of the data table. Then stage one or more additional collisions using other masses. Measure the velocities before and after each collision.

a) Enter the measured values in the data table.

3. Organize a table for recording the momentum of each object before and after each of the above collisions.

a) Prepare a table similar to the following example in your log:

Momentum of Object before and after Collisions
Momentum = Mass × Velocity

Before the Collision		After the Collision
Momentum of Object 1 kg (m/s)	Momentum of Object 2 kg (m/s)	Momentum of Combined Objects 1 and 2 kg (m/s)

b) Calculate the momentum of each object before and after each of the above collisions and enter each momentum value in the table.

c) Calculate and compare the total momentum before each collision to the total momentum after each collision.

d) Allowing for minor variations due to errors of measurement, write in your log a general conclusion about how the momentum before a collision compares to the momentum afterward.

FOR YOU TO READ

The Law of Conservation of Momentum

In this activity, you investigated another conservation principle that is a hallmark of physics—the conservation of momentum. If you sum all of the momenta before a collision or explosion, you know that the sum of all the momenta after the collision will be the same.

If the momentum before a collision is 500 kg·m/s, then the momentum after the collision is 500 kg·m/s. A football player stops to catch a pass. The player is not moving and therefore has momentum equal to zero. If an opponent that has a momentum of 500 kg·m/s then hits the player, both players move off with (a combined) 500 kg·m/s of momentum. Any time you see a collision in sports, you can explain that collision using the conservation of momentum.

Conservation of momentum is an experimental fact. Physicists have compared the momenta before and after a collision between pairs of objects ranging from railroad cars slamming together to subatomic particles impacting one another at near the speed of light. Never have any exceptions been found to the statement, "The total momentum before a collision is equal to the total momentum after the collision if no external forces act on the system." This statement is known as the Law of Conservation of Momentum. In all collisions between cars and trucks, between protons and protons, between planets and meteors, the momentum before the collision equals the momentum after.

A single cue ball hits a rack of 15 billiard balls and they all scatter. It would seem like everything has changed. Physicists have discovered that in this collision, as in all collisions and explosions, nature does keep at least one thing from changing—the total momentum. The sum of the momenta of all of the billiard balls immediately after the collision is equal to the momentum of the original cue ball. Nature loves momentum. Irrespective of the changes you visually note, the total momentum undergoes no change whatsoever. The objects may move in new directions and with new speeds, but the momentum stays the same. There aren't many of these conservation laws that are known.

Conservation of momentum can be shown to emerge from Newton's Laws. Newton's Third Law states that if object *A* and object *B* collide, the force of object *A* on *B* must be equal and opposite to the force of object *B* on *A*.

$$F_{A \text{ on } B} = -F_{B \text{ on } A}$$

The negative sign shows mathematically that the equally sized forces are in opposite directions.

Newton's Second Law states that $F = ma$. Also, acceleration, a, equals the change in velocity divided by the change in time ($a = \Delta v/\Delta t$):

$$m_B a_B = -m_A a_A$$

$$m_B \frac{\Delta v_B}{\Delta t} = -m_A \frac{\Delta v_A}{\Delta t}$$

$$m_B \frac{(v_f - v_i)_B}{\Delta t} = -m_A \frac{(v_f - v_i)_A}{\Delta t}$$

Since the change in time must be the same for both objects (*A* acts on *B* for as long as *B* acts on *A*), then Δt can be eliminated from both sides of the equation.

Combining the initial velocities (v_i) on one side of the equation and the final velocities (v_f) on the other side of the equation:

$$m_A v_{iA} + m_B v_{iB} = m_A v_{fA} + m_B v_{fB}$$

$$(m_A v_A)_{before} + (m_B v_B)_{before} =$$
$$(m_A v_A)_{after} + (m_B v_B)_{after}$$

Newton's Laws have yielded the conservation of momentum. The momentum of object *A* *before* the collision plus the momentum of object *B* *before* the collision equals the momentum of object *A* *after* the collision plus the momentum of object *B* *after* the collision

This equation not only works in one-dimensional collisions, but works equally well in the extraordinarily complex two-dimensional collisions of billiard balls and three-dimensional collisions of bowling.

Solving conservation of momentum problems is easy. Calculate each object's momentum before the collision. Calculate each object's momentum after the collision. The totals before the collision must equal the total after the collision.

There are a variety of collisions involving two objects. In each collision, momentum is conserved and the same equation is used. The equation gets simpler when one of the objects is at rest and has zero momentum. You may wish to draw two sketches for each collision—one showing each object before the collision and one showing each object after the collision. By writing the momenta you know directly on the sketch, the calculations become easier.

Collision Type 1: One moving object hits a stationary object and both stick together and move off at the same speed:

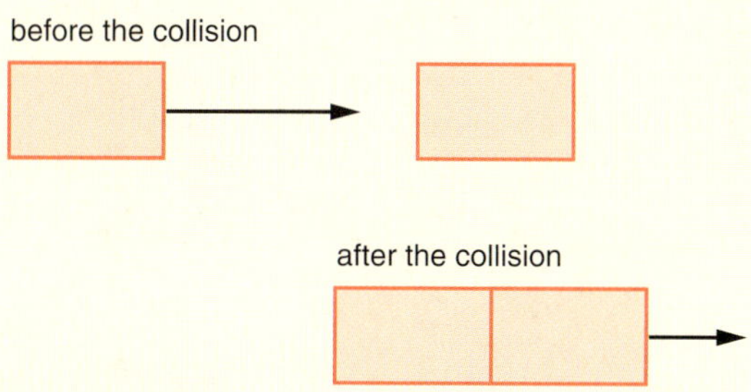

Collision Type 2: Two stationary objects explode and move off in opposite directions.

Collision Type 3: One moving object hits a stationary object. The first object stops, and the second object moves off.

Collision Type 4: One moving object hits a stationary object, and both move off at different speeds.

Collision Type 5: Two moving objects collide, and both objects move at different speeds after the collision.

Collision Type 6: Two moving objects collide, and both objects stick together and move off at the same speed.

Sample Problem 1

A 75.00-kg ice skater is moving to the east at 3.00 m/s toward his 50.00-kg partner, who is moving toward him (west) at 1.80 m/s. If he catches her up and they move away together, what is their final velocity?

Strategy: This is a problem involving the Law of Conservation of Momentum. The momentum of an isolated system before an interaction is equal to the momentum of the system after the interaction. As you are working through this problem, remember that the v in this expression is velocity and that it has direction as well as magnitude. Make east the positive direction, and then west will be negative.

Givens:

$$m_b = 75.00 \text{ kg}$$
$$m_g = 50.00 \text{ kg}$$
$$v_b = 3.00 \text{ m/s}$$
$$v_g = -1.80 \text{ m/s}$$

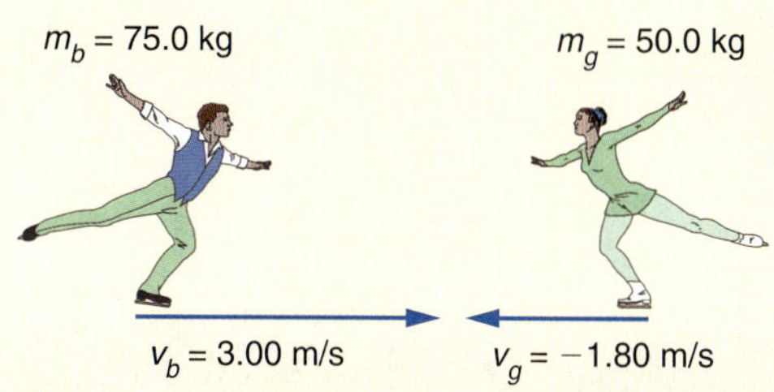

Solution:

$(m_b v_b)_{before} + (m_g v_g)_{after} = [(m_b + m_g)v_{bg}]_{after}$

$(75.00\text{ kg})(3.00\text{ m/s}) + (50.00\text{ kg})(-1.80\text{ m/s}) = (75.00\text{ kg} + 50.00\text{ kg})v_{bg}$

$$v_{bg} = \frac{225\text{ kg·m/s} - 90.0\text{ kg·m/s}}{125.00\text{ kg}}$$

$$= 1.08\text{ m/s}$$

Sample Problem 2

A steel ball with a mass of 2 kg is traveling at 3 m/s west. It collides with a stationary ball that has a mass of 1 kg. Upon collision, the smaller ball moves away to the west at 4 m/s. What is the velocity of the larger ball?

Strategy: Again, you will use the Law of Conservation of Momentum. Before the collision, only the larger ball has momentum. After the collision, the two balls move away at different velocities.

Givens:

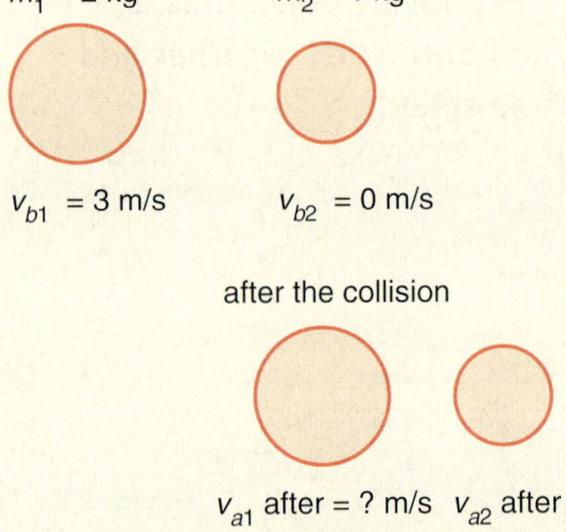

$m_1 = 2\text{ kg}$ $\quad v_{b2} = 0\text{ m/s}$

$m_2 = 1\text{ kg}$ $\quad v_{a2} = 4\text{ m/s}$

$v_{b1} = 3\text{ m/s}$

Solution:

$(m_1 v_1)_b + (m_2 v_2)_b = (m_1 v_1)_a + (m_2 v_2)_a$

$(2\text{ kg})(3\text{ m/s}) + (1\text{ kg})(0\text{ m/s}) = (2\text{ kg})v_{a1} + (1\text{ kg})(4\text{ m/s})$

$6\text{ kg·m/s} = (2v_{a1})\text{ kg} + 4\text{ kg·m/s}$

$v_{a1} = 1\text{ m/s}$

Reflecting on the Activity and the Challenge

The Law of Conservation of Momentum is a very powerful tool for explaining collisions in sports and other areas. The law works even when, as often happens in sports, one of the objects involved in a collision "bounces back," reversing the direction of its velocity and, therefore, its momentum, as a result of a collision. When describing a collision between a bat and ball, or a collision between two people, you can describe how the total momentum is conserved.

Physics To Go

1. A railroad car of 2000 kg coasting at 3.0 m/s overtakes and locks together with an identical car coasting on the same track in the same direction at 2.0 m/s. What is the speed of the cars after they lock together?

2. In a hockey game, an 80.0-kg player skating at 10.0 m/s overtakes and bumps from behind a 100.0-kg player who is moving in the same direction at 8.00 m/s. As a result of being bumped from behind, the 100.0-kg player's speed increases to 9.78 m/s. What is the 80.0-kg player's velocity (speed and direction) after the bump?

3. A 3-kg hard steel ball collides head-on with a 1-kg hard steel ball. The balls are moving at 2 m/s in opposite directions before they collide. Upon colliding, the 3-kg ball stops. What is the velocity of the 1-kg object after the collision? (Hint: Assign velocities in one direction as positive; then any velocities in the opposite direction are negative.)

4. A 45-kg female figure skater and her 75-kg male skating partner begin their ice dancing performance standing at rest in face-to-face position with the palms of their hands touching. Cued by the start of their dance music, both skaters "push off" with their hands to move backward. If the female skater moves at 2.0 m/s relative to the ice, what is the velocity of the male skater? (Hint: The momentum before the skaters push off is zero.)

5. A 0.35-kg tennis racquet moving to the right at 20.0 m/s hits a 0.060-kg tennis ball that is moving to the left at 30.0 m/s. The racquet continues moving to the right after the collision, but at a reduced speed of 10.0 m/s. What is the velocity (speed and direction) of the tennis ball after it is hit by the racquet?

6. A stationary 3-kg hard steel ball is hit head-on by a 1-kg hard steel ball moving to the right at 4 m/s. After the collision, the 3-kg ball moves to the right at 2 m/s. What is the velocity (speed and direction) of the 1-kg ball after the collision? (Hint: Direction is important.)

7. A 90.00-kg hockey goalie, at rest in front of the goal, stops a puck (m = 0.16 kg) that is traveling at 30.00 m/s. At what speed do the goalie and puck travel after the save?

8. A 45.00-kg girl jumps from the side of a pool into a raft (m = 0.08 kg) floating on the surface of the water. She leaves the side at a speed of 1.10 m/s and lands on the raft. At what speed will the girl-raft system begin to travel across the pool?

9. Two cars collide head on. Initially, car A (m = 1700.0 kg) is traveling at 10.00 m/s north and car B is traveling at 25.00 m/s south. After the collision, car A reverses its direction and travels at 5.00 m/s while car B continues in its initial direction at a speed of 3.75 m/s. What is the mass of car B?

10. A proton ($m = 1.67 \times 10^{-27}$ kg) traveling at 2.50×10^5 m/s collides with an unknown particle initially at rest. After the collision the proton reverses direction and travels at 1.10×10^5 m/s. Determine the change in momentum of the unknown particle.

11. You shoot a 0.04-kg bullet moving at 200.0 m/s into a 20.00-kg block initially at rest on an icy pond.
    a) What is the velocity of the bullet-block combination?
    b) The coefficient of friction between the block and the ice is 0.15. How far would the block slide before coming to rest?

12. Write a 15- to 30-s voice-over that highlights the conservation of momentum in a sport of your choosing.

Activity 9 Circular Motion

GOALS

In this activity you will:

- Understand that a centripetal force is required to keep a mass moving in a circular path at constant speed.
- Understand that a centripetal acceleration accompanies a centripetal force, and that, at any instant, both the acceleration and force are directed toward the center of the circular path.
- Apply the equation for circular motion.
- Understand that centrifugal force is the reaction to centripetal force.

To avoid becoming too dizzy, limit your spins while standing to about four.

What Do You Think?

Racecars can make turns at 150 mph.

- **What forces act on a racecar when it moves along a circular path at constant speed on a flat, horizontal surface?**

Record your ideas about this question in your *Active Physics* log. Be prepared to discuss your responses with your small group and the class.

For You To Do

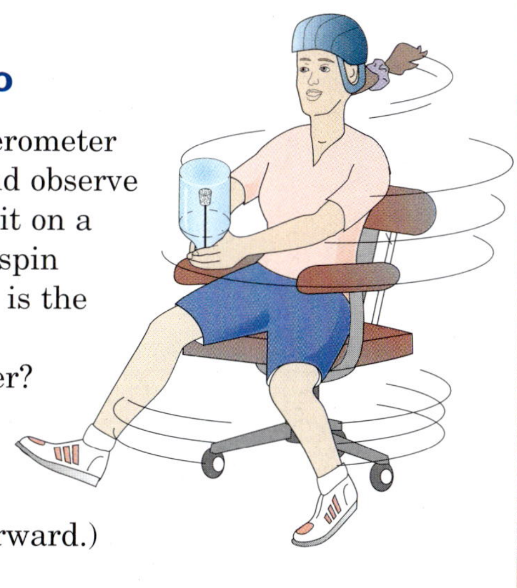

1. Hold an accelerometer in your hands and observe it as you either sit on a rotating stool or spin around while standing. What is the direction of the acceleration indicated by the accelerometer? (You can find out how the cork indicates acceleration by holding it and noting its behavior as you accelerate forward.)

a) Make a sketch in your log to simulate a snapshot photo taken from above as the accelerometer was moving along a circular path. Show the circular path, the accelerometer "frozen" at one instant, the cork "frozen" in leaning position, and an arrow to represent the velocity of the accelerometer at the instant represented by your sketch.

2. Review in your textbook and your log how you used a force meter to apply a constant force to objects to cause the objects to accelerate in **Activity 2**.

a) Based on the results of **Activity 2**, write a brief statement in your log that summarizes how the amount and direction of acceleration of an object depends on amount and direction of the force acting on the object.

3. Start a ball rolling across the floor. While it is rolling, catch up with the ball and use the force meter to push exactly sideways, or perpendicular, to the motion of the ball with a fixed amount of force. Carefully follow alongside the ball and, as will be necessary, keep adjusting the direction of push so that it is always perpendicular to the motion of the ball.

a) Make a top view sketch in your log that shows:

- a line to represent the straight-line path of the ball before you began pushing sideways on the ball
- a dashed line to represent the straight-line path on which the ball would have continued moving if you had not pushed sideways on it
- a line of appropriate shape to show the path taken by the ball as you pushed perpendicular to the direction of the ball's motion with a constant amount of force.

b) When you pushed on the ball exactly sideways to its motion, did you cause the ball to move either faster or slower? Explain your answer.

c) Assuming that friction could be eliminated to allow the ball to continue moving at constant speed, describe what you would need to do to make the ball keep moving on a circular path.

d) If you stop pushing on the ball, how does the ball move? Try it, and use a sketch and words in your log to describe what happens.

4. Review each of the items listed below. Copy each item into your log and write a statement to discuss how each item is related to an object moving along a circular path. If an item does not apply to circular motion, explain why.
   a) Galileo's Principle of Inertia
   b) Newton's First Law of Motion
   c) Newton's Second Law of Motion

Physics Words

centripetal acceleration: the inward radial acceleration of an object moving at a constant speed in a circle.

FOR YOU TO READ

The Unbalanced Force Required for Circular Motion

During the above activities you saw two things that are related by Newton's Second Law of Motion, $F = ma$. First, the accelerometer showed that when an object moves in a circular path there is an acceleration that at any instant is toward the center of the circle. This acceleration has a special name, **centripetal acceleration**. The word centripetal means "toward-the-center"; therefore, centripetal acceleration refers to acceleration toward the center of the circle when an object moves in a circular path.

You also saw that a centripetal force, a toward-the-center force, causes circular motion. When a centripetal force is applied to a moving object, the object's path curves; without the centripetal force, the object follows the tendency to move in a straight line. Therefore, a centripetal force, when applied, is an unbalanced force, meaning that it is not "balanced off" by another force.

Newton's Second Law seems to apply to circular motion just as well as it applies to accelerated motion along a straight line, but with a strange "twist." It is a clearly correct application of $F = ma$ to say that a centripetal force, F, causes a mass, m, to experience an acceleration, a. However, the strange part is that when an object moves along a circular path at constant speed, acceleration is happening with no change in the object's speed. The force changes the direction of the velocity.

Velocity describes both the amount of speed and the direction of motion of an object. Thinking about the velocity of an object moving with constant speed on a circular path, it is true that the velocity is changing from one instant to the next not in the amount of the velocity, but with respect to the direction of the velocity. The diagram shows an object moving at constant speed on a circular path. Arrows are used to represent the velocity of the object at several instants during one trip around the circle.

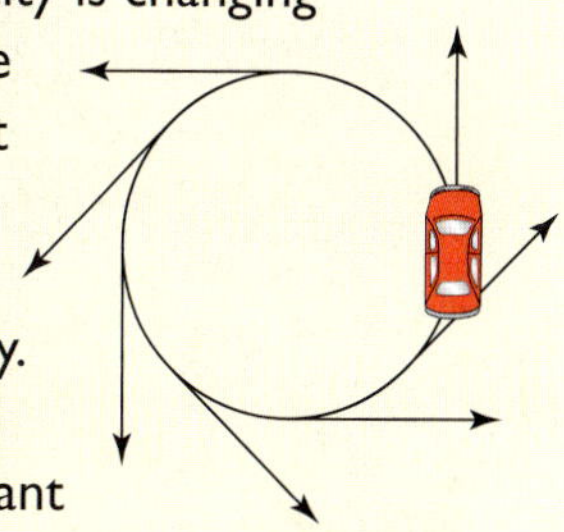

→

Physicists have shown that a special form of Newton's Second Law governs circular motion:

$$F_C = ma_C = \frac{mv^2}{r}$$

where F_C is the centripetal force in newtons,
m is the mass of the object moving on the circular path in kilograms,
a_C is the centripetal acceleration in m/s^2,
v is the velocity in m/s, and
r is the radius of the circular path in meters.

Sample Problem

Find the centripetal force required to cause a 1000.0-kg automobile travelling at 27.0 m/s (60 miles/hour) to turn on an unbanked curve having a radius of 100.0 m.

Strategy: This problem requires you to find centripetal force. You can use the equation that uses Newton's Second Law to calculate F_C.

Givens:

$m = 1000.0$ kg

$y = 27.0$ m/s

$r = 100.0$ m/s

Solution:

$$F_C = \frac{mv^2}{r}$$

$$= \frac{(1000.0\text{ kg }(27.0\text{ m/s})^2}{100.0\text{ m}}$$

$$= \frac{(1000.0\text{ kg} \times 730\text{ m}^2/\text{s}^2)}{100.0\text{ m}}$$

$$= 7300\text{ N}$$

If the force of friction is less than the above amount, the car will not follow the curve and will skid in the direction in which it is travelling at the instant the tires "break loose."

Reflecting on the Activity and the Challenge

Both circular motion and motion along curved paths that are not parts of perfect circles are involved in many sports. For example, both the discus and hammer throw events in track and field involve rapid circular motion before launching a projectile. Track, speed skating, and automobile races are done on curved paths. Whenever an object or athlete is observed to move along a curved path, you can be sure that a force is acting to cause the change in direction. Now you are prepared to provide voice-over explanations of examples of motion along curved paths in sports, and in many cases you perhaps can estimate the amount of force involved.

Physics To Go

1. For the car used as the example in the **For You To Read**, what is the minimum value of the coefficient of sliding friction between the car tires and the road surface that will allow the car to go around the curve without skidding? (Hint: First calculate the weight of the car, in newtons.)

2. If you twirl an object on the end of a string, you, of course, must maintain an inward, centripetal force to keep the object moving in a circular path. You feel a force that seems to be pulling outward along the string toward the object. But the outward force that you detect, called the "centrifugal force," is only the reaction to the centripetal force that you are applying to the string. Contrary to what many people believe, there is no outward force acting on an object moving in a circular path. Explain why this must be true in terms of what happens if the string breaks while you are twirling an object.

3. A 50.0-kg jet pilot in level flight at a constant speed of 270.0 m/s (600 miles per hour) feels the seat of the airplane pushing up on her with a force equal to her normal weight, 50.0 kg × 10 m/s^2 = 500 N. If she rolls the airplane on its side and executes a tight circular turn that has a radius of 1000.0 m, with how much force will the seat of the airplane push on her? How many "g's" (how many times her normal weight) will she experience?

4. Imagine a video segment of an athlete or an item of sporting equipment moving on a circular path in a sporting event. Estimate the mass, speed, and radius of the circle. Use the estimated values to calculate centripetal force and identify the source of the force.

5. Below are alternate explanations of the same event given by a person who was not wearing a seat belt when a car went around a sharp curve:

   a) "I was sitting near the middle of the front seat when the car turned sharply to the left. The centrifugal force made my body slide across the seat toward the right, outward from the center of the curve, and then my right shoulder slammed against the door on the passenger side of the car."

   b) "I was sitting near the middle of the front seat when the car turned sharply to the left. My body kept going in a straight line while, at the same time due to insufficient friction, the seat slid to the left beneath me, until the door on the passenger side of the car had moved far enough to the left to exert a centripetal force against my right shoulder."

   Are both explanations correct, or is one right and one wrong? Explain your answer in terms of both explanations.

6. People seem to be fascinated with having their bodies put in a state of circular motion. Describe an amusement park ride based on circular motion that you think is fun, and describe what happens to your body during the ride.

7. a) Explain why football players fall on a wet field while changing directions during a play.

   b) Include the concepts centripetal force and Newton's Laws in a revised explanation.

   c) In a new revision, make the explanation exciting enough to include in your sports video voice-over.

PHYSICS AT WORK

Dean Bell

TELEVISION PRODUCER USES SPORTS TO TEACH MATH AND PHYSICS

Dean Bell is an award-winning filmmaker and television writer, director, and producer. His show *Sports Figures* is a highly acclaimed ESPN educational television series designed to teach the principles of physics and mathematics through sports. His approach has been to tell a story, pose a problem, and then follow through with its mathematical and scientific explanation. "But always," he says, "you must make it fun. It has to be both educational and entertaining."

Dean began his career as a filmmaker after college. He landed the apprentice film editor's position on a Woody Allen film. From there, he worked his way up in the field, from assistant editor, to editor and finally writer, director, and producer.

"I've always been a fan of educational TV," he states, "although I never thought that was where my career would take me. It's one of life's little ironies that I've ended up producing this type of show. You see, my father worked in scientific optics and was very science oriented. He was always delighted in finding out how things worked, and was even on the Mr. Wizard TV show a few times."

Dean writes the script for each segment, working together with top educational science consultants. "We spend a day coming up with ideas and then researching each subject thoroughly. Our job is to illustrate the relationship between a sports situation and the related mathematical or physics principles.

"At the end of the day," says Dean, "it really is nice to be working on a show that means something and that is so worthwhile. I'm still getting ahead in my career as a film and TV producer, but now I'm also an educator."

Chapter 1 Assessment

Your big day has arrived. You will be meeting with the local television station to audition for a job as a "physics of sports" commentator. Whether you will get the job will be decided on the quality of your voice-over.

With what you learned in this chapter, you are ready to do your science commentary on a short sports video. Choose a videotape from a sports event, either a school event or a professional event. Each of you will be responsible for producing your own commentary, whether or not you worked in cooperative groups during the activities. You are not expected to give a play-by-play description, but rather describe the rules of nature that govern the event. Your viewers should come away with a different perspective of both sports and physics. You may produce one of the following:

- **a written script**
- **a live narrative**
- **a video soundtrack**
- **an audiocassette**

Review the criteria by which your voice-over dialogue or script will be evaluated. Your voice-over should:

- **use physics principles and terms correctly**
- **have entertainment value**

After reviewing the criteria, decide as a class the point value you will give to each of these criteria:

- **How important is the physics content? How many physics terms and principles should be illustrated to get the minimum credit? The maximum credit?**
- **What value would you place on the entertainment aspect? How do you fairly assess the excitement and interest of the broadcast?**

Physics You Learned

Galileo's Principle of Inertia

Newton's First Law of Motion

Newton's Second Law of Motion

Newton's Third Law of Motion

Weight

Center of mass; center of gravity

Friction between different surfaces

Momentum

Law of Conservation of Momentum

Centripetal acceleration

Centripetal force

Chapter 2

SAFETY

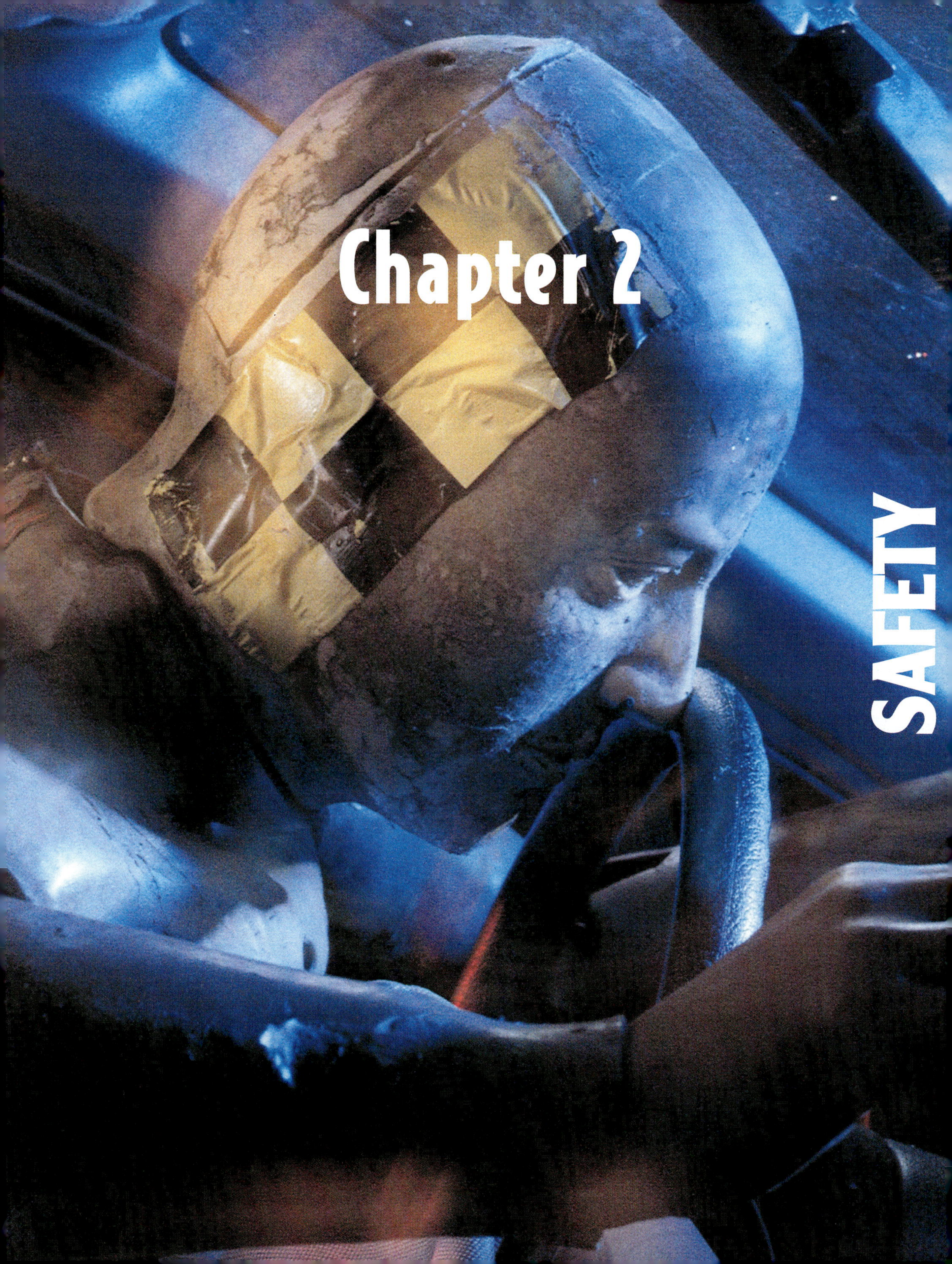

Scenario

Probably the most dangerous thing you will do today is travel to your destination. Transportation is necessary, but the need to get there in a hurry, and the large number of people and vehicles, have made transportation very risky. There is a greater chance of being killed or injured traveling than in any other common activity. Realizing this, people and governments have begun to take action to alter the statistics. New safety systems have been designed and put into use in automobiles and airplanes. New laws and a new awareness are working together with these systems to reduce the danger in traveling.

What are these new safety systems? You are probably familiar with many of them. In this chapter, you will become more familiar with most of these designs. Could you design or even build a better safety device for a car or a plane? Many students around the country have been doing just that, and with great success!

Challenge

Your design team will develop a safety system for protecting automobile, airplane, bicycle, motorcycle, or train passengers. As you study existing safety systems, you and your design team should be listing ideas for improving an existing system or designing a new system for preventing accidents. You may also consider a system that will minimize the harm caused by accidents.

Your final product will be a working model or prototype of a safety system. On the day that you bring the final product to class, the teams will display them around the room while class members informally view them and discuss them with members of the design team. During this time, class members will ask questions about each others products. The questions will be placed in envelopes provided to each team by the teacher. The teacher will use some of these questions during the oral presentations on the next day.

The product will be judged according to the following three parts:

1. The quality of your safety feature enhancement and the working model or prototype.
2. The quality of a five-minute oral report that should include:
   - **the need for the system**
   - **the method used to develop the working model**
   - **the demonstration of the working model**
   - **the discussion of the physics concepts involved**
   - **the description of the next-generation version of the system**
   - **the answers to questions posed by the class**
3. The quality of a written and/or multimedia report including:
   - **the information from the oral report**
   - **the documentation of the sources of expert information**
   - **the discussion of consumer acceptance and market potential**
   - **the discussion of the physics concepts applied in the design of the safety system**

Criteria

You and your classmates will work with your teacher to define the criteria for determining grades. You will also be asked to evaluate your own work. Discuss as a class the performance task and the points that should be allocated for each part. A starting point for your discussions may be:

- **Part 1 = 40 points**
- **Part 2 = 30 points**
- **Part 3 = 30 points**

Since group work is made up of individual work, your teacher will assign some points to each individual's contribution to the project. If individual points total 30 points, then parts 1, 2 and 3 must be changed so that the total remains at 100.

Activity 1 Response Time

GOALS

In this activity you will:

- Identify the parts of the process of stopping a car.
- Measure reaction time.
- Wire a series circuit.

What Do You Think?

Many deaths that occur on the highway are drivers and passengers in vehicles that did not cause the accident. The driver was not able to respond in time to avoid becoming a statistic.

- **How long would it take you to respond to an emergency?**

Record your ideas about this question in your *Active Physics* log. Be prepared to discuss your responses with your small group and the class.

For You To Do

1. To stop a car, you must move your foot from the gas pedal to the brake pedal. Try moving your right foot between imaginary pedals.

a Estimate how long it takes to move your foot between the imaginary pedals. Record your estimate.

2. The first step in stopping a car happens even before you move your foot to the brake. It takes time to see or hear something that tells you to move your foot. Test this by having a friend stand behind you and clap. When you hear the sound, move your foot between imaginary pedals.

a) Estimate how long it took you to respond to the loud noise. Record your estimate.

3. Create a simple electric circuit to test your response time. Your group will need a battery in a clip, two switches, a flashlight bulb in a socket, and connecting wires. Connect the wires from one terminal of the battery to the first switch, then to the second switch, to the light bulb, and back to the battery.

Have your teacher approve your circuit before proceeding to Step 4.

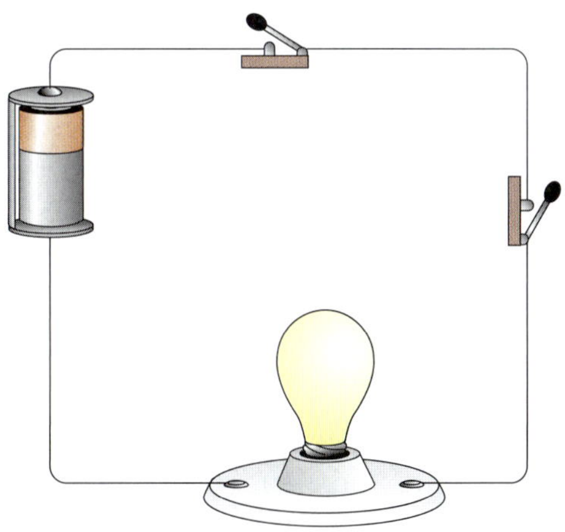

4. Close one switch while the other is open. Close the other switch. Take turns turning the light off and on with each person operating only one switch.

a) Record what happens in each case.

5. Try to keep the light on for exactly one second, then five seconds. You can estimate one second by saying "one thousand one."

a) How quickly do you think you can turn the light off after your partner turns it on? The time the bulb is lit is your response time. Record an estimate of your response time in your log.

6. Find your response time using the electric circuit.
   - a) How could you improve the accuracy of the measurement?
   - b) How would repeating the investigation improve the accuracy?
7. Test your response time with the other equipment set up in your classroom. Use a standard reaction time meter, such as one used in driver education. You will need to follow the directions for the model available in your class.

   - a) Record your response time.
8. Use two stopwatches. One person starts both stopwatches at the same time, and hands one to her lab partner. When the first person stops her watch, the lab partner stops his. The difference in the two times is the response time.
   - a) Record your response time.
9. Use a centimeter ruler. Hold the centimeter ruler at the top, between thumb and forefinger, with zero at the bottom. Your partner places thumb and forefinger at the lower end, but does not touch the ruler. Drop the ruler. Your partner must stop the ruler from falling by closing thumb and forefinger.
   - a) The position of your partner's fingers marks the distance the ruler fell while her nervous system was responding. Record the distance in your log.
   - b) The graph at the top of the next page shows the relationship between the distance the ruler fell and the time it took to stop it. Use the graph to find and record your response time.

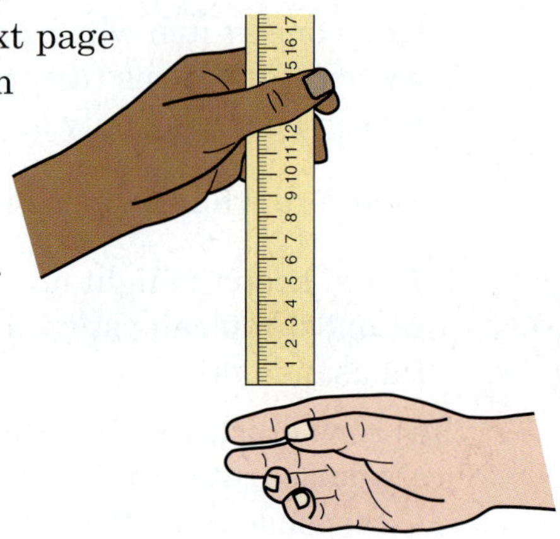

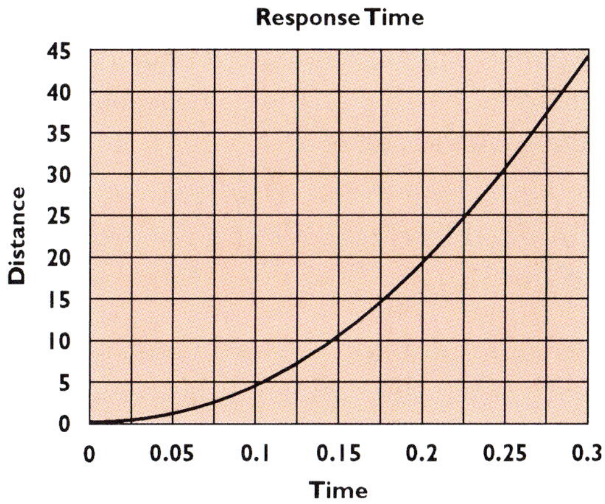

10. Compare the measures of your response obtained from each strategy.
    a) Explain why they were not all the same.
    b) What measure do you think best reports your response time? Why?

11. Compare the measures you obtained with those of other students.
    a) Record the results for the fastest, slowest, and average response times.
    b) Why do you think response times vary for people of the same age? Discuss this with your group and then record your answer.

Reflecting on the Activity and the Challenge

The amount of time people require before they can act has a direct impact on their driving. It takes time to notice a situation and more time to respond. A person who requires a second to respond to what he or she sees or hears is more likely to have an accident than someone who responds in half a second. Your **Chapter Challenge** is to design and build an improved safety device for a car. You may be able to design a car that helps drivers to stay alert and helps them become more aware of their surroundings. Anything that you can do to decrease a driver's response time will make the car safer.

Physics To Go

1. Test the response time of some of your friends and family with the centimeter ruler. Bring in the results from at least three people of various ages.
2. How do the values you found in **Question 1** compare with those you obtained in class? What do you think explains the difference, if any?
3. Take a dollar bill and fold it in half lengthwise. Have someone try to catch the dollar bill between his or her forefinger and middle finger. Most people will fail this task.
   a) Explain why it is so difficult to catch the dollar bill.
   b) Repeat the dollar bill test, letting them catch it with their thumb and forefinger.
   c) Explain why catching it with thumb and forefinger may have been easier. Try to include numbers in your answer such as the length of the dollar, the time for the dollar to fall, and average response time.
4. Does a racecar driver need a better response time than someone driving around a school? Explain your answer, giving examples of the dangers each person encounters.
5. Apply what you learned from this activity to describe how knowing your own response time can help you be a safer driver.

Stretching Exercises

1. Build a device with a red light and a green light. If the red light turns on, you must press one button and measure the response time. If the green light turns on, you must press a second button and measure the response time. Have your teacher approve your design before proceeding. How do response times to this "decision" task compare with the response times measured earlier?

2. Use the graph for response time to construct a response-time ruler with the distance measurement converted to time. You can now read response times directly.

3. Do you think some groups of people have better or worse response times than others? Consider groups such as basketball players, video game players, taxi drivers, or older adults. Plan an investigation to collect data that will help you find an answer. Include in your plan the number of subjects, how you will test them, and how you will organize and interpret the data collected. Have your teacher approve your plan before you proceed.

Activity 2 Speed and Following Distance

GOALS

In this activity you will:

- Define speed.
- Identify constant and changing speeds.
- Interpret distance-time and speed-time graphs.
- Contrast average and instantaneous speeds.
- Calculate the distance traveled at constant speed.

What Do You Think?

In a rear-end collision, the driver of the car in back is always found at fault.

- **What is a safe distance between your car and the car in front of you?**
- **How do you decide?**

Record your ideas about these questions in your *Active Physics* log. Be prepared to discuss your responses with your small group and the class.

For You To Do

1. A strobe photo is a multiple-exposure photo in which a moving object is photographed at regular time intervals. The strobe photo below shows a car traveling at 30 mph.

a) Copy the sketch in your log.

2. Think about the difference between the motion of a car traveling at 30 mph and one traveling at 45 mph.

a) Draw a sketch of a strobe photo, similar to the one above, of a car traveling at 45 mph.

b) Are the cars the same distance apart? Were they farther apart or closer together than at 30 mph?

c) Draw a sketch for a car traveling at 60 mph. Describe how you decided how far apart to place the cars.

3. The following sketch shows a car traveling at different speeds.

a) Copy the sketch in your log. Mark where the car is traveling fast, where it is traveling slowly, and where it is traveling at a constant speed. How did you know?

4. A sonic ranger connected to a computer will produce a graph that shows an object's motion. Use the sonic ranger setup to obtain the following graphs to print or sketch in your log.

Make sure the path of motion is clear of any hazards.

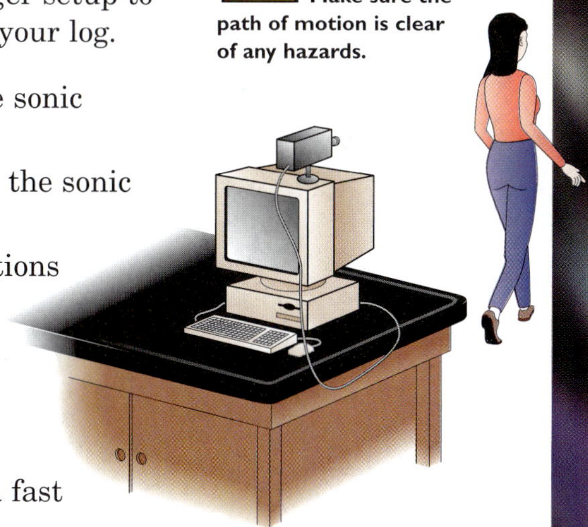

a) Sketch a graph of a person walking toward the sonic ranger at a normal speed.

b) Sketch a graph of a person walking away from the sonic ranger at a normal speed.

c) Sketch a graph of a person walking both directions at a very slow speed.

d) Sketch a graph of a person walking both directions at a fast speed.

5. Predict what the graph will look like if you walk toward the system at a slow speed and away at a fast speed. Test your prediction.

a) Record your prediction in your log.

b) Based on your measurements, how accurate was your prediction?

6. Repeat any of the motions in **Steps 4** or **5** for a more thorough analysis.
   a) From your graph, determine the total distance you walked.
   b) How long did it take to walk that distance?
   c) Divide the distance you walked by the time it took. This is your average speed in meters per second (m/s).

Physics Words

speed: the change in distance per unit time; speed is a scalar, it has no direction.

PHYSICS TALK

Speed

The relationship between **speed**, distance, and time can be written as:

$$\text{Speed} = \frac{\text{Distance traveled}}{\text{Time elapsed}}$$

If your speed is changing, this gives your average speed. Using symbols, the same relationship can be written as:

$$v_{av} = \frac{\Delta d}{\Delta t}$$

where v_{av} is average speed

Δd is change in distance or displacement.

Δt is change in time or elapsed time.

Sample Problem 1

You drive 400 mi. in 8 h. What is your average speed?

Strategy: You can use the equation for average speed.

$$v_{av} = \frac{\Delta d}{\Delta t}$$

Givens:

$$\Delta d = 400 \text{ mi.}$$

$$v_{av} = \frac{\Delta d}{\Delta t}$$

$$= \frac{400 \text{ mi.}}{8 \text{ h}}$$

$$= 50 \text{ mph (miles per hour)}$$

Your average speed is 50 mph. This does not tell you the fastest or slowest speed that you traveled. This also does not tell you how fast you were going at any particular moment.

Sample Problem 2

Elisha would like to ride her bike to the beach. From car trips with her parents, she knows that the distance is 30 mi. She thinks she can keep up an average speed of about 15 mph. How long will it take her to ride to the beach?

Strategy: You can use the equation for average speed.

$$v_{\text{av}} = \frac{\Delta d}{\Delta t}$$

However, you will first need to rearrange the terms to solve for elapsed time.

$$\Delta t = \frac{\Delta d}{v_{av}}$$

Solution:

$$\Delta t = \frac{\Delta d}{v_{av}}$$

$$= \frac{30 \text{ mi.}}{15 \text{ mph}}$$

$$= 2 \text{ h}$$

FOR YOU TO READ

Representing Motion

One way to show motion is with the use of strobe photos. A strobe photo is a multiple-exposure photo in which a moving object is photographed at regular time intervals. The sketches you used in **Steps 1, 2,** and **3** in **For You To Do** are similar to strobe photos. Here is a strobe photo of a car traveling at the average speed of 50 mph.

Another way to represent motion is with graphs. The graph below shows a car traveling at the average speed of 50 mph.

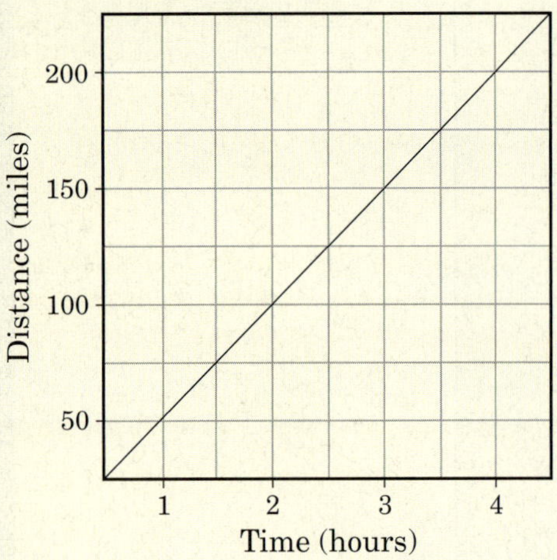

Kilometers and Miles

Highway signs and speed limits in the USA are given in miles per hour, or mph. Almost every other country in the world uses kilometers to measure distances. A kilometer is a little less than two-thirds of a mile. Kilometers per hour (km/h) is used to measure highway driving speed. Shorter distances, such as for track events and experiments in a science class, are measured in meters per second, m/s.

You will use mph when working with driving speeds, but meters per second for data you collect in class. The good news is that you do not need to change measures between systems. It is important to be able to understand and compare measures.

To help you relate the speeds with which you are comfortable to the data you collect in class, the chart below gives *approximate* comparisons.

School zone	25 mph	40 km/h	11 m/s
Residential street	35 mph	55 km/h	16 m/s
Suburban interstate	55 mph	90 km/h	25 m/s
Rural interstate	75 mph	120 km/h	34 m/s

Reflecting on the Activity and the Challenge

You now know how reaction time and speed affect the distance required to stop. You should be able to make a good argument about tailgating as part of the **Chapter Challenge**. If your car can be designed to limit tailgating or to alert drivers to the dangers of tailgating, it will add to improved safety.

Physics To Go

1. Describe the motion of each car moving to the right. The strobe pictures were taken every 3 s (seconds).

   a) 

   b) 

2. Sketch strobe pictures of the following:

   a) A car starting at rest and reaching a final constant speed.
   b) A car traveling at a constant speed then coming to a stop.

3. For each graph below, describe the motion of the car:

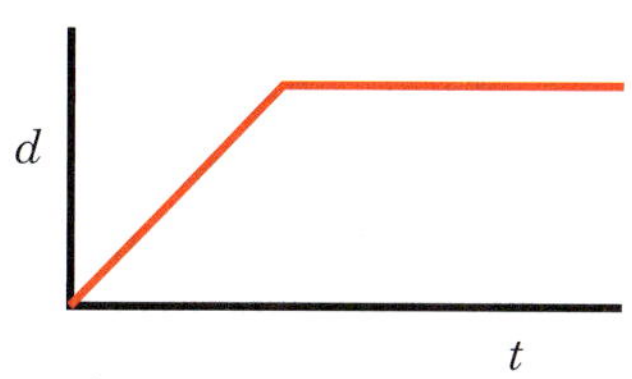

a)

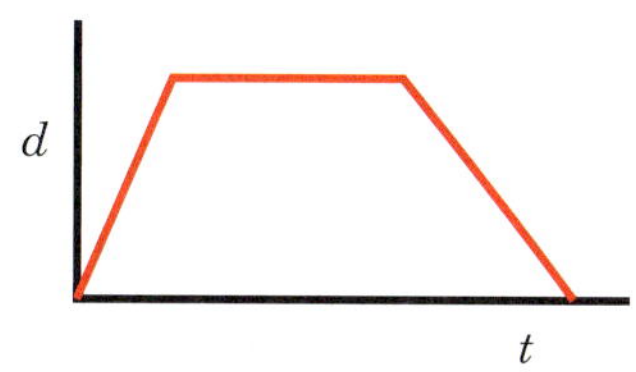

b)

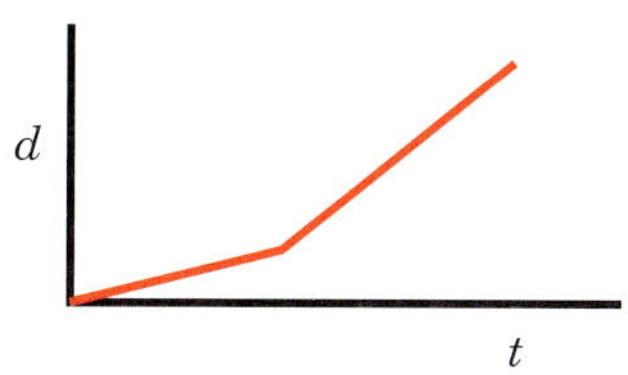

c)

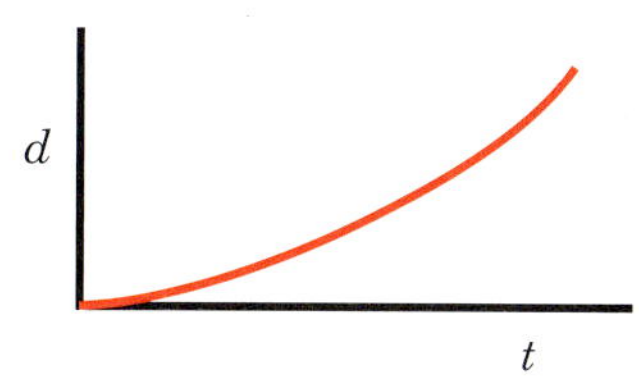

d)

4. A racecar driver travels at 110 m/s (that's almost 250 mph) for 20 s. How far has the driver traveled?

5. A salesperson drove the 215 miles from New York City to Washington, DC, in 412 hours.
   a) What was her average speed?
   b) How fast was she going when she passed through Baltimore?

6. If you planned to walk to a park that was 5 miles away, what average speed would you have to keep up to arrive in 2 hours?

7. Use your average response time from **Activity 1** to answer the following:
   a) How far does your car travel in meters during your response time if you are moving at 55 mph (25 m/s)?
   b) How far does your car travel during your response time if you are moving at 35 mph (16 m/s)? How does the distance compare with the distance at 55 mph?
   c) Suppose you are very tired and your response time is doubled. How far would you travel at 55 mph during your response time?

8. According to traffic experts, the proper following distance you should leave between your car and the vehicle in front of you is two seconds. As the vehicle in front of you passes a fixed point, say to yourself "one thousand one, one thousand two." Your car should reach the point as you complete the phrase. How can the experts be sure? Isn't two seconds a measure of time? Will two seconds be safe on the interstate highway?

9. You calculated the distance your car would move during your response time. Use that information to determine a safe following distance at:
   a) 25 mph
   b) 55 mph
   c) 75 mph

10. Apply what you learned in this activity to write a convincing argument that describes why following a car too closely (tailgating) is dangerous. Include the factors you would use to decide how close counts as "tailgating."

Stretching Exercises

Measure a distance of about 100 m. You can use a football field or get a long tape or trundle wheel to measure a similar distance. You also need a watch capable of measuring seconds. Determine your average speed traveling that distance for each of the following:

a) a slow walk
b) a fast walk
c) running
d) on a bicycle
e) another method of your choice

Activity 3 Accidents

GOALS

In this activity you will:

- Evaluate your own understandings of safety.
- Evaluate the safety features on selected vehicles.
- Compare and contrast the safety features on selected vehicles.
- Identify safety features in selected vehicles.
- Identify safety features required for other modes of transportation (in-line skates, skateboards, cycling, etc.).

What Do You Think?

Chances are you will not be able to avoid being in an accident at some time in the future.

- **How can you protect yourself from serious injury, or even death, should an accident occur?**
- **What do you think is the greatest danger to you or the people in an accident?**

Record your ideas about these questions in your *Active Physics* log. Be prepared to discuss your responses with your small group and the class.

For You To Do

1. Many people think that they know the risks involved with day-to-day transportation. The "test" below will check your knowledge of automobile accidents. The statements are organized in a true and false format. Record a T in your log for each statement you believe is true and an F if you believe the statement is false. Your teacher will supply the correct answers for discussion at the end of the activity.

   a) More people die because of cancer than automobile accidents.

   b) Your chances of surviving a collision improve if you are thrown from the car.

   c) The fatality rate of motorcycle accidents is less than that of cars.

   d) A large number of people who are belted into their cars are killed in a burning or submerged car.

   e) If you don't have a child restraint seat, you should place the child in your seat belt with you.

   f) You can react fast enough during an accident to brace yourself in the car seat.

   g) Most people die in traffic accidents during long trips.

   h) A person not wearing a seat belt in your car poses a hazard to you.

   i) Traffic accidents occur most often on Monday mornings.

   j) Male drivers between the ages of 16 and 19 are most likely to be involved in traffic accidents.

   k) Casualty collisions are most frequent during the winter months due to snow and ice.

   l) More pedestrians than drivers are killed by cars.

   m) The greatest number of roadway fatalities can be attributed to poor driving conditions.

   n) The greatest number of females involved in traffic accidents are between the ages of 16 and 20.

   o) Unrestrained occupant casualties are more likely to be young adults between the ages of 16 and 19.

2. Calculate your score. Give yourself two points for a correct answer, and subtract one point for an incorrect answer. You might want to match your score against the descriptors given below.

   21 to 30 points: Expert Analyst

   14 to 20 points: Assistant Analyst

   9 to 13 points: Novice Analyst

   8 points and below: Myth Believer

   a) Record your score in your log. Were you surprised about the extent of your knowledge? Some of the reasons behind these facts will be better understood as you continue to travel through this chapter.

Obtain permission from the cars' owners before proceeding.

3. Survey at least three different cars for safety features. The list on the next page will allow you to evaluate the safety features of each of the cars. Place a check mark in the appropriate square.

   Number 1 indicates very poor or nonexistent, 2 is minimum standard, 3 is average, 4 is good, and 5 is very good.

   For example, when rating air bags: a car with no air bags could be given a 1 rating, a car with only a driver-side air bag a 2, a car with driver and passenger side air bags a 3, a car with slow release driver and passenger-side air bags a 4, and a car which includes side-door air bags to the previous list a 5. You may add additional safety features not identified in the chart. Many additional features can be added!

   a) Copy and complete the table in your log.

   b) Which car would you evaluate as being safest?

Car Tested: Make and Model _____	Year _____				
Safety Feature	Rating				
Padded front seats	1	2	3	4	5
Padded roof frame	1	2	3	4	5
Head rests	1	2	3	4	5
Knee protection	1	2	3	4	5
Anti-daze rear-view mirror that brakes on impact	1	2	3	4	5
Child proof safety locks on rear doors	1	2	3	4	5
Padded console	1	2	3	4	5
Padded sun visor	1	2	3	4	5
Padded doors and arm rests	1	2	3	4	5
Steering wheel with padded rim and hub	1	2	3	4	5
Padded gear level					
Padded door pillars					
Air bags					

Reflecting on the Activity and the Challenge

Serious injuries in an automobile accident have many causes. If there are no restraints or safety devices in a vehicle, or if the vehicle is not constructed to absorb any of the energy of the collision, even a minor collision can cause serious injury. Until the early 1960s, automobile design and construction did not even consider passenger safety. The general belief was that a heavy car was a safe car. While there is some truth to that statement, today's lighter cars are far safer than the "tanks" of the past.

The safety survey may have provided ideas for constructing a prototype of a safety system used for transportation. If it has, write down ideas in your log that have been generated from this activity.

Physics To Go

1. Review and list all the safety features found in today's new cars. As you compile your list, write next to each safety feature one or more of the following designations:

   F: effective in a front-end collision.

   R: effective in a rear-end collision.

   S: effective in a collision where the car is struck on the side.

   T: effective when the car rolls over or turns over onto its roof.

2. Make a list of safety features that could be used for cycling.

3. Make a list of safety features that could be used for in-line skating.

4. Make a list of safety features that could be used for skateboarding.

5. Ask family members or friends if you may evaluate their car. Discuss and explain your evaluation to the car owners. Record your evaluation and their response in your log.

Stretching Exercises

1. Read a consumer report on car safety. Are any cars on the road particularly unsafe?
2. Collect brochures from various automobile dealers. What new safety features are presented in the brochures? How much of the advertising is devoted to safety?

Activity 4 Life (and Death) before Seat Belts

GOALS

In this activity you will:

- Understand Newton's First Law of Motion.
- Understand the role of safety belts.
- Identify the three collisions in every accident.

What Do You Think?

Throughout most of the country, the law requires automobile passengers to wear seat belts.

- **Should wearing a seat belt be a personal choice?**
- **What are two reasons why there should be seat belt laws and two reasons why there should not?**

Record your ideas about these questions in your *Active Physics* log. Be prepared to discuss your responses with your small group and the class.

Perform the activity outside of traffic areas. Do not obstruct paths to exits. Do not leave carts lying on the floor.

For You To Do

1. In this activity, you will investigate car crashes where the driver or passenger does not wear seat belts. Your model car is a laboratory cart. Your model passenger is molded from a lump of soft clay. With the "passenger" in place, send the "car" at a low speed into a wall.

Safety

a) Describe, in your log, what happens to the "passenger."

2. Repeat the collision at a high speed. Compare and contrast this collision with the previous one.

a) Compare and contrast requires you to find and record at least one similarity and one difference. A better response includes more similarities and differences.

3. You can conduct a more analytical experiment by having the cart hit the wall at varying speeds. Set up a ramp on which the car can travel. Release the car on the ramp and observe as it crashes into the wall. Repeat the collision for at least two ramp heights.

a) Record the heights of the ramp and describe the results of the collision. Describe the collision by noting the damage to the "passenger."

Physics Words

Newton's First Law of Motion: an object at rest stays at rest and an object in motion stays in motion unless acted upon by an unbalanced, external force.

inertia: the natural tendency of an object to remain at rest or to remain moving with constant speed in a straight line.

PHYSICS TALK

Newton's First Law of Motion

Newton's First Law of Motion (also called the Law of **Inertia**) is one of the foundations of physics. It states:

An object at rest stays at rest, and an object in motion stays in motion unless acted upon by a net external force.

There are three distinct parts to Newton's First Law.

Part 1 says that objects at rest stay at rest. This hardly seems surprising.

Part 2 says that objects in motion stay in motion. This may seem strange indeed. After looking at the collisions of this activity, this should seem clearer.

Part 3 says that Parts 1 and 2 are only true when no force is present.

FOR YOU TO READ

Three Collisions in One Accident!

Arthur C. Damask analyzes automobile accidents and deaths for insurance companies and police reports. This is how Professor Damask describes an accident:

Consider the occupants of a conveyance moving at some speed. If the conveyance strikes an object, it will rapidly decelerate to some lower speed or stop entirely; this is called the first collision. But the occupants have been moving at the same speed, and will continue to do so until they are stopped by striking the interior parts of the car (if not ejected); this is the second collision. The brain and body organs have also been moving at the same speed and will continue to do so until they are stopped by colliding with the shell of the body, i.e., the interior of the skull, the thoracic cavity, and the abdominal wall. This is called the third collision.

Newton's First Law of Motion explains the three collisions:

- First collision: the car strikes the pole; the pole exerts the force that brings the car to rest.
- Second collision: when the car stops, the body keeps moving; the structure of the car exerts the force that brings the body to rest.
- Third collision: the body stops, but the heart and brain keep moving; the body wall exerts the force that brings the heart and brain to rest.

Even with all the safety features in our automobiles, some deaths cannot be prevented. In one accident, only a single car was involved, with only the driver inside. The car failed to follow the road around a turn, and it struck a telephone pole. The seat belt and the air bag prevented any serious injuries apart from a few bruises, but the driver died. An autopsy showed that the driver's aorta had burst, at the point where it leaves the heart.

Reflecting on the Activity and the Challenge

In this activity you discovered that an object in motion continues in motion until a force stops it. A car will stop when it hits a pole but the passenger will keep on moving. If the car and passenger have a large speed, then the passenger will continue moving with this large speed. The passenger at the large speed will experience more damage from the fast-moving cart.

Have you ever heard someone say that they can prevent an injury by bracing themselves against the crash? They can't! Restraining devices help provide support. Without a restraining system, the force of impact is either absorbed by the rib, skull, or brain.

Use Newton's First Law of Motion to describe your design. How will your safety system protect passengers from low speed and higher speed collisions?

Physics To Go

1. Describe how Newton's First Law applies to the following situations:

   • You step on the brakes to stop your car.

   (Sample answer: You and the car are moving forward. The brakes apply a force to the tires and stop them from rotating. Newton's law states that an object in motion will remain in motion unless a force acts upon it. In this case, the force is friction between the ground and the tires. You remain in motion since the force that stopped the car did not stop you.)

   • You step on the accelerator to get going.

   • You turn the wheel to go around a curve. (Hint: You keep moving in a straight line.)

   • You step on the brakes, and an object in the back of the car comes flying forward.

2. Give two more examples of how Newton's First Law applies to vehicles or people in motion.
3. According to Newton's First Law, objects in motion will continue in motion unless acted upon by a force. Using Newton's First Law, explain why a cart that rolls down a ramp eventually comes to rest.
4. The skateboard, shown in the picture to the right, strikes the curb. Draw a diagram indicating the direction in which the person moves. Use Newton's First Law to explain the direction of movement.

5. Explain, in your own words, the three collisions during a single crash as described by Professor Damask in **For You To Read**.
6. Use the diagrams below to compare the second and third collisions described by Professor Damask with the impact of a punch during a boxing match.

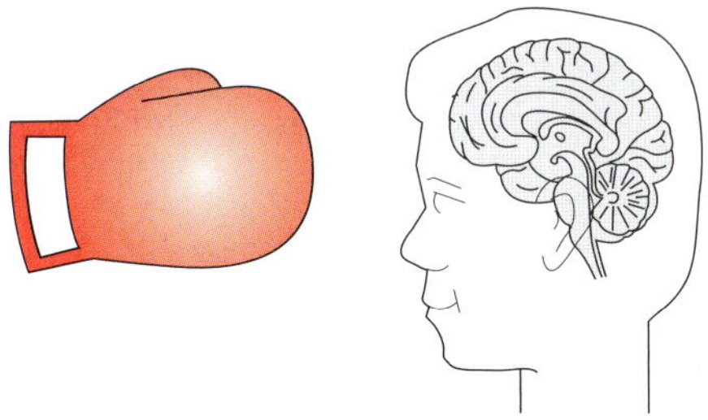

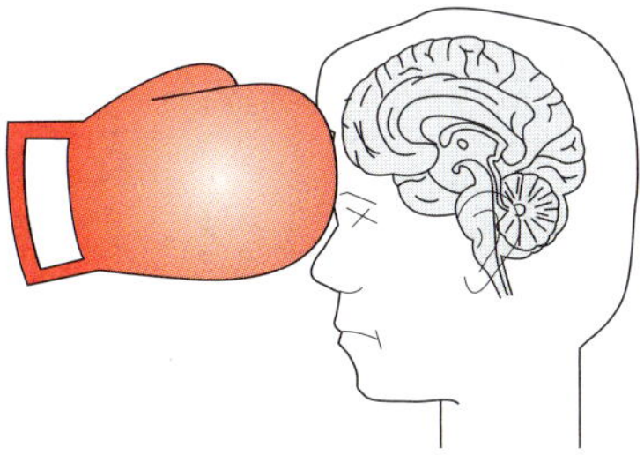

7. When was the law instituted requiring drivers to wear seat belts?

Stretching Exercises

1. Determine what opinions people in your community hold about the wearing of seat belts. Compare the opinions of the 60+ years old and 25 to 59 year old groups with that of the 15 to 24 year old group. Survey at least five people in each age group: Group A = 15 to 24 years, Group B = 25 to 59 years, and Group C = 60 years and older. (Survey the same number of individuals in each age group.) Ask each individual to fill out a survey card.

A sample questionnaire is provided below. You may wish to eliminate any question that you feel is not relevant. You are encouraged to develop questions of your own that help you understand what attitudes people in your community hold about wearing seat belts. The answers have been divided into three categories: 1 = agree; 2 = will accept, but do not hold a strong opinion; and 3 = disagree. Try to keep your survey to between five and ten questions.

Age group:	Date of Survey:		
Statement	Agree	No strong opinion	Disagree
1. I believe people should be fined for not wearing seat belts.	1	2	3
2. I wouldn't wear a seat belt if I didn't have to.	1	2	3
3. People who don't wear seat belts pose a threat to me when they ride in my car.	1	2	3
4. I believe that seat belts save lives.	1	2	3
5. Seat belts wrinkle my clothes and fit poorly so I don't wear them.	1	2	3

2. Make a list of reasons why people refuse to wear seat belts. Can you challenge these opinions using what you have learned about Newton's First Law of Motion?

Activity 5 Life (and Fewer Deaths) after Seat Belts

GOALS

In this activity you will:

- Understand the role of safety belts.
- Compare the effectiveness of various wide and narrow belts.
- Relate pressure, force and area.

What Do You Think?

In a collision, you cannot brace yourself and prevent injuries. Your arms and legs are not strong enough to overcome the inertia during even a minor collision. Instead of thinking about stopping yourself when the car is going 30 mph, think about catching 10 bowling balls hurtling towards you at 30 mph. The two situations are equivalent.

- **Suppose you had to design seat belts for a racecar that can go 200 mph. How would they be different from the ones available on passenger cars?**

Record your ideas about this question in your *Active Physics* log. Be prepared to discuss your responses with your small group and the class.

For You To Do

1. In this activity you will test different materials for their suitability for use as seat belts. Your model car is, once again, a laboratory cart; your model passenger is molded from a lump of soft clay. Give your passenger a seat belt by stretching a thin piece of wire across the front of the passenger, and attaching it on the sides or rear of the car.

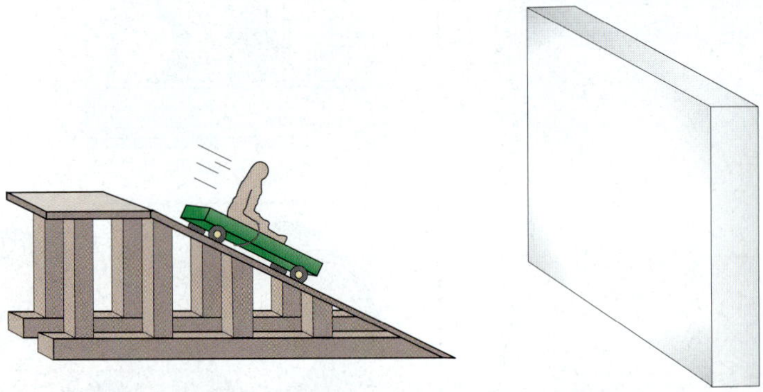

Perform the activity outside of traffic areas. Do not obstruct paths to exits. Do not leave carts lying on the floor.

2. Make a collision by sending the car down a ramp. Start with small angles of incline and increase the height of the ramp until you see significant injury to the clay passenger.
   a) In your log, note the height of the ramp at which significant injury occurs.

3. Use at least two other kinds of seat belts (ribbons, cloth, etc.). Begin by using the same incline of ramp and release height as in **Step 2**.
   a) In your log, record the ramp height at which significant injury occurs to the "passenger" using the other kinds of seat belt material.

4. Crash dummies cost $50,000! Watch the video presentation of a car in a collision, with a dummy in the driver's seat. You may have to observe it more than once to answer the following questions:
   a) In the collision, the car stops abruptly. What happens to the driver?
   b) What parts of the driver's body are in the greatest danger? Explain what you saw in terms of the law of inertia (Newton's First Law of Motion).

Physics Words

force: a push or a pull that is able to accelerate an object; force is measured in newtons; force is a vector quantity.

pressure: force per surface area where the force is normal (perpendicular) to the surface; measured in pascals.

FOR YOU TO READ

Force and Pressure

When you repeated this experiment accurately each time, the **force** that each belt exerted on the clay was the same each time that the car was started at the same ramp height. Yet different materials have different effects; for example, a wire cuts far more deeply into the clay than a broader material does.

The force that each of the belts exerts on the clay is the same. When a thin wire is used, all the force is concentrated onto a small area. By replacing the wire with a broader material, you spread the force out over a much larger area of contact.

The force per unit area, which is called **pressure**, is much smaller with a ribbon, for example, than with a wire. It is the pressure, not the force, that determines how much damage the seat belt does to the body. A force applied to a single rib might be enough to break a rib. If the same force is spread out over many ribs, the force on each rib can be made too small to do any damage. While the total force does not change, the pressure becomes much smaller.

PHYSICS TALK

Pressure is the force per unit area:

$$P = \frac{F}{A}$$

where F is force in newtons (N)

A is area in meters squared (m^2)

and P is pressure in newtons per meter squared (N/m^2).

Force can be measured using a spring scale.

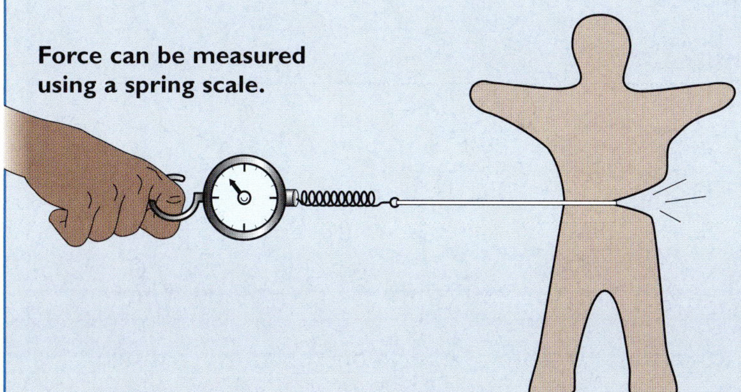

Sample Problem

Two brothers have the same mass and apply a constant force of 450 N while standing in the snow. Brother A is wearing snow shoes that have a base area of 2.0 m². Brother B, without snowshoes, has a base area of 0.1 m². Why does the brother without snowshoes sink into the snow?

Strategy: This problem involves the pressure that is exerted on the snow surface by each brother. You can use the equation for pressure to compare the pressure exerted by each brother.

Givens:

$$F = 450 \text{ N}$$
$$A_1 = 2.0 \text{ m}^2$$
$$A_2 = 0.1 \text{ m}^2$$

Solution:

Brother A	Brother B
$P = \frac{F}{A}$	$P = \frac{F}{A}$
$= \frac{450 \text{ N}}{2.0 \text{ m}^2}$	$= \frac{450 \text{ N}}{0.1 \text{ m}^2}$
$= 225 \text{ N/m}^2$ or 230 N/m^2	$= 4500 \text{ N/m}^2$

The pressure that Brother B exerts on the snow is much greater.

Reflecting on the Activity and the Challenge

In this activity you gathered data to provide evidence on the effectiveness of seat belts as restraint systems. The material used for the seat belt and the width of the restraint affected the distortion of the clay figure. By applying the force over a greater area, the pressure exerted by the seat belt during the collision can be reduced.

It is important to note that not every safety restraint system will be a seat belt or harness, but that all restraints attempt to reduce the pressure exerted on an object by increasing the area over which a force is applied.

How will your design team account for decreasing pressure by increasing the area of impact? Think about ways that you could test your design prototype for the pressure created during impact. Your presentation of the design will be much more convincing if you have quantitative data to support your claims. Simply stating that a safety system works well is not as convincing as being able to show how it reduces pressure during a collision.

Physics To Go

1. Use Newton's First Law to describe a collision with the passenger wearing a seat belt during a collision.

2. What is the pressure exerted when a force of 10 N is applied to an object that has an area of:

   a) 1.0 m^2?

   b) 0.2 m^2?

   c) 15 m^2?

   d) 400 cm^2?

3. A person who weighs approximately 155 lb. exerts 700 N of force on the ground while standing. If his shoes cover an area of 400 cm^2 (0.0400 m^2), calculate:

   a) the average pressure his shoes exert on the ground

   b) the pressure he would exert by standing on one foot

4. For comparison purposes, calculate the pressure you exert in the situations described below. Divide your weight in newtons, by the area of your shoes. (To find your weight in newtons multiply your weight in pounds by 4.5 N/lb. You can approximate the area of your shoes by tracing the soles on a sheet of centimeter squared paper.)
   a) How much pressure would you exert if you were standing in high heels?
   b) How much pressure would you exert while standing on your hands?
   c) If a bed of nails contains 5000 nails per square meter, how much force would one nail support if you were to lie on the bed? With this calculation you can now explain how people are able to lie on a bed of nails. It's just physic!

5. Describe why a wire seat belt would not be effective even though the force exerted on you by the wire seat belt is identical to that of a cloth seat belt.

6. Do you think there ought to be seat belt laws? How does not using seat belts affect the society as a whole?

7. Conduct a survey of 10 people. Ask each person what percentage of the time they wear a seat belt while in a car. Be prepared to share your data with the class.

Stretching Exercises

The pressure exerted on your clay model by a thin wire can be estimated quite easily. Loop the wire around the "passenger," and connect the wire to a force meter.

a) Pull the force meter hard enough to make the wire sink into the model just about as far as it did in the collision.
b) Record the force as shown on the force meter (in newtons).
c) Estimate the frontal area of the wire—its diameter times the length of the wire that contacts the passenger. Record this value in centimeters squared (cm^2).
d) Divide the force by the area. This is the pressure in newtons per centimeter squared (N/cm^2).

Activity 6 Why Air Bags?

GOALS

In this activity you will:

- Model an automobile air bag.
- Relate pressure to force and area.
- Demonstrate that the force of an impact can be reduced by spreading it out over a longer time.

What Do You Think?

Air bags do not take the place of seat belts. Air bags are an additional protection. They are intended to be used with seat belts to increase safety.

- **Why are air bags effective?**
- **How does the air bag protect you?**

Record your ideas about these questions in your *Active Physics* log. Be prepared to discuss your responses with your small group and the class.

For You To Do

1. You will use a large plastic bag or a partially inflated beach ball as a model for an air bag. Impact is provided by a heavy steel ball, or just a good-sized rock, dropped from a height of a couple of meters.

   Gather the equipment you will need for this activity. Your problem is to find out how long it takes the object to come to rest. What is the total time duration from when the object first touches the air bag until it bounces back?

2. With a camcorder, videotape the object striking the air bag from a given height, such as 1.5 m.

   a) Record the exact height, from which you dropped the object.

3. Play the sequence back, one frame at a time. Count the number of frames during the time the object is moving into the air bag—from the moment it first touches the bag until it comes to rest, before bouncing. Each frame stands for $\frac{1}{30}$ s. (Check your manual.)

   a) In your log, record the number of frames and calculate how long it takes for the object to come to rest.

   If a camcorder is not available, the experiment may be performed, although less effectively, by attaching a ticker-tape timer to the falling object.

   After the object is dropped, with the object still attached, stretch the tape from the release position to the air bag. Mark the dot on the tape that was made just as the object touched the air bag.

   Now push the object into the air bag, about as far as it went just before it bounced. Mark the tape at the dot that was made as the object came to rest. The dots should be close together for a short interval at this point.

   Now count the time that passed between the two marks you made. (You must know how rapidly dots are produced by your timer.)

Set up the activity in an area clear of obstruction. Arrange for containment of the dropped object.

4. Repeat **Steps 2** and **3**, but this time drop the ball against a hard surface, such as the floor. Keep the height from which the object is dropped constant.
   a) Record how long it takes for the object to come to rest on a floor.
5. Choose two other surfaces and repeat **Steps 2** and **3**.
   a) Record how long it takes the object to come to rest each time.
   b) In your log, list all the surfaces you tested in the order in which you expect the most damage to be done to a falling object, to the least damage.
   c) Is there a relationship between the time it takes for the object to come to rest and the potential damage to the object landing on the surface? Explain this relationship in your log.

PHYSICS TALK

Force and Impulse

Newton's First Law states that an object in motion will remain in motion unless acted upon by a net external force. In this activity you were able to stop an object with a force. In all cases the object was traveling at the same speed before impact. Stopping the object was done quickly or gradually. The amount of damage is related to the time during which the force stopped the object. The air bag was able to stop the object with little damage by taking a long time. The hard surface stopped the object with more damage by taking a short time.

Physicists have a useful way to describe these observations. An **impulse** is needed to stop an object. That impulse is defined as the product (multiplication) of the force applied and the time that the force is applied.

Physics Words

impulse: the product of force and the interval of time during which the force acts; impulse results in a change in momentum.

$$\text{Impulse} = F\Delta t$$

where F is force in newtons (N)

Δt is the time interval during which the force is applied in seconds (s).

Impulse is calculated in newton seconds (Ns).

An object of a specific mass and a specific speed will need a definite impulse to stop. Any forces acting for enough time can provide that impulse.

If the impulse required to stop is 60 Ns, a force of 60 N acting for 1 s has the required impulse. A force of 10 N acting for 6 s also has the required impulse.

Force F	Time Interval Δt	Impulse $F\Delta t$
60 N	1 s	60 Ns
10 N	6 s	60 Ns
6000 N	0.01 s	60 Ns

The greater the force and the smaller the time interval, the greater the damage that is done.

Sample Problem

A person requires an impulse of 1500 Ns to stop. What force must be applied to the person to stop in 0.05 s?

Strategy: You can use the equation for impulse and rearrange the terms to solve for the force required.

$$\text{Impulse} = F\Delta t$$

$$F = \frac{\text{Impulse}}{\Delta t}$$

Givens:

$$\text{Impulse} = 1500 \text{ Ns}$$

$$\Delta t = 0.05 \text{ s}$$

Solution:

$$F = \frac{\text{Impulse}}{\Delta t}$$

$$= \frac{1500 \text{ Ns}}{0.05 \text{ s}}$$

$$= 30{,}000 \text{ N}$$

Reflecting on the Activity and the Challenge

People once believed that the heavier the automobile, the greater the protection it offered passengers. Although a heavy, rigid car may not bend as easily as an automobile with a lighter frame, it doesn't always offer more protection.

In this activity, you found that air bags are able to protect you by extending the time it takes to stop you. Without the air bags, you will hit something and stop in a brief time. This will require a large force, large enough to injure you. With the air bag, the time to stop is longer and the force required is therefore smaller.

Force and impulse must be considered in designing your safety system. Stopping an object gradually reduces damage. The harder a surface, the shorter the stopping distance and the greater the damage. In part this provides a clue to the use of padded dashboards and sun visors in newer cars. Understanding impulse allows designers to reduce damage both to cars and passengers.

Physics To Go

1. If an impulse of 60 Ns is required to stop an object, list in your log three force and time combinations (other than those given in the **Physics Talk**) that can stop an object.
2. A person weighing 130 lb. (60 kg) traveling at 40 mph (18 m/s) requires an impulse of approximately 1000 Ns to stop. Calculate the force on the person if the time to stop is:

   a) 0.01 s
   b) 0.10 s
   c) 1.00 s

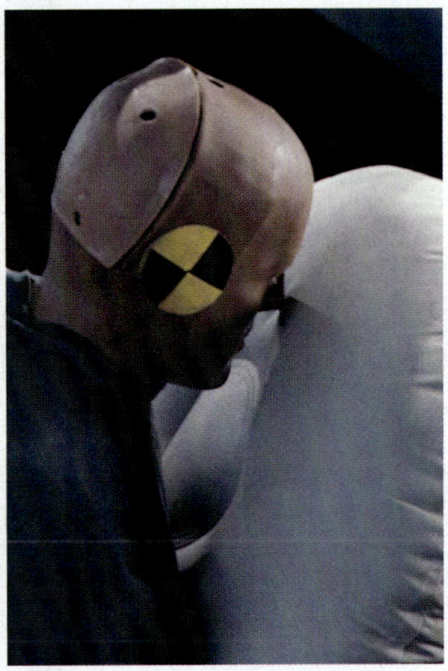

3. Explain in your log why an air bag is effective. Use the terms force, impulse, and time in your response.
4. Explain in your log why a car hitting a brick wall will suffer more damage than a car hitting a snow bank.
5. There are several other safety designs that employ the concept of spreading out the time interval of a force. Describe in your log how the ones listed below perform this task:

   a) the bumper
   b) a collapsible steering wheel
   c) frontal "crush" zones
   d) padding on the dashboard

6. There are many other situations in which the force of an impulse is reduced by spreading it out over a longer time. Explain in your log how each of the actions below effectively reduces the force by increasing the time. Use the terms force, impulse, and time in your response.

   a) catching a hard ball
   b) jumping to the ground from a height
   c) bungee jumping
   d) a fireman's net

7. The speed of airplanes is considerably higher than the speed of automobiles. How might the design of a seat belt for an airplane reflect the fact that a greater impulse is exerted on a plane when it stops?
8. Airplanes have seat belts. Should they also have air bags?

Activity 7 Automatic Triggering Devices

GOALS

In this activity you will:

- Design a device that is capable of transmitting a digital electrical signal when it is accelerated in a collision.

What Do You Think?

An air bag must inflate in a sudden crash, but must not inflate under normal stopping conditions.

- **How does the air bag "know" whether to inflate?**

Record your ideas about this question in your *Active Physics* log. Be prepared to discuss your responses with your small group and the class.

For You To Do

Inquiry Investigation

1. Form engineering teams of three to five students. Meet with your engineering team to design an automatic air bag triggering device using a knife switch, rubber bands, string, wires, and a flashlight bulb. Other materials may also be supplied by you or your teacher.

Safety

Be sure to receive your teacher's approval before using any material.

2. The design parameters are as follows:

- The device must turn a flashlight bulb on, or turn it off. This will be interpreted as the trigger signal.
- The device must not trigger if the car is brought to a sudden stop from a slow speed.
- The device must trigger if the sudden stop is from a high speed.
- The car containing the device must be released down a ramp. The car will then strike a wall at the bottom of the ramp.
- The battery and bulb must be attached to the car along with the triggering device. The bulb does not have to remain in the final on or off state, but it must at least flash to show triggering.

3. Follow your teacher's guidelines for using time, space, and materials as you design your triggering device.

4. Demonstrate your design team's trigger for the class.

FOR YOU TO READ

Impulse and Changes in Momentum

It takes an unbalanced, opposing force to stop a moving car. **Newton's Second Law** lets you find out how much force is required to stop any car of any mass and any speed.

The overall idea can be shown using a concept map.

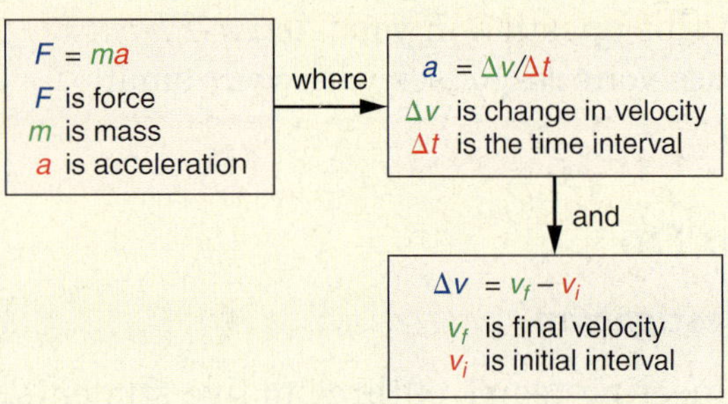

If you know the mass and can determine the acceleration, you can calculate the force using Newton's Second Law:

$$F = ma$$

A moving car has a forward **velocity** of 15 m/s. Stopping the car gives it a final velocity of 0 m/s.

$$\begin{aligned}\text{The change in velocity} &= v_{final} - v_{initial} \\ &= 0 - 15 \text{ m/s} \\ &= -15 \text{ m/s.}\end{aligned}$$

Any change in velocity is defined as **acceleration**.

In this case the change in velocity is –15 m/s. If the change in velocity occurs in 3 s, the acceleration is –15 m/s in 3 s, or –5 m/s every second, or –5 m/s^2. You can look at this as an equation:

$$a = \frac{\Delta v}{\Delta t} = \frac{v_f - v_o}{\Delta t} = \frac{0 \text{ m/s} - 15 \text{ m/s}}{3 s} = -5 \text{ m/s}^2$$

If the car had stopped in 0.5 s, the change in speed is identical, but the acceleration is now:

$$a = \frac{\Delta v}{\Delta t} = \frac{v_f - v_o}{\Delta t} = \frac{0 \text{ m/s} - 15 \text{ m/s}}{0.5 \ s} = -30 \text{ m/s}^2$$

Newton's Second Law informs you that unbalanced outside forces cause all accelerations. The force stopping the car may have been the frictional force of the brakes and tires on the road, or the force of a tree, or the force of another car. Once you know the acceleration, you can calculate the force using Newton's Second Law. If the car has a mass of 1000 kg, the unbalanced force for the acceleration of –5 m/s *every* second would be –5000 N. The force for the larger acceleration of –30 m/s *every* second would be –30,000 N. The negative sign tells you that the unbalanced force was opposite in direction to the velocity.

The change in velocity, acceleration, and force give a complete picture.

There is another, equivalent picture that describes the same collision in terms of **momentum** and impulse.

Any moving car has momentum. Momentum is defined as the mass of the car multiplied by its velocity $p = mv$.

The impulse/momentum equation tells you that the momentum of the car can be changed by applying a force for a given amount of time.

Impulse = change in momentum

↓

$\Delta p = F\Delta t$ → where → $\Delta p = M\Delta v$

Using the impulse/momentum approach explains something different about a collision. Consider this question, "why do you prefer to land on soft grass rather than on hard concrete?" Soft grass is preferred because the force on your legs is less when you land on soft grass. Let's find out why using Newton's Second Law and then by using the impulse/momentum relation.

→

Newton's Second Law explanation: Whether you land on concrete or soft grass, your change in velocity will be identical. Your velocity may decrease from 3 m/s to 0 m/s. On concrete, this change occurs very fast, while on soft grass this change occurs in a longer period of time. Your acceleration on soft grass is smaller because the change in velocity occurred in a longer period of time.

$$a = \frac{\Delta v}{\Delta t}$$

When the change in the period of time gets larger, the denominator of the fraction gets larger and the value of the acceleration gets smaller.

When landing on grass, Newton's Second Law then tells you that the force must be smaller because the acceleration is smaller for an identical mass. $F = ma$. Smaller acceleration on grass requires a smaller force. Smaller forces are easier on your legs and you prefer to land on soft grass.

Momentum/impulse explanation: Whether you land on concrete or soft grass, your change in momentum will be identical. Your velocity will decrease from 3 m/s to 0 m/s on either concrete or grass.

$$F\Delta t = \Delta p$$

You can get this change in momentum with a large force over a short time or a small force over a longer time.

If your mass is 50 kg, the amount of your change in momentum may be 150 kg m/s, when you decrease your velocity from 3 m/s to 0 m/s. There are many forces and associated times that can give this change in the value of the momentum.

If you could land on a surface that requires 3 s to stop, it will only require 50 N. A more realistic time of 1 s to stop will require a larger force of 150 N. A hard surface that brings you to a stop in 0.01 s requires a much larger force of 15,000 N.

On concrete, this change in the value of the momentum occurs very fast (a short time) and requires a large force. It hurts. On soft grass this change in the value of the momentum occurs in a longer time and requires a small force: it is less painful and is preferred.

Change in value of momentum	Force	Change in Time Δt	$F\Delta t$
150 kg m/s	50 N	3 s	150 kg m/s
150 kg m/s	75 N	2 s	150 kg m/s
150 kg m/s	150 N	1 s	150 kg m/s
150 kg m/s	1500 N	0.1 s	150 kg m/s
150 kg m/s	15,000 N	0.01 s	150 kg m/s

Physics To Go

1. How do impulse and Newton's First Law (the Law of Inertia) play a role in your air bag trigger design?
2. Imagine a device where a weight is hung from a string within a metal can. If the weight hits the side of the can, a circuit is completed. How do impulse and the Law of Inertia work in this device?
3. In cars built before 1970, the dashboard was made of hard metal. After 1970, the cars were installed with padded dashboards like you find in cars today. In designing a safe car, it is better to have a passenger hit a cushioned dashboard than a hard metal dashboard.
   a) Explain why the padded dashboard is better using Newton's Second Law.
   b) Explain why the padded dashboard is better using impulse and momentum.
4. Why would you prefer to hit an air bag during a collision than the steering wheel?
   a) Explain why the air bag is better using Newton's Second Law.
   b) Explain why the air bag is better using impulse and momentum.
5. Explain why you bend your knees when you jump to the ground.
6. Catching a fast ball stings your hand. Why does wearing a padded glove help?
7. When a soccer ball hits your chest, you can stiffen your body and the ball will bounce away from you. In contrast, you can "soften" your body and the ball drops to your feet. Explain how the force on the ball is different during each play.

Physics Words

Newton's Second Law of Motion: if a body is acted upon by an external force, it will accelerate in the direction of the unbalanced force with an acceleration proportional to the force and inversely proportional to the mass.

velocity: speed in a given direction; displacement divided by the time interval; velocity is a vector quantity, it has magnitude and direction.

acceleration: the change in velocity per unit time.

momentum: the product of the mass and the velocity of an object; momentum is a vector quantity.

Stretching Exercises

1. How does a seat belt "know" to hold you firmly in a crash, but allow you to lean forward or adjust it without locking? Write your response in your log.
2. Go to a local auto repair shop, junk yard, or parts supply store. Ask if they can show you a seat belt locking mechanism. How does it work? Construct a poster to describe what you have learned.

Activity 8 The Rear-End Collision

GOALS

In this activity you will:

- Evaluate from simulated collisions, the effect of rear-end collisions on the neck muscles.
- Understand the causes of whiplash injuries.
- Understand Newton's Second Law of Motion.
- Understand the role of safety devices in preventing whiplash injury.

What Do You Think?

Whiplash is a serious injury that is caused by a rear-end collision. It is the focus of many lawsuits, loss of ability to work, and discomfort.

- **What is whiplash?**
- **Why is it more prominent in rear-end collisions?**

Record your ideas about these questions in your *Active Physics* log. Be prepared to discuss your responses with your small group and the class.

For You To Do

1. You will use two pieces of wood to represent the torso (the trunk of the body) and the head of a passenger. Attach a small piece of wood (about 1" x 2" x 2") to a larger piece of wood (about 1" x 3" x 10") with some duct tape acting like a hinge between the two pieces.

   a) Make a sketch to show your passenger. Label what each part of the model passenger represents.

2. Set up a ramp against a stack of books about 40 cm high, as shown in the diagram below. Place the wooden model passenger at the front of a collision cart positioned about 50 cm from the end of the ramp. Release a second cart from a few centimeters up the ramp.

   a) In your log record what happens to the head and torso of the wooden model.

Perform the activity outside of traffic areas. Do not obstruct paths to exits. Do not leave carts lying on the floor.

3. With the first cart still positioned about 50 cm from the end of the ramp, release the second cart from the top of the ramp.

   a) Describe what happens to the head of the model passenger in this collision.

   b) Use Newton's First Law of Motion to explain your observations.

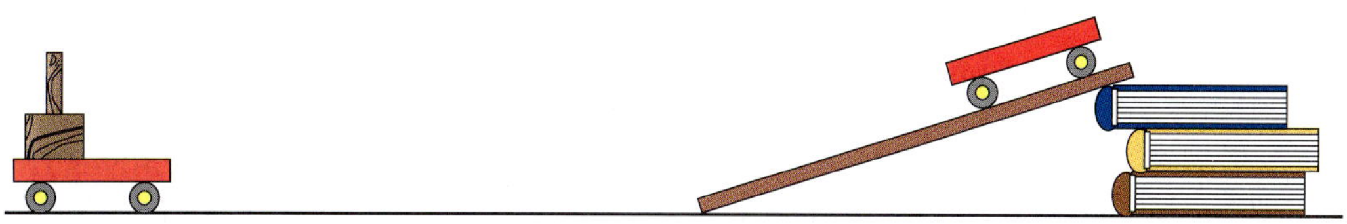

4. The duct tape represents the neck muscles and bones of the vertebral column. How large a force do the neck muscles exert to keep the head from flying off the body, and to return the head to the upright position? To answer this question, begin by estimating the mass of an average head.

   a) Estimate and record in your log the mass of an average human head. The mass would be close to the mass of a filled water container of the same size.

5. Mark off a distance about 30-cm long on the lab table or the floor. Obtain a piece of wood and attach it to a spring scale. Pull the wooden mass with the spring scale over the distance you marked.

a) In your log record the force required to pull the mass and the time it took to cover the distance.

b) Repeat the step, but vary the time required to pull the mass over the distance. Record the forces and the times in your log.

c) Use your observations to complete the following statement:

The shorter the time (that is, the greater the acceleration) the ______ the force required.

6. The ratio of the mass of the wood to the estimated mass of the head is the same as the ratio of the forces required to pull them.

a) Use the following ratio to calculate how large a force the neck muscles exert to keep the head from flying off the body, and to return the head to the upright position under different accelerations.

$$\frac{\text{mass of head}}{\text{mass of wood}} = \frac{\text{force to move head}}{\text{force to move wood}}$$

7. Whiplash is a serious injury that can be caused by a rear-end collision. The back of the car seat pushes forward on the torso of the driver and the passengers and their bodies lunge forward. The heads remain still for a very short time. The body moving forward and the head remaining still causes the head to snap backwards. The neck muscles and bones of the vertebral column become damaged. The same muscles must then snap the head back to its place atop the shoulders.

a) What type of safety devices can reduce the delay between body and head movement to help prevent injury?

b) What additional devices have been placed in cars to help reduce the impact of rear-end collisions?

FOR YOU TO READ

Newton's Second Law of Motion

Newton's First Law of Motion is limited since it only tells you what happens to objects if net force acts upon them. Knowing that objects at rest have a tendency to remain at rest and that objects in motion will continue in motion does not provide enough information to analyze collisions. Newton's Second Law allows you to make predictions about what happens when an external force is applied to an object. If you were to place a collision cart on a level surface, it would not move. However, if you begin to push the cart, it will begin to move.

Newton's Second Law states:

If a body is acted on by a force, it will accelerate in the direction of the unbalanced force. The acceleration will be larger for smaller masses. The acceleration can be an increase in speed, a decrease in speed, or a change in direction.

Newton's Second Law of Motion indicates that the change in motion is determined by the force acting on the object, and the mass of the object itself.

Analyzing the Rear-End Collision

This activity demonstrated the effects of a rear-end collision. Newton's First Law and Newton's Second Law can help explain the "whiplash" injury that passengers suffer during this kind of collision.

Imagine looking at the rear-end collision in slow motion. Think about all that happens.

1. A car is stopped at a red light. This is the car in which the driver is going to be injured with whiplash. The driver is at rest within the car.
2. The stopped car gets hit from the rear.
3. The car begins to move. The back of the seat pushes the driver forward and his torso moves with the car. The driver's head is not supported and stays back where it is.
4. The neck muscles hold the head to the torso as the body moves forward. The muscles then "whip" the head forward. The head keeps moving until it gets ahead of the torso. The head is stopped by the neck muscles. The muscles pull the head back to its usual position. Ouch!

Let's repeat the description of the collision and insert all of the places where Newton's First Law applies. Newton's First Law states that *an object at rest stays at rest and an object in motion stays in motion unless acted upon by an unbalanced, outside force.*

1. A car is stopped at a red light. This is the car in which the driver is going to be injured with whiplash. The driver is at rest within the car. *Newton's First Law: an object (the driver) at rest stays at rest.*

2. The stopped car gets hit from the rear.
3. The car begins to move. The back of the seat pushes the driver forward and his torso moves with the car. *Newton's First Law: an object (the driver's torso) at rest stays at rest unless acted upon by an unbalanced, outside force.* The driver's head is not supported and stays back where it is. *Newton's First Law: an object (the driver's head) at rest stays at rest.*
4. The neck muscles hold the head to the body as the body moves forward. The muscles then "whip" the head forward. *Newton's First Law: an object (the head) at rest stays at rest unless acted upon by an unbalanced, outside force.* The head keeps moving until it gets ahead of the torso. *Newton's First Law: an object (the head) in motion stays in motion.* The head is stopped by the neck muscles. *Newton's First Law: an object (the head) in motion stays in motion unless acted upon by an unbalanced, outside force.* The muscles pull the head back to its usual position. *Newton's First Law: an object at rest stays at rest unless acted upon by an unbalanced, outside force.* Ouch!

Let's repeat the description of the collision and insert all of the places where Newton's Second Law applies. Newton's Second Law states that *all accelerations are caused by unbalanced, outside forces, $F = ma$.* An acceleration is any change in speed.

1. A car is stopped at a red light. This is the car in which the driver is going to be injured with whiplash. The driver is at rest within the car.
2. The stopped car gets hit from the rear.
3. The car begins to move. *Newton's Second Law: the car accelerates because of the unbalanced, outside force from the rear; $F = ma$.* The back of the seat pushes the driver forward and his torso moves with the car. *Newton's Second Law: the torso accelerates because of the unbalanced, outside force from the back of the seat; $F = ma$.* The driver's head is not supported and stays back where it is.
4. The neck muscles hold the head to the torso as the body moves forward. The muscles then "whip" the head forward. *Newton's Second Law: the head accelerates because of the unbalanced force of the muscles; $F = ma$.* The head keeps moving until it gets ahead of the torso. The head is stopped by the neck muscles. *Newton's Second Law: the head accelerates (slows down) because of the unbalanced force from the neck muscles; $F = ma$.* The muscles pull the head back to its usual position. *Newton's Second Law: the head accelerates because of the unbalanced force from the rear; $F = ma$.* Ouch!

Newton's Second Law informs you that all accelerations are caused by *unbalanced, outside* forces. It does not say that all forces cause accelerations. An object at rest may have many forces acting upon it. When you hold a book in your hand, the book is at rest. There is a force of gravity pulling the book down. There is a force of your hand pushing the book up. These forces are equal and opposite. The "net" force on the book is zero because the two forces balance each other. There is no acceleration because there is no "net" force.

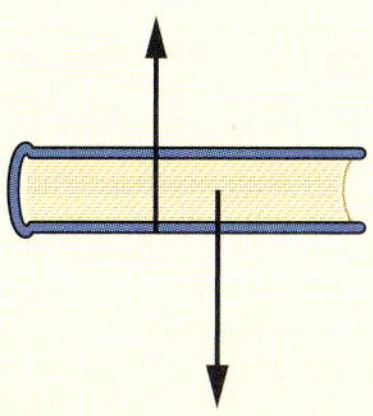

Both forces act through the center of the book. They are shifted a bit in the diagram to emphasize that the upward force of the hand acts on the bottom of the book and the downward force of gravity acts on the middle of the book.

As a car moves down the highway at a constant speed, there are forces acting on the car but there is no acceleration. This indicates that the net force must be zero. The force of the engine on the tires and road moving the car forward must be equal and opposite to the force of the air pushing the car backward. These forces balance each other in this case, where the speed is not changing. There is no net force and there is no acceleration. The car stays in motion at a constant speed. A similar situation occurs when you push a book across a table at constant speed. The push is to the right and the friction is to the left, opposing motion. If the forces are equal in size, there is no net force on the book and the book does not accelerate—it moves with a constant speed.

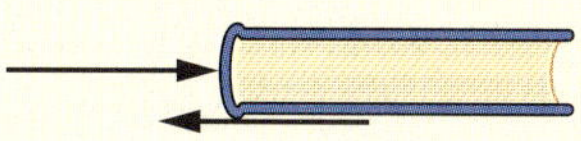

Reflecting on the Activity and the Challenge

The vertebral column becomes thinner and the bones become smaller as the column attaches to the skull. The attachment bones are supported by the least amount of muscle. Unfortunately, the smaller bones, with less muscle support, make this area particularly susceptible to injury. One of the greatest dangers following whiplash is the damage to the brainstem. The brainstem is particularly vital to life support because it regulates blood pressure and breathing movements. Consider how your safety device will help prevent whiplash following a collision. What part of the restraining device prevents the movement of the head?

Physics To Go

1. Why are neck injuries common after rear-end collisions?
2. Explain why the packages in the back move forward if a truck comes to a quick stop.
3. As a bus accelerates, the passengers on the bus are jolted toward the back of the bus. Indicate what causes the passengers to be pushed backward.
4. Why would the rear-end collision demonstrated by the laboratory experiment be most dangerous for someone driving a motorcycle?
5. Would headrests serve the greatest benefit during a head-on collision or a rear-end collision? Explain your answer.
6. A cork is attached to a string and placed in a jar of water as shown by the diagram to the right. Later, the jar is inverted.
   a) If the glass jar is pushed along the surface of a table, predict the direction in which the cork will move.
   b) If you place your left hand about 50 cm in front of the jar and push it with your right hand until it strikes your left hand, predict the direction in which the cork will move.

Be sure the outside of the jar is dry so it does not slip out of your hands.

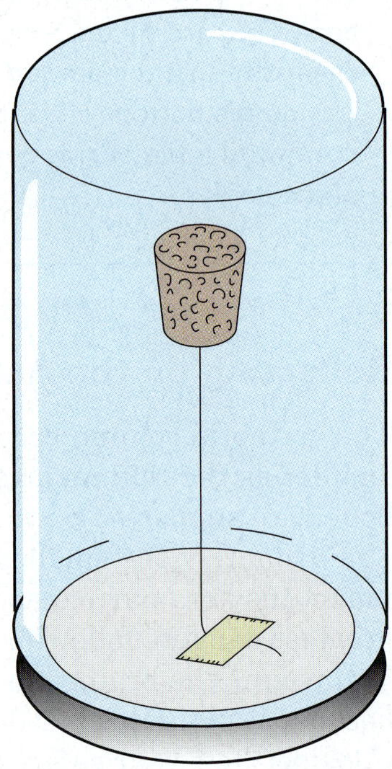

Activity 9 Cushioning Collisions (Computer Analysis)

GOALS

In this activity you will:

- Apply the concept of impulse in the analysis of automobile collisions.
- Use a computer's motion probe (sonic ranger) to determine the velocity of moving vehicles.
- Use a computer's force probe to determine the force exerted during a collision.
- Compare the momentum of a model vehicle before the collision with the impulse applied during the collision.
- Explore ways of using cushions to increase the time that a force acts during a primary collision.

What Do You Think?

The use of sand canisters around bridge supports and crush zones in cars are examples of technological systems that are designed to minimize the impact of collisions between a car and a stationary object or another car.

- **How do these technological systems reduce the impact of the primary collision?**

Record your ideas about this question in your *Active Physics* log. Be prepared to discuss your responses with your small group and the class.

For You To Do

1. In this investigation you will be using a force probe that is attached to a computer to determine the effectiveness of different types of cushions for a toy vehicle. Release a toy car at

Safety

Perform the activity outside of traffic areas. Do not obstruct paths to exits. Do not leave carts lying on the floor.

the top of a ramp and measure the force of impact as it strikes a barrier at the bottom. A sonic ranger can be mounted on the ramp to measure the speed of the toy car prior to the collision. Open the appropriate computer files to prepare the sonic ranger to graph velocity vs. time and the force probe to graph force vs. time.

2. Mount the sonic ranger at the bottom of a ramp and place the force probe against a barrier about 10 cm from the bottom of the ramp, as shown in the diagram. Attach an index card to the back of the car, to obtain better reflection of the sound wave and improve the readings of the sonic ranger.

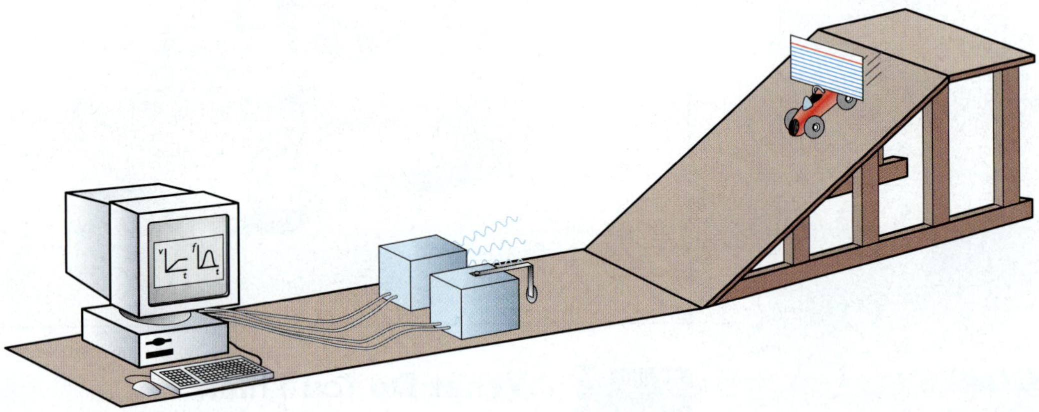

3. Conduct a few runs of the car against the force probe to ensure that the data collection equipment is working properly.

4. Attach a cushioning material to the front of the car. Conduct a number of runs with the same type of cushioning. Make sure that the car is coasting down the same slope from the same position each time.

a) Make copies of the velocity vs. time and force vs. time graphs that are displayed on the computer.

5. Repeat **Step 4** using other types of cushioning materials.

a) Record your observations in your log.

6. Use the graphical information you obtained in this activity to answer the following:
   - a) Compare the force vs. time graphs for the cushioned cars with those for the cars without cushioning.
   - b) Compare the areas under the force vs. time graphs for all of the experimental trials.
   - c) Compute the momentum of the car (the product of the mass and the velocity) prior to the collision and compare it with the area under the force vs. time graphs.
   - d) Summarize your comparisons in a chart.
   - e) How can impulse be used to explain the effectiveness of cushioning systems?
   - f) Describe the relationship between impulse ($F\Delta t$) and the change in momentum ($m\Delta v$).

PHYSICS TALK

Change in Momentum and Impulse

Momentum is the product of the mass and the velocity of an object.

$$p = mv$$

where p is the momentum,

m is the mass,

and v is the velocity.

Change in momentum is the product of mass and the change in velocity.

$$\Delta p = m\Delta v$$

Impulse = change in momentum

$$F\Delta t = m\Delta v$$

Sample Problem

A vehicle has a mass of 1500 kg. It is traveling at 15.0 m/s. Calculate the change in momentum required to slow the vehicle down to 5.0 m/s.

Strategy: You can use the equation for calculating the change in momentum.

$$\Delta p = m\Delta v$$

Recall that the Δ symbol means "the change in." If you know the final and initial velocities you can write this equation as:

$$\Delta p = m(v_f - v_i)$$

where v_f is the final velocity and

v_i is the initial velocity.

Givens:

$m = 1500$ kg

$v_f = 5.0$ m/s

$v_i = 15.0$ m/s

Solution:

$$\Delta p = m\,(v_f - v_i)$$

$$= 1500 \text{ kg } (5.0 \text{ m/s} - 15.0 \text{ m/s})$$

$$= 1500 \text{ kg } (-10.0 \text{ m/s})$$

$$= -15{,}000 \text{ kg} \cdot \text{m/s}$$

Reflecting on the Activity and the Challenge

What you learned in this activity better prepares you to defend the design of your safety system. The principles of momentum and impulse must be used to justify your design. Previously, you discovered objects with greater mass are more difficult to stop than smaller ones. You determined that increasing the velocity of objects also makes them more difficult to stop. Objects that have a greater mass or greater velocity have greater momentum.

Linking the two ideas together allows you to begin examining the relationship between momentum and impulse. For a large momentum change in a short time, a large force is required. A crushed rib cage or broken leg bones often result. The change in the momentum can be defined by the impulse on the object.

What device will you use to increase the stopping time for the **Chapter Challenge** activity? Make sure that you include impulse and change in momentum in your report. Your design features must be supported by the principles of physics.

Physics To Go

1. Helmets are designed to protect cyclists. How would the designer of helmets make use of the concept of impulse to improve their effectiveness?
2. The Congress of the United States periodically reviews federal legislation that relates to the design of safer cars. For many years, one regulation was that car bumpers must be able to withstand a 5 mph collision. What was the intent of this regulation? The speed was later lowered to 3 mph. Why? Should it be changed again?
3. If a car has a mass of 1200 kg and an initial velocity of 10 m/s (about 20 mph) calculate the change in momentum required to:

   a) bring it to rest

   b) slow it to 5 m/s (approximately 10 mph)

4. If the braking force for a car is 10,000 N, calculate the impulse if the brake is applied for 1.2 s. If the car has a mass of 1200 kg, what is the change in velocity of the car over this 1.2 s time interval?
5. A 1500-kg car, traveling at 5.0 m/s after braking, strikes a power pole and comes to a full stop in 0.1 s. Calculate the force exerted by the power pole and brakes required to stop the car.
6. For the car described in **Question 5**, explain why a breakaway pole that brings the car to rest after 2.8 s is safer than the conventional power pole.
7. Write a short essay relating your explanation for the operation of the cushioning systems to the explanation of the operation of the air bags.
8. Explain why a collapsible steering wheel is able to help prevent injuries during a car crash.
9. Compare and contrast the two force vs. time graphs shown.

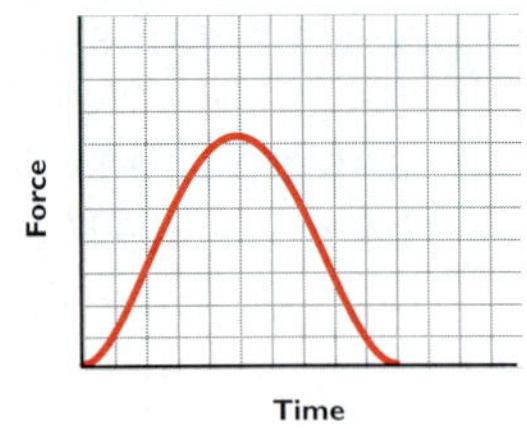

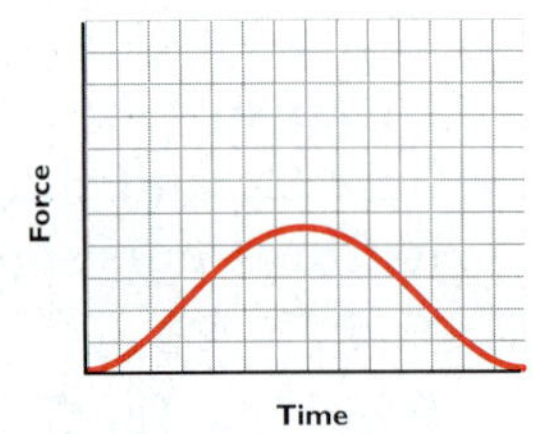

Stretching Exercises

Package an egg in a small container so that the egg will not break upon impact. Your teacher will provide the limitations in the construction of your package. You may be limited to two pieces of paper and some tape. You may be limited to a certain size package or a package of a certain weight. Bring your package to class so that it can be compared in a crash test with the other packages.

(Hint: Place each egg in a plastic bag before packaging to help avoid a messy cleanup.)

PHYSICS AT WORK

Mohan Thomas

DESIGNING AUTOMOBILES THAT SAVE LIVES

Mo is a Senior Project Engineer at General Motors North American Operation's (NAO) Safety Center and his responsibilities include making sure that different General Motors vehicles meet national safety requirements. Several of the design features that Mo has helped to develop have been implemented into vehicles that are now out on the road.

"This is how it works," he explains. "An engineer for a vehicle comes to us here at the Safety Center and requests technical assistance with design features to help them meet the side impact crash regulations required by the government. You have to analyze the physical forces of an event, which involves one car hitting another car on the side and then the door smashing into the driver," he continues. "We'll study the velocity, acceleration, momentum, and inertia in an event, as well as the materials used in the vehicle itself."

"The initial energy of an impact from one vehicle on another," states Mo, "has to be managed by the vehicle that's getting hit. Our goal is to manage the energy in such a way that the occupant in the vehicle being hit is protected. You take the forces that are coming into the vehicle and you redirect them into areas around the occupant. The framework of the car, therefore, is very important to the design, as well as energy-absorbing materials used in the vehicle."

Mo grew up in Chicago, Illinois, and has always enjoyed math and science, but he was also interested in creative writing. He wanted to combine math and science with creative work and has found that combination in the design work of engineering. "The nice part of being at the Safety Center," states Mo, "is that you know that you are contributing to something meaningful. The bottom line is that the formulas and problems that we are working on are meant to save people's lives."

Chapter 2 Assessment

Your design team will develop a safety system for protecting automobile, airplane, bicycle, motorcycle or train passengers. As you study existing safety systems, you and your design team should be listing ideas for improving an existing system or designing a new system for preventing accidents. You may also consider a system that will minimize the harm caused by accidents.

Your final product will be a working model or prototype of a safety system. On the day that you bring the final product to class, the teams will display them around the room while class members informally view them and discuss them with members of the design team. At this time, class members will generate questions about each others' products. The questions will be placed in envelopes provided to each team by the teacher. The teacher will use some of these questions during the oral presentations on the next day. The product will be judged according to the following:

1. The quality of your safety feature enhancement and the working model or prototype.
2. The quality of a 5-minute oral report that should include:

- **the need for the system**
- **the method used to develop the working model**
- **demonstration of the working model**
- **discussion of the physics concepts involved**
- **description of the next-generation version of the system**
- **answers to questions posed by the class**

3. The quality of a written and/or multimedia report including:

- **the information from the oral report**
- **documentation of the sources of expert information**
- **discussion of consumer acceptance and market potential**
- **discussion of the physics concepts applied in the design of the safety system**

Criteria

Review the criteria that were agreed to at the beginning of the chapter. If they require modification, come to an agreement with the teacher and the class.

Your project should be judged by you and your design team according to the criteria before you display and share it with your class. Being able to judge the quality of your own work before you submit it is one of the skills that will make you a "treasured employee"!

Physics You Learned

Newton's First Law of Motion (inertia)

Pressure (N/m^2)

Pressure = $\frac{\text{force}}{\text{area}}$ $(P = \frac{F}{a})$

Distance vs. time relationships

Time interval =
time$_{(final)}$ – time$_{(initial)}$ $(\Delta t = t_f - t_i)$

Impulse (N × *time*),

Impulse =
force × time interval (Impulse = $F \times \Delta t$)

Stopping distance

Newton's Second Law of Motion, constant acceleration, net force
Force = mass × acceleration
$F = ma$

Momentum = mass × velocity $(p = mv)$

Conservation of momentum

Change in momentum is affected by mass and change in velocity

Change in momentum =
mass × change in velocity $(\Delta p = m\, \Delta v)$

Impulse =
change in momentum $(F\Delta t = m\Delta v)$

Chapter 3

THE TRACK & FIELD CHAMPIONSHIP

Scenario

High school students across the country compete in track and field events locally. Some of the best competitors go on to compete at the state or national levels. The world's largest such event is the Penn Relays, which includes competitors at the high school, college, and professional levels. Your school has been invited to send a team to the Penn Relays (or some other big competition). Everyone is excited and nervous about the event. The team has read that some professional athletes hire physicists and other science professionals to help them improve their performances. Your school's track team cannot afford professional physicists but hopes that students studying *Active Physics* may be able to provide some assistance. Can the *Active Physics* students use their understanding to help the track team improve its performance?

Challenge

Your challenge is to write a physics manual about track and field training for your high-school team to help improve its performance. The manual should:

- **help students compare themselves with their competitors**
- **include a description of physics principles as they relate to track events**
- **provide specific techniques to improve performance**

Criteria

How will the manual be graded? What qualities should a good manual have? Discuss these issues in small groups and with your class. You may decide that some or all of the following qualities should be graded in your track and field manual:

- **physics principles**
- **inclusion of charts**
- **past records**
- **relevant equations**
- **definitions**
- **specific techniques**

Any advice you give should be understandable to athletes who have not studied physics. You can describe any activities you have done to explain how you know that the technique works, but you should not tell so much about each activity that the reader becomes bored.

Once you list the criteria for judging the track and field manuals, you should also decide how many points should be given for each criterion. If the group is going to hand in one manual, you must also agree on the way in which individuals will receive their grade. One method may be that the individual contributions toward the project receive 75% of the individual grade, and the group grade provides the remaining 25% of the individual grade. You should discuss different strategies and choose the one that is best suited to your school.

Activity 1 Running the Race

GOALS

In this activity you will:

- Measure distance in meters along a straight line.
- Use a stopwatch to measure the amount of time for a running person to travel a measured distance.
- Use measurements of distance and time to calculate the average speed of a running person.
- Compare the average speeds of a running person during segments of a run.
- Compare the average speeds of different persons running along the same path.
- Infer factors which affect the average speed of a running person.

What Do You Think?

Very few people in the world can run 100 meters in less than 10 seconds.

- **How can you measure a runner's speed?**
- **Does running twice as far take twice as much time?**

Record your ideas about these questions in your *Active Physics* log. Be prepared to discuss your responses with your small group and the class.

For You To Do

1. Your teacher will indicate the location of a "track." Place a mark on the track at the 0-, 5-, 15-, and 10-m positions.

2. Place students with stopwatches at the 5-, 10-, 15-, and 20-m marks to serve as timers. Each timer should measure the amount of time for each runner to reach the timer's assigned mark.

3. Have someone serve as starter, saying "Ready, Set, Go!" for each runner. All watches should be started on the "Go" signal, and each watch should be stopped as the runner passes the timer's assigned mark. Measure the time for three runners.

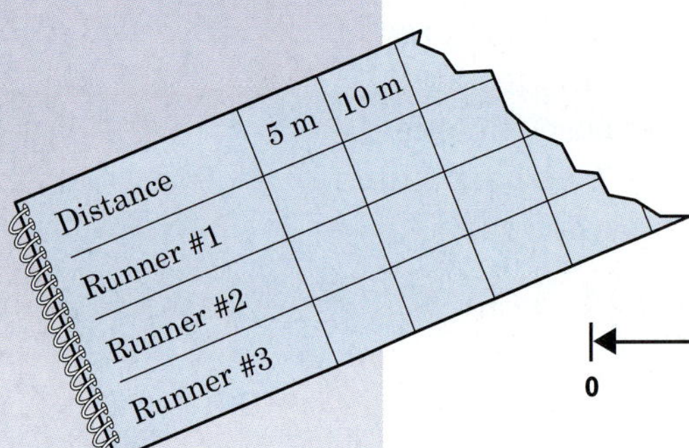

a) In your log, record the time it takes each runner to reach the 5-, 10-, 15-, and 20-m positions.

4. Calculate the amount of time taken to run each 5-m interval. To calculate the time taken to run from the 5- to 10-m mark, you will need to subtract the time at the 5-m mark from the time at the 10-m mark.

a) Use a table similar to the following to write the results in your log.

5. Does running twice as far take twice as much time?

Distance	0-5 m	5-10 m	10-15 m	15-20 m
Runner #1 Time				
Speed				
Runner #2 Time				
Speed				
Runner #3 Time				
Speed				

a) Use data from the 20-m dashes to explain your answer in your log.

6. Calculate the average speed during each 5-m interval. Use the equation:

$$\text{Average speed} = \frac{\text{Distance traveled}}{\text{Time taken}}$$

a) Record the average speed during each 5-m interval in the table in your log.

7. Use your data to answer the following questions:

   a) In which distance interval did each runner have the greatest average speed? Circle the interval in your log for each runner.

   b) Was the interval of greatest speed the same or different for different runners?

   c) Which runner holds the record for the fastest 5-m interval?

   d) Describe each runner's total dash in terms of speed during distance intervals.

   e) Estimate the amount of time taken for each runner to reach maximum speed.

8. Calculate the average speed of each runner for the entire dash.

   a) Record the average speed in your log.

   b) If the three dashes had been an Olympic time trial, which runner would have won? What was the winning speed?

9. Use data from each runner's dash.

   a) Write suggestions in your log for the runners to improve their performances.

PHYSICS TALK

Average Speed

The relationship between average **speed**, distance, and time can be written as:

$$\text{Average speed} = \frac{\text{Distance}}{\text{Time}}$$

This relationship can be rearranged as follows:

$$\text{Distance} = \text{Average Speed} \times \text{Time}$$

$$\text{Time} = \frac{\text{Distance}}{\text{Average Speed}}$$

Physics Words

speed: the change in distance per unit time; speed is a scalar; it has no direction.

FOR YOU TO READ

The **Scenario** for this chapter mentioned that the Penn Relays is the world's largest track and field event. It includes athletes at the high school, college, and professional levels. Examine the record times of male and female athletes in running events at the Penn Relays in the table to the right.

Many students in your class may be able to run 100 m in less than 15 s. Some members of your school's track and field team might be able to do the same in under 12 s. Do a calculation to show yourself that even a slow person who could run 100 m in 15 s could, theoretically, break the Penn Relays record for the 1500-m race if it is true that "running 15 times farther takes 15 times as much time." It will be necessary to convert the Penn Relays record given in minutes and seconds into seconds. (Example: A time of 4:08 for the mile equals 248 s.)

Obviously, not only speed but also stamina is involved in winning a race. An athlete needs to be able to run at high speed and have the stamina, or strength, to keep the speed as high as possible for the entire race. Both speed and stamina can be improved through training and knowledge. Athletes who hold records are those who run both fast and smart.

Penn Relays Record Times		
Distance (meters)	Time–Men (minutes:seconds)	Time–Women (minutes:seconds)
100	0:10.47	0:11.44
200	0:21.07	0:23.66
400	0:45.49	0:52.33
800	1:48.8	2:05.4
1500	3:49.67	4:24.0
mile	4:08.7	4:49.2
3000	8:05.8	9:15.3
5000	15:09.36	16:59.5

Reflecting on the Activity and the Challenge

You know from measurements of your classmates' running that the person who travels the entire distance of a race in the least time wins. The winner also has the highest average speed for the overall race. You also know that the speed of each runner in your class changed during the run.

Search for patterns in the distances and times of men and women who hold records in the Penn Relays. Do the speeds vary with the distance of the events? What information could you give your school's track and field team about the Penn Relays right now? What further information do you need?

Physics To Go

1. a) Calculate the average speed of the male who holds the Penn Relays record for the 1500-m run.
   b) From the data you gathered, are there students in your class who can reach the same speed as the male 1500-m record holder?
   c) Do you think the fastest student in your class could run the 1500 m in record time?
2. Calculate the average speeds of women who hold Penn Relays records in the 100-, 200-, 400-, and 1500-m runs. What is the pattern of speeds?
3. Find some record times for your school's track team in running events and compare them with the Penn Relays record times. Also compare the average speeds for races of equal distance for your school with those for the Penn Relays.
4. Is it fair to compare data for high school running events with data for world records? Why or why not?
5. How do the amounts of time taken by a champion 100-m runner to travel the first and last 50 m of a sprint compare? What is the basis for your answer?
6. Track coaches often measure "split times" for runners during races. For example, a runner's time might be measured every 100 m during a 400-m race. How could "splits" be useful for helping runners improve both speed and stamina?

Activity 2 Analysis of Trends

GOALS

In this activity you will:

- Sketch a best fit curve or trend line on a graph on which points have been plotted.
- Extrapolate graphs to predict trends beyond available data.
- Create graphs of data presented in tabular form.
- Calculate the average speed of a runner given distance and time.
- Calculate the time required for a runner to travel a specified distance given the runner's average speed.
- Interpret information presented in graphical form.

What Do You Think?

Current trends indicate that women will start outrunning men in 65 years.

- **Is it useful to compare track records over many years of time?**
- **Can future track records be predicted based on past performances?**

Record your ideas about these questions in your *Active Physics* log. Be prepared to discuss your responses with your small group and the class.

For You To Do

1. Look at the graph "Speed Versus Year: Men's Olympic 400-m Dash." The average speed of runners is shown on the vertical axis, and the year in which the race was run is shown on horizontal axis. Take a moment to be sure you understand that the plotted points show a 100-year history of the speeds of male athletes in the Olympic 400-m dash.

Speed Versus Year:
Men's Olympic—400-m Dash

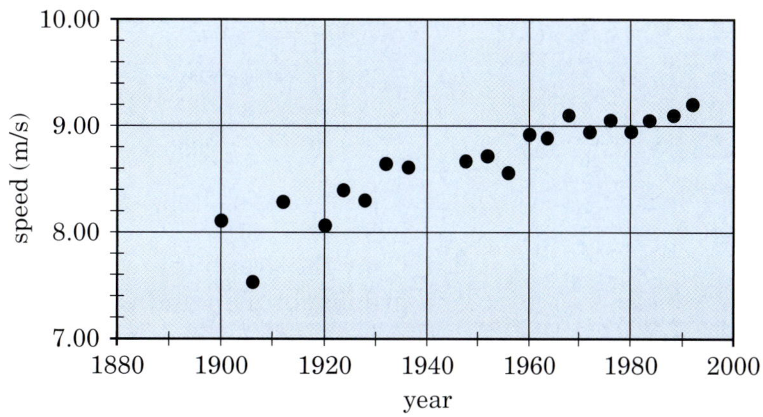

a) Copy the graph into your log. Sketch either a straight line or a smooth, curved line through the data points to show what you think is the shape of the graph. Do this for the points plotted from 1896 to 1996.

b) Comment in your log about what you see as the trend of average speed in the 400-m dash over the past 100 years.

c) Make the best guess you can to sketch how the graph would continue to the year 2020. This process of going beyond the data is called extrapolation. Try it.

d) According to your extrapolation, what will be the speed of the winning runner in the men's Olympic 400-m dash in the year 2020?

2. Look at the graph "Speed Versus Distance: Men and Women, Penn Relays." Notice that the average speed of runners is shown on the vertical axis and the distance of the race is shown on the horizontal axis. Be sure you understand that there are two sets of plotted points, one for men and one for women. Also, be sure you understand that the plotted points show how average speed varies with distance of the race for both men and women.

Speed Versus Distance:
Men and Women, Penn Relays Records

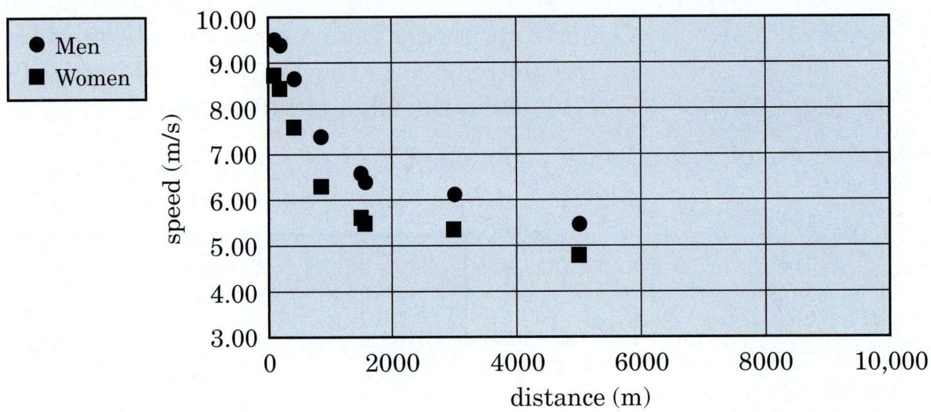

a) Copy the graph into your log. Sketch the shapes of the graphs for men and women by connecting the plotted points with either a straight line or a curved line, according to your choice.

b) Comment on the trends of average speed for both men and women as the distance of races gets longer and longer.

c) Extrapolate from the graphs to predict what the record speed at the Penn Relays would be for 10,000-m races for men and women.

d) Try to use extrapolation to find a race distance for which men and women would run at the same average speed. Comment on your attempt.

3. The kind of data analysis you performed in this activity can be made much easier with a computer or graphing calculator. Basic distance and time data from track records can be entered into a computer spreadsheet software program or a graphing calculator. The computer or calculator can then be instructed to calculate speed from distance versus time data and to display graphs. The graphs can be analyzed to show trend lines, extrapolations, and other information. Your teacher or someone else familiar with computers and calculators may be able to help you use such devices as tools for data analysis.

Reflecting on the Activity and the Challenge

Your plan to help your school's team at the Penn Relays will only be as good as your knowledge of current performances in track and field. Also, knowledge of trends in how performances are changing with time in various events is also important.

Research shows that women are improving their track performances about twice as fast as men. This is because more and more women are participating in running as a sport than ever before. Certainly, you will want to consider the possibility of either a male or female athlete bringing home a medal from the Penn Relays for your school. Data can help you decide which athletes at your school will have the best chance.

Perhaps it's time to get some data about the performance of athletes at your school so that you can begin forming a strategy to help your team. Your teacher will help you with this.

Physics To Go

1. Get distance and time data for the best performances of your school's boys' and girls' track teams for races of various distances.
   a) Calculate the average speed for each event.
   b) Plot the average speeds as data points on the Penn Relays graph used in **For You To Do.** Be sure to keep male and female data separate.
   c) Connect the points to show the shapes of the graphs for males and females.
   d) For which race events does the graph for males at your school come closest to touching the graph for men at the Penn Relays? What does this mean? Do the same analysis for females at your school and the Penn Relays.

2. In **For You To Do** you extrapolated from a graph to predict the speed of the winner of the men's 400-m dash in the Olympics in the year 2020. Predict the winning time for that race. (Hint: See **Physics Talk** in **Activity 1**.)

3. In **For You To Do** you extrapolated from two graphs to predict the average speeds of men and women who would win if a 10,000-m run were held at the Penn Relays. Predict the time of the winning man and woman.

4. A runner had an average speed of 10.14 m/s for 9.86 s. Calculate the distance the runner traveled.

5. The table gives the years and the winning times for the women's 400-m dash in the Olympics.
   a) Calculate the winning speeds.
   b) Plot a speed vs. year graph for the data.
   c) What is the trend of the average speeds over the past years?
   d) From your graph extrapolate what the winning speed will be in 2020.

Year	Time (seconds)
1964	52.00
1968	52.00
1972	51.08
1976	49.29
1980	48.88
1984	48.83
1988	48.65
1992	48.83

Activity 3 Who Wins the Race?

GOALS

In this activity you will:

- Measure short time intervals in arbitrary units.
- Measure distances to the nearest millimeter.
- Use a record of an object's position vs. time to calculate the object's average speed during designated position and time intervals.
- Measure and describe changes in an object's speed.
- Relate changes in the speed of an object traveling on a complex sloped track to the shape of the track.

What Do You Think?

The fastest human can't go as fast as a car traveling at 25 miles per hour.

Who wins the race?

- **The runner with the highest finishing speed?**
- **The runner with the highest average speed?**
- **The runner with the greatest top speed?**

Take a few minutes to write answers to these questions in your *Active Physics* log. Discuss your answers with your small group to see if you agree or disagree with others. Be prepared to discuss your group's ideas with the entire class.

For You To Do

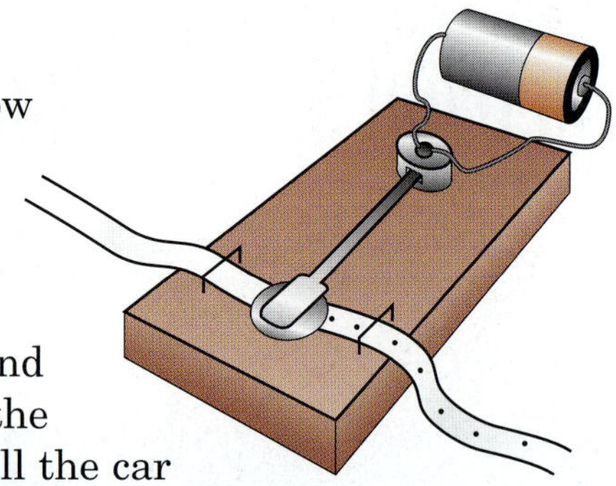

1. Your teacher will show you how to use a ticker-tape timer to record a toy car's position and the time it takes to move. Thread a piece of paper tape about 1 m long in the timer, and attach one end of the tape to the car. Turn on the timer, and pull the car at a nearly constant speed so that the tape is dragged completely through the timer.

Use only the tape provided. Do not substitute other paper.

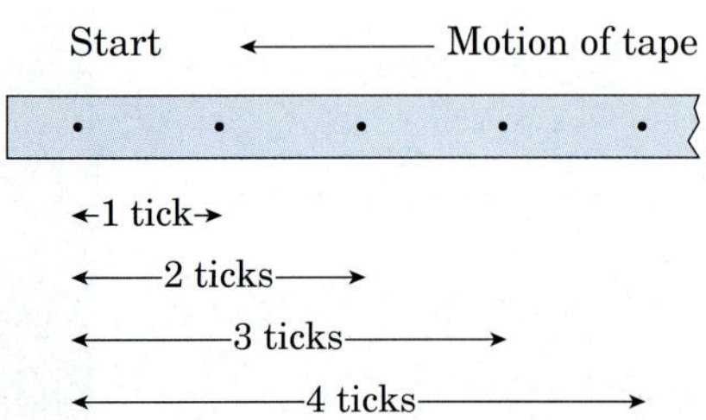

2. Examine the pattern of dots that the timer makes on the tape. Assume that the timer makes dots that are separated by equal amounts of time. The amount of time from one dot to the next will be called a "tick." Obviously, a tick in this case is some small fraction of a second. You will use the tick as a unit of time in this activity and not worry about converting it to seconds.
   a) Do you agree that the distance from one dot to the next on the tape is the distance that the car travelled during one tick of time? Check with your teacher if you have difficulty with this. When you understand this concept, record it in your log.
   b) Do you agree that if you measured the distance from one dot to the next in a unit such as centimeters, you would know the car's speed during that part of the motion in "centimeters per tick"? (Remember, average speed equals distance divided by, or per unit of, time.) When you understand this concept, record it in your log.
   c) Is the spacing between the dots about the same all along the tape, or does the spacing vary?
   d) What would it mean if the spacing stays about the same? If the spacing varies?
   e) Find the part of the tape that shows the beginning of the car's motion where the dots are far enough apart to be seen clearly, and mark one dot as the first dot for analysis. Also locate the last dot made before the tape left the timer at the end of the motion. Measure the distance from the first clear dot to the last dot in centimeters (to the nearest $\frac{1}{10}$ cm, or 1 mm). Also, count the number of tick time intervals (the total number of spaces between dots) from the first to the last dot. Record these measurements in your log.

f) Use the data from **Part (e)** to calculate the average speed of the car in “centimeters per tick.” Record your work in your log.

3. Make another “run” with the car, using another paper tape, but this time try to pull the car to make it go faster and faster along its path.

a) Compare the pattern of dots on the tape for this run to when you tried to pull the car at constant speed. In your log, explain differences between the two records of motion in terms of position, speed, and time.

b) Choose a section near the middle of the “faster and faster” tape that is 5-tick intervals long (count 5 spaces between dots, not 5 dots) and mark the beginning and end of the 5-tick time interval. Measure the distance between the first and last dots of the interval. Calculate the average speed during the interval in “centimeters per tick.” Record your work.

c) How would the average speed compare if you were to measure it for a similar interval earlier in the run? Later? How do you know? Write your answer in your log.

4. Now let’s race! Your “runner” is going to race on each of four inclines as shown. You are going to use the techniques learned above to analyze each race.

Track 1: **This is a simple incline 1.00 m long, supported by a meter stick along its length to keep the track from sagging. One end is raised so that h = 0.10 m.**

Track 2: **This time the incline is 0.40 m long, with a 0.60-m level section for the remaining part. As for track 1, h = 0.10 m.**

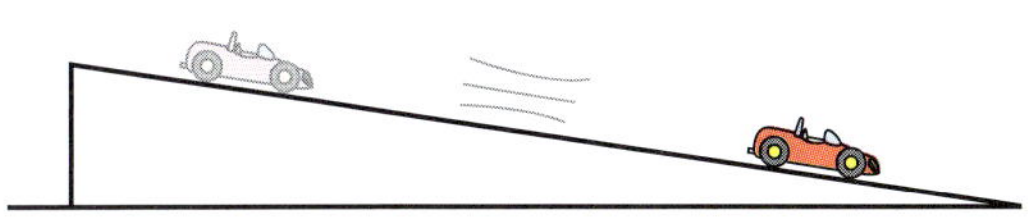

Track 3: **The setup is the same as for track 1, except that h = 0.15 m.**

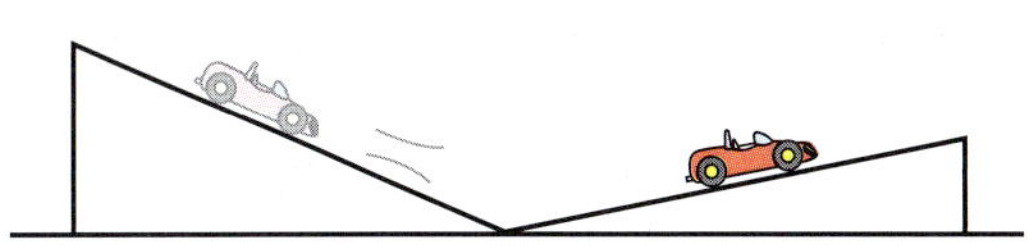

Track 4: **Set the first height at 0.20 m. At about 0.50 m along the track, slope the track up so that the second height is 0.10 m. Use a manila folder to smooth the transition between the inclines.**

a) Predict which track will produce the winning result if your car is released and allowed to run 1 m along each track, starting from the top. Record your prediction in your log.

5. Run your toy car on each track. In each case:

Allow the force of gravity to do the pulling by simply releasing the car at the start of the run.

Have the car pull at least 1 m of tape through the timer. This may require adding a "leader" to the 1-m tape length to allow for the distance between the car and the timer at the start of the race.

At the beginning of each run, the timer should be started and then the car should be released.

After each run, mark the track number on the tape, mark the first clear dot made at the beginning of the run, and place a mark on the tape 1 m (100 cm) beyond the first clear dot.

a) For each race, explain in your log how you will analyze the tapes to measure the final speed, the top speed, and the overall average speed in centimeters per tick.

b) Make the necessary measurements, and do the calculations to fill in the speed values in a table similar to the one below.

Track Number	Final Speed	Top Speed	Average Speed
1			

c) Which track produced the winning run in the big race? How can you tell? Explain your answer in your log.
If a photogate timer is available, this is a preferable way in which to measure the velocities. If not, try to minimize the friction of the timer from the experiment.

Reflecting on the Activity and the Challenge

Now you can see that it is the details of what happens during a race that determines who wins. The distance of a race and the time taken to run it do not reveal what a champion does along the way to win races consistently.

Speed within most races varies. The sprinters who get up to top speed quickly and maintain their speed throughout a race often win. Those who start quickly and "fade" at the end of the race often lose.

Helping athletes at your school analyze their performances in terms of speed during parts of a race will be needed if they are to compete with the best runners.

Physics To Go

1. Describe a procedure that you could use to convert one "tick" of the timer used in this activity into seconds of time. How could you find out how many "ticks" equal one second?
2. What would the spacing of dots look like for a ticker-tape timer record of an object that is slowing down in its motion?
3. From what you observed and measured during this activity, describe how the speed of a toy car behaves as it travels:
   a) on a straight ramp that slopes downward
   b) on a level surface when the car already has some speed at the beginning
   c) on a straight ramp that slopes upward
4. Aisha and Bert are running at constant speeds, Aisha at 9.0 m/s and Bert at 8.5 m/s. They both cross a "starting line" at the same time. The "finish line" is 100 m away.
   a) How long does it take Aisha to finish the race?
   b) How long does it take Bert to finish the race?
   c) Where is Bert when Aisha crosses the finish line?
   d) By how many meters does Aisha finish ahead of Bert?
5. The Penn Relays women's high school record for the 1500-m run is 4 min, 24.0 s. The women's high school record for the mile (1609 m) run at the Penn Relays is 4 min, 49.2 s. In which race did the record holder have the greatest average speed in meters/second?
6. Salina runs the 200-m race for the school's track team. She runs the first 100 m at 9.0 m/s. Then she hears her classmates cheer, "GO, Salina, GO!" and runs the final 100 m at 10.0 m/s.
   a) Calculate the time for Salina to run the first 100 m.
   b) Calculate the time for Salina to run the final 100 m.
   c) Calculate Salina's average speed for the entire race.

Activity 4 Understanding the Sprint

GOALS

In this activity you will:

- Calculate the average speed of a runner given distance and time.
- Produce a histogram showing the average speeds of a runner during segments of a race; analyze changes in the runner's speed.
- Produce a graph of distance versus time from split time data for a runner.
- Estimate the slope of a distance versus time graph at specified times.
- Recognize that the slope of a distance versus time graph at a particular time represents the speed at that time.

What Do You Think?

It was not believed to be humanly possible to run a mile in less than four minutes until Roger Bannister of England did it in 1954.

- **How much time does it take to get "up to speed"?**

Take a few minutes to write an answer to this question in your *Active Physics* log. Discuss your answer with your small group to see if you agree or disagree. Be prepared to discuss your group's ideas with the class.

For You To Do

1. Carl Lewis established a world record for the 100-m dash at the World Track and Field Championships held in Tokyo, Japan, in 1991. The times at which he reached various distances in the race (his "split times," or "splits") are shown in the table below.

Distance (meters)	0.0	10.0	20.0	30.0	40.0	50.0	60.0	70.0	80.0	90.0	100.0
Time (seconds)	0.00	1.88	2.96	3.88	4.77	5.61	6.45	7.29	8.13	9.00	9.86

a) In your log, copy and complete the table to the right. Use subtraction to calculate the time taken by Carl Lewis to run each 10 m of distance during the race.

b) Calculate Lewis's average speed during each 10 m of the race. The values of the time interval and the average speed have been entered in the table for the first 10 m of the dash.

Distance (meters)	Time Interval (seconds)	Average Speed (meters/second)
0.0–10.0	1.88	5.32
10.0–20.0		
80.0–90.0		
90.0–100.0		

2. Use the data you created for the above table to make a bar graph to give you a visual display of Carl Lewis's average speed during each 10 m of his world-record 100-m dash. Use a piece of graph paper set up as shown to the right.

Carl Lewis's World Record 100-m Dash Average Speed, 10-m Intervals

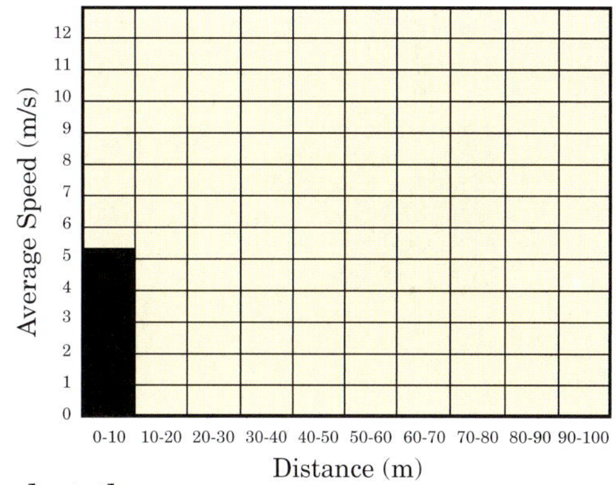

a) Tape or copy the bar graph in your log.

3. Analyze the bar graph to answer these questions:

a) At what position in the dash did Lewis reach top speed? How close can you state that position to the nearest meter? to the nearest 10 m? Explain your answer.

b) How well did Carl Lewis keep his top speed once he reached it? Did he seem to be getting tired at the end of the race? Give evidence for your answers.

c) Can you tell how fast Lewis was going at an exact position in the race, such as at 15.0 m or 20.0 m? Why or why not?

d) It took 9.86 s for Lewis to run the entire 100 m. Calculate his overall average speed. Draw a horizontal line across the bar graph at an appropriate height to represent the average speed for the entire race. Compare the height of each bar on the graph with the height of the line. Explain what the comparisons mean.

4. Use the "splits" given at the start of this activity to plot a graph of Carl Lewis's position versus time.

a) On a piece of graph paper, make a vertical distance scale from 0 to 100 m and a horizontal time scale from 0 to 10 s. Plot each position at the appropriate time and connect the points to show what you think is the shape of the graph. Tape or copy the graph in your log.

5. Compare the distance versus time graph with the bar graph of speed versus distance.

   a) When the distance versus time graph is curving early in the run, do the bars on the bar graph change in height or are they fairly steady in height? What does this comparison mean? When the graph is climbing in a straight line, what is happening to the heights of the bars on the bar graph? What does this comparison mean?

   b) Someone said, "The slope, or steepness, of a distance versus time graph at any instant is the speed at that instant." Do you believe this statement? Why or why not?

   c) Describe the slope, or steepness, of the distance versus time graph each second (1.00 s, 2.00 s, 3.00 s, and so on) during Lewis's record run.

Reflecting on the Activity and the Challenge

In this activity you saw two different ways to analyze in detail the speed of a runner during a race. If you had split-time information for runners in sprint events for your school's track team, you could help the runners find out, for example, if they are "letting up" at the end of a race or how rapidly they are reaching top speed.

With more knowledge about details of their performances, your school's runners may find that they can improve parts of their races.

PHYSICS TALK

$$\text{Speed} = \frac{\text{Distance traveled}}{\text{Time elapsed}}$$

On a distance versus time graph, the speed is equal to the slope of the graph.

Physics To Go

1. If you were to design a track for a toy car to run as in **Activity 3** to simulate Carl Lewis's 100-m dash, what shape would you design for the track?
2. For long distances, humans can run at a constant speed of about 6 m/s, pigs at about 4 m/s, and horses at 20 m/s. Sketch a distance versus time graph with three lines showing a person, a pig, and a horse running for 100 s.
3. Sketch a distance versus time graph for a person who is not moving at all.
4. In a 2 × 100-m relay race, Joan ran the first 100 m at a speed of 5 m/s and then Rami ran the next 100 m at a speed of 10 m/s. What was the average speed for the entire relay race? (Hint: It is not 7.5 m/s.)
5. Do you think it is possible for a runner to keep increasing speed for an entire race as a strategy to win?
6. Examine what excellent runners do in long-distance races. On the right is a chart of Eamonn Coghlan's split time and total time every 200 m in a mile (1609-m) race.
   a) Use the total times listed to calculate Coghlan's average speed for distances of 200, 400, 800, and 1000 m distances during his run.
   b) Compare Coghlan's average speeds for distances of 200, 400, 800, and 1000 m with the average speeds of Penn Relays record holders at the same distances. (Penn Relays record average speeds for various distances are listed in **Activity 1**.) What patterns do you see in the comparison? At what distances was Coghlan's speed getting closest to world-record speed?
   c) Use Coghlan's split times to identify the distance interval (0–200 m, 200–400 m, 400–600 m, and so on) when he had the greatest average speed. Also identify the interval of lowest average speed.
   d) You found out in this activity that Carl Lewis slowed down slightly near the end of his record 100-m dash. Did Eamonn Coghlan do the same thing near the end of his mile run? Use data in your answer.
   e) How could you use data about Coghlan's performance to give advice to members of your school's track team who enter long-distance events?

Distance (m)	Split Time (s)	Total Time (s)
0	0.00	0.00
200	29.23	29.23
400	29.87	59.10
600	30.09	89.19
800	30.25	119.44
1000	29.88	149.32
1200	29.90	179.22
1400	29.38	208.60
1600	29.55	238.15

Activity 5 Acceleration

GOALS

In this activity you will:

- Understand the definition of acceleration.
- Understand meters per second per second as the unit of acceleration.
- Use an accelerometer to detect acceleration.
- Use an accelerometer to make semi-quantitative comparisons of accelerations.
- Distinguish between acceleration and deceleration.

A cork attached to a string floats in a liquid-filled bottle.

Use only plastic bottles to construct accelerometers.

What Do You Think?

Accelerating out of the starting blocks is important if a runner is going to win a race.

- **If all runners in a dash have equal top speeds and none "fades" at the end of the race, what determines who wins?**
- **How is response time a factor in determining who wins the race?**

Write your answers to these questions in your *Active Physics* log. Be prepared to discuss your ideas with your small group and other members of your class.

For You To Do

1. In this activity, you will use an "accelerometer," a device for measuring acceleration. There are many kinds of accelerometers. The diagram on the left shows a "cork accelerometer."

Explore how the accelerometer works by holding it in your hands in front of you so that you can look down to see the top of the cork in the bottle and liquid. Hold the accelerometer upright so that the cork is centered.

2. Take the accelerometer for a walk. Observe any movement of the cork, especially as you start from a resting position and speed up (accelerate). Walk at a fairly constant speed, and then slow down to a stop (decelerate). Try it a few times—starting, walking, and stopping at normal rates.

   a) Record your observations of the cork's movement.

3. Repeat the above walk and observe what happens if you start faster, if you walk faster at a constant speed, and if you stop faster.

   a) Record your observations in your log.

4. Repeat the walk in **Step 2,** but walk backward.

a) Record your observations in your log.

5. Use your observations to answer the following questions in your log.

Describe the *amount* and the *direction* the cork leans in each of the following situations:

a) standing at rest

b) low acceleration while walking forward

c) high acceleration while walking forward

d) low constant speed while walking forward

e) high constant speed while walking forward

f) high deceleration (slowing down) while walking forward

g) low deceleration while walking forward

6. Someone said, "Deceleration while walking forward is the same as acceleration while walking backward."

a) Do you agree or disagree? Use your observations of the accelerometer for your answer.

7. Set up a system to take the accelerometer for a ride on a cart that is being pulled by a falling weight, as shown in the diagram to the right. Hang a weight on the string and allow it to pull the cart as you observe the accelerometer. Record your answers to the questions below in your log.

a) Does the cart appear to accelerate? Does the accelerometer tell you that it's accelerating? How can you tell?

b) Does the accelerometer show that the acceleration is constant or changing? How can you tell?

Keep area under the falling mass clear.

8. Repeat **Step 7** using a larger weight to pull the cart. Answer the following questions in your log:

a) How does the acceleration for the large weight compare with the acceleration for the small weight? How can you tell?

b) Which produced a more steady, constant acceleration: using the falling weights or walking with the accelerometer? What evidence do you have for your answer?

Physics Words

acceleration: the change in velocity per unit time.

vector: a quantity that has both magnitude and direction.

PHYSICS TALK

Acceleration

The relationship between **acceleration,** speed, and time can be written as:

$$\text{Acceleration} = \frac{\text{Change in speed}}{\text{Time interval}}$$

Using symbols, the same relationship can be written as:

$$a = \frac{\Delta v}{\Delta t}$$

where

a = acceleration,

Δv = change in speed, and

Δt = the time interval, or change in time, for the change in speed to happen.

Sample Problem

A sprinter at the start of a race increases speed from 0 m/s to 5.0 m/s as the clock runs from 0 s to 2.0 s. Find the sprinter's acceleration.

Strategy: You can use the equation for acceleration.

$$a = \frac{\Delta v}{\Delta t}$$

You can the write this equation to show the change in speed from the initial to final speed and the time interval.

$$a = \frac{v_{final} - v_{initial}}{t_{final} - t_{initial}}$$

Givens:

$v_{initial} = 0$ m/s

$v_{final} = 5.0$ m/s

$t_{initial} = 0$ s

$t_{final} = 2.0$ s

Solution:

$$a = \frac{v_{final} - v_{initial}}{t_{final} - t_{initial}}$$

$$= \frac{5.0 \text{ m/s} - 0 \text{ m/s}}{2.0 \text{ s} - 0 \text{ s}}$$

$$= \frac{5.0 \text{ m/s}}{2.0 \text{ s}}$$

$$= 2.5 \text{ (m/s)/s or } 2.5 \text{ m/s}^2$$

Mathematically, meters per second per second is equal to meters per second squared. Therefore, when you use "m" and "s" as abbreviations for meters and seconds, you can shorten the unit of acceleration to m/s^2. The following ways of stating the unit of acceleration are the same:

(m/s)/s $\quad$ m/s^2

When speaking about acceleration, you can describe the unit as "meters per second squared" or "meters per second every second."

FOR YOU TO READ

Scalars and Vectors

Acceleration is defined as a change in velocity with respect to time. In everyday language, you probably use the words velocity and speed interchangeably. In physics, the two words have precise and distinct meanings.

Speed is a measure of how fast something is moving. A runner's speed may be 8 m/s or 9 m/s. Velocity is a measure of how fast and in what direction something is moving. A runner's velocity may be 8 m/s north or 9 m/s east.

Acceleration is a change in velocity with respect to time. Velocity can change when the speed changes or when the direction changes (or if both change). The accelerometer in the activity indicated acceleration when the speed of the cart changed. The accelerometer will also indicate acceleration when the cart changes direction. You can hold the accelerometer in your hand. As you rotate, you will notice that the cork leans toward the center of the circle indicating that there is an acceleration.

Quantities in physics that have magnitude (e.g., 8 m/s) and direction (e.g., east) are called **vectors.** Velocity is an example of a vector →

quantity. Acceleration and **displacement** are also vectors. Displacement is a vector because a person can run 50 m north or 50 m toward the goal.

Some quantities don't have direction. You may be 17 years old. The temperature may be 66°. There may be 29 students in your physics class. None of these values have direction and are referred to as **scalars.** Distance and speed are also scalars. A distance of 1200 m or 3 cm does not have any direction. A speed of 22 m/s also has no direction and is a scalar. (This is why physicists mean different things by velocity and speed.)

The wear and tear on your shoes or the tires of your car can be a measure of the distance you have traveled. The odometer of your car measures the distance your car has traveled, in all directions. The speedometer on your car measures the speed of your car, while traveling in any direction.

Imagine you drive 30 mi. to the east at 30 mph and then drive 30 mi. west on your way back home at 30 mph. Your total distance traveled is 60 mi. and your average speed is 30 mph for two hours. Your total displacement will be zero and your average velocity will be zero. Your total displacement is zero because you ended up in the same position as you began. Your average velocity can be zero because you traveled at 30 mph east for the first hour and then you traveled 30 mph west (or –30 mph) for the second hour.

A runner in the 1600-m race covers a distance of 1600 m. Her displacement is 0 m if she ends up in the same place that she began. (This is the case if the race is run on an oval track.)

If she runs the 1600 m in 300 s, then you can calculate her average speed.

$$\text{Average speed} = \frac{\text{total distance}}{\text{total time}} = \frac{600\ \text{m}}{300\ \text{s}} = 2\ \text{m/s}$$

You can also calculate her average velocity, which is in sharp contrast to her speed.

$$\text{Average velocity} = \frac{\text{total displacement}}{\text{total time}} = \frac{0\ \text{m}}{300\ \text{s}} = 0\ \text{m/s}$$

The introduction of vector and scalar quantities may seem to make things more complicated than they need to be. As you analyze more complex situations, like throwing a javelin or high jumping, the use of vectors will make these descriptions easier. Mathematics is only introduced in physics when it simplifies analysis of a situation and makes things clearer.

Scalar Quantities	Vector Quantities
Distance	Displacement
Speed	Velocity
Time	Acceleration
Mass	Force (mass x acceleration)
	Momentum (mass x velocity)
Energy	
Work	

Reflecting on the Activity and the Challenge

You now know a lot more about acceleration. Since a race always begins with a speed of zero and ends with runners in motion, acceleration is part of every race. Depending on the distance of a race, acceleration may go up, go down, or disappear several times during the race. One runner may have more acceleration than another runner at the start of a race, or one runner's acceleration may change during a race. Do the athletes on your school's track team know about acceleration? If not, maybe they should, and perhaps you can help them.

Physics To Go

1. Is there anything in nature that has constant acceleration?
2. If Carl Lewis were to carry a cork accelerometer during the start of a sprint, describe what the accelerometer would do.
3. Did Carl Lewis accelerate for the entire 100 m of his world-record dash? Explain his pattern of accelerations.
4. If you are running, getting tired, and slowing down, are you accelerating? Explain your answer.
5. When you throw a ball straight up in the air, what is the direction of its acceleration while it is going up? While it is coming down?
6. What additional tips could you give your school's track team as a result of this activity?

Physics Words

scalar: a quantity that has magnitude, but no direction.

displacement: the difference in position between a final position and an initial position; it depends only on the endpoints, not on the path; displacement is a vector; it has magnitude and direction.

7. Which of the following terms represent scalar quantities and which are vector quantities? Explain your choice for each.

   a) force
   b) displacement
   c) distance
   d) acceleration

8. A student walks 3 blocks south, 4 blocks west, and 3 blocks north. What is the displacement of the student?
9. A person travels 6 m north, 4 m east, and 6 m south. What is the total displacement?
10. If a woman runs 100 m north and then 70 m south, what is her total displacement?
11. Which of the following statements about the movement of an object with zero acceleration are true and which are false? For each statement, explain why you indicated it to be true or false.

    a) The object may be speeding up.
    b) The object may be in motion.
    c) The object must be at rest.
    d) The object may be slowing down.

12. Carl Lewis began at rest and reached a speed of 9.26 m/s after 1.96 s of the race. Calculate his acceleration, assuming that it was constant during this time.
13. Objects in free fall that are not affected by air resistance have acceleration due to gravity of 9.8 m/s^2. Calculate the final velocity that a falling object will have after:

    a) 1.0 s b) 2.0 s c) 3.0 s d) 4.0 s e) 5.0 s

14. The javelin reaches a speed of 30.0 m/s within 0.6 s. What is the acceleration of the javelin?
15. During the 100-m run, Carl Lewis decreased his speed late in the race. His speed decreased from 11.9 m/s to 11.5 m/s in 0.87 s. What was his acceleration during that time interval?
16. A pole-vaulter manages to get over a 5.0-m bar. The acceleration due to gravity is 9.8 m/s^2.

    a) When he falls from that height, how fast will he be going after 0.5 s?
    b) What will be his average speed after 0.5 s?
    c) How far will he have fallen after 0.5 s?

Activity 6 Measurement

GOALS

In this activity you will:

- Calibrate the length of a stride.
- Measure a length by pacing and with a meter stick.
- Identify sources of error in measurement.
- Evaluate estimates of measurements as reasonable or unreasonable.
- Measure various objects and calculate the error in each measurement.

What Do You Think?

In about 200 B.C. Eratosthenes calculated the circumference of the Earth. He used shadows cast by the Sun in two cities and a measurement of the distance between the two cities. The distance between the cities was found by pacing.

- **Two people measure the length of the same object. One reports a length of 3 m. The other reports a length of 10 m. Has one of them goofed? Why do you think so?**
- **What if the measurements were 3 m and 3.1 m?**

Record your ideas about these questions in your *Active Physics* log. Be prepared to discuss your responses with your small group and the class.

For You To Do

1. Select a cleared distance along the floor of the cafeteria, gym, hall, corridor, or classroom, or a paved area away from traffic out of doors.

2. Measure the length of your stride using a meter stick. Finding the length of your stride is an example of calibration—making a scale for a measuring instrument.

   a) Record your measurement in your log.

3. Count off the number of strides it takes you to cover the selected distance.

   a) Record this in your log.

4. Use the number of strides you took and the length of your stride to compute the distance.

   a) Record your calculations.

5. List the results of the measurements made by the entire class on the board.

   a) Do all the measurements agree?

   b) Why do you think there are differences among the measurements made by different students? List as many different sources of error as you can.

   c) Suggest a way of improving your measurements.

6. Measure the distance with a meter stick.

   a) Record your measurement in your log. Make a list of all the class measurements on the board.

   b) Can you develop a system that will produce measurements all of which agree exactly or will there always be some difference in measurements? Justify your answers.

7. Physicists identify two kinds of errors in measurement. Errors that can be corrected by calculation are called systematic errors. For example, if you made a length measurement starting at the 1 cm mark on a ruler, you could correct your measurement by subtracting 1 cm from the final reading on the ruler.

Errors that come from the act of measuring are called random errors. No measurement is perfect. When you measure something you make an approximation. Random errors exist in any measurement. Scientists provide an estimate of the size of the random errors in their data.

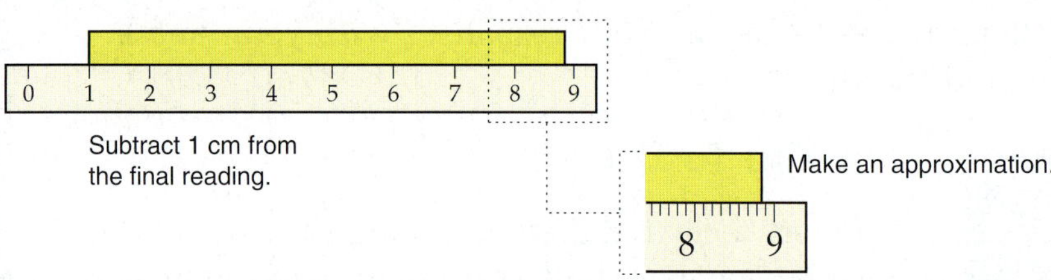

   a) Classify the sources of error you have listed as systematic or random.

   b) Estimate the size of each error.

8. Sometimes a precise measurement is not needed. A good estimate will do. What is a good estimate? Use your common sense and prior knowledge to judge if an estimate is a "good," or reasonable, one. Determine if each is reasonable. Explain your answers.

> Example:
>
> - A single-serving drink container holds 5 kg of liquid.
>
> Use common sense and mental math to see if this is a good estimate. One kilogram of water takes up about 1 L of volume. Five kilograms of water would take up 5 L. Most drinks are like water in their density. A 5-L container is much bigger than a single serving.

a) A college football player has a mass of 100 kg.

b) A high school basketball player is 4-m tall.

c) Your teacher works 1440 min every day.

d) A poodle has a mass of 60 kg.

e) Your classroom has a volume of 150 m^3.

f) The distance across the school grounds is 1 km.

FOR YOU TO READ

Measurement and Track Records

Measurements are crucial in all sporting events. As you learned in this activity all measurements have some error associated with them. These measurement errors can bring into question whether a world record was actually broken.

Every four years, the summer Olympics take place in a different city. Each city builds a track to be 400 m. How accurately built is the 400-meter track? Could it be 1 cm longer in one city and 1 cm shorter in another city? If so, then a runner in the first Olympics may be running an extra distance of 2 cm on every lap. If the race is a 1600-m race, then one runner may run a total of 8 cm further. It takes at least 0.01 s to run 8 cm. If a runner "beats" the world record by only 0.01 s, it may have been because she was running on a shorter track, not because she was faster.

Reflecting on the Activity and the Challenge

A measurement is never exact. When you make a measurement, you estimate. All measurements have random errors. You can try to minimize these errors but you cannot eliminate them. One decision you must make is how accurate a measurement you really need.

In this **Chapter Challenge**, you will be writing a track and training manual. You may want the runners to know that the inside lane is shorter than the outside lane. You may also want to inform them that some tracks are very slightly different than other tracks. It's easy to tell who was faster when two runners compete side by side. It's more difficult to tell who was faster when one runner competed on a different track in a different city.

Physics To Go

1. Get a meter stick and centimeter ruler. Find the length of five different-sized objects, such as a door (height), a table top, a large book, a pencil, and a postage stamp.
   a) Which measuring tool is best for measuring each object?
   b) Calculate the error in each measurement. What kind of errors are these?
2. Pace off the size of a room. Estimate your accuracy. Then check your accuracy with a meter stick or tape measure.
3. Give an estimated value about which you and someone else would agree. Then give an estimated value about which you and someone else would not agree.
4. An Olympic swimming pool is 50 m long. Do you think the pool is built to an accuracy of 1 m (49 m to 51 m) or 1 cm (49.99 m to 50.01 m)?
5. An oil tanker is said to hold 5 million barrels of oil. Do you think this measurement is accurate? How accurate?
6. Choose 5 food products. How accurate are the measurements on their labels?
7. Are these estimates reasonable? Explain your answers.
   a) A 2-L bottle of soft drink is enough to serve 12 people at a meeting.
   b) A mid-sized car with a full tank of gas can travel from Boston to New York City without having to refuel.

Activity 7 Increasing Top Speed

GOALS

In this activity you will:

- Calculate the average speed of a runner given distance and time.
- Measure the frequency of strides of a running person.
- Measure the length of strides of a running person and calculate speed.
- Recognize that either the equation (Speed = Distance / Time) or the equation (Speed = Stride Frequency × Stride Length) may be used to calculate a runner's speed with equivalent results.
- Infer ways in which stride length and stride frequency can be adjusted by a runner to increase speed.

What Do You Think?

A cheetah can reach a top speed of 60 miles per hour (about 30 m/s).

- **What can a runner do to increase top speed?**

Record your ideas about this question in your *Active Physics* log. Be prepared to discuss your responses with your small group and the class.

For You To Do

1. Watch the video of a runner.

2. Use information from the video to answer the following questions. Record your data and show your calculations in your log.

a) Use the total distance traveled and the total time to calculate the runner's speed in yards per second.

$$\text{Speed (yards/second)} = \frac{\text{Distance (yards)}}{\text{Time (seconds)}}$$

(Yards/second are used since the running is being done on a football field.)

b) Count the number of strides taken by the runner during the entire run. Use the number of strides and the total time to calculate the runner's stride frequency in strides per second.

$$\text{Stride frequency (strides/second)} = \frac{\text{Number of strides}}{\text{Time (seconds)}}$$

c) Calculate the average length of one stride for the runner. To do so, measure the length of several single strides and then calculate the average length per stride. The unit for your answer will be yards/stride.

d) Calculate the runner's speed using the following "new" equation for speed. Your answer will be in yards/second. Speed = Stride Frequency × Stride Length.

3. Compare the results of using the two equations for calculating speed.

a) Do the equations agree on the runner's speed? How good is the agreement?

b) How could you explain any difference in results?

4. You will test the "new" equation at a "track." Your teacher will show you where to set up the track. Place marks at intervals of 0.75-m along a 12-m track.

5. Starting from the "zero" mark, walk so that you step on each mark. Count the number of strides to complete the walk, and use a stopwatch to measure the total time.

a) Record your data in your log.

b) Calculate your speed using the distance/time equation.

c) Calculate your stride frequency.

$$\text{Stride frequency (strides/second)} = \frac{\text{Number of strides}}{\text{Time (seconds)}}$$

d) Calculate your speed using the stride frequency and stride length.

Speed = Stride Frequency × Stride Length.

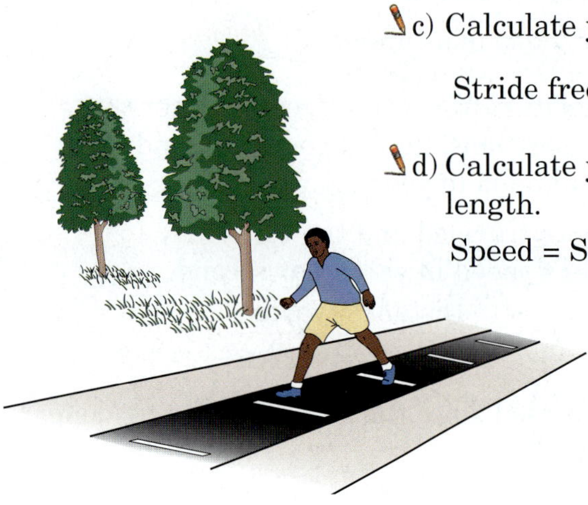

6. What happens to your speed if you change your stride length?

   a) Record what you think will happen to the speed if you decrease the length of the stride.

7. Mark the track at 0.50-m intervals and walk again, stepping on each mark. Again measure the time and count the number of strides.

   a) Record the data in your log.

   b) Calculate your speed using both equations.

   c) How well do your speeds calculated by both methods on this track compare?

8. Compare your performances on the two different tracks.

   a) Was your speed on the track which had 0.75-m stride lengths about the same as, or different from, your speed on the track which had 0.50-m intervals? Why?

   b) If you must step on each mark, what would you need to do to make your speed on the second track equal to your speed on the first track? Express your answer in numerical terms.

PHYSICS TALK

Calculating Speed Using Frequency and Length

If it is really true that Speed = Frequency × Stride Length, then it should also be true that the relationship produces an answer that has a unit of speed, such as meters/second. If frequency is measured in strides/second and if length is measured in meters/stride, then

$$\frac{\text{strides}}{\text{second}} \times \frac{\text{meters}}{\text{stride}} = \frac{\text{meters}}{\text{second}}$$

The equation produces an answer that has a unit of speed.

Reflecting on the Activity and the Challenge

Distance runners learn from their coaches how to make conscious changes in stride length and frequency to improve their performances. To increase speed, the "trick" is to increase one—either frequency or stride length—without decreasing the other. Good runners know how to increase either, or even both, when needed in a race. Think of experiments that you could do with members of your school's track team to help them to learn to use frequency and stride length to win races.

Physics To Go

1. A runner's stride length is 2.0 m and her frequency is 1.8 strides/s.

   a) Calculate her average speed.
   b) What would be her time for a 200-m race?
2. A runner maintains a constant speed of 6.0 m/s. If his stride length is 1.5 m, what is his stride frequency?

3. If the runner in **Question 2** increases his stride length to 1.6 m without changing his stride frequency, what will be his new speed?

4. If a marching band has a frequency of 2.0 strides/s and if each stride length is 0.65 m, what is the band's marching speed?

5. The graphs on the next page are reproduced from an article that reported the characteristics of runners who are 30 or more years of age.

   Write a statement that explains the information on the two graphs. Include your own inferences about what happens to stride frequency with increasing age.

6. The running events called "hurdles" present special problems involving stride length and frequency. In both 100- and 400-m races, runners must jump over hurdles placed at regular intervals along the track. Discuss special techniques that hurdlers must develop to make sure that they are ready to jump when they reach each hurdle.

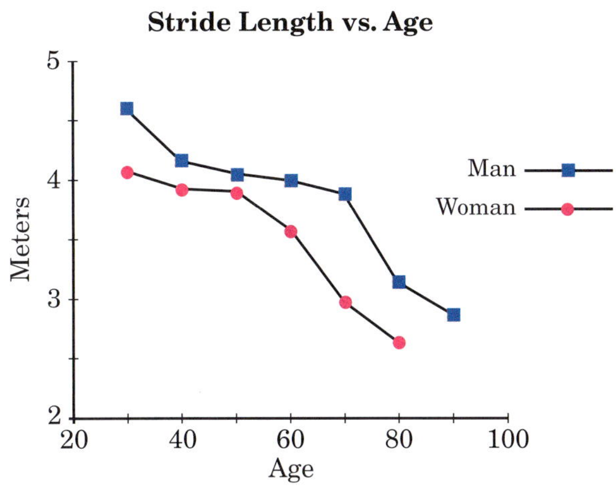

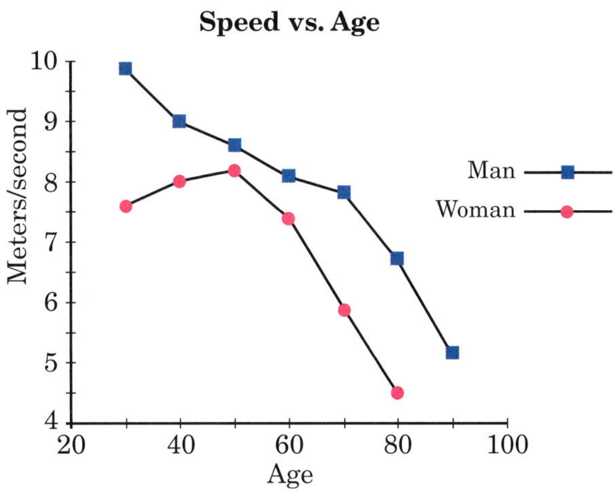

"Sprint Stride Kinematics."
Nancy Hamilton in *Log of Applied Biomechanics*, Volume 9, 1993, pp. 15–26

Stretching Exercises

Common sense suggests that physical characteristics, such as the length of a person's legs, may affect stride length. Do long-legged people have greater stride length? Can long-legged people maintain the same stride frequency when running as people with shorter legs? If so, do long-legged people have higher running speeds? Ask some classmates or friends in your neighborhood to serve as volunteers for an experiment designed to identify and test the effects of leg length, or other body characteristics, on stride length and frequency during walking and/or running.

Activity 8 Projectile Motion

GOALS

In this activity you will:

- Understand and correctly apply the term "free fall."
- Understand and correctly apply the term "projectile."
- Understand and correctly apply the term "trajectory."
- Understand and correctly apply, within physics context, the term "range."
- Observe that all projectiles launched horizontally from the same height strike a level surface in equal times, regardless of launch speed (including zero launch speed).
- Understand that the vertical and horizontal components of a projectile's motion operate independent of one another.
- Infer factors which affect the range of a projectile.
- Infer the shape of a projectile's trajectory.
- Infer ways in which stride length and stride frequency can be adjusted by a runner to increase speed.

What Do You Think?

Some track and field events involve launching things into the air such as a shot put, a javelin, or even one's body in the case of the long jump.

- **What determines how far an object thrown into the air travels before landing?**

Record your ideas about this question in your *Active Physics* log. Be prepared to discuss your response with your small group and the class.

For You To Do

1. Place one coin at the edge of a table with about half of the coin hanging over the edge. Place another coin flat on the table and use a finger to shoot this coin across the tabletop to strike the first coin. Aim "off center" so that the coin at the edge of the table drops straight down and the projected coin leaves the edge of the table with some horizontal speed. Repeat the event as many times as needed to record your answers to the following question in your log:

a) Which coin hits the floor first? (Hearing is the key to observation here, although you may also wish to rely on sight.)

Make sure the path is clear before launching the coins.

2. Vary the speed of the projected coin.

a) How does its speed affect the amount of time for either coin to fall to the floor?

b) How does its speed affect how far across the floor the projected coin lands?

3. Use a box, stack of books, or a different table or countertop to vary the height.

a) Which coin hits the floor first?

b) How does increasing the height affect how far the projected coin travels horizontally as it falls?

4. Your teacher will supervise an activity in which one student sits on a chair that is moving at constant speed. While the chair is moving, the student will throw a ball straight up into the air and try to catch it when it comes down. The class will stand in a line along the track to observe the event, prepared to mark the horizontal distance (range) the ball travels from release to catch.

a) In your log write your prediction of what you think will happen.

b) Write in your log what you observed about the ball's trajectory (shape of the ball's path) and the ball's approximate range (horizontal distance) for trials in which you varied the speed of the chair and the launching speed of the ball.

c) According to your observations, what factors affect the range of the ball?

d) According to your observations, what is the shape of the trajectory?

PHYSICS TALK

Projectiles and Trajectories

Physicists often work with objects that have been launched into the air in a state of "free fall." In a free fall, the main force acting on the object while it is in the air is the downward pull of the Earth's gravity.

An object launched into the air is called a **projectile**. Examples of projectiles are a javelin, a shot put, or a broad jumper. The path that the projectile follows when launched into the air is called the **trajectory**.

Physics Words

projectile: an object traveling through the air.

trajectory: the path followed by an object that is launched into the air.

FOR YOU TO READ

Vector Components

By observing two falling coins and by tossing a ball in a moving chair, you gained evidence of two very important aspects of how thrown objects move in space. Since the shot put, the javelin, the hammer, and even the high jumper are objects thrown in the air, these two observations are crucial to helping the track team improve its performance.

The horizontally thrown coin and the dropped coin hit the ground at the same time, when there is no air resistance. Under careful observations, you find that this is always true – the horizontal motion of the coin does not affect its downward motion. If you were to take a picture of the coin every tenth of a second, you would observe the two coins as shown on the following page:

Both coins fall the same amount in each tenth of a second. The vertical motions are identical. The projected coin kept moving to the right, but its vertical motion was identical to the dropped coin.

Similarly, the projected coin has a constant speed to the right, when there is no air resistance. The vertical motion does not affect this constant horizontal speed.

At any point in its motion, the projected coin is moving down and to the right. You can draw its velocity at any time. After a short time, it has a small vertical speed and its constant horizontal speed. You can add these as vectors as you did with forces.

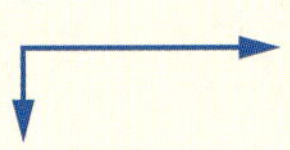

To add these two velocity vectors, you "complete the rectangle" and draw the diagonal. This diagonal is the resultant velocity vector.

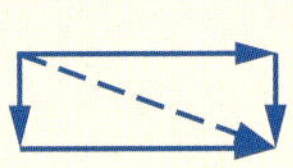

An alternative way to add these two vectors is to put them "tip to tail." By sliding one vector over (maintaining its length and direction), the resultant is then drawn from the tail of the first vector to the tip of the second vector.

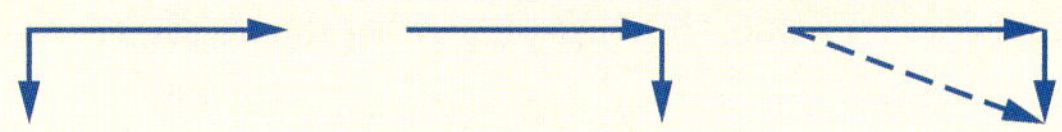

As you notice, the resultant vector is identical in length and direction, independent of whether you added the two vectors by the "rectangle" or "tip-to-tail" method.

If you look at the velocity some time later, you notice that the coin is moving faster in the vertical direction but continues horizontally at the same speed. You can add the vectors to determine what happens to the resultant vector.

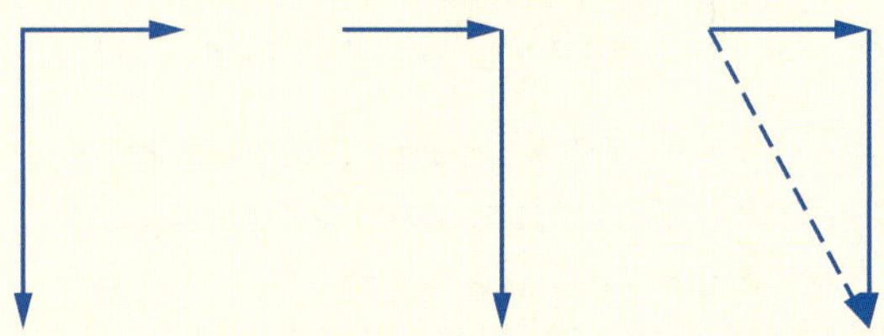

The resultant or total velocity has gotten larger and its direction has changed. The total velocity is in a more vertical direction.

If you took the velocity at any one point in the path, you could also use that resultant velocity vector to find the horizontal and vertical components. You draw the resultant velocity vector to the correct size and in the correct direction. Next, you draw horizontal and vertical axes from the tail of the vector. Then you draw a line from the tip of the vector to the horizontal axis. By doing this, you can see the horizontal and vertical vectors that can add together to produce this resultant.

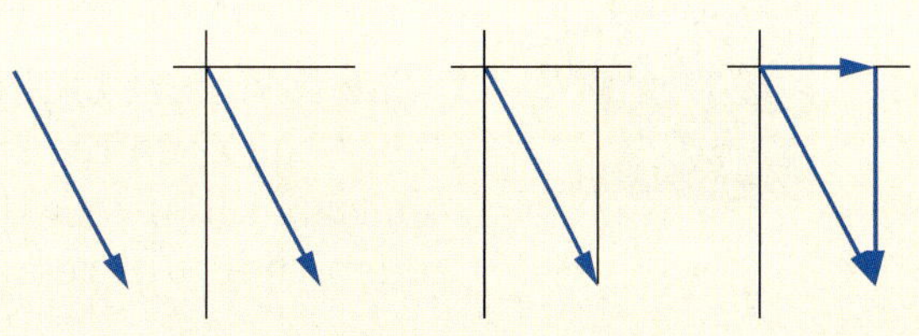

If you were to take numerous velocity vectors representing the path of the object, you would notice two things. First, that the horizontal velocity components would always be equal. The second is that the vertical component increases as time goes on.

Sample Problem

a) A javelin is thrown at 20.0 m/s at an angle of 30° with respect to the horizontal. What is its velocity in the *x*-direction?

b) If the javelin were thrown at 40°, what is its velocity in the *x*-direction?

c) How far does each javelin travel after 3.0 s?

Strategy: You can solve the first two parts by drawing vector diagrams to scale and finding the *x*-components. In **Part (c)** you can find how far each javelin traveled by using the relationship:

Distance = velocity x time

$d = vt$

Solution:

a) The first vector must be 20 units long at an angle of 30°. (The scale is 1 unit = 1 m/s.)

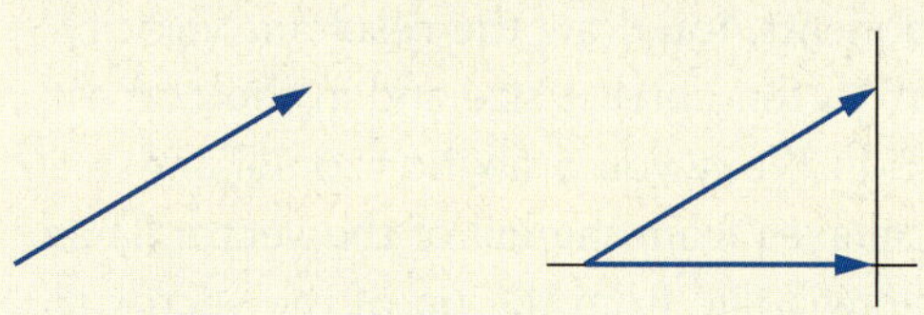

Measuring the *x*-component and using the scale, you find the *x*-component is 17.3 m/s.

b) The second vector is also 20 units long at an angle of 40°.

Measuring the *x*-component and using the scale, you find the *x*-component is 15.3 m/s.

c) The first trajectory has a horizontal velocity component of 17.3 m/s for 3.0 s. Its distance is:

$$\begin{aligned} d &= vt \\ &= (17.3 \text{ m/s})(3.0 \text{ s}) \\ &= 52 \text{ m} \end{aligned}$$

The second trajectory has a horizontal velocity component of 15.3 m/s for 3.0 s. Its distance is:

$$\begin{aligned} d &= vt \\ &= (15.3 \text{ m/s})(3.0 \text{ s}) \\ &= 46 \text{ m} \end{aligned}$$

Reflecting on the Activity and the Challenge

The first part of this activity (two falling coins) demonstrated that the time required for a coin to fall is independent of the horizontal speed. If two long jumpers rise to the same height, then they will remain in the air for identical times.

The second part of this activity (the rolling chair) showed that the faster the chair is moving, the farther the ball will travel horizontally. If a long jumper is able to increase horizontal speed, then the jumper will travel farther.

To maximize the distance an object travels, you should try to maximize the horizontal speed and maximize the height it can rise. How can you use these conclusions from the activity to improve the performance of your broad jumper?

Physics To Go

1. Draw a sketch of your two coins leaving the table. Show where each coin is at the end of each tenth of a second. Remember to emphasize that they both hit the ground at the same time.

2. Repeat the sketch of the two coins leaving the table, but this time have one of the coins moving at a very high speed.

3. It is said that a bullet shot horizontally and a bullet dropped will both hit the ground at the same time. Draw sketches of this (the bullet is like a very, very fast-moving coin).

4. a) Survey your friends and family members to find out which they think will hit the ground first, a bullet that is dropped, or a fast-moving bullet.
   b) Explain why you think people may believe that the two coins hit the ground at the same time, but that they have a more difficult time believing the same fact about bullets.

5. Use evidence from your observations of the two coins in this activity to prove that a 100 mile/hour pitch thrown horizontally by a major league player will hit the ground in the same amount of time as a 10 mile/hour pitch thrown horizontally from the same height by a child.

6. Use evidence from your observations of the ball and chair in this activity to show the truth of the statement, "A projectile's horizontal motion has no effect on its vertical motion, and vice versa."

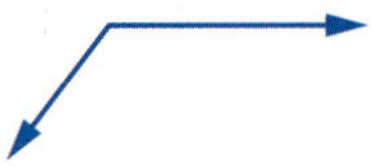

7. The diagram to the left shows two forces acting on a point at the same time. Draw the vector that represents their resultant.

8. Above a flat horizontal plane, an arrow, *A*, is shot horizontally from a bow at a speed of 50 m/s. A second arrow, *B*, is dropped from the same height and at the same instant as *A* is fired. Neglecting air friction, how does the time *A* takes to strike the plane compare to the time *B* takes to strike the plane?

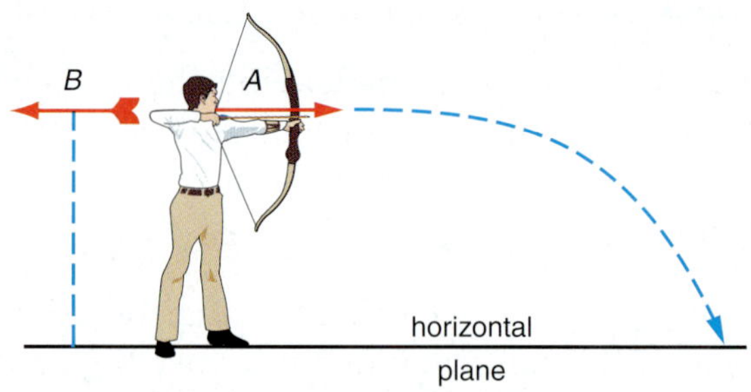

9. A swimmer jumps into a river and swims directly for the opposite shore at 2.0 km/h. The current in the river is 3.0 km/h. What is the swimmer's velocity relative to the shore?

10. A javelin is thrown at 15 m/s at an angle of 35° to the horizontal.

    a) What is its velocity in the x-direction?
    b) How far has the javelin traveled in 2.0 s?

11. A shot put is released at 12 m/s at an angle of 40° to the horizontal.

    a) What is its velocity in the x-direction?
    b) How far has the shot put traveled in 0.5 s?

12. Write a note to your school's track coach describing how the information you learned in this activity could help the team's long-jump athletes.

Activity 9 The Shot Put

GOALS

In this activity you will:

- Measure the acceleration due to gravity.
- Apply the equation Speed $= g\Delta t$ to calculate the speed attained by an object which has fallen freely from rest for a time interval Δt.
- Understand that the average speed at a specified time of an object which has fallen freely from rest is equal to half of the speed attained by the object at the specified time.
- Apply the equation (Distance = Average Speed x Δt) to calculate the distance travelled by an object which has fallen freely from rest for a time interval Δt.
- Use mathematical models of free fall and uniform speed to construct a physical model of the trajectory of a projectile.
- Use the motion of a real projectile to test a physical model of projectile motion.
- Use a physical model of projectile motion to infer the effects of launch speed and launch angle on the range of a projectile.

What Do You Think?

A world record shot put of 23.12 m was set by Randy Barnes of the USA in 1990.

- **Will a higher or lower launch point of a projectile increase range?**
- **Will a particular launch angle increase range?**
- **Will a greater launch speed of a projectile increase range?**

Record your ideas about these questions in your *Active Physics* log. Be prepared to discuss your group's responses with your small group and the class.

For You To Do

1. Your teacher will provide you with a method of measuring the acceleration caused by the Earth's gravity for objects in a condition of free fall. One simple recommended method uses a picket fence and a photogate timer attached to a computer. The picket fence is dropped and the computer measures the time between black slats of the fence. The computer then displays the acceleration

due to gravity. A second method uses a ticker-tape timer and a mass. This second method requires more class time for the analysis of data.

a) In your log, describe the procedure, data, calculations, and the value of the acceleration of gravity obtained. The acceleration of gravity is used often, so it is given its own symbol, "g."

2. In your log, make a table similar to the following. Some data already have been calculated and entered in the table to help you get started.

Time of Fall (seconds)	Speed at End of Fall (m/s)	Average Speed of Fall (m/s)	Distance (m)
0.0	0	0.0	
0.1	1	0.5	
0.2	2		
0.3			
0.4			
0.5			

a) Calculate and record in the table the speed reached by a falling object at the end of each 0.10 s of its fall.

Example:

Assume $g = 10 \text{ m/s}^2$

Speed = acceleration × time

Speed at the end of 0.2 s = $10 \text{ m/s}^2 \times 0.2 \text{ s} = 2 \text{ m/s}$

b) Calculate and record the average of all of the speeds the object has had at the end of each 0.10 s of its fall. Since the object's speed has increased uniformly from zero to the speed calculated above at the end of each 0.10 s of the fall, the average speed will be one-half of the speed reached at the end of each 0.10 s of falling.

Example:

$$\text{Average speed} = \frac{\text{Zero + speed at end of time interval}}{2}$$

$$\text{Average speed during 0.2 s of fall} = \frac{(0 \text{ m/s} + 2 \text{ m/s})}{2}$$

$$= 1 \text{ m/s}$$

(Significant digits have not been used for clarity.)

c) Calculate and record the distance the object has fallen at the end of each 0.10 s of its fall. To do this, use the familiar equation Distance = Average speed × time.

Example:

The average speed during 0.2 s of falling is 1 m/s.

$$\begin{aligned} \text{Distance} &= \text{Average speed} \times \text{time} \\ &= 1 \text{ m/s} \times 0.2 \text{ s} \\ &= 0.2 \text{ m} \end{aligned}$$

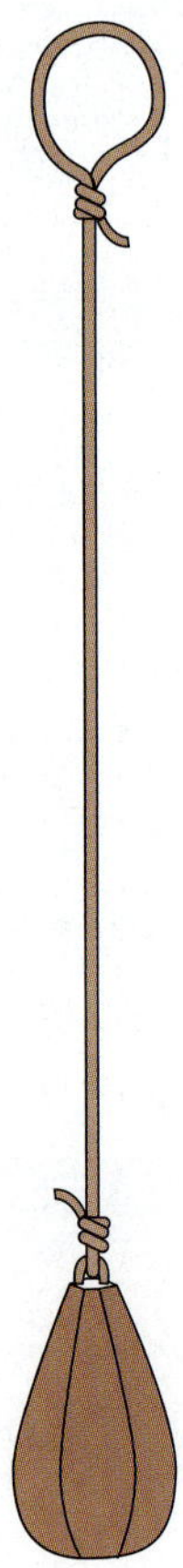

3. The table you have completed is a mathematical model of an object falling freely from rest. Now you will change the mathematical model into a physical model. Your teacher will assign your group a particular row in the data table for information about the falling object.

   Assemble a string and weight, as shown in the diagram, for the distance of fall assigned to your group.

4. Place a small label on the weight, showing your group's name and the time of fall assigned to your group.

5. Your teacher will place a horizontal row of pins labeled 0.0 s, 0.1 s, 0.2 s, etc. along the top edge of a chalkboard in your classroom. The times noted on the labels correspond to the instants for which you calculated distances of fall in the table. The horizontal distance from one pin to the next in the row is 40 cm. The 40-cm spacing of the pins is a model of the positions an object would have every 0.10 s if it traveled along the horizontal row of pins at constant speed.

a) Calculate the horizontal speed by dividing the distance traveled, 40 cm, during each time interval, by 0.1 s. (Dividing a number by 0.1 is equivalent to multiplying the number by 10). Show your calculation and the result in your log.

b) Hang your string and weight assembly from the pin corresponding to the time assigned to your group. Place a small mark on the chalkboard at the bottom end of the string and weight assembly.

6. A volunteer from the class should draw a smooth curve which connects the marks on the chalkboard. A volunteer should try to match the path, a trajectory, by throwing a tennis ball horizontally from the point corresponding to time = 0.0 s. To match the trajectory, the ball will need to be thrown horizontally at the speed calculated in **Step 5 (a)** above. This may require a few practice tries.

a) Write your observations in your log.

7. Create a "mirror image" of the trajectory by moving the 0.1–0.5 s pins to positions 40 cm to the left of the 0.0 s pin. Hang the string and weight assemblies, mark the chalkboard and connect the points to create the second half of an "arch-shaped" model of a trajectory.

8. A volunteer should try to throw a ball to match this trajectory. Have another person prepared to catch the ball.

a) What conditions seem to be necessary to match the trajectory? Write your observations in your log.

b) When a volunteer is able to match the trajectory, the class should agree upon and give the volunteer instructions to test, one at a time, the effects of launch speed and launch angle on the range of the projectile. Write your observations in your log.

9. Your teacher will show you a "portable" version of the row of pins used in **Step 5** above. It is different in two ways: one additional time interval of 0.1 s has been added (the row is 40 cm longer), and a string to show the distance of fall, 1.8 m, at a time of 0.6 s has been added.

10. Rest the end of the stick corresponding to 0.0 s on the tray at the bottom of the chalkboard while inclining the stick at an angle of 30°.

    a) Is the path indicated by the bottom ends of the string and weights assemblies a "true" trajectory? Have a volunteer try to match it. Record your observations.

    b) Repeat for angles of 45°, 60° and other angles of interest. Record your observations (it may be necessary to rest the lower end of the model on the floor to prevent the upper end from hitting the ceiling of the room).

    c) What angle of inclination predicts the greatest range for the projectile? Record your observation.

    d) Incline the stick to 90° (straight up)—do this outdoors if the ceiling is not high enough. What is being modeled in this case? Record your thoughts.

FOR YOU TO READ

Modeling Projectile Motion

The activities you have just completed demonstrate that a projectile has two motions that act at the same time and do not affect one another. One of the motions is constant speed along a straight line, corresponding to the amount of launch speed and its direction. The second motion is downward acceleration at 10 m/s^2 caused by Earth's gravity, which takes effect immediately upon launch. The trajectory of a projectile becomes simple to understand when these two simultaneous motions are kept in mind.

This activity also demonstrates the main thing that scientists do: create models to help you understand how things in nature work. In this activity you saw how two kinds of models, a mathematical model (the table of times, speeds, and distances during falling) and a physical model (the evenly spaced strings of calculated lengths) correspond to reality when a ball is thrown. For a scientific model to be accepted, the model must match reality in nature. By that requirement, the models used in this activity were good ones.

Technology as a Tool

Trajectories of projectiles can be modeled using a computer or graphing calculator. Your teacher, or someone else familiar with such devices, may be able to help you find computer software or enter equations into a calculator. These tools will allow you to manipulate variables such as launch angle, launch speed, launch height, and range to enhance your ability to explore and understand projectile motion.

In this activity, you analyzed the motion of a shot put by looking at the horizontal motion and the vertical motion independently.

You were able to find the vertical distance traveled by first finding the average velocity and then multiplying that average velocity by the time. The vertical distance traveled can be found in one step by using the equation

$$y = \frac{1}{2}at^2$$

where a is the acceleration due to gravity (9.8 m/s^2 on Earth).

The value of 9.8 m/s^2 is often rounded up to be 10 m/s^2, as it was in the **For You To Do** section.

You now have a means to analyze two-dimensional motion mathematically. The analysis of two-dimensional motion begins with the recognition that the horizontal and vertical components are independent of one another, as you discovered in the activity. The horizontal velocity always remains the same. The vertical velocity always increases with time.

	Horizontal Component	**Vertical Component**
Position	Position $x = v_x t$ where v_x is the horizontal component of the velocity, t is the time, and x is the horizontal displacement.	$y = \frac{1}{2}at^2$ where a is the acceleration due to gravity (9.8 m/s^2 on Earth), t is the time, and y is the vertical displacement.
Velocity	The horizontal velocity is a constant. There is no net force in the horizontal direction. With no force, there is no acceleration.	$v_y = at$ where a is the acceleration due to gravity (9.8 m/s^2 on Earth), t is the time and v_y is the vertical velocity.
Acceleration	No acceleration in the x-direction.	Acceleration due to gravity in the y-direction.

During a broad jump the athlete runs and then travels in a parabola (a bowl-shaped curve). The faster she runs, the faster is her horizontal velocity. She must jump in the air to get height so she can stay in the air longer. She does this without slowing down the horizontal velocity. It's easier to analyze her motion by looking at the second half of her trajectory.

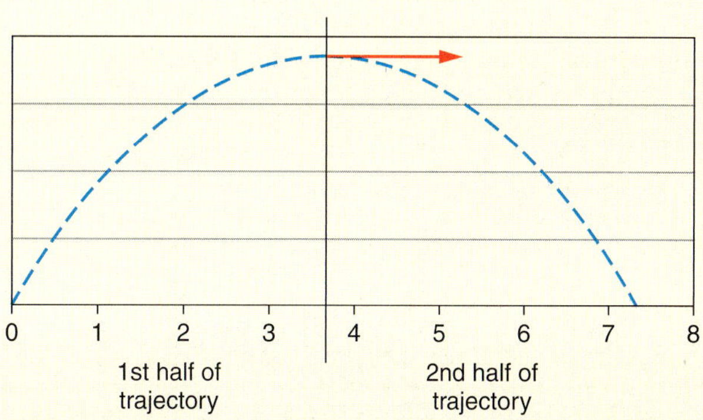

By focusing on the second half of the trajectory, the jumper appears to jump horizontally from a certain height. Suppose the height is 1.8 m. Look at the following sample problem.

Sample Problem

A broad jumper jumps horizontally from a height of 1.8 m with a horizontal velocity of 6.0 m/s. Where will she land?

Strategy:

Step 1: Use the vertical motion information to determine the time in the air.

If she fell straight down from 1.8 m or jumped horizontally from 1.8 m, her vertical motion would be identical. Using the equation $y = \frac{1}{2}at^2$, you can find the time she is in the air.

Solution:

If she were in the air 1.0 s, the vertical distance would be

$$y = \frac{1}{2}at^2$$

$$y = \frac{1}{2}(9.8 \text{ m/s}^2)(1.0 \text{ s})(1.0 \text{ s})$$

$$= 4.9 \text{ m}$$

Since she falls from only 1.8 m, she must be in the air for less than 1 s. You can try some other times:

Time (s)	Distance (m)
0.3	0.441 or 0.4
0.4	0.784 or 0.8
0.5	1.225 or 1.2
0.6	1.764 or 1.8
0.7	2.401 or 2.4

To fall from 1.8 m, it would take a bit less than 0.6 s.

If you use algebra, you can rearrange the equation to solve for time.

$$y = \frac{1}{2}at^2$$

$$t = \sqrt{\frac{2y}{a}}$$

Solving for time:

$$t = \sqrt{\frac{2d}{a}}$$

$$= \sqrt{\frac{2(1.6 \text{ m})}{9.8 \text{ m/s}^2}}$$

$$= 0.57 \text{ s}$$

Strategy:

Step 2: If she has a horizontal velocity of 6.0 m/s and she is in the air for 0.57 s, where will she land? Her horizontal motion can be found by recognizing that distance equals velocity times time.

Solution:

$$x = v_x t$$

$$= (6.0 \text{ m/s})(0.57 \text{ s})$$

$$= 3.42 \text{ m or } 3.4 \text{ m}$$

The jumper moves 3.4 meters on the way down. Her total jump is twice this value since she moves horizontally on the way up as well.

$$x \text{ total} = 6.8 \text{ m}$$

You can solve lots of problems like this. A set of problems was solved where the jumper left the ground with the same total velocity but changed the angle. What was found was that the longest jump occurred when the athlete left the ground at an angle of 45°.

Let's see if this makes sense. If the athlete jumps straight up, she maximizes her time in the air but has no horizontal velocity. She will be in the air a long time, but won't go anywhere horizontally. If the athlete jumps straight out at a very small angle, she has a large horizontal component, but is not in the air very long. If she leaves the ground at 45°, she is in the air for quite some time and still has a large horizontal velocity. This angle of 45° gives the maximum range.

Reflecting on the Activity and the Challenge

The information learned about projectile motion in this activity applies not only to the shot put but to any track and field event that involves throwing things into the air (including the self-launching of a human body as in the hurdles, long jump, or high jump). It has been reported that one Olympian who competed in the shot put increased his range in that event by nearly 4 meters, based on suggestions made by a physicist. You are now a physicist specializing in projectile motion. How will you help your school's track team?

Physics To Go

1. If the launching and landing heights for a projectile are equal, what angle produces the greatest range? Why?
2. Compared to a launch angle of 45°, what happens to the amount of time a projectile is in the air if the launch angle is:
   a) Greater than 45°?
   b) Less than 45°?

3. For a constant launch speed, what angle produces the same range as a launch angle of:
   a) 30°?
   b) 15°?

4. If you launch a projectile from a high building, the angle for greatest range is less than 45°. Explain why this is true.

5. If a shot putter releases the projectile 2 m above level ground, what angle produces the greatest range? 45°? More than 45°? Less than 45°? Why? How could you find the exact angle?

6. Analysis of performances of long jumpers has shown that the typical launch angle is about 18°, far less than the angle needed to produce maximum range. Why do you think this occurs?

7. You are familiar with Carl Lewis as a medal-winning sprinter. But he is also an Olympic gold medalist in the long jump. Why do you think he's successful in both events?

8. The diagram shows a ball thrown toward the east and upward at an angle of 30° to the horizontal. Point X represents the ball's highest point.
   a) What is the direction of the ball's acceleration at point X? (Ignore friction.)
   b) What is the direction of the ball's velocity at point X?

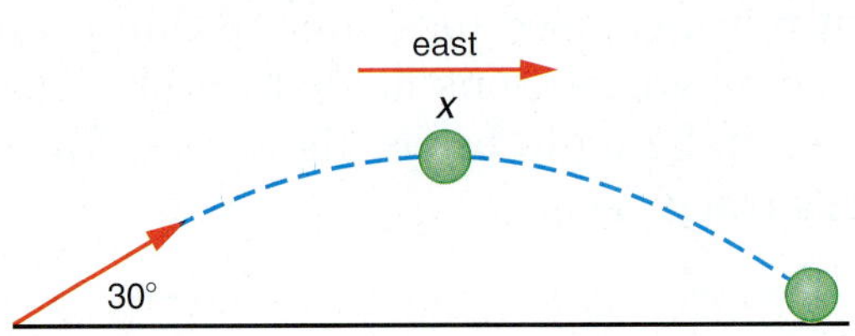

9. The diagram at the top of the next page shows a baseball being hit with a bat. Angle θ represents the angle between the horizontal and the ball's initial direction of motion. Which value of θ would result in the ball traveling the longest horizontal distance?

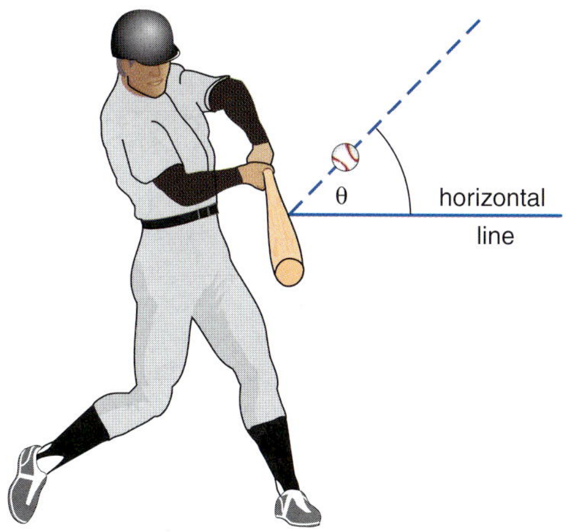

10. Four balls, each with mass m and initial velocity v, are thrown at different angles by a ball player. Neglecting air friction, which angular direction produces the greatest projectile height?

11. An object is thrown horizontally off a cliff with an initial velocity of 5.0 m/s. The object strikes the ground 3.0 s later.
    a) What is the vertical speed of the object as it reaches the ground?
    b) What is the horizontal speed of the object 1.0 s after it is released?
    c) How far from the base of the cliff will the object strike the ground?

12. The diagram below shows a ball projected horizontally with an initial velocity of 20.0 m/s east, off a cliff 100 m high.
    a) During the flight of the ball, what is the direction of its acceleration?

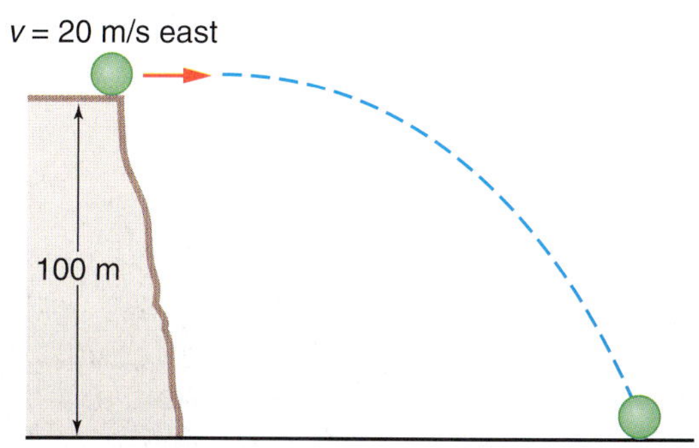

b) How many seconds does the ball take to reach the ground?

Activity 10 Energy in the Pole Vault

GOALS

In this activity you will:

- Understand and apply the equation Kinetic Energy = $1/2\ mv^2$.
- Understand and apply the equation Gravitational Potential Energy = mgh.
- Recognize that restoring forces are active when objects are deformed.
- Understand and apply the equation Potential Energy Stored in a Spring = $1/2kx^2$.
- Understand and measure transformations among kinetic and potential forms of energy.
- Conduct simulations of transformations of energy involved in the pole vault.

What Do You Think?

You would need a fence more than 20 feet high to keep the world champion pole vaulter (6.14-m record) out of your yard.

- **If good vaulters have a 25-foot pole, why can't they vault 25 feet?**
- **What factors do you think limit the height vaulters have been able to attain?**

Take a few minutes to write answers to these questions in your *Active Physics* log. Compare your answers with the answers given by others in your group. Be prepared to represent your group's ideas to the entire class.

For You To Do

1. Carefully clamp a ruler in a vertical position so that the clamp is near the bottom end and the top end extends a few centimeters above the edge of a tabletop. Tape a pen or pencil to the surface of the ruler near the top end of the ruler so that the writing end of the pen extends to one side of the top end of the ruler. If the top end of ruler moves as it is bent, the pen will move with it.

2. Set up a ramp as shown in the diagram. Three different starting points on a ramp will be used to roll a ball across the tabletop at three different speeds. Each time the ball rolls, it will strike the ruler near the top end, causing the ruler to bend. A marking surface held in contact with the tip of the pen or pencil will be used to measure the deflection.

3. Roll the ball at low, medium and high speeds to bend the ruler. In each case, measure the amount of deflection of the end of the ruler as indicated by the length of the pen mark.

   a) If the rolling ball represents the running vaulter and the ruler represents the pole in the model, how does the amount of bend in the pole depend on the vaulter's running speed? Record your data and response in your log.

4. Carefully clamp a ruler flat-side down to a tabletop so that two-thirds of the ruler's length extends over the edge of the table.

5. Place a penny on the top surface of the ruler at the outside end.

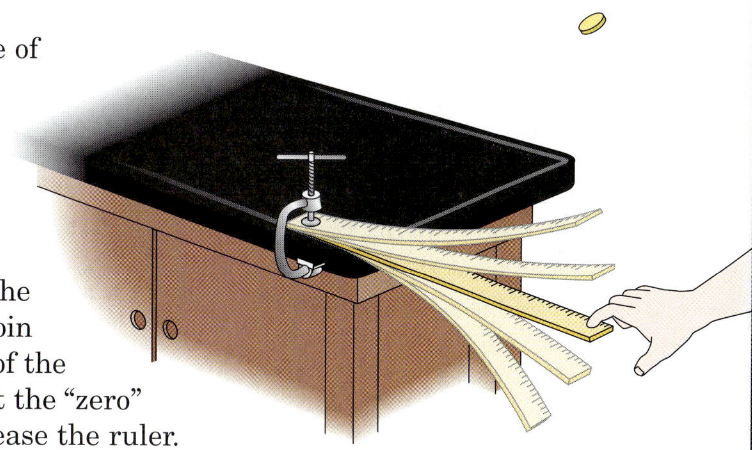

6. Use a second ruler to measure a 1 cm downward deflection of the outside end of the clamped ruler; that is, bend the ruler downward. Prepare to measure the maximum height to which the coin flies upward using the position of the coin when the ruler is relaxed at the "zero" vertical position of the coin. Release the ruler.

   a) Record in your log the height to which the coin travels.

   b) Repeat **Step 6** above for ruler deflections of 2 cm and 3 cm, and in each case, record the maximum height of the "vaulted" coin.

   c) How does the height to which the coin travels seem to be related to the amount of deflection of the ruler? (Remember, the coin is a projectile, in this case.)

Do not deflect the ruler excessively.

FOR YOU TO READ

Conservation of Energy in the Pole Vault

The pole vault is a wonderful example of the Law of Conservation of Energy. The forms of energy are changed, or transformed, from one to another during a vault, but, in principle, the total amount of energy in the system of the vaulter and the pole remains constant. Food energy provides muscular energy for the vaulter to run, gaining an amount of **kinetic energy.** Some of the vaulter's kinetic energy is used to catapult the vaulter with an initial speed upward and the remaining kinetic energy is converted into work done on the pole to cause it to store an amount of **spring potential energy.** As the bent pole straightens, its potential energy is delivered to the vaulter to increase the vaulter's gravitational potential energy.

In making measurements of the ruler's deflection and the height of the coin, you continued your study of the conservation of energy – the most important principle of science. Richard Feynman, an American physics giant of the 20th century, provides a story that may help you to understand energy conservation. In his story, a child plays with 28 blocks. Every day the child's mother counts the blocks and always finds the total to be 28. On one occasion, she only finds 27 blocks, but then realizes that one block is hidden in a box. By finding the mass of the box and its contents, she can determine that 1 block must be inside, if she knew the mass of the empty box. On another day, she finds only 25 blocks, but can see that the water in a pail is higher than expected. By measuring the height difference, and knowing something about the original height of the water and the volume of a block, she determines that 3 blocks are below the surface of the water. On a third day, she finds 30 blocks! She then remembers that a friend came over to visit and decides that 2 of the blocks belong to the friend. Feynman equates "counting the blocks" with measuring the total energy. There were 28 blocks and there will always be 28 blocks. If there are 28 units of energy, then there will always be 28 units of energy.

In the activity, the energy came from deflecting the ruler. The muscles in your arm applied a force over a distance and deflected the ruler and the ruler gained spring potential energy. As you let go of the ruler it tossed the coin into the air. The ruler lost some of its energy and the moving coin gained kinetic energy. The coin rose up in the air, slowing down along the way. The kinetic energy was transformed as the coin rose, and there was a gain in gravitational potential energy. As the coin descended, the gravitational potential energy became kinetic energy once again.

If you were to measure the energy, you might find that the work done by your muscles in deflecting the ruler may have been equal to, say three joules. If that were so, then the three joules of work would have created three joules of spring potential energy that would have become three joules of kinetic energy that would then have become three joules of gravitational potential energy. The energy comes from external work and this energy then remains constant.

You can calculate the work done by your muscles by multiplying the force applied and the distance over which the force was applied.

$$W = F \cdot d$$

The dot between the F and the d is to signify multiplication of the force and displacement when they are (at least partly) in the same direction.

There are also equations that can help you to measure the energies.

Spring potential energy:

$$PE_{spring} = \frac{1}{2}kx^2$$

where k is the spring constant, and x is the amount of bending.

Gravitational potential energy:

$$PE_{grav} = mgh$$

where m is the mass of the object, g is the acceleration due to gravity, h is the height through which the object is lifted.

Kinetic energy:

$$KE = \frac{1}{2}mv^2$$

where m is the mass of the moving object and v is the speed of the object.

All of the energies are expressed in joules. Energy is a scalar. It has no direction.

A second example of energy transformations that you may be familiar with is in the sport of archery. In archery, you pull on the bowstring and the arrow flies through the air. If you shoot the arrow straight up, you can look at energy conservation in the following way: Your arm does work on the bow by pulling on the string and stretching the bow. The bow now has spring potential energy. The string is let go. The bow loses its spring potential energy and the arrow

gains kinetic energy. If the arrow goes up, the kinetic energy of the arrow now becomes gravitational potential energy. As the arrow descends, the gravitational potential energy becomes kinetic energy. If the arrow sticks up in the ground and comes to rest, all of the kinetic energy transfers to the ground and becomes heat energy. These ideal transformations of energies occur when there are no external forces, like air resistance, depleting energy from the original amount of energy.

The force to stretch a string is a Hooke's Law force.

$$F = kx$$

This is Hooke's Law. It says that the larger the stretch of a spring, the larger the force required. When you pull a bowstring, you need almost no force to pull it a tiny bit. As you stretch it, the force you must apply gets greater. The average force will be halfway between the zero force (to start the stretch) and the final force for the last stretch. The final force is kx. The initial force is 0. The average force is:

$$\bar{F} = \frac{kx + 0}{2} = \frac{1}{2}kx$$

The total stretch of the spring is x. The work done is:

$$W = F \cdot d = \left(\frac{1}{2}kx\right)x = \frac{1}{2}kx^2$$

which is the expression for the spring's potential energy.

You can also calculate the work done to lift an object of mass m up through a distance h. Here the applied force must be equal to the weight of the object mg.

$$W = F \cdot d = mgh$$

which is the expression for the gravitational potential energy.

Also, you can calculate the work done in accelerating an object from an initial velocity v_i to a final velocity v_f with a constant force.

$$\begin{aligned} W &= F \cdot d \\ &= mad \\ &= \frac{1}{2}mv_f^2 - \frac{1}{2}mv_i^2 \end{aligned}$$

which is the expression for the change in kinetic energy.

The work equations and the energy equations are identical. This is where the statement "energy is the ability to do work" originates.

Sample Problem 1

Your teacher gives you a pop-up toy. When you push down on it, it sticks to the desk for a moment and then pops into the air.

a) If the toy has a mass of 100.0 g and leaps 1.20 m off the table, how much potential energy does it have at its point of maximum height? (Use $g = 9.80\ \text{m/s}^2$.)

Strategy: The toy has a type of energy that depends on its position in the Earth's gravitational field. You can use this "potential" energy to do some work or to change to some other form. So you need to use the formula for gravitational potential energy.

Givens:

$m = 100.0\ \text{g} = 0.100\ \text{kg}$

$h = 1.20\ \text{m}$

Solution:

$$\begin{aligned} PE_{grav} &= mgh \\ &= (0.100\ \text{kg})(9.80\ \text{m/s}^2)(1.20\ \text{m}) \\ &= 1.18\ \text{J} \end{aligned}$$

b) When the toy jumps off the desk, with what speed does it leave?

Strategy: At the point where it jumps off the desk, the toy has its maximum amount of kinetic energy. This is what becomes the potential energy at the peak. Because energy is conserved, these two values will be equal—kinetic energy at the bottom equals the potential energy at the peak.

Givens:

$$PE_{grav} = 1.18\ \text{J}$$

Solution:

$$PE_{grav} = KE$$

$$KE = \frac{1}{2}mv^2$$

You can use algebra to rearrange the equation to solve for v.

$$v = \sqrt{\frac{KE}{\frac{1}{2}m}}$$

$$= \sqrt{\frac{1.18\ \text{J}}{\frac{1}{2}(0.100\ \text{kg})}}$$

$$= 4.90\ \text{m/s}$$

c) If you push the toy down 2.00 cm to make it stick to the desk, what is the spring constant of the spring in the toy?

Strategy: Where did the kinetic energy to make the toy leap off the desk come from? It came from doing work on the spring and storing a different type of potential energy: elastic potential energy. How much elastic potential energy did the toy store? 1.18 J! So again you can use conservation of energy to solve this problem!

Givens:

$x = 2.00\ \text{cm} = 0.0200\ \text{m}$

Solution:

$$PE_{spring} = \frac{1}{2}kx^2$$

You can use algebra again to rearrange the equation to solve for k.

$$k = \frac{PE_{spring}}{\frac{1}{2}x^2}$$

$$= \frac{1.18\ \text{J}}{\frac{1}{2}(0.0200\ \text{m})^2}$$

$$= 5900\ \text{N/m}$$

This sounds like a really large value for such a little toy, but remember that this is newtons per meter and the spring in the toy is only compressed a few centimeters.

d) What force was needed to compress the spring the 2.00 cm?

Strategy: Now that you know the compression and the spring constant, it is possible to find the amount of force required to press down on the spring.

Givens:

$k = 5900\ \text{N/m}$
$x = 0.0200\ \text{m}$

Solution:

$$F = kx$$
$$= (5900\ \text{N/m})(0.0200\ \text{m})$$
$$= 118\ \text{m or } 120\ \text{N}$$

Sample Problem 2

At what height, above the surface of Earth, could a tennis ball ($m = 57$ g) be dropped to give it the same kinetic energy it has when traveling at 45 m/s? (Neglect air resistance.)

Strategy: As an object is dropped from a height, its gravitational potential energy decreases and its kinetic energy increases. Assume that you are looking for the vertical position, which will yield a speed of 45 m/s the instant *before* the ball touches the ground. The problem can be solved in one step using the conservation of energy ($PE_{grav} = KE$).

Givens:

$m = 57\ \text{g} = 0.057\ \text{kg}$
$v = 45\ \text{m/s}$
$g = 9.8\ \text{m/s}^2$

→

Solution:

$$PE_{grav} = KE$$

$$mgh = \frac{1}{2}mv^2$$

You can use algebra to rearrange the equation to solve for *h*. Notice that in solving the problem in one step, you do not have to take into account the mass of the ball.

$$h = \frac{v^2}{2g}$$

$$= \frac{(45 \text{m/s})^2}{2 \times 9.8 \text{ m/s}^2}$$

$$= \frac{2025 \text{ m}^2\text{s}^2}{19.6 \text{ m/s}^2}$$

$$= 103.3 \text{ m or } 100 \text{ m}$$

When the ball is traveling at 45 m/s, it has a kinetic energy of 58 J. (You can calculate this:)

$$\frac{1}{2}mv^2 = \frac{1}{2}(0.57 \text{ kg})(45 \text{ m/s}^2)$$

If the tennis ball were positioned at a location 100 m above Earth, the gravitational potential energy of the ball would also equal 58 J.

Reflecting on the Activity and the Challenge

In this activity you were told that throughout the event of pole vaulting, energy changes from one form to another, but the total amount of energy in the system at all instants remains the same. (A small amount of energy may be lost from the system of the vaulter and pole by making a dent in the end of the pit, which stops the pole or by generating heat in the pole as it bends.)

This will be important for the athletes at your school to know. They will need to understand what they must do to gain the greatest height in a pole vault. By examining the equations for kinetic and potential energy, they will also be able to appreciate why there is a limit to the height that somebody can vault. The faster someone runs, the more kinetic energy they have. This kinetic energy makes the pole bend. The more kinetic energy there is, the more bend in the pole is expected. Potential energy stored in the bent pole will be transformed to increase the vaulter's gravitational potential energy. The more bend in the pole, the higher the vaulter goes. One key to success in the pole vault is to have the most kinetic energy. Another key to success is to bend the pole as much as possible. You may be able to use this knowledge to help a pole vaulter in your school improve performance.

Physics Words

kinetic energy: the energy an object possesses because of its motion.

spring potential energy: the internal energy of a spring due to its compression or stretch.

Physics To Go

1. Describe the energy transformations in the shot put.
2. Describe the energy transformations in the high jump.
3. Assume that a vaulter is able to carry a vaulting pole while running as fast as Carl Lewis in his world record 100-m dash. Also assume that all of the vaulter's kinetic energy is transformed into gravitational potential energy. What vaulting height could that person attain? (Hint: Use the equation $\frac{1}{2}mv^2 = mgh$.)
4. Why doesn't the length of the pole alone determine the limit of vaulting height?
5. Some poles lose a significant amount of energy to heat as they flex. Use the Law of Conservation of Energy to explain how this would affect performance.
6. The women's pole vault world record as of spring 1997 was 4.55 m, set by Emma George. What do you estimate was Emma's speed prior to planting the pole? Use conservation of energy for your prediction.
7. Sergei Bubka held the world record for the pole vault as of spring 1997 at 6.14 m. How did Sergei's speed compare with Emma George's speed?
8. A 2.0-kg rock is dropped off a 100-m high cliff.
   a) Without using kinematics, calculate the speed the rock would be going when it got to the bottom of the cliff.
   b) Can you do this if you do not know the mass of the rock?
9. A bow is strung with a bowstring that has a spring constant of 1500 N/m.
   a) If you pull it back 25 cm, how much work are you doing on the string?
   b) If the string is pushing against an arrow that has a mass of 0.10 kg, how fast would the arrow be going when it left the bow?

10. An exercise spring has a spring constant of 315 N/m.
    a) How much work is required to stretch the spring 30 cm?
    b) What force is needed to stretch the spring 30 cm?

11. A toy car (m = 0.04 kg) is released from rest and slides down a frictionless track 1 m high. At the bottom of the track it slides along a horizontal portion until it hits a spring (k = 18 N/m). The spring is attached to an immovable object. What is the maximum compression of the spring?

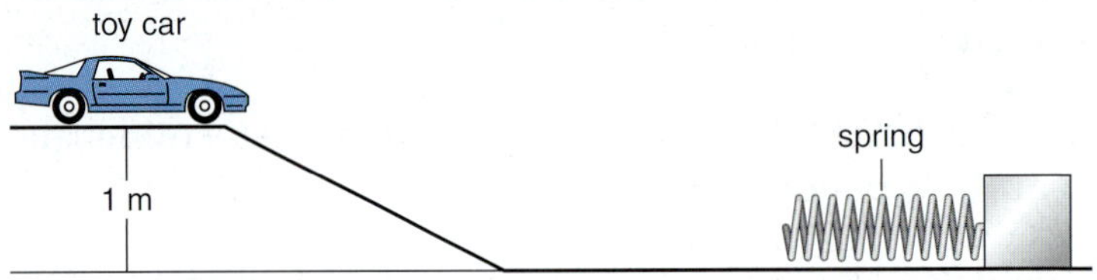

12. A roller coaster is poised at the top of a hill 50 m high.
    a) How fast would it be going when it went over the top of the next hill on the track, which is only 30 m high?
    b) From a practical point of view, why is it advantageous that this ride is mass independent?

13. A 40-g bullet leaves the muzzle of a gun at a speed of 300 m/s.
    a) What is the kinetic energy of the bullet as it leaves the barrel?
    b) If the barrel of the gun is 12 cm long, what is the force acting on the bullet while it is in the barrel?

14. A super-ball (m = 30 g) is dropped from a height of 3 m.
    a) At what vertical position is $PE_{\text{grave}} = KE$?
    b) What is the speed of the ball at this position?

15. A water-balloon (m = 300 g) is launched horizontally from a platform 2 m above the ground with a slingshot. The slingshot (k = 60 N/m) is stretched 40 cm before launch. How far from the platform will the balloon strike the ground?

PHYSICS AT WORK

Erv Hunt

Coach to the Champions

In 1998, Erv Hunt began his 25th season as the head coach of the University of California at Berkeley's track and field team. He has brought more than just a reputation as one of the foremost track and field coaches in the United States to his job. He has brought to his team the experience of having served as the US men's head track and field coach at the 1996 Olympic Games in Atlanta.

"We have a much better understanding of the science of running now, the physics of it, than we use to," states Erv, "and that is one of the reasons runners have improved over the years." Erv believes that it's important when training to understand pace work and intervals during the different stages of a race. He begins each training season with his team in a slow and relaxed manner. "I want the athletes to know what it feels like to run correctly without tensing up," he says, "then they can slowly pick up the pace. I work on breaking down their running techniques, creating strategies, at different intervals and seeing what works with a sense of control. Too often runners begin to lose control mentally and physically when they're running fast. I want them to have the same sense of control at fast paces that they have during the slower paces."

To be around and train some of the great athletes during his career has been an incredible experience for Erv. "For someone to reach that level of excellence makes them somewhat a different breed of person," and Erv has had the privilege to work with some of the best.

Chapter 3 Assessment

In this chapter you studied physics as it relates to track and field events. Now you have the knowledge to write a physics manual about track and field training for your high school team to help improve its performance. The manual should:

- **help students compare themselves with the competition**
- **include a description of physics principles as they relate to track events**
- **provide specific techniques to improve performance**

Review and remind yourself of the grading criteria that was agreed on by the class at the beginning of the chapter. You may have decided that some or all of the following qualities should be graded in your track and field manual:

- **physics principles**
- **inclusion of charts**
- **past records**
- **relevant equations**
- **definitions**
- **specific techniques**

Any advice you give should be understandable to athletes who have not studied physics. You can describe any activities you have done to explain how you know that the technique works, but you should not tell so much about each activity that the reader becomes bored.

Physics You Learned

Distance, Time, Speed

$$\text{Speed} = \frac{\text{Distance traveled}}{\text{Time taken}}$$

Speed vs. Time graphs

Histograms

Average speed, Instantaneous speed

Distance vs. Time graphs

Acceleration

Accelerometers

Speed = Frequency × Stride Length

Trajectory motion

Horizontal and Vertical motion

Optimum angle for trajectories

Potential energy

Kinetic energy

Chapter 4

THRILLS AND CHILLS

4

Thrills and Chills

Scenario

You are excited and scared as you sit back into the seat. You pull the safety restraints into place. The next thing you know, you are beginning a slow but steady ascent into the sky. Then, with a sudden jolt, you reach the top. This is where the thrill or nightmare begins. You hurtle down the track at ever-increasing speeds. You are flung against one side of your seat as you scream around a curve. You shriek as you hang upside down, fortunately, firmly secured to your seat. All the time, your stomach has no idea where it is or you are. Finally, you come to rest where you began. What a ride! Want to go again?

Roller coasters have been enjoyed for many years. However, the roller coaster that may appeal to you, may not appeal to your parents or other friends and relatives.

Challenge

A roller coaster, called the Terminator Express, has been designed for an amusement park. Your challenge is to take the roller coaster design and modify it for a select group of riders. For instance, you may decide that you will modify the roller coaster so that young children can experience the thrill of a roller coaster in a safe and non-threatening way. You may prefer to design the roller coaster for adults that are a bit squeamish about the big hills and sharp turns. They want to experience the thrill of a roller coaster ride, but are ready to pass on the death-defying action. On the other hand, you may choose to design a roller coaster for daredevils that are ready to handle any thrill you can provide. You may also wish to design a roller coaster for people who are physically challenged or visually impaired. However, in each situation, you must ensure that the roller coaster is safe and nobody will be in danger.

Criteria

You will make the design based on whatever group you choose. Your design will have to meet certain criteria. All designs must have certain components including:

- **at least two hills**
- **one horizontal curve**

In addition, you will have to provide evidence that the ride is safe. Safety data will include the height, speed, and acceleration of the roller coaster at five designated locations. Finally, you will have to calculate the work required to get the roller coaster rolling.

You will present your design as both a model and a written poster or report.

Make a list of criteria for each of the required elements of the roller coaster. What should a roller coaster design that deserves an "A" look like? Are there other elements that you think should be included in a roller coaster grade? How much should each element be worth? Discuss these questions with your small group and the class.

Activity 1 The Big Thrill

GOALS

In this activity you will:

- Be able to draw and interpret a top view and a side view of a roller coaster ride.
- Conclude that thrills in roller coaster rides come from accelerations and changes in accelerations.
- Define acceleration as a change in velocity with respect to time and to recognize the units of acceleration.
- Be able to measure and calculate velocity and acceleration.

What Do You Think?

The tallest wooden roller coaster has a height of about 66 m (218 ft.). The tallest steel roller coaster is 128 m (420 ft.) high. This is as tall as a 40-story high rise!

- **Which part of the roller coaster produces the loudest screams? Why?**

Record your ideas about these questions in your *Active Physics* log. Be prepared to discuss your response with your small group and the class.

For You To Do

Part A: Sketch of the Roller Coaster

1. Sketch a roller coaster with a first hill of 15 m that quickly descends to 6 m and then turns to the left in a big circle (radius of 10 m) and then descends back to the ground.

2. Compare your roller coaster design to those drawn by other people on your design team.

   a) Which sketch do you like the best? Provide three reasons why you prefer that sketch.

3. Create two sketches for the same roller coaster. The first sketch should be a side view. The second sketch should be a view from the sky.

   a) What are the advantages of having two sketches?

4. Below is the roller coaster that has been designed by the professional team that is asking for your help. It is called the Terminator Express. There are two views of the roller coaster. The first view is a side view. The second view is a view from the sky.

   a) Sketch the side and top view in your *Active Physics* log.

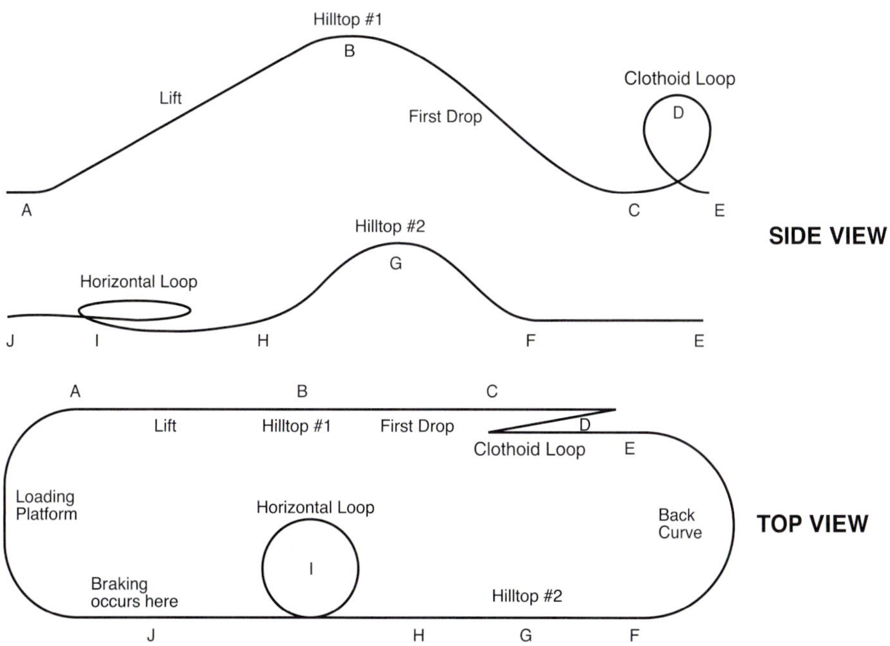

5. The roller coaster car begins from the loading platform and then rises along the lift. It arrives at the top of hilltop #1 and then makes its first drop. It then goes into a vertical loop. This clothoid loop (it has a big radius at the bottom and a small radius at the top) allows the riders to be safely upside down. The car then goes along the back curve, rises over hilltop #2 and then swings into a horizontal loop. The brakes are applied and the roller coaster comes to a stop.

   Have one team member read this description as you move your finger along the roller coaster track.

   Repeat the procedure with the top view.

The rider must be wearing a bicycle helmet.

Part B: Roller Coaster Fun

1. You will now blindfold someone in your group in order to observe the thrilling parts of a roller coaster ride. The blindfolded person will sit in a chair with wheels.

   a) Write down the safety concerns when one of your team members is blindfolded and you will be pushing him or her. What could go wrong? How can you prevent this?

2. Push on the chair of the blindfolded team member. While the person is moving give the blindfolded team member another push. Continue the pushing with pushes from different directions.

   a) Notice when the blindfolded team member smiles or laughs or exhibits some emotion. Is it when the push comes from the back or the front or the side?

3. A person's velocity is a measure of the speed and direction of the person. The person's velocity may have been 1.2 m/s north or 1.5 m/s toward the door or 1.0 m/s toward the window. In each case there is a magnitude (1.0 m/s, 1.2 m/s, 1.5 m/s) and a direction (north, toward the door, toward the window).

   a) Was the velocity responsible for the "rider's" reactions? Did the blindfolded person react more when they were going faster as they were gently pushed or pulled or when they were moving in a certain direction?

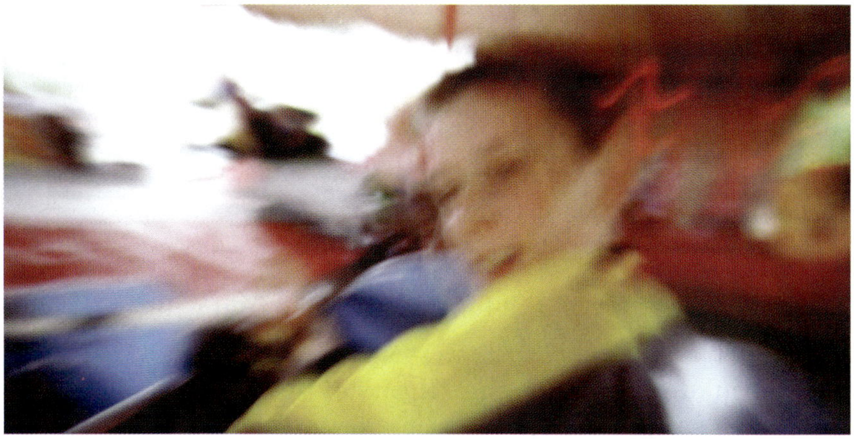

4. The change in a person's velocity over time is referred to as acceleration. Suppose a person were moving at 1.5 m/s toward the door and someone pushed him or her and made the person move at 1.3 m/s toward the window. There is acceleration because there was a change in speed and a change in direction.

   Acceleration also occurs if the person changed velocity from 1.1 m/s north to 1.4 m/s north. Here the acceleration is due to a change in speed.

   There would also have been acceleration if the person changed velocity from 1.3 m/s east to 1.3 m/s south. Here the acceleration is due to a change in the direction, with no change in speed.

   a) Was acceleration responsible for the reactions of the blindfolded person? Did they react more when they accelerated?

5. Acceleration is a change in velocity in a specific time. The change from 1.1 m/s north to 1.5 m/s north may have taken 1 s.

   The change in velocity is 0.4 m/s in one second. There are a number of ways in which this can be stated:

   0.4 m/s (meters per second) in one second

   0.4 m/s every second

   0.4 m/s per second

   0.4 (m/s)/s

   0.4 m/s^2

Be prepared to stop the ball at the end of the track so it does not roll onto the floor. If the ball rolls onto the floor pick it up right away. If you cannot find the ball, tell your teacher right away.

Part C: Measuring Velocities and Calculating Accelerations

1. The value 1.5 m/s north is a velocity. The velocity 1.5 m/s tells you that the object can travel 1.5 m in 1 s.

   a) If an object were moving across the table, what instruments would you need for measurements to determine if the object were traveling at 1.5 m/s?

2. Place a track flat on the top of your table. You will set a steel ball moving along the track.

   a) Calculate and record the velocity of the object using a ruler and a stopwatch. The equation for calculating velocity is

   $$v = \frac{\Delta d}{\Delta t}$$

   where v is the velocity,

   d is the displacement, and

   t is the time.

   The symbol Δ (delta) signifies "change in."

   (Remember: a velocity must have a direction.)

3. Decrease the speed of the steel ball.

   a) Calculate and record its velocity again.

4. Your teacher will demonstrate the use of a photogate timer. The timer starts when an object breaks the beam. The timer stops when the beam is no longer broken. The elapsed time can be measured very accurately.

   a) To determine the velocity of the steel ball, what additional information would you (or the computer) have to know?

5. A large steel ball with a diameter of 6 cm passes through a photogate. The elapsed time recorded on the photogate timer is 2 s.

   a) Calculate the speed of the ball. (Since the speed is requested, you do not have to worry about the direction of motion. Speed is a scalar—it has no direction. Velocity is a vector—it has direction.)

6. Use the photogate timer to help you find the speed of the steel ball traveling down the track.

a) Record the speed of the steel ball traveling down the track.

7. Create a slope for a steel ball to travel down. Have a ball travel down the slope.

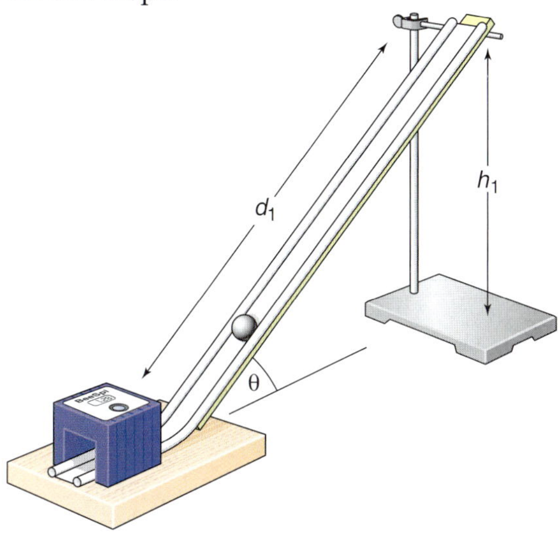

a) Measure and record the speeds at two different points.

b) Calculate the acceleration of the ball from the two speed measurements. (You will also have to measure the time between velocity measurements.) Acceleration is the change in velocity with respect to time. The equation to calculate acceleration is

$$a = \frac{\Delta v}{\Delta t}$$

where a is the acceleration,

v is the velocity, and

t is the time.

The symbol Δ (delta) signifies "change in," and can be calculated by subtraction.

Part D: Acceleration on the Roller Coaster—Pulling g's

1. On a roller coaster, you often feel heavier or lighter as you whip around curves. The accelerations take a toll on your body. This is often called "pulling g's." The Terminator Express has a number of places where you will be "pulling g's." Try to imagine a ride on the roller coaster shown.

a) Copy the drawing below into your *Active Physics* log. Indicate where the passengers might feel light, and where they might feel heavy. Also indicate on the diagram where you think the coaster is speeding up and where it is slowing down.

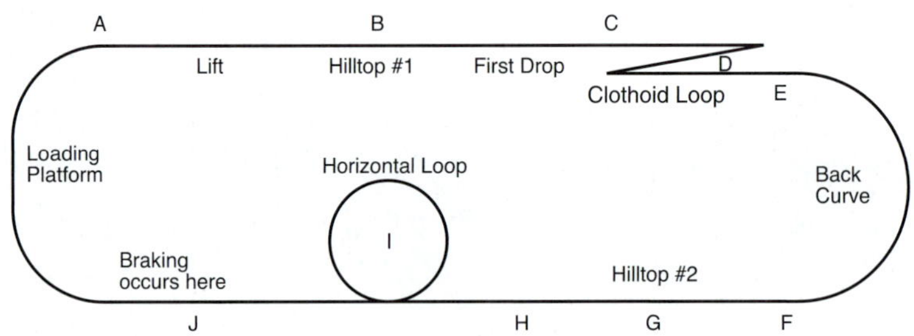

TOP VIEW

Physics Words

scalar: a quantity that has magnitude, but no direction.

vector: a quantity that has both magnitude and direction.

PHYSICS TALK

Measuring Velocity and Acceleration

In this activity you were introduced to some terms that you would need to understand in order to redesign the Terminator Express.

Distance is measured with a piece of string or a tape measure along a path. The unit used is usually a meter. For example, an object traveled 3 m. Distance is a **scalar** quantity. It has no direction.

Displacement is measured with a meter stick with direction included. It depends only on the endpoints, not on the path. An object traveled 3.5 m east. Displacement is a **vector**. It has magnitude (3.5 m) and direction (east).

In the diagram to the right, the curve represents the path of an object. The straight line represents the displacement.

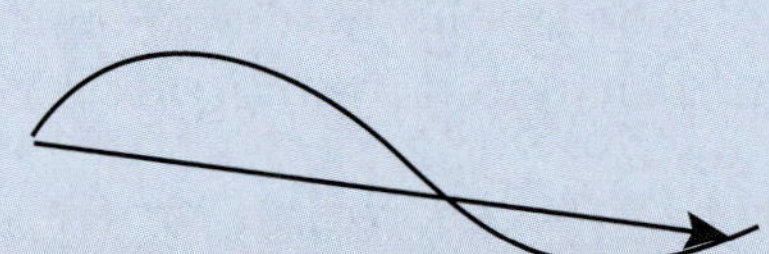

Time is measured with a stopwatch or other type of watch or clock.

Speed is the change in distance per unit time. The object's speed may be 4 m/s. This means that the object moves 4 m every second. Speed is a scalar. It has no direction.

Velocity is the change in displacement per unit time. The object's velocity may be 4 m/s south. Velocity is a vector. It has magnitude (4 m/s) and direction (south).

The equation to calculate velocity is

$$v = \frac{\Delta d}{\Delta t}$$

where v is the velocity,

d is the displacement, and

t is the time.

The symbol Δ (delta) signifies "change in."

For a person walking one lap around a city block, the distance is equal to the perimeter of the city block. The speed is equal to this distance divided by the time to complete the walk. The displacement equals 0 (since the person ended up where she started) and the velocity equals 0 as well.

Acceleration is the change in velocity per unit time. An object's acceleration may be 5 m/s every second. This means that the object changes its speed by 5 m/s every second. The speed will increase from 0 m/s to 5 m/s to 10 m/s to 15 m/s with each change requiring one second. 5 m/s every second is also written as 5 m/s^2 (five meters per second squared).

The equation to calculate acceleration is

$$a = \frac{\Delta v}{\Delta t}$$

where a is the acceleration,

v is the velocity, and

t is the time.

The symbol Δ (delta) signifies "change in."

Physics Words

speed: the change in distance per unit time; speed is a scalar; it has no direction.

velocity: speed in a given direction; displacement divided by the time interval; velocity is a vector quantity; it has magnitude and direction.

acceleration: the change in velocity per unit time.

Reflecting on the Activity and the Challenge

Velocity and acceleration are a big part of roller coaster fun. Traveling at a big speed is not enough to give a Big Thrill. The thrills come from accelerating around the curves and along the straight segments. This acceleration changes your speed and your direction as you ride along the path of the coaster. Additional thrills come from changes in acceleration. More rapid changes require greater accelerations. In designing your variation of the Terminator Express, you will want to ensure that the speeds and accelerations are right for your riders. You are required to have hills and turns, but the loop may be too much for your riders.

Physics To Go

1. Draw a top view and a side view of a roller coaster with the following characteristics: The roller coaster car begins from the loading platform and then rises along the lift. It arrives at the top of hilltop #1 and makes its first drop. It then climbs hill #2 which is half the height of hill #1. The car then goes along the back curve, rises over hilltop #3, and swings into a horizontal loop. The coaster then comes out of the loop onto a level plane. The brakes are applied and the roller coaster comes to a stop.

2. Identify where the biggest thrill will be in the Terminator Express roller coaster. Explain why this will be The Big Thrill.

3. Speed doesn't produce thrills. Living on the Earth, you already have a big speed.

   a) The Earth makes a complete revolution once every 24 h. City A is close to the Equator and travels a large circumference in 24 h. City B is close to the Arctic Circle and travels a small circumference in 24 hours. Which city has the greater speed?
   b) The circumference of (path around) the Earth's Equator is about 40,000 km. It requires one day or 24 h to complete one revolution. Calculate the speed you are traveling on Earth.
   c) Why don't you get a big thrill going at such a high speed?

4. A roller coaster rider changes from a speed of 4 m/s to 16 m/s in 3 s. Calculate the magnitude of the acceleration of the ride.

5. Identify the following situations as one of distance, displacement, speed, velocity, or acceleration. Indicate which is a vector and why.

   a) a car traveling at 50 km/h
   b) a student riding a bike at 4 m/s toward home
   c) a roller coaster ride whips around a left turn at 5 m/s
   d) a roller coaster car rises 12 m
   e) a train ride takes you 150 km

6. Suppose you were designing a roller coaster for young pre-school children.
   a) Describe two changes you would make to the Terminator Express roller coaster. Explain why you would make these changes.
   b) Draw the top and side view of the roller coaster with these additional changes.

7. A cart is 10 cm long. It travels through a photogate in 2 s. Calculate the cart's speed.

8. A second cart is 5 cm long. If it were traveling at the same speed as the cart in **Question 7**, what would the photogate timer record as the elapsed time?

Stretching Exercise

Investigate roller coasters on the Internet. Which are the most modern? What are some innovations in newer roller coasters? What features from historic coasters have been retained? Compare wooden and steel coasters.

Activity 2 What Goes Up and What Comes Down

GOALS

In this activity you will:

- Measure the speed of an object at the bottom of a ramp.
- Recognize that the speed at the bottom of a ramp is dependent on the initial height of release of the cart and independent of the angle of incline of the ramp.
- Complete a graph of speed versus height of the ramp.
- Define and calculate gravitational potential energy and kinetic energy.
- State the conservation of energy.
- Relate the conservation of energy to a roller coaster ride.

What Do You Think?

The steepest angle of descent on a wooden roller coaster is 70°. The steepest angle of descent on a steel roller coaster is 90°.

Two roller coaster slides are shown in the illustration below.

- **Which roller coaster will give the bigger thrill? Why?**

Record your ideas about these questions in your *Active Physics* log. Be prepared to discuss your response with your small group and the class.

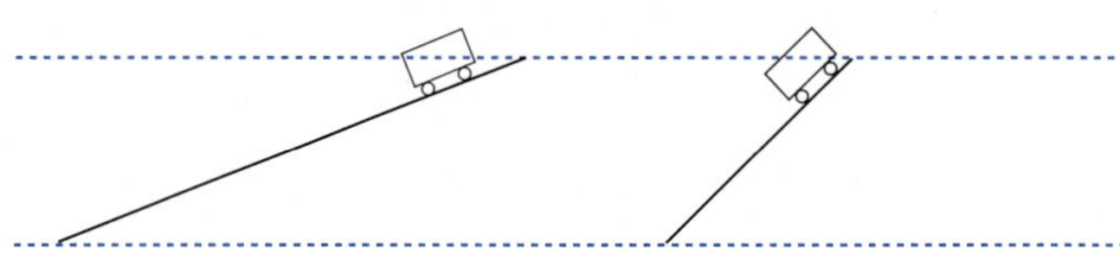

For You To Do

In this activity, you will investigate the speed at the bottom of a roller coaster slide. You will use a steel ball and a track to determine if a pattern exists between the placement of the ball and its speed at the bottom of the track. A pattern for speed will allow you to predict the speed for a new roller coaster.

1. The basic setup for this inquiry investigation is a track and a steel ball (or cart). You can measure distances to the nearest tenth of a centimeter with a ruler and speeds with a photogate timer.

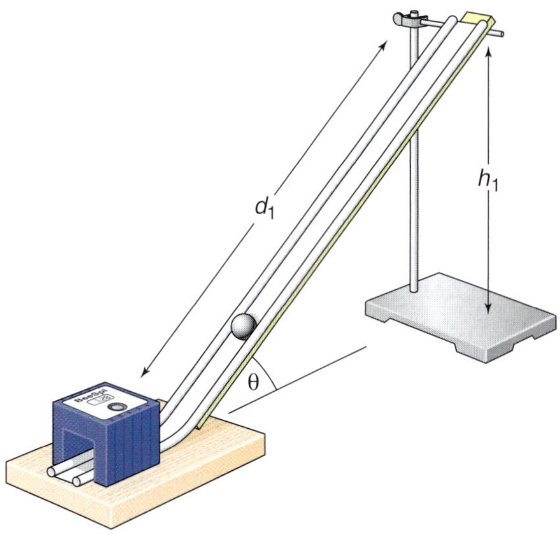

Be prepared to stop the ball at the end of the track so it does not roll onto the floor. If the ball rolls onto the floor pick it up right away. If you cannot find the ball, tell your teacher right away.

2. Your first step is to determine the speed of the steel ball at the bottom of the incline when the ball is placed at different points along the track. You should not vary the angle of the track. It will be useful to record the speed as a function of both the distance (d) and the height (h). (Although you can find the height if you know the distance, it's easier to make the extra measurement than to do the calculation using trigonometry.)

a) Complete the data table in your *Active Physics* log for at least four different positions.

Angle of track =

Height (m)	Distance (m)	Speed at bottom (m/s)

3. Change the angle of the track and repeat the investigation. First use the same heights and then use the same distances. This will require eight measurements of speed.

a) Create two data charts, height and speed, and distance and speed. Complete the measurements.

4. Review the data from the first and second tracks.

a) Is there a pattern between the heights and the speed or between the distances and the speed or both? Describe the pattern.

5. Change the angle of the track again.

a) Construct a data chart where you can predict the speeds from the distances OR heights that you have chosen.

b) Add a column to your chart so that it now has one column for predicted speeds and one for measured speeds. Complete the measurements for your chart.

c) How good were your predictions?

6. Complete the same investigation using a curved track. Measuring the distance along a curved track will require some ingenuity. Measuring the height is similar to the straight track.

a) Construct a data table that includes both predicted speeds and measured speeds.

b) Write a summary statement comparing the speed at the bottom of a curved track and the speed at the bottom of a straight track.

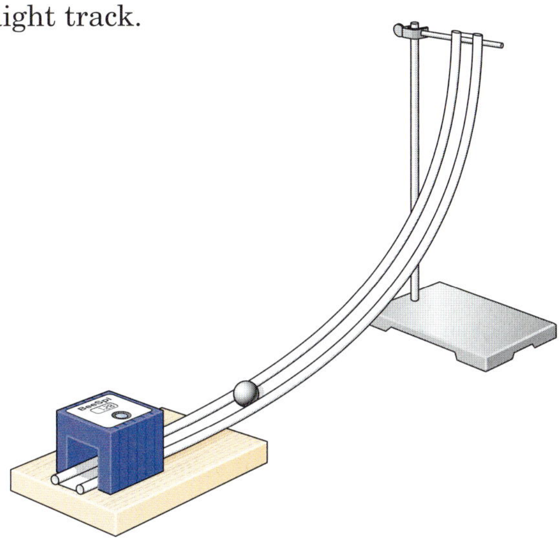

7. Complete a similar investigation using a pendulum. Measuring the distance the pendulum bob travels will require some ingenuity. Measuring the height is similar to the straight track.

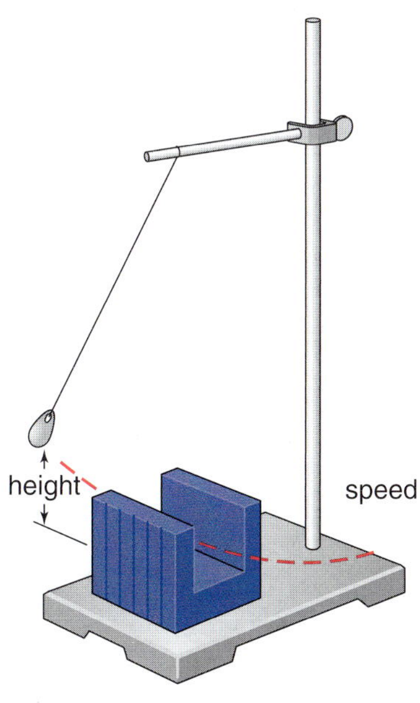

a) Construct a data table that includes both predicted speeds and measured speeds.

b) Write a summary statement comparing the speed at the bottom of the swing for a pendulum and the speed at the bottom of a straight track.

8. A valuable way in which to analyze data is with a graph. Take any data set (straight track, curved track or pendulum) and construct two graphs.

a) The first graph should have height on the x-axis and speed at the bottom on the y-axis.

b) The second graph should have height on the x-axis and the speed squared on the y-axis. This will require you to calculate v^2 for each speed.

9. Graphs with curves are difficult to interpret. It's hard to tell if the curve is part of a circle, ellipse, hyperbola, parabola, or none of these. Graphs with straight lines are much easier. The equation for all straight lines is

$$y = mx + b$$

where m is the slope of the graph and

b is the y-intercept.

In your straight-line graph, $(\text{speed})^2$ is the y-variable and height is the x-variable. You can now write an equation for the graph.

$$y = mx + b$$

Since the graph intersects the origin, the value of the y-intercept, b, is 0.

The equation for the graph becomes

$$y = mx + 0 \text{ or } y = mx.$$

Substituting for the variables in your graph:

$$(\text{speed})^2 = \text{slope (height) or}$$

$$v^2 = mh$$

a) Calculate the slope of your graph and record its value.

b) Compare the value of your slope with those of other groups in the class.

FOR YOU TO READ

Gravitational Potential Energy and Kinetic Energy

By varying the slope of the incline and measuring speeds, you were able to find that the speed at the bottom of a track is determined not by the length of the incline, but by the initial height of the incline. Two carts sliding down inclines will have the same final speed if they both start from the same height.

The carts shown in the diagram will have identical speeds at the bottom of the inclines. The second one will get there sooner, but will arrive with the same speed as the first one. (You are assuming that friction is so small that you can ignore it.)

This connection between the height and the speed was valid for different inclines, for curved tracks, and for a pendulum.

The concept of energy can be used to describe this relationship. In your activity, the steel ball or cart at the top of the incline has **gravitational potential energy** (GPE). (Gravitational potential energy is the energy an object has as a result of its position in a gravitational field.) The cart at the bottom of the hill has **kinetic energy** (KE). (Kinetic energy is the energy an object possesses because of its motion.) GPE is dependent on the height of the cart above the ground. The KE is dependent on the speed of the cart. A larger change in GPE is associated with a larger KE.

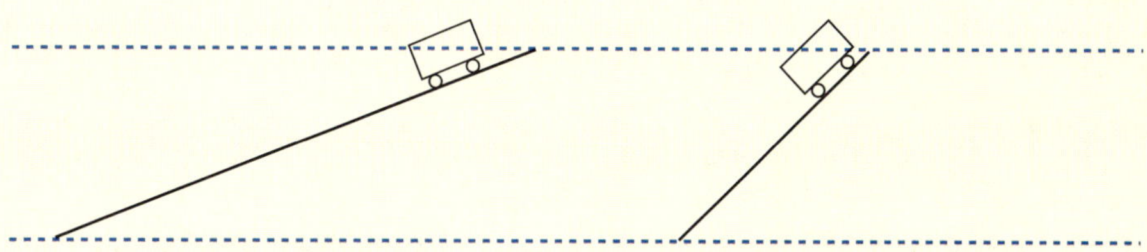

→

Physics Words

gravitational potential energy: the energy a body possesses as a result of its position in a gravitational field.

kinetic energy: the energy an object possesses because of its motion.

The equation for gravitational potential energy is

$$GPE = mgh$$

where m is the mass of the cart,

g is the acceleration due to gravity, and

h is the height above the ground.

You could just as easily have defined h as the height above the lab table. You will only concern yourself with the change in height. In this activity, the mass of the steel ball or cart and the acceleration due to gravity remained the same throughout the observations.

The equation for kinetic energy is

$$KE = \frac{1}{2} mv^2$$

where m is the mass of the cart, and

v is the velocity of the cart.

The unit for energy is a **joule** (symbol, J). Both GPE and KE are measured in joules. The chart below shows some calculations for a roller coaster car of mass 200 kg and an initial height of 20 m. Notice that at the top of the roller coaster there are lots of joules of GPE and zero joules of KE. At the bottom of the incline, there are zero joules of GPE and lots of joules of KE. At the two other positions listed, there are some joules of GPE and some joules of KE.

Without knowing the velocity, it would seem that you could not calculate the KE. However, the sum of the GPE and KE must always be 40,000 J. This is because that was the total energy at the beginning. When the roller coaster was at a height of 20 m, there was no movement and 0 J of KE. All of the energy at this point was the 40,000 J of GPE. This 40,000 J becomes very important for this roller coaster. The sum of GPE and KE must always be 40,000 J at any point on the roller coaster. (There is, of course, in real life, some loss of energy to the

Mass of car = 200 kg and g = 10 m/s^2 (approximate value)			
Position of car (height) (m)	**GPE (J) = mgh**	**KE (J) = 1/2 mv^2**	**GPE + KE (J)**
Top (20 m)	40,000	0	40,000
Bottom (0 m)	0	40,000	40,000
Halfway down (10 m)	20,000	20,000	40,000
Three-quarters way down (5 m)	10,000	30,000	40,000

environment due to friction that must be taken into consideration. However, we will neglect that for now.)

At the bottom of the roller coaster (see line 2 on the chart), there are 0 J of GPE. To total 40,000 J, there must be 40,000 J of kinetic energy KE at the bottom.

Halfway down (see line 3 on the chart), the KE must equal 20,000 J, so that the sum of the GPE (20,000 J) and the KE (20,000 J) once again equals 40,000 J.

Three-quarters of the way down (see line 4 on the chart), the KE must equal 30,000 J. The sum of the GPE (10,000 J) and the KE (30,000 J) once again equals 40,000 J.

Given any height, you can determine the GPE and then determine the KE. In this roller coaster, the GPE and KE must equal 40,000 J.

In a higher roller coaster, the GPE and KE might equal 60,000 J. You can still calculate the GPE at any height and then find the corresponding KE.

A word of caution is necessary when discussing the conservation of energy. The sum of the GPE and KE only remains the same if there are no losses of energy due to friction, sound, or other outside sources and no additions of energy from motors.

The conservation of energy provides a way to find the kinetic energy if you know the change in height of the roller coaster. However, it also allows you to find the speed of the roller coaster. Using algebra, you can calculate the speed.

$$\text{Energy (at the bottom)} = \text{Energy (at the top)}$$

$$\text{KE (bottom)} + \text{GPE (bottom)} = \text{KE (top)} + \text{GPE (top)}$$

$$\text{KE (bottom)} + 0 = 0 + \text{GPE (top)}$$

$$\text{KE (bottom)} = \text{GPE (top)}$$

$$\frac{1}{2}mv^2 = mgh$$

$$v^2 = 2gh$$

where g is the acceleration due to gravity.

In the above equation, the *m* cancelled out. This means that the speed is independent of the mass of the car. (It doesn't matter whether the roller coaster car has 2 or 4 passengers.)

Since $g = 9.8\ \text{m/s}^2$ then $v^2 = 2gh$. You can see why your graph of v^2 versus *h* was a straight line. The slope of the line should equal: $2 \times 9.8\ \text{m/s}^2 = 19.6\ \text{m/s}^2$.

Physics Words

joule: the SI unit for work and all other forms of energy; one joule (1 J) of work is done when a force of one newton moves an object one meter in the direction of the force.

PHYSICS TALK

Calculating Kinetic Energy from Gravitational Potential Energy

The gravitational potential energy is defined as GPE = mgh

The kinetic energy is defined as KE = $\frac{1}{2}mv^2$

In a system like your roller coaster, the sum of the GPE and KE is constant.

$$\text{GPE} + \text{KE} = \text{constant}$$

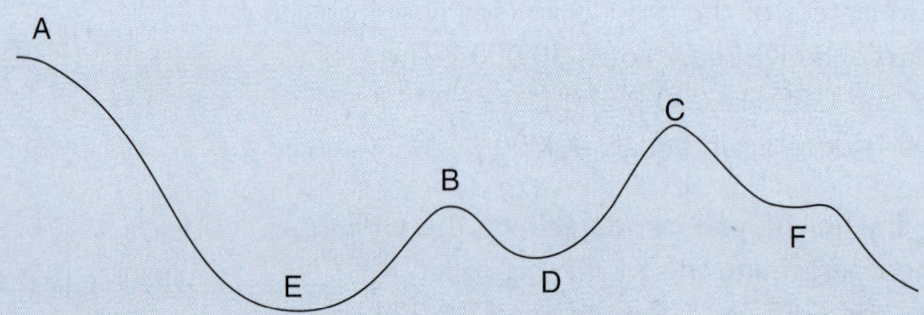

In the roller coaster above, if the GPE at point A is 30,000 J and the KE at point A is 0 J, then the total energy at point A and each and every other point on the roller coaster is 30,000 J. The total energy is also 30,000 J at points B, C, D, E, and F and every point in between. Since points B and F have the same height, the roller coaster car must also have the same GPE. That implies that the roller coaster car also has the same KE and is therefore going at the same speed at both points.

The height determines the GPE. The total energy at every point is the same. If you know the GPE, you can easily find the KE. The KE informs you about the speed.

Reflecting on the Activity and the Challenge

In designing a roller coaster, it is necessary to know how the speed of the roller coaster will vary. Knowing that the sum of the GPE and KE is constant is crucial in finding the speed at each point on the ride. How to calculate the GPE and KE will also be important when you want to ensure safety. You cannot let the roller coaster fall off the tracks, nor can you build a roller coaster that injures people. If your roller coaster begins on top of a tall hill, the height of that hill will determine the speed for each and every point along the roller coaster.

Physics To Go

1. For which track is the speed the greatest at the bottom? (Assume no friction.)

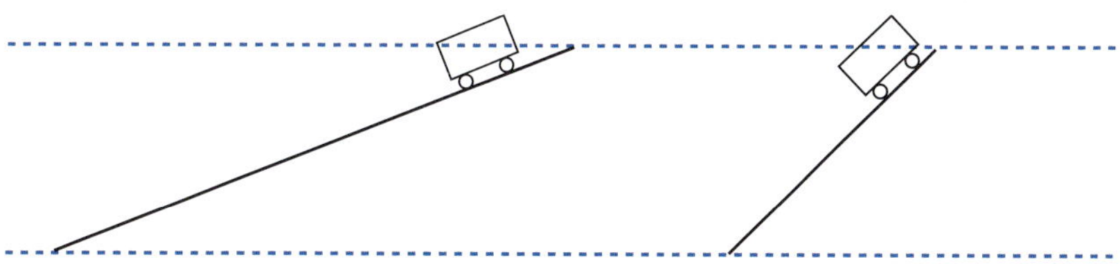

2. For which position is the speed the greatest at the bottom?

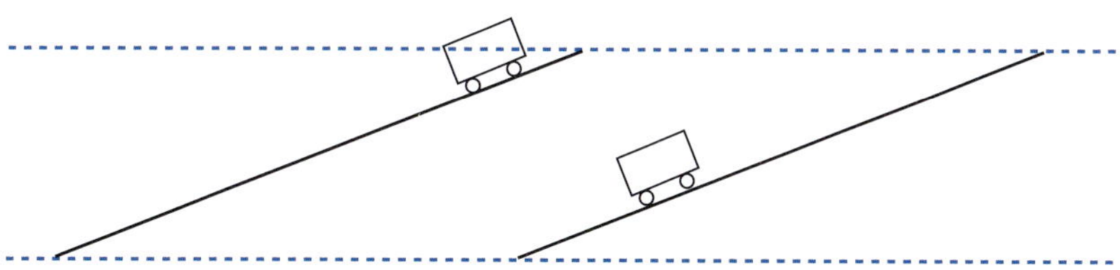

3. State the conservation of energy as it applies to roller coasters. Include in your statement GPE, KE, mgh, and $\frac{1}{2}mv^2$.

4. Complete the chart below for a roller coaster.

Mass of car = 200 kg and g = 10 m/s² (approximate)			
Position of car (height) (m)	**GPE (J) = *mgh***	**KE (J) = $\frac{1}{2}mv^2$**	**GPE +KE (J)**
Top (30 m)	60,000	0	
Bottom (0 m)			
Halfway down (15 m)			
Three-quarters way down (7.5 m)			

5. Complete the chart below for a roller coaster.

Mass of car = 300 kg and g = 10 m/s² (approximate)			
Position of car (height) (m)	**GPE (J) = *mgh***	**KE (J) = $\frac{1}{2}mv^2$**	**GPE +KE (J)**
Top (25 m)			
Bottom (0 m)			
Halfway down (10 m)			
Three-quarters way down (5 m)			

6. A pendulum is lifted to a height of 0.75 m. The mass of the bob is 0.2 kg.
   a) Calculate the GPE at the top.
   b) Find the KE at the bottom.
   c) At what position will the GPE and the KE be equal?

7. What do you think now? Two roller coaster slides are shown below. Which will give the biggest thrill? Why?

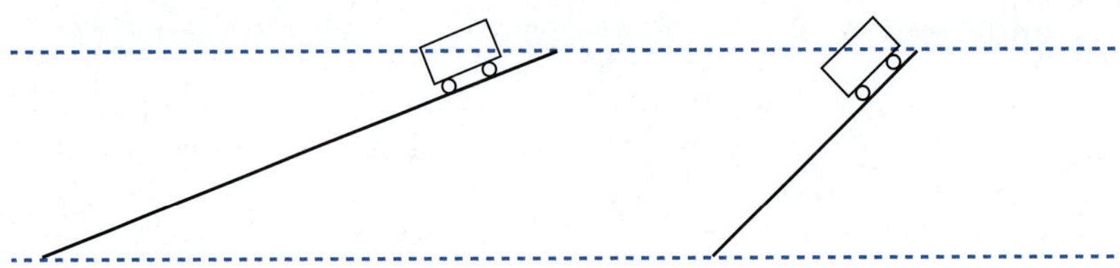

8. A roller coaster ride in the early morning only has 6 passengers. In the afternoon it has 26 passengers. Will the speed of the roller coaster change with more passengers aboard? Explain your answer.

9. Below is a side view of a roller coaster that starts from rest at position A.

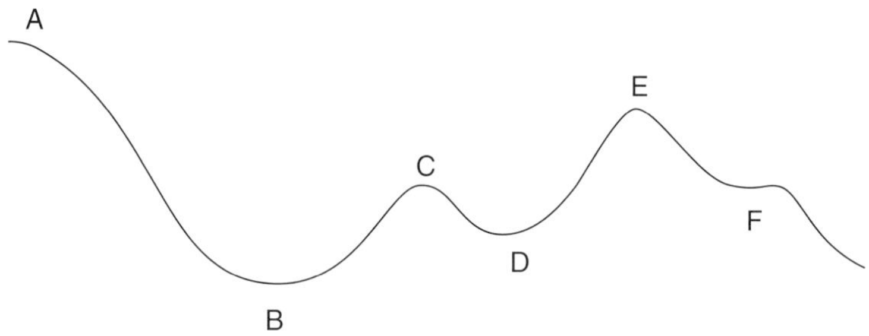

a) At which point is the roller coaster car traveling the fastest? Explain.
b) At which two points is the roller coaster car traveling at the same speed? Explain.
c) Is the roller coaster car traveling faster at E or D? Explain.

10. Below is a side view of a roller coaster that starts from rest at position A.

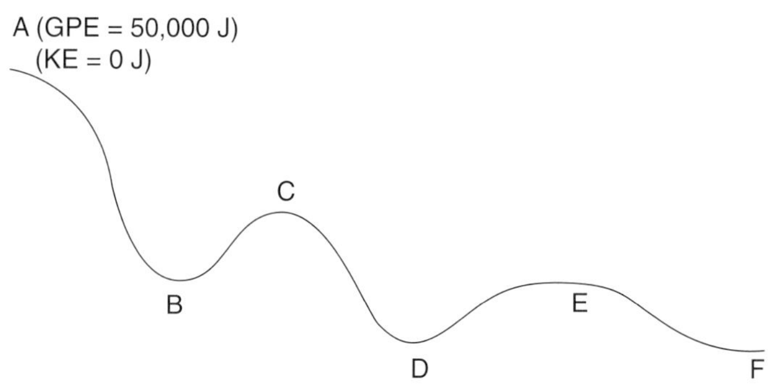

a) Determine plausible values for the GPE and KE at points B, C, D, E, and F.
b) At which two places is the roller coaster traveling at the same speed. Explain using GPE, KE, and speed in your explanation.

11. Complete this chart for a modified Terminator Express roller coaster. You may wish to try this on a spreadsheet.

Mass of car = 200 kg and g = 10 m/s^2 (approximate)					
Location	**Height (m)**	**Speed (m/s^2)** $v = \sqrt{2gh}$	**GPE (J)** mgh	**KE (J)** $\frac{1}{2}mv^2$	**Total energy (J) GPE + KE**
Top					
Bottom of hill					
Top of 1st hill					
Top of loop					
Horizontal loop					

Stretching Exercise

As a skateboarder practices on the vert, there are constant changes in the gravitational potential energy GPE and the kinetic energy KE. Research the size of the vert and report back on how the conservation of energy plays an integral part in this sport. You may also wish to make measurements of skateboarders in the vert.

Activity 3 More Energy

GOALS

In this activity you will:

- Measure the kinetic energy of a pop-up toy.
- Calculate the spring potential energy from the conservation of energy and using an equation.
- Recognize the general nature of the conservation of energy with heat, sound, chemical, and other forms of energy.

What Do You Think?

The concept of a "lift hill" for a roller coaster was developed in 1885. This was the initial hill that began a roller coaster ride. A chain or a cable often pulled up the train to the top of this hill.

- **How does the roller coaster today, get up to its highest point?**
- **Does it cost more to lift the roller coaster if it is full of people?**

Record your ideas about these questions in your *Active Physics* log. Be prepared to discuss your response with your small group and the class.

For You To Do

Eye protection must be worn during this activity. Have team members step back before the toy is released.

1. Everybody loves those little pop-up toys. You press the plunger, place it on a table, and "pop!" it flies into the air. In this inquiry investigation, you will determine the kinetic energy, KE, of the pop-up toy when it leaves the ground.

   Play with the toy to get a sense of how high it jumps.

   a) What is the approximate height of a jump?

   b) How consistent is the pop-up toy from one jump to the next?

2. Discuss among your group two distinct methods you can use to determine the KE when the pop-up toy leaves the ground. One method will require the photogate timer. The second method will require a meter stick to measure the height of the jump.

   a) Record your two methods in your *Active Physics* log. Since another team may want to understand what you have done, be quite careful to list all the steps. Indicate how all measurements are completed, and what is recorded or calculated.

3. Complete the investigation using both methods.

   a) Record your results. If during the experiment you changed your procedure, you should also record any changes here.

   b) Compare the KE determinations from the two methods.

Be sure the nickels are secure on the toy. Retape after every two to three trials.

4. Tape some nickels to the top of the pop-up toy in order to approximately double its mass. The mass of a nickel is approximately 5 g or 0.005 kg. (A dime is *not* 10 g; a quarter is *not* 25 g; and a penny is *not* 1 g.)

5. Repeat the investigation and find the KE of the pop-up toy as it leaves the ground.

   a) Why do you think that the heavier pop-up toy behaved differently? Use the terms GPE and KE in your explanation.

6. Answer the following questions in your *Active Physics* log:

   a) What is the toy's KE and GPE when it sits on the table?
   b) What happens to the toy's GPE and KE as it rises from the table?
   c) If the total energy of the toy is conserved, where does this KE and GPE come from as it rises?
   d) Where is the toy when its KE and when its GPE is greatest?

7. The pop-up toy had KE and GPE as it rose above the table. While it was sitting there, it also had spring potential energy, SPE. This SPE was converted to KE which then became GPE as the pop-up toy ascended. Using the concept of conservation of energy from the last activity, you notice that before popping up, there was all SPE. Just after popping up, there was all KE. When reaching the highest point, there was all GPE. The total energy at all other points was the same as the total SPE before popping, the total KE just after popping or the total GPE at its peak. You can show this in a chart. Notice that total energy is conserved. You now have spring potential energy SPE in addition to GPE and KE.

   a) Complete the chart in your log with other reasonable values for SPE, KE, GPE and the sum in the respective columns.

Position above table (m)	SPE (J)	KE (J)	GPE (J)	SPE + KE + GPE (J)
At rest on table: height = 0 m	20	0	0	20
Just after popping: height = 0 m	0	20	0	20
At peak: height = 0.30 m	0	0	20	20
1/2 the way up: height = 0.15 m			10	
With the spring only partially opened: height = 0 m				
Some other position: height = ? m				

FOR YOU TO READ

Conservation of Energy

Kaitlyn, Hannah, and Nicole share an apartment. Hannah keeps a bowl by the door filled with quarters that she can use for the washer and dryer at the laundromat. On Tuesday, Hannah counts her money and finds that she has 24 quarters or $6.00 in quarters. This is just the right amount for her laundry on Saturday. On Wednesday morning, Nicole comes rushing up to the apartment because she needs some quarters for the parking meter. She takes three quarters from the bowl and replaces them with six dimes and three nickels. The total money in the bowl is still $6.00. On Wednesday afternoon, Kaitlyn needs to buy a fifty-cent newspaper from the machine that takes all coins but pennies. Kaitlyn takes two quarters from Hannah's bowl and replaces these coins with fifty pennies. The total in the bowl is still $6.00.

Wednesday night, Hannah comes home and notices that her bowl is filled with quarters, pennies, nickels, and dimes. She knows that it still adds up to the $6.00 that was there in the morning, but also knows that she cannot do her laundry without all the money in quarters. Her roommates agree to exchange all the coins with quarters the next day.

The money in the bowl could represent the energy in a system. The total amount of energy may have been 600 J. As the coins in the bowl change from quarters to pennies to dimes and nickels and back to quarters, the energy in the system can vary from kinetic energy to gravitational potential energy to spring potential energy in any combinations.

If Kaitlyn had taken the two quarters and not replaced them with pennies, then the total money would be less. The loss in money due to Kaitlyn would have resulted in that money being somewhere else. In some systems, energy is lost as well. A bouncing ball does not get to the same height in each successive bounce. Some of the energy of the ball becomes sound energy and heat energy. These can be measured and will

show that the energy left the system but did not disappear. In the pop-up toy and the roller coaster, the total energy can be GPE, KE, SPE, but the sum of the energies must always be the same.

As you followed the changes in Hannah's bowl of money, you knew that there were ways to measure the total amount of money. Fifty pennies is identical in value to two quarters. Scientists look for all the energies in a system. There is electrical energy, light energy, nuclear energy, sound energy, heat energy, chemical energy, and others. Each one is able to be calculated using measurements. All the energies are measured in joules. The total number of joules must always remain the same.

In a real roller coaster, the roller coaster has all its energy as GPE (gravitational potential energy) as it sits on the highest hill. Most of this energy becomes KE (kinetic energy) as the roller coaster is released. Some small amount of the energy is converted to heat and a smaller part to sound.

Where does the roller coaster get all of that GPE that drives the rest of the ride? Something has to pull the roller coaster up to the top of the hill. The energy to pull the roller coaster is usually electric. The electrical energy comes from a power plant (which burns oil, gas, coal, or uses nuclear energy or water's potential energy as it comes crashing down) or from a local generator that may use gasoline.

After the cars are pulled to the top of the hill, the roller coaster is a closed system. Energy no longer enters the system. The total energy of the roller coaster remains the same except for losses due to heat and sound. In introductory physics, you can usually ignore heat and sound in the first analysis.

Physics Words

spring potential energy: the internal energy of a spring due to its compression or stretch.

PHYSICS TALK

Calculating Spring Potential Energy

In this activity you extended the conservation of energy principle to include the **spring potential energy**. It is possible to calculate the spring potential energy.

The equation for spring potential energy is

$$\text{SPE} = \frac{1}{2}kx^2$$

where k is the spring constant and

x is the amount of stretch or compression of the spring.

A spring that is difficult to compress or stretch will have a large spring constant k. That spring will "pack" more SPE for an identical compression than a spring that is easy to compress.

The total energy of a spring toy that can jump into the air is the sum of the SPE, the GPE, and the KE. Once the spring is compressed, you can consider it a closed system and the sum of these three energies, GPE, KE, and SPE must remain constant.

$$\text{GPE} + \text{KE} + \text{SPE} = \text{constant}$$

$$mgh + \frac{1}{2}mv^2 + \frac{1}{2}kx^2 = \text{constant.}$$

Reflecting on the Activity and the Challenge

There are other energies—heat, sound, chemical, etc. In your analysis of the roller coaster, you may decide to ignore heat and sound, but you had better mention this in your report. In the actual construction, it will be important to take into account that a small amount of energy is being dissipated (lost).

The roller coaster uses electrical energy to get the cars to the top of the hill. This is similar to using the chemical energy of your body to compress the pop-up toy so that you can watch it jump.

Once the energy is in the spring of the pop-up toy, you can observe it as a closed system. The SPE becomes KE, which becomes GPE. In the same way, once the cars are on top of the hill, the GPE can become KE.

Physics To Go

1. Complete the chart with other reasonable values for SPE, KE, GPE, and the sum in the respective columns.

Position above table (m)	SPE (J)	KE (J)	GPE (J)	SPE + KE + GPE (J)
At rest on table: height = 0 m	25			
Just after popping: height = 0 m				
At peak: height = 0.60 m				
1/2 the way up: height = 0.30 m				
With the spring only partially opened: height = 0 m				
Some other position: height = ? m				

2. How would the chart values in **Question 1** change if some extra mass were attached to the pop-up toy?

3. You throw a ball into the air and catch it on the way down. Beginning with the chemical energy in your muscles, describe the energy transformations of the ball.

4. Why can the second hill of the roller coaster not be larger than the first hill?

5. Why does the roller coaster not continue forever and go up and down the hills over and over again?

6. A roller coaster of mass 300 kg ascends to a height of 15 m. How much electrical energy was required to raise the cars to this height?

7. A roller coaster has a mass of 400 kg and a speed of 15 m/s.
   a) What is the KE of the roller coaster?
   b) What will be the GPE of this roller coaster at its highest point?
   c) How high can the roller coaster go with this much energy?

8. A ball is thrown upward from Earth's surface. While the ball is rising, is its gravitational potential energy increasing, decreasing, or remaining the same?

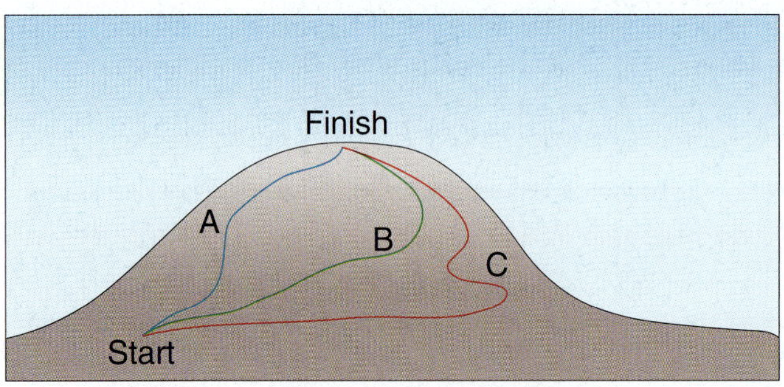

9. Three people of equal mass climb a mountain using paths A, B, and C shown in the diagram.

Along which path(s) does a person gain the greatest amount of gravitational potential energy from start to finish: A only, B only, C only, or is the gain the same along all paths?

Activity 4 Your "At Rest" Weight

GOALS

In this activity you will:

- Distinguish between mass and weight.
- Calculate weight in newtons.
- Measure the effect of weight on the stretch of a spring.
- Graph the relationship between weight and stretch of a spring.
- Use a spring to create a scale and explain how Newton's Second Law is used in the creation of the scale.
- Calculate spring forces using Hooke's Law.

What Do You Think?

A canary and an elephant have enormous differences in weight. The elephant may weigh more than 10,000 times as much as the canary.

- **Can you use the same scale to weigh a canary and an elephant?**
- **How does a bathroom scale work?**

Record your ideas about these questions in your *Active Physics* log. Be prepared to discuss your response with your small group and the class.

For You To Do

Part A: Mass and Weight

1. There is a ride at the amusement park today in which all you do is drop straight down. If you were to record your motion, you would find that your speed increases by 9.8 m/s every second. This value of 9.8 m/s every second is the acceleration due to gravitation. All objects near the surface of the Earth fall at this same rate of change of velocity with respect to time.

   You have Galileo to thank for this insight. As the story goes, he dropped two objects from the leaning tower of Pisa and observed them hitting the ground at the same time. The story may not be true, but Galileo did perform many experiments with balls rolling down inclined planes. The "dropping experiment" has been repeated many, many times with very precise equipment and with the effects of air resistance minimized or eliminated.

   a) If you were at the leaning tower of Pisa and dropped a baseball and a bowling ball at the same time, which would hit the ground first? Explain your answer.

b) If you dropped a baseball and a piece of paper, which would hit the ground first? Explain your answer.

c) How would you modify the statement, "All objects fall at the same acceleration" to account for your observation of a baseball and a piece of paper?

2. Produce at a supermarket is often priced by weight. Apples may cost 79 cents per pound and watermelon may cost 22 cents per pound.

a) What is a "pound"?

b) How would you define weight?

3. In physics, **weight** is defined as the force of the Earth's gravity on an object. The weight is the mass of an object multiplied by the acceleration due to gravity. Large masses are heavy and small masses are light. In the metric system, the mass of an object is measured in kilograms. The acceleration due to gravity is measured as 9.8 m/s every second or 9.8 m/s^2. Using this, you can calculate the weight of a student with a mass of 50 kg.

Physics Words

weight: the vertical, downward force exerted on a mass as a result of gravity.

$$\text{weight} = (\text{mass}) \times (\text{acceleration due to gravity})$$

$$w = ma_g$$

$$= (50 \text{ kg})(9.8 \text{ m/s}^2)$$

$$= 490 \text{ kg}\cdot\text{m/s}^2$$

$$= 490 \text{ N (newtons)}$$

The unit for weight is a newton (N). Weight is a force and has the same units as force. Recall that force is numerically equal to mass times acceleration ($F = ma$). Weight is the force due to gravity where the acceleration is the acceleration due to gravity. This can be written using symbols as a_g or g. $F = ma_g$ or $F = mg$.

a) Compare the weights of a gymnast with a mass of 40 kg and a football player with a mass of 110 kg.

4. The newton is a good metric unit for weight. However, you may wish to compare this to a pound, with which you are more familiar. Each kilogram weighs 2.2 lb. A 220-lb. football player has a mass of 100 kg. The weight of 100 kg, according to the equation $w = mg$, is 980 N.

a) Find the weight in newtons of a bowling ball that weighs 11 lb.

b) Find the weight in newtons of a $\frac{1}{4}$-lb burger (a meat patty that has a weight of $\frac{1}{4}$-lb).

5. The weight of the $\frac{1}{4}$-lb burger is close to 1 N. In a country that uses metric measurements, a burger restaurant could call their $\frac{1}{4}$-lb burger a "Newton Burger." You can use this as an approximate way to determine how much something weighs in newtons if you know the weight in pounds. A 50-lb. person has the equivalent weight of 200 $\frac{1}{4}$-lb burgers. Therefore, the 50-lb. person has an approximate weight of 200 N.

a) Find the approximate weight in newtons of a roller coaster car that weighs 1500 lb.

Part B: The Properties of Springs

1. You can use a set of masses to determine the properties of springs. Make a spring from a piece of paper. Take a piece of $8\frac{1}{2} \times 11$ paper and fold it like an accordion.

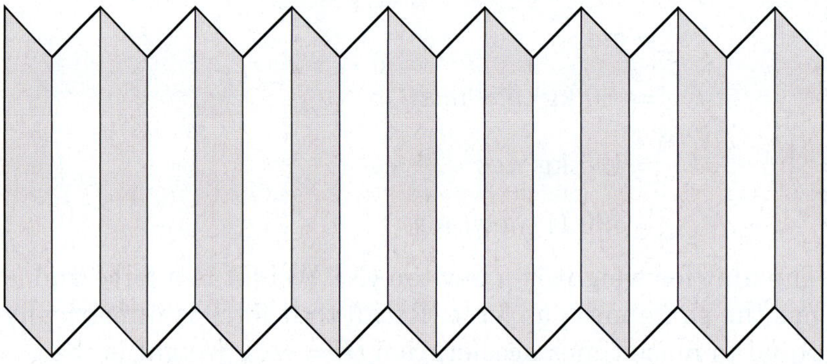

2. With a small force, stretch the paper spring slightly.

a) Record what happens when you release it.

3. With a small force, compress the paper spring slightly.

a) Record what happens when you release it.

4. Does the paper spring return to its original size and shape as the force of the stretch increases? Try it.

   a) Record your observations.

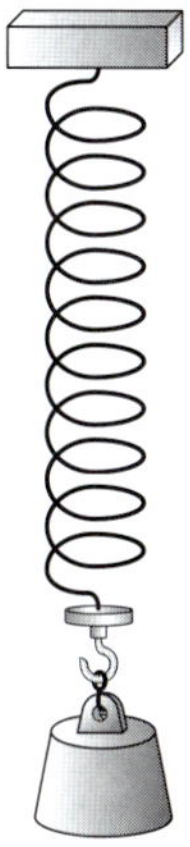

5. A metal spring has the same properties as the paper spring. The metal spring is better able to restore itself to its original shape than the paper spring. However, the metal spring can also be stretched past its load limit and you should be careful not to do this.

6. The stretch of a metal spring can be measured precisely. Secure the spring vertically. You are now ready to measure its stretch with a given set of weights.

   You will have to convert from grams to kilograms to newtons. If a 100-g mass is used, this is equivalent to 0.1 kg. The mass of 0.1 kg has a weight of 0.98 N. This can be written as a single equation:

$$
\begin{aligned}
w &= mg \\
&= (100\ \text{g})\left(\frac{1\ \text{kg}}{1000\ \text{g}}\right)(9.8\ \text{m/s}^2) \\
&= 0.98\ \text{kg}\cdot\text{m/s}^2 \\
&= 0.98\ \text{N}
\end{aligned}
$$

   Notice that you converted grams to kilograms by multiplying by 1 kg/1000 g. In math class, you learned that you could always multiply a number by 1 and not change its value. For instance, $27 \times 1 = 27$. Since 1 kg is equal to 1000 g, the fraction 1 kg/1000 g has an equivalent numerator and denominator and the fraction equals 1. When you use this fraction in the equation, the 100 g gets converted to 0.1 kg. The gram units "drop out" and you are left with kilogram units. This dimensional analysis was done because you want the mass in kilograms.

Be careful when placing and removing masses that the spring does not snap anyone. Have one person hold the bottom and top of the spring as another person adds or removes each mass.

a) In your *Active Physics* log, create a data table with four columns. Label the first three columns mass, weight, and stretch of spring. The fourth column will be left blank for the time being.

7. Measure the stretch of the spring for different weights.

a) Record your measurements in the data table.

b) Plot a graph with the weights on the x-axis and the stretch of the spring on the y-axis.

c) From your graph, predict what the stretch would be for a weight that you have not tried but between the weights that you have measured. This type of prediction from a graph is called interpolation.

8. Test your prediction by measuring the stretch of the spring for that weight.

a) How good was your prediction?

9. Repeat the investigation for a second spring that looks different than the first. The spring may have larger or smaller coils or the coils may be closer together or further apart. You should have a new data table, a new graph, and a new interpolation.

a) Describe how the springs differ in physical appearance.

b) Describe how the springs differ in terms of the data chart and corresponding graph.

10. Invent a graph for a third spring.

a) Sketch the invented spring's graph.

b) Write a description of a spring that would have such data. The description should include the ease or difficulty of stretching the spring.

11. Return to your first data chart. Divide the weight by the stretched distance for each measurement.

a) Record these values in the fourth column.

b) What do you notice about these calculated values?

c) Repeat for the calculations for the second data chart.

d) What value might your invented spring have for this column?

Part C: The Spring as a Weighing Machine

1. The spring stretches a different amount for each weight. Create a scale for weighing objects using one of the two springs that you have previously used. A scale has a spring and an arrow that points to the weight of the hanging object.

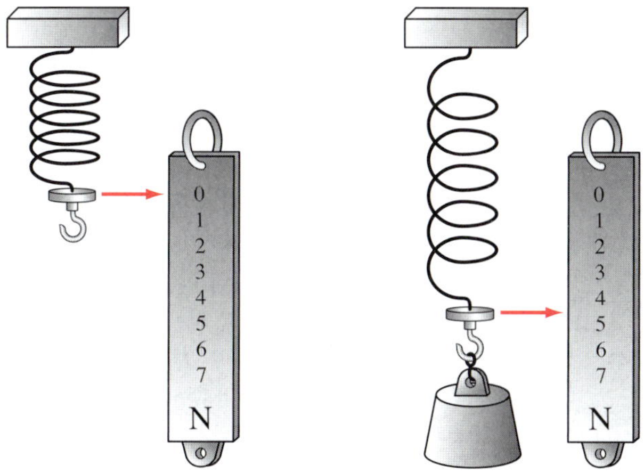

2. Choose three known masses. Measure their weight on your scale.

a) Record your values.

3. Choose two objects of unknown weight. Measure their weight on your scale.

a) Record the object and its weight.

FOR YOU TO READ

Hooke's Law Describes the Restoring Force a Spring Exerts

Many springs have the property that the stretch of the spring is directly proportional to the force applied to it. This means that if you double the force, the stretch of the spring doubles. If you triple the force, the stretch of the spring triples. And if you make the force 2.7 times larger, the stretch of the spring is 2.7 times as large.

Springs that behave in this way are said to obey Hooke's Law. Sir Robert Hooke discovered this property of springs. The law explains very simply and precisely what restoring force a spring exerts if it is stretched. The more you stretch a spring, the larger the restoring force by the spring. You can describe this relationship in words or with a graph or with a mathematical equation. The equation for Hooke's Law is:

$$F = -kx$$

where F is the force,

x is the displacement, and

k is the spring constant.

The spring constant k is an indication of how easy or difficult it is to stretch or compress a spring.

The negative sign in the equation indicates that the pull by the spring is opposite to the direction it is stretched or compressed. Stretch a spring down and it tries to pull up. Stretch a spring to the right and it tries to pull to the left. Compress a spring and it tries to push.

You can calculate the value of the spring constant k by measuring the force exerted by the spring and the stretch of the spring.

Sample Problem 1

A 3.0-N weight is suspended from a spring. The spring stretches 2.0 cm. Calculate the spring constant.

Strategy: If a 3.0-N weight is suspended at rest from the spring, the spring must be applying a force of 3.0 N. If the spring were applying a force of less than 3.0 N, the weight would accelerate down. If the spring were applying a force of more than 3.0 N, the weight would accelerate up. When the force of gravity on the mass is 3.0 N down and the spring exerts a force of 3.0 N up, then the mass has no net force on it and it remains at rest once the friction brings it to rest. In this activity, you always took your measurements when the mass had stopped moving.

Givens:

$F = 3.0\text{ N}$

$x = 2.0\text{ cm}$

Solution:

$$F = kx$$

$$k = \frac{F}{x}$$

$$= \frac{3.0\text{ N}}{2.0\text{ cm}}$$

$$= 1.5\text{ N/cm}$$

With a set of data points for the weight and the stretch, you would find that all values of k are the same or constant. And so k is the spring constant.

As you determined in the activity, you can also record the data on a graph. The graph can also be used to determine the spring constant k.

If the data are graphed so that the force is on the y-axis and the stretch is on the x-axis, then the spring constant will be the slope of the graph. You can see this if you compare the equations for Hooke's Law and the equation for a straight line (when the y-intercept is zero).

From here on, the negative sign in the equation will be omitted. In mathematics, you would say that you are calculating the absolute value of F using the absolute value of the stretch x. You must always keep in mind that the direction of the force is opposite to the direction of the stretch.

Hooke's Law: $F = kx$

Straight line: $y = mx + b$ (where $b = 0$)

Since the force is on the y-axis and the stretch x is on the x-axis, the slope m of the straight-line graph is k.

If the graph is constructed so that the force is on the x-axis and the stretch is on the y-axis, then the slope will be the reciprocal of the spring constant $1/k$. You can see this by comparing the equations for Hooke's Law and the equation for a straight line. In this case, we will write the equation for the straight line as $Y = MX + B$ so that the X-axis won't be confused with the stretch x, which is being placed on the Y-axis.

Hooke's Law: $F = kx$

$$x = \left(\frac{1}{k}\right)F$$

Straight line: $Y = MX$

Since the force is on the x-axis and the stretch x is on the y-axis, the slope m of the straight-line graph is $1/k$.

→

Sample Problem 2

Weights are hung from a spring and the stretch is measured. The data collected is shown in the graph below. Calculate the spring constant from the graph.

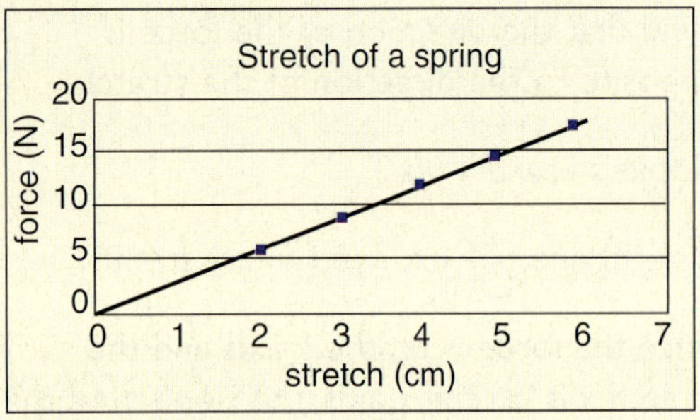

Strategy: Since the force is on the *y*-axis and the stretch is on the *x*-axis, you can compare the equations for a straight line and Hooke's Law.

Hooke's Law: $F = kx$

Straight line: $y = mx$

The slope of the graph will be equal to the spring constant *k*.

Givens:

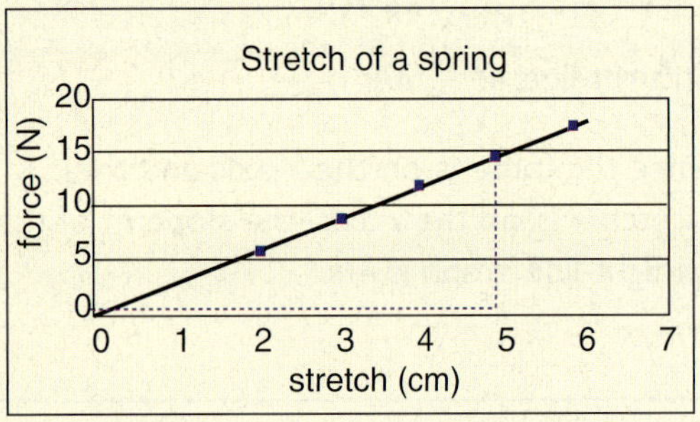

Solution:

$$\text{Slope} = \frac{\text{rise}}{\text{run}}$$

$$= \frac{15\text{N}}{5.0\text{ cm}}$$

$$= 3.0\text{ N/cm}$$

Stretch and Compress

You began the activity by both compressing and stretching a paper spring. You then made measurements on a stretched spring. Conducting an investigation with a compressed spring would produce similar results.

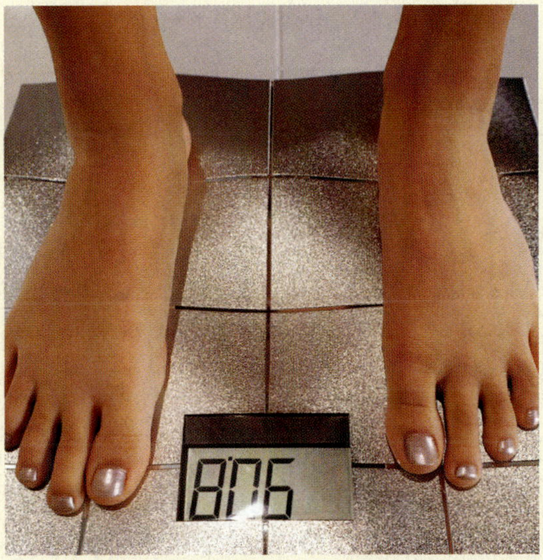

Many bathroom scales work by compressing the spring. Inside the bathroom scale is a spring. When you step on the scale, the spring compresses just enough to provide a force equal to your weight. The more weight, the more compression of the spring is required. The top of the spring is connected to a scale that has been calibrated.

As the spring gets compressed, the arrow points to a different number corresponding to the compression and force of the spring.

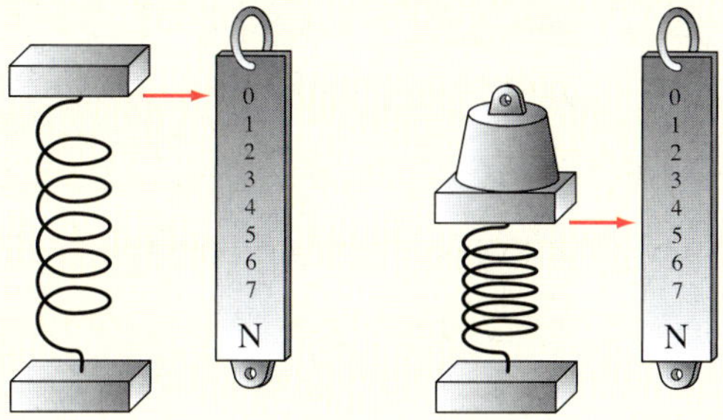

The scale does not read the weight of the object. The scale reads the compression of the spring. Of course, under normal circumstances the compression of the spring provides a force equal to your weight. You can then say that the scale reads your weight.

PHYSICS TALK

Hooke's Law

The restoring force of a spring is proportional to the stretch of a spring. The larger the stretch of a spring, the larger the force exerted by the spring.

The mathematical description of the force of a spring is called **Hooke's Law**. The equation is:

$$F = -kx$$

where F is the force,

x is the displacement, and

k is the spring constant.

The spring constant k is an indication of how easy or difficult it is to stretch or compress a spring. The negative sign in the equation indicates that the pull by the spring is opposite to the direction it is stretched or compressed. The spring has a tendency to return to its original shape.

→

Physics Words

Hooke's Law: the distance of stretch or compression of a spring is directly proportional to the force applied to it.

A person steps on a scale. The scale moves a bit and then comes to rest. There are two forces acting on the person at rest. There is the force of gravity pulling down on the person. There is also the force of the spring pushing up on the person. These two forces are equal. The net force on the person is zero and the person remains at rest.

The sum of the forces = ma (Newton's Second Law)

The sum of the forces = 0 (The acceleration is zero)

Force due to gravity + force due to the spring = 0

$$mg + (-kx) = 0$$

Reflecting on the Activity and the Challenge

Part of the fun of the roller coaster is the sensation you get as your weight appears to change at the peaks and valleys of the ride. In order to understand these apparent weight changes, you must understand how a scale works. All spring scales are based on Hooke's Law. The stretch (or compression of a spring) is directly proportional to the weight pulling or pushing on the spring. The more pull, the more is the stretch. Mathematically, you can write Hooke's Law as $F = kx$ where F is the weight (Force of Gravity) and x is the stretch. (The minus sign has been left out, but this requires you to remember that the force is in the opposite direction from the stretch.) The spring constant k is constant for a specific spring. The scale measures the compression of the spring. This force, when everything is at rest is equal to your weight. When the roller coaster starts moving, the scale will read many different values. You may want to include the weight changes in your design for your roller coaster. To do this, you may want to describe how a spring scale works.

Physics To Go

1. Objects near the surface of the Earth accelerate at a rate of 9.8 m/s every second. Based on this information, how fast will an object be going after it has fallen each of the following lengths of time?

   a) 2 s
   b) 5 s
   c) 10 s

2. Objects near the surface of the Moon accelerate at a rate of 1.6 m/s every second. Based on this information, how fast will an object be going after it has fallen each of the following lengths of time?

   a) 2 s
   b) 5 s
   c) 10 s

3. Calculate the weight of the following objects:

   a) a football player with a mass of 100 kg
   b) a toddler with a mass of 10 kg
   c) an adult with a mass of 60 kg

4. Use the approximation that a $\frac{1}{4}$-lb burger is one newton. Write down the approximate weights in newtons of the following objects:

   a) a 130-lb student
   b) a 1000-lb roller coaster cart
   c) a 50-lb child

5. Weights were hung from a spring and the stretch of the spring was measured. The data chart is given to the right.

   a) Graph the data with the weight on the x-axis and the stretch on the y-axis.
   b) Draw the best-fit line through the data points. (Do not connect the dots; draw the best-fit line.)
   Find the slope of the graph.
   What is the meaning of the slope?

Weight (n)	Stretch (cm)
0.0	0.0
0.3	2.0
0.7	4.6
1.2	8.0
2.0	13.0
2.4	16.2
3.1	21.0

c) Invent a graph for a second spring. Sketch the invented spring's graph. Write a description of a spring that would have such data. The description should include the ease of difficulty of stretching the spring.

6. When Robert Hooke first described the relationship that has come to be known as Hooke's Law, he wrote "as the force, so the stretch." Explain in a full sentence or two what Hooke meant by this. (Hooke wrote this as a footnote in Latin with the letters all mixed up. This allowed him to keep his discovery a secret for a while.)

7. A weight of 12 N causes a spring to stretch 3.0 cm. What is the spring constant k of the spring?

8. Two springs have spring constants of 10.0 N/cm and 15.0 N/cm. Which spring is more difficult to stretch?

9. Calculate the spring constant k from the graph of a stretched spring below.

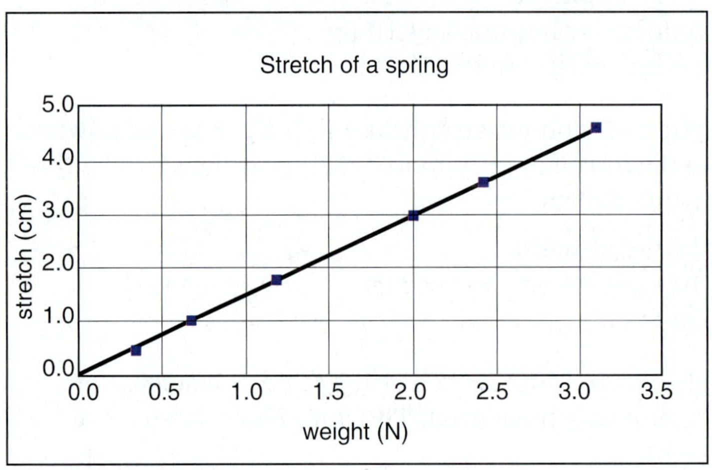

10. Describe how a bathroom scale works.

Stretching Exercise

Get permission to take apart a bathroom scale. Investigate the parts. Create sketches to explain how the scale works and the function of all of the parts. When your explanation is complete, put the scale back together.

Activity 5 Weight on a Roller Coaster

GOALS

In this activity you will:

- Recognize that the weight of an object remains the same when the object is at rest or moving (up or down) at a constant speed.
- Explore the change in apparent weight as an object accelerates up or down.
- Analyze the forces on a mass at rest, moving with constant velocity, or accelerating by drawing the appropriate force vector diagrams.
- Mathematically predict the change in apparent weight as a mass accelerates up or down.

What Do You Think?

As the roller coaster moves down that first hill, up the second hill, and then over the top, you feel as if your weight is changing. In roller coaster terms this is called airtime. It is the feeling of floating when your body rises up out of the seat.

- **Does your weight change when you are riding on a roller coaster?**
- **If you were sitting on a bathroom scale, would the scale give different readings at different places on the roller coaster?**

Record your ideas about these questions in your *Active Physics* log. Be prepared to discuss your response with your small group and the class.

For You To Do

In this activity, you will investigate the weight changes you feel when you are on a roller coaster. You will use the spring scale for your observations. However, you will explain what you observe with both the spring scale and the bathroom scale.

Part A: Moving the Mass at a Constant Speed

1. Hang a mass from the spring scale and note the force. When the mass is not moving the acceleration equals 0 and therefore the force of gravity pulling the mass down and the force exerted by the spring pulling the mass up must be equal. The force of the spring equals the force of gravity. You could also just say, "the spring is measuring the weight."

   a) Record the weight of the mass in your log.

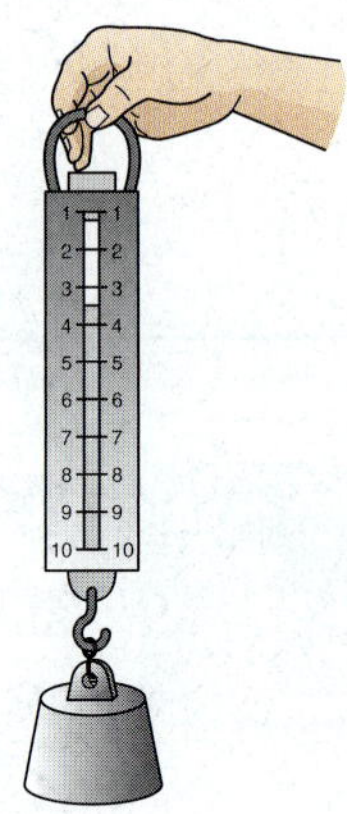

2. With your arm extended down, move the mass up until your arm is as high as you can reach. Once you start the mass moving, you want to keep lifting it at a constant speed. Your lab partners will try their best to read the spring scale during the time that the mass is moving at constant speed. Ignore the readings when you first start moving the mass and when you stop it. You will return to those observations later.

You may have to repeat this a few times so that you can lift the mass at a constant speed so that your lab partners can observe it.

a) Record your finding in your log.

3. The observations in **Step 2** may have been difficult for you to make accurately. The spring scale should have displayed the same reading when the scale moved at constant speed as it did when it was suspended at rest. The same result will occur in an elevator. If you are on a bathroom scale, the scale will compress and display your weight when the elevator is at rest. It will display the same weight when the elevator is moving at a constant speed between floors.

   Repeat the observation with the spring scale moving down at a constant speed. Is the weight once again the same? Please remember, you are only interested in the weight reading as the mass descends at a constant speed, not when you first get it to move.

   a) Record your findings.

4. **Newton's First Law** states that "An object at rest remains at rest and an object in motion remains in motion, unless acted upon by an unbalanced force." **Newton's Second Law** states that "An accelerating object must have a net force acting upon it: $F = ma$." When the mass is being lifted at constant speed, there is no acceleration since acceleration is defined as a change in speed with respect to time. If there is no acceleration, then there is no net force.

   a) Draw a box in your log and draw arrows to show the forces on the box when it is not accelerating.

5. In physics, the arrows you drew are called vectors. Check your drawing to see if the arrows (vectors) have these features:

Physics Words

Newton's First Law of Motion: an object at rest stays at rest and an object in motion stays in motion unless acted upon by an unbalanced, external force.

Newton's Second Law of Motion: if a body is acted upon by an external force, it will accelerate in the direction of the unbalanced force with an acceleration proportional to the force and inversely proportional to the mass.

a) Was the weight vector drawn down? Why did you draw it this way?

b) Was the force of the spring vector drawn up? Why did you draw it this way?

c) Were there any other force vectors? What were they representing?

d) Were the weight vector and the spring vector equal in length? The size of the vectors implies the size of the force. If the forces are equal, then the lengths of the vectors should be the same. If needed, change your force diagram so that the vectors represent the size of the forces.

6. Now consider the box if it were moving down.

a) Draw a second box with the force vectors when the box is moving down at a constant speed. Provide an explanation using Newton's First Law and Newton's Second Law (similar to **Step 4**) as a rationale for your diagram.

Part B: Accelerating the Mass

1. It's now time to return to the scale readings when you first start moving the mass. With your arm extended down, accelerate the mass up until your arm is as high as you can reach. Your lab partners will try their best to read the spring scale during the time that the mass starts to move. Once again, you may have to repeat this a few times so that you can lift as others observe.

a) Record your observation in your *Active Physics* log.

b) Use Newton's Second Law ($F = ma$) to make sense of the observation in your log.

c) Draw a box representing the mass and draw the force vectors acting on the box as it first begins to move.

2. Check your drawing to see if the force vectors have these features:

a) Was the weight vector drawn down? Why did you draw it this way?

b) Was the force of the spring vector drawn up? Why did you draw it this way?

c) Were there any other force vectors? What were they representing?

d) Were the weight vector and the spring vector equal in length? The size of the vectors implies the size of the force.

3. Since the box is accelerating up, the force of the spring must have been larger than the force of gravity. Newton's Second Law indicates that acceleration up requires a net force up. In your force-vector diagram, the vector representing the spring scale should be larger than the force of the gravity vector.

a) If required, modify your diagram.

4. The spring scale displays a value larger than the mass's weight on Earth. Suppose you were standing on a bathroom scale in an elevator. The elevator begins to move up.

a) How would the reading on the bathroom scale compare to your weight at rest?

5. Predict what would happen to the scale reading when the mass stops moving upward.

a) Record your prediction in your log.

b) Repeat the observations for the moments when the mass stops its upward motion. Describe your observation in your log.

6. Begin with the mass high in the air and suspended by the spring scale.

a) If you begin to lower the mass, in which direction is the acceleration?

b) Draw a force-vector diagram that has a net force in the direction of the acceleration.

c) Do you predict that the scale will read a value higher or lower than it does when the mass is at rest? Record your prediction.

d) Try it out. Record your observations in your log.

7. As a summary of what changes occur to the spring scale reading (or the bathroom scale reading in an elevator), complete the following chart in your log. Some responses are provided.

	Acceleration (up, down, zero)	Scale reading (larger, smaller, equal to weight)
1. Elevator at rest on bottom floor	Zero	Equal
2. Elevator starts moving up	Up	Larger
3. Elevator moves up at constant speed		
4. Elevator comes to rest on top floor		
5. Elevator is at rest at top floor		
6. Elevator begins to move down	Down	Smaller
7. Elevator moves down at constant speed		
8. Elevator comes to rest on bottom floor		
9. Elevator at rest on bottom floor		

8. Riding in an elevator is similar to riding in a roller coaster. Although the physics is the same, the elevator ride does not have the excitement of a roller coaster.

a) Compare elevator rides and roller coasters by providing three similarities and three differences.

FOR YOU TO READ

Forces Acting during Acceleration

As the roller coaster moves you about, you feel in your stomach and from the car seat that things are happening. These, however, are more than just feelings. These changes can be measured. You can use physics to explain these changes.

If you were sitting on a scale on a level roller coaster, at rest or moving with a constant velocity, the scale reading would be equal to your weight. The force of the Earth pulling on you (your weight), shown as a blue vector in the diagram, would be equal to the force of the compressed spring within the bathroom scale, shown as a red vector. A force diagram shows this.

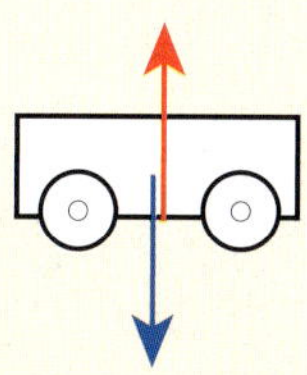

If the person on the roller coaster weighs 600 N, then the bathroom scale would have to be providing a force of 600 N. Any smaller and the person would accelerate down. Any larger and the person would accelerate up. When you first sit on the scale, the compression is too little and you do move down. The spring compresses and provides a larger force but you continue to move down. You go past the compression you need and the spring then pushes up. You go back up and down and up and down and continue this movement until the spring's force is exactly equal to your weight.

As the elevator or roller coaster starts moving up, there is acceleration up. (Remember that acceleration is a change in velocity with respect to time.) For you to accelerate up, there must be an unbalanced force pushing you up. Since you are in contact with the bathroom scale then it must be the bathroom scale that is pushing you up. (Yes, the elevator or roller coaster is pushing on the scale, but in physics you only have to worry about the forces on you for this situation.)

The scale reading will be greater than your weight. The force of the Earth pulling on you (your weight), shown as a blue vector, would be less than the force of the compressed spring within the bathroom scale, shown as a red vector. A force diagram shows this.

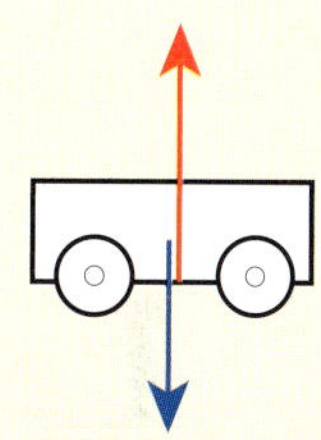

If the person on the roller coaster weighs 600 N, then the bathroom scale would have to be providing a force of greater than 600 N.

Another way of looking at the situation is to look first at the forces. According to the vector diagram, the force of the scale is larger than the force of gravity. The net force is therefore up and the object will accelerate up according to Newton's Second Law, $F = ma$.

PHYSICS TALK

Calculating Acceleration

You can calculate the acceleration of an elevator if you know the weight of the object and measure the force of the spring scale during the acceleration. Assume that a person on a scale weighs 600.0 N and the force of the scale on the person is 700.0 N up.

$$\text{Net force} = 700.0\ \text{N} - 600.0\ \text{N}$$
$$= 100.0\ \text{N}$$

If the weight of the object is 600.0 N, you can calculate its mass:

$$\text{Weight} = ma_g \quad \text{where } a_g = 9.8\ \text{m/s}^2.$$
$$600.0\ \text{N} = m\ (9.8\ \text{m/s}^2)$$
$$m = 61\ \text{kg}$$

Knowing the mass and the net force, you can calculate the acceleration using Newton's Second Law:

$$F = ma$$
$$100.0\ \text{N} = (61\ \text{kg})\ a$$
$$a = 1.6\ \text{m/s}^2$$

Similarly, you can calculate the reading of the spring scale if you know the acceleration of the elevator.

Sample Problem

An elevator at the top floor begins to descend with an acceleration of 2.0 m/s^2. What will a bathroom scale read if a 50.0 kg person is standing on the scale during the acceleration?

Strategy: Since the elevator is accelerating down, the net force must be down. The vector forces would look like this:

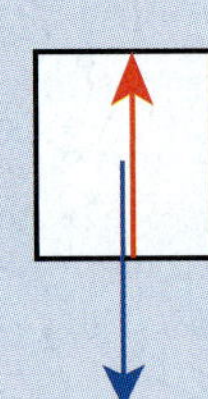

Newton's Second Law states that

$$\Sigma F = ma$$

where ΣF means "the vector sum of the forces" or "the net force."

Since the weight is greater than the force of the spring and in the opposite direction, you can write this as:

Weight – force by spring = ma

Givens:

$m = 50.0$ kg

$a = 2.0$ m/s^2

Solution:

Weight = ma_g

= (50.0 kg)(9.8 m/s^2)

= 490 N

Weight – force by spring = ma

490 N – F_s = (50.0 kg)(2.0 m/s^2)

F_s = 390 N

The scale would read 390 N instead of the person's weight, which is 490 N.

F_s = 390 N

If the elevator accelerates up, the scale reads a value higher than the weight of 490 N. There is really no limit to the upward acceleration though the person would become unconscious if the acceleration were greater than eight times the acceleration due to gravity. This has been experimentally determined by test pilots.

If the elevator accelerates down, the scale reads a value lower than the weight of 490 N.

There is a lower limit on the weight. If the elevator accelerates down at $9.8\,\mathrm{m/s^2}$ (e.g., the cable broke and the elevator is in free fall), the scale would not push at all and its reading would be 0 N.

$$\Sigma F = ma$$

$$\text{Weight} - \text{force by spring} = ma$$

$$(50.0\text{ kg})(9.8\,\mathrm{m/s^2}) - F_s = (50.0\text{ kg})(9.8\,\mathrm{m/s^2})$$

$$490\text{ N} - F_s = 490\text{ N}$$

$$F_s = 0\text{ N}$$

This is what is experienced when someone skydives out of a plane and has not yet pulled the cord on the parachute. It is also felt for a few moments in the amusement park ride where you are in free fall.

Reflecting on the Activity and the Challenge

A ride in an elevator is a lot like a ride in a roller coaster. The elevator moves up and down, but at too slow a pace and with too little acceleration to provide a great deal of excitement. When the elevator is at rest or moving up or down at constant velocity, your weight readings are identical. That's because at rest or moving at a constant velocity requires no net force.

The force of the scale up on you is equal to the force of your weight down. The bathroom scale denotes the value of the force up on you. When objects accelerate (change their velocities), there is a net force. When the elevator accelerates up, you also accelerate up. This is because the Earth pulls down on you less than the scale pushes you up. The scale reads a larger weight than before. You also feel as if you weigh more. When the elevator accelerates down, you also accelerate down. This is because the force of the scale up on you is less than the force of your weight down. The scale reads a smaller weight than before. You also feel as if you weigh less. If the elevator cable were to break, you would have only the force of your weight pulling you down. The scale would not push up and you and the weight reading would be zero. You would be "weightless." Roller coasters, like elevators, have sections where the acceleration is up or down. At these locations, people sense that they weigh more or less. When you design your roller coaster, you may want to take into consideration how large an acceleration your sample population would enjoy.

Physics To Go

1. The vector diagram shows a block of wood that can move up and down. The red vector represents the force pushing up on the block. The blue vector represents the weight of the block.
   a) Could the block be at rest?
   b) Could the block be moving up at a constant speed?
   c) Could the block be accelerating down?
   d) If you wrote "no" for any of the above questions, draw a force diagram for that description.

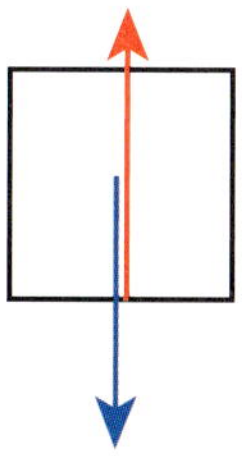

2. Complete the following chart in your log. Some responses are provided.

	Acceleration (up, down, zero)	Scale reading (larger, smaller, equal to weight)
1. Elevator at rest on bottom floor	Zero	Equal
2. Elevator starts moving down		
3. Elevator moves down at constant speed		
4. Elevator comes to rest on bottom floor		
5. Elevator is at rest at bottom floor		
6. Elevator begins to move up	Up	Larger
7. Elevator moves up at constant speed		
8. Elevator comes to rest on top floor		
9. Elevator at rest on top floor	Zero	Equal

3. A student usually weighs 140 lbs. On an elevator, the person is surprised to find that the scale only reads 137 lbs. for a few moments. Describe the movement of the elevator.

4. A person in an elevator at rest weighs 600 N. The elevator is about to move from the 2nd floor to the 5th floor. When it first starts to move, what will the passenger observe about his weight?

5. An elevator at the top floor begins to descend with an acceleration of 1.5 m/s^2. A person is standing on a bathroom scale in the elevator.

a) Will the bathroom scale's reading increase or decrease once the elevator starts?

b) What will a bathroom scale read if a 50-kg person is standing on the scale during the acceleration?

6. A 50-kg student is on a scale in the elevator.

a) What will be the scale reading when the elevator is at rest?

b) What will be the scale reading when the elevator accelerates up at a rate of 2 m/s^2?

c) What will be the scale reading as the elevator travels up at a constant speed?

7. Explain the meanings of the three sketches below. Specifically, why is there a different scale reading for the same student in each elevator?

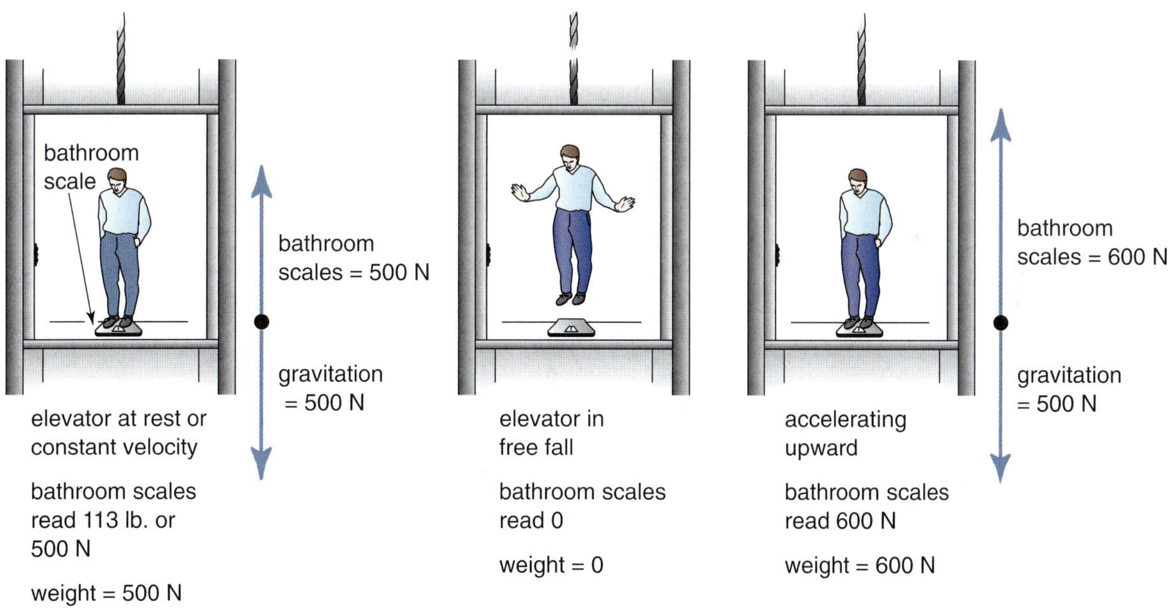

Stretching Exercise

Use a digital camera or projector and record some of the activities while riding in an elevator. Go with an adult to a place where there is an elevator that moves up and down several floors.

Activity 6 On the Curves

GOALS

In this activity you will:

- Recognize that an object in motion remains in motion unless acted upon by a force – Newton's First Law.
- Explain how a force toward a fixed center will allow a car to travel in circular motion.
- Describe how the centripetal force is dependent on the speed and the radius of the curve and the mass of the object.
- Solve problems using the equation for centripetal force
- Recognize that safety considerations limit the acceleration of a roller coaster to below 4 g.

What Do You Think?

The first looping coaster was built in Paris, France. It had about a 4-m (13 ft.) wide loop. One of the largest loops today is about 35 m (120 ft.) wide.

- **Why don't you fall out of the roller coaster car when it goes upside down during a loop-the-loop?**

Record your ideas about this question in your *Active Physics* log. Be prepared to discuss your response with your small group and the class.

For You To Do

In this activity, you will explore the behavior of the roller coaster on horizontal curves where you rip across the side and on vertical curves where you find yourself upside down. The more you understand about the requirements of curves on roller coasters, the more freedom you will have in designing your roller coaster.

Part A: Moving on Curves

1. A battery-operated car can move by turning on the switch. Investigate the toy car's motion under different circumstances.

   a) Let the car go. Describe its motion in your *Active Physics* log.

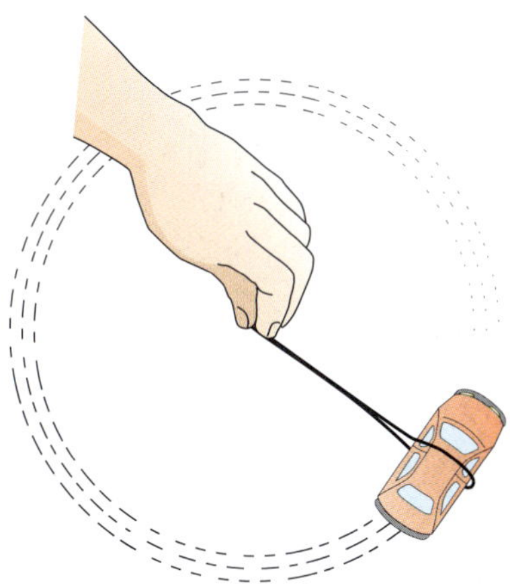

Be sure to pick up the car from the floor when it is not in use so it does not present a hazard for people walking in the room.

   b) Attach a string to the side of the car and hold the other end of the string fixed to a point on the floor. Describe the motion of the car.

   c) Predict what will happen to the motion of the car when your end of the string is let go. Provide a reason for your prediction.

   d) Test your prediction and record your observations in your log.

2. Your investigation with the toy car demonstrates that a force is needed for circular motion. This is big stuff. Whenever you see anything moving in a circle, you should remind yourself

of the movement of the car. Without the force of the string, the car moves in a straight line. If something moves in a circle, there must be a force that keeps it moving in a circle.

a) What force kept the toy car moving in a circle?

b) In which direction must this force point?

3. There is no string that keeps a real car moving around a curve. However, if the car is to move around the curve, there must be a force pointing toward the center of the curve. Imagine the road surface covered with slick ice. The car would not "make the curve" but would keep moving in a straight line and go off the road. It wouldn't matter which way you pointed the wheels—no turning.

a) What is the force that keeps a real car moving around a curve?

4. In a roller coaster, there are horizontal curves similar to those on the road.

a) Sketch the coaster moving around such a curve.

b) Draw an arrow showing the velocity of the car. This is the direction it would go if there were no force.

c) Draw an arrow showing the direction of the force that keeps the coaster moving in a circle.

5. There are two orientations of the roller coaster car you will investigate as it travels in a horizontal circle. The passengers can be sitting up as they would in an automobile with the wheels of the roller coaster down. They could also be on their sides with the wheels of the roller coaster facing away from the center of the circle. In each of these orientations, the force moving the car in a circle will be toward the center.

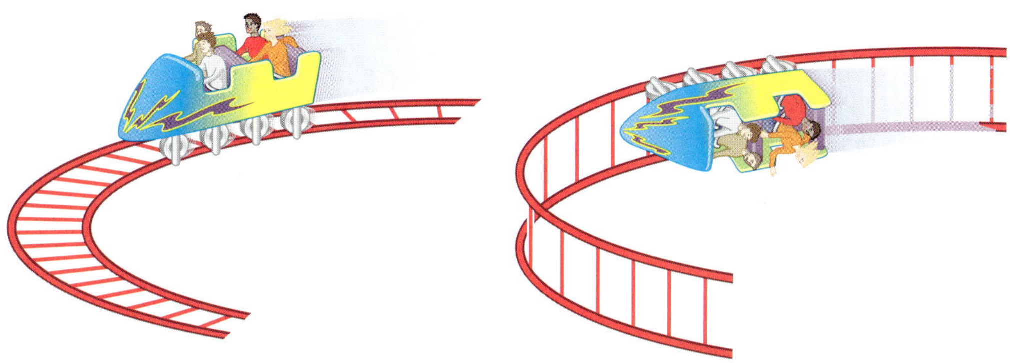

a) Is that the way that you drew it in **Step 4**? Make any changes necessary to your diagram.

b) Identify what force causes the roller coaster to move in the circle in each of these two cases. To help you identify the forces, you may want to imagine the circumstances with which the roller coaster would not "hold the turn" but would move off in a straight line.

6. The roller coaster may also do a loop-the-loop as it travels in a vertical circle. If the loop were a perfect circle, as illustrated below, there would always have to be a force toward the center of circle.

a) Make a sketch of the loop in your *Active Physics* log.

b) Draw the force vectors showing the direction of the force for the positions of the roller coaster noted.

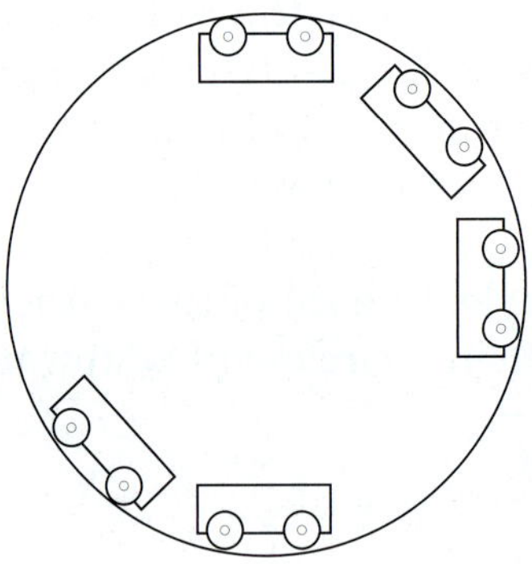

7. The gravitational force F_g is acting on the car at all times. To move in a circle, there must be a force toward the center of the circle. At the top of the circle, the gravitational force is toward the center of the circle. For a car at the bottom of the circle, the gravitational force is in the opposite direction to what is required for circular motion. The only other force at the bottom of the loop is the track pushing up on the car. This upward force must be responsible for the car moving in a circle.

a) Draw the gravitational force and the track's force on the car when the car is at the bottom of the loop and moving in a circle

8. Check your force diagram for the car at the bottom.

a) Is the gravitational force (or weight) vector pointing down?

b) Is the force of the track pointing up?

c) Is the force of the track pointing up larger than the weight vector pointing down?

9. The force of the track on the car is called the normal force F_N. (It is called the normal force because it is "normal" or "perpendicular" to the track.) This normal force must be present on the car when it rounds the loop at the bottom.

a) Could the normal force at the top of the loop be zero?

Part B: How Much Force is Required?

1. Your teacher will supply you with a rubber stopper and an attached string. You should wear safety glasses to protect your eyes in case the string should break or your partner accidentally loses grip of the string. You will use the stopper and string as a qualitative way to investigate the force needed to move something in a circle.

   Twirl the stopper at a slow speed in a horizontal circle. Increase the speed. Observe the force that your fingers are applying to the string as you increased the speed.

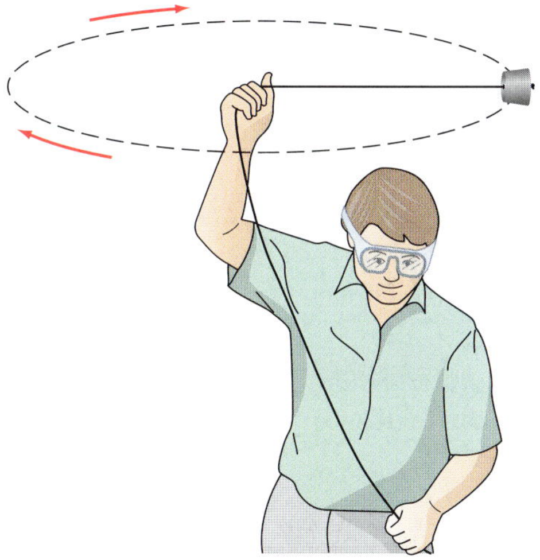

Everyone must be wearing safety glasses. Stand clear of anything that is glass.

   a) Write down a description of what you observed concerning the speed of the stopper and the force on your fingers.

2. Now twirl a string with two or three rubber stoppers attached.

   a) Compare the force that your fingers applied to a string with one rubber stopper and a string with more than one cork.

   b) In **Steps 1** and **2** did you keep the speeds and radii (length of string) of the twirls identical? Why is this important?

   c) Write down your observations about the force of your fingers and the mass of the stoppers.

3. Twirl one stopper, but this time change the length of the string.
   a) Write down a description of what you observed concerning the length of the string and the force on your fingers.
   b) What properties of the twirling stopper must you hold constant if you wish to compare only how changes in length affect the required force?

4. Transfer your observations about the cork on the string to your area of interest for the **Chapter Challenge**—roller coasters. To keep a roller coaster moving in a circle requires a force toward the center of the circle. The wheels on the track, the surface of the track, or the force of gravity can supply this force.
   a) How does the required force change when the speed of the roller coaster changes?
   b) How does the required force change when the mass of the roller coaster changes?
   c) How does the required force change when the radius of the curve changes?
   d) If the speed of a roller coaster were increased, how might you strengthen the track to provide the additional force required?

5. In physics, scientists often look at "limiting cases" to help them understand a concept better. A limiting case is the most extreme case that you may imagine. For instance, analyze the limiting cases for a roller coaster going around a horizontal curve that is not banked. If the car's speed got larger and larger and larger, the limiting case would be an infinite speed.
   a) If the cart were going at an infinite speed, how large a frictional force would be required by the track to keep the cart moving in the curve? Write your response down in your *Active Physics* log.
   b) The other limiting case is a car with zero speed. How large a frictional force would the track have to provide if the cart were moving very, very, very, slowly around a curve? Write your response down in your *Active Physics* log.

6. A roller coaster car in a vertical loop requires a force toward the center of the loop. At the top of the loop, the car requires a force toward the center of the loop, which is straight down. This force can be supplied by the force of gravity and by the normal force of the track.

a) In the limiting case, where the car is traveling as fast as possible, how much force will be required to keep it moving in the circle? Since the force of gravity is mg and doesn't change its value, what produces the very large force?

b) In the limiting case, where the car is moving as slowly as it can while still moving in a circle, the force of gravity is the only force acting on the car. Describe what would happen if the car moved faster than this. Where would the extra force come from? If the car moved slower than this, what would happen? (Hint: It would need a smaller force, but the force of gravity can't get smaller.) Would the car be able to travel in the circle?

c) Describe how the construction of a roller coaster track in a vertical loop is impacted by the speed of the roller coaster.

Part C: Is There an Equation?

1. The success of physics in describing the world is due to the discovery that mathematics can describe events precisely, accurately, and concisely. The equation for circular motion is

$$F_c = \frac{mv^2}{R}$$

where F_c is the centripetal force,

m is the mass,

v is the velocity, and

R is the radius.

This equation accurately describes your observations. (You can see how the equation is derived in the **Physics Talk** section.)

The centripetal force, F_c, on the left side of the equation is the force required to move something in a circle. It is always directed toward the center of the circle. (Reminder: when something moves in a circle a force is required. Remember the toy car with the string attached. The string always supplied a force toward the center of the circle and the car moved in a circle.)

On the right side of the equation are variables that can change when objects move in circles.

a) If the mass increases on the right side of the equation, then the right side of the equation gets larger. What happens to the F_c? Describe in your log how this agrees with your observations.

b) If the velocity increases on the right side of the equation, what happens to the F_c? Describe in your log how this agrees with your observations.

c) The equation tells you more about how the change in velocity affects the force than you could determine from your qualitative exercise. The equation says that the force increases as the square of the velocity, v^2. If the velocity triples, then v^2 is nine times as large. Tripling the velocity requires nine times the force. If the velocity quadruples (4 times as large), then v^2 is sixteen (4×4) times as large. And, if the velocity increases by a factor of 10, then v^2 is 100 times as large.

A roller coaster car going with twice the speed around a banked curve needs a stronger track. Write down in your log how much stronger the track must be for a doubling of the speed.

d) If the radius of the curve increases on the right side of the equation, then the right side of the equation gets smaller since the R is in the denominator of the fraction. What happens to the F_c?

e) Complete the following sentence in your log: The gentler the curve, the _______ the force required to keep the car moving along the curve. If the curve is tight (R is very small) then a _______ force is required.

2. The limiting case of the large curve is where the curve's radius is so very large that the curve and a straight line are hardly distinguishable. On a straight path, no force is required.

a) Describe in your log how this agrees with your observations of the cork on a string.

PHYSICS TALK

Calculating Centripetal Force

Newton's First Law states that an object at rest stays at rest and an object in motion stays in motion unless acted upon by an unbalanced force. A car with no net force will be at rest or travel at a constant speed in a straight line. Newton's Second Law states that accelerations require forces: $F_{net} = ma$. Acceleration is defined as a change in velocity with respect to time.

$$a = \frac{\Delta v}{\Delta t}$$

This change in velocity can be a change in speed (the car can go faster or slower) or a change in direction (the car can move in a curve.) Newton's Second Law states that any acceleration requires a force.

Acceleration due to a change in speed is easy to calculate. When a car accelerates from 10 m/s north to 30 m/s north, its change in velocity is 20 m/s north. If the change occurred in 4 s, the acceleration is (20 m/s)/4 s = 5 m/s^2.

For acceleration due to a change in direction, the calculation is a bit more difficult. A car traveling at 10.0 m/s north that then travels 10.0 m/s east after 4.0 s also has acceleration. To determine the change in velocity, you have to calculate the change in the velocity by subtracting vectors $v_f - v_i$, where v_f is the final velocity and v_i is the initial velocity. This is mathematically equivalent to adding $v_f + (-v_i)$.

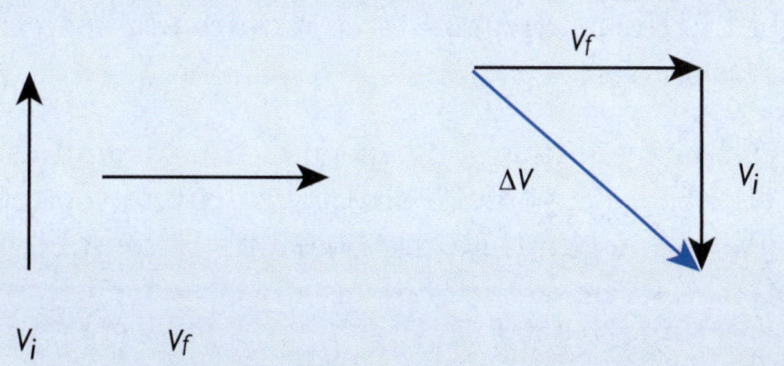

The magnitude of this change in velocity Δv can be found with the Pythagorean Theorem and is equal to 14 m/s. You can see that the direction is 45° south of east. The acceleration is $\frac{\Delta v}{\Delta t}$ = (14 m/s)/4 s = 3.5 m/s^2 at an angle of 45° south of east.

Looking at the vector diagram, you can see that this makes sense. To change from moving north to moving east the car needs a push in the southeast direction. If these were two velocities of a car moving in a circle, this would be average acceleration during one-quarter revolution. The direction at the midpoint would be directly toward the center of the circle.

There is another way to calculate the acceleration without using vectors and without the need for the time. You can use an equation derived from an analysis of the circular motion.

$$a = \frac{v^2}{R}$$

If you know the speed of the roller coaster car and the radius of the circle, you can directly calculate the required acceleration. Knowing the mass, you can find the required force by using Newton's Second Law, $F = ma$.

Sample Problem

A roller coaster car moving at 12.0 m/s swings into a horizontal turn with a radius of curvature equal to 20.0 m.

a) What is the acceleration of the roller coaster?
b) If the mass of the passengers and car total 300 kg, what is the centripetal force required to keep the car on its tracks?

Strategy: Since you know the speed of the car and the radius of the circle, you can directly calculate the required acceleration. You can then use Newton's Second Law to calculate the force.

Givens:

$$v = 12.0 \text{ m/s}$$
$$R = 20.0 \text{ m}$$
$$m = 300.0 \text{ kg}$$

Solution:

$$a = \frac{v^2}{R} = \frac{(12.0 \text{ m/s})^2}{20.0 \text{ m}} = 7.2 \text{ m/s}^2$$

$$F = ma = (300.0 \text{ kg})(7.2 \text{ m/s}^2) = 2200 \text{ N}$$

This force will have to be supplied by the track to the wheels of the coaster.

Physics Words

centripetal force: a force directed towards the center that causes an object to follow a circular path.

FOR YOU TO READ

Centripetal Force

The fun of a roller coaster is the fun of whipping around the turns and flipping upside down. All objects moving in circles require a force toward the center of the circle. In a roller coaster moving around the curve, this force is the track on the wheels of the car when the wheels are down during the curve.

In a roller coaster curve where the car tilts vertically and the wheels face the outside of the circle, the force toward the center is the force of the wall holding the track. This is called a normal force F_N because it is normal (perpendicular) to the wall.

In any circular motion, the force that keeps the object moving in a circle is called the centripetal force. The toy car moving in a circle had a **centripetal force** that was the force of tension in the attached string. A car moving around a curve has the force of friction between the tires and road as the centripetal force. The roller coaster car rounding a turn on its side may have the force of the track as the centripetal force. The centripetal force is not an additional force. It is the name given to a force like friction, tension, gravity, or the normal force when that force causes an object to move in a circle.

When the roller coaster is in a vertical loop, the direction of the centripetal force is always changing to ensure that the force vector always points to the center of the circular track. Pay particular attention to how this is phrased. Although the force is always toward the center, the direction is always changing since in the circle, the force may be toward the left or the right or up, but still point toward the center.

In the vertical loop, this centripetal force can be either the gravitational force, the normal force of the track on the car or a combination of the two. When it is a combination of the two, you must add the forces as vectors. At the bottom of the circle, the normal force points toward the center of the circle (upward) while the gravitational force points downward. The vector sum of these two forces must be toward the center of the circle. You can therefore conclude that the normal force is larger than the gravitational force. The normal force corresponds to your apparent weight, as it did in the elevator activity. This is why you feel as if you weigh more at the bottom of the loop of the roller coaster.

At the top of the loop-the-loop, the gravitational force and the normal force both act downward, toward the center of the loop. The sum of these two vectors provides the required centripetal force. How much of the normal force is required will depend on the mass and velocity of the car.

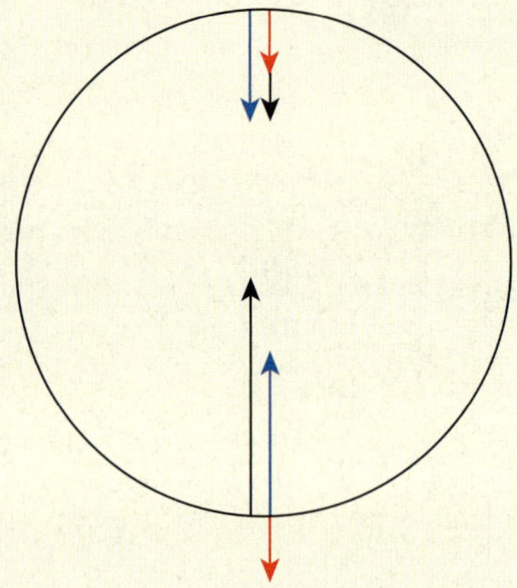

The blue force vectors show the required centripetal force to keep the car moving in a circular path at the top and bottom of the loop. Notice that the size of the vectors is different since the car has a larger speed at the bottom of the loop. The directions are different because the centripetal force must always point toward the center of the circle.

The red force vector represents the force of gravity or weight of the cars. Both weight vectors are identical because the weights of the roller coaster car are identical at the top and bottom.

The black vector represents the normal force of the track on the car. The sum of the normal force plus the weight must be equal to the required centripetal force. At the top, the normal force is small since the weight contributes to the centripetal force. If the speed decreases, the required centripetal force would be less and less. There comes a point where the gravitational force (weight) would be all that is required to keep the car moving in a circle. In that case, the normal force is zero. In this special situation where no normal force is required, you could actually have a small gap in the top of the track and the car would continue to move in a circle.

At the bottom of the roller coaster, the car would need a normal force of the track on the car that would be greater than the weight since the weight is not providing any help for the required centripetal force.

This is summarized in the following tables.

The force of gravity pulling you down is used to move you in a circle. You are moving down but you are also moving across. If the speed of the roller coaster were not sufficient, you would require a centripetal force less than the gravitational force to move you in a vertical circle. The gravitational force would cause you to fall. It's your large speed that keeps you from falling when you are upside down.

Apparent Weight and the Roller Coaster Ride

You discovered earlier that an elevator ride could give you a sense of weight changes during accelerations. In the roller coaster

Fast-moving roller coaster			
	Required centripetal force	**Force of gravity (weight)**	**Normal force (the force of the track on the car)**
At the top of the loop	5000 N	1000 N	4000 N
At the bottom of the loop	9000 N	1000 N	10,000 N

Slow-moving roller coaster			
	Required centripetal force	**Force of gravity (weight)**	**Normal force (the force of the track on the car)**
At the top of the loop	2100 N	1000 N	1100 N
At the bottom of the loop	6100 N	1000 N	7100 N

loop-the-loop, the passenger will also experience changes in apparent weight. The normal force is an indication of the apparent weight, as it was in the elevator. A passenger on the roller coaster feels lighter at the top of the roller coaster. This is similar to the elevator because in both cases you feel lighter because acceleration is down. A passenger on the roller coaster feels heavier at the bottom of the roller coaster. Once again, this is similar to the elevator because in both cases you feel heavier because acceleration is up.

In the slow-moving roller coaster in the chart, the apparent weight (normal force) at the top may only be 1100 N, while the apparent weight (normal force) at the bottom may be 7100 N.

There are three locations that you can use to summarize the discussion on forces and weight. On a level track with the cart moving at constant speed, the sum of the forces must be zero.

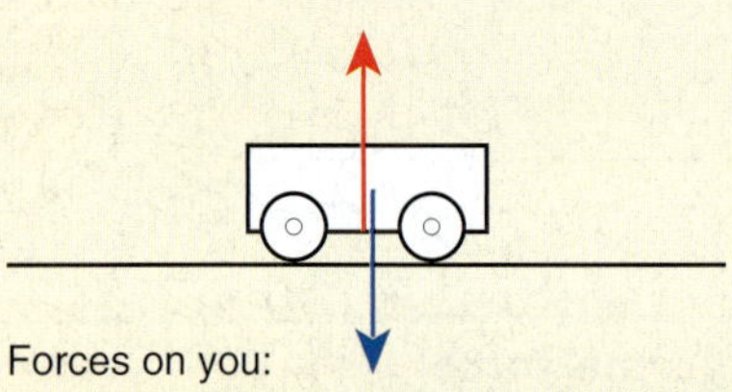

Forces on you:
The force of the seat on you = 500 N (apparent weight).
The force of gravity on you is 500 N (weight).

At the bottom of the hill or loop, there must be a net force toward the center of the circle to keep you moving in a circular path.

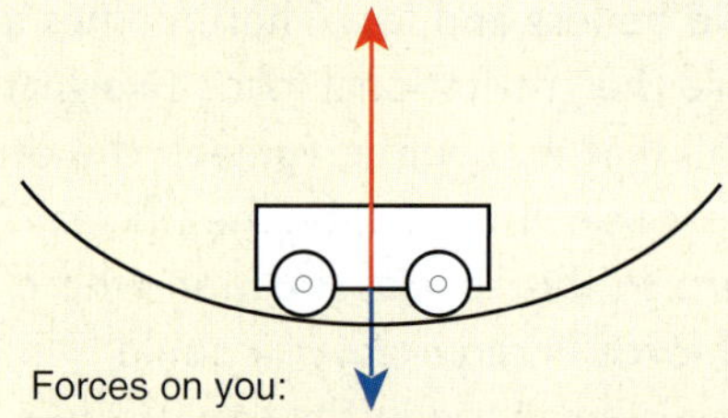

Forces on you:
The force of the seat on you = 1000 N.
The force of gravity on you is 500 N (weight).
You feel as if you weigh 1000 N (apparent weight).

At the top of the hill or loop, there must be a net force toward the center of the circle to keep you moving in a circular path.

Forces on you:
The force of the seat on you = 100 N.
The force of gravity on you is 500 N (weight).
You feel as if you weigh 100 N (apparent weight).

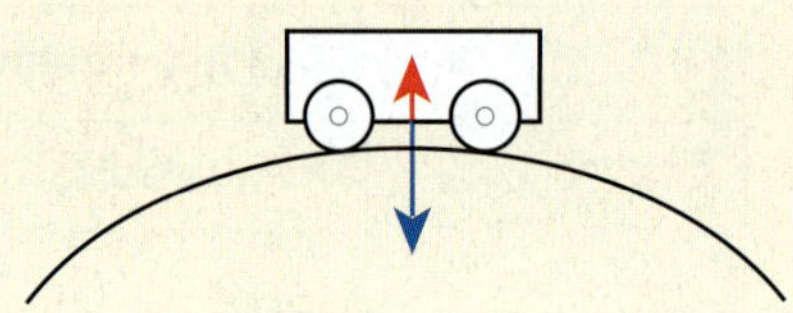

Roller coasters do not use loops that are circular. They use a clothoid loop (it has a big radius at the bottom and a small radius at the top). In this way, at the top of the loop the roller coaster is moving in a small circle, while at the bottom it is moving in a larger circle. This is done to ensure that the roller coaster car can make the turn at the top but not gain so much speed at the bottom that the person at the bottom would be pulling more than 4 g's.

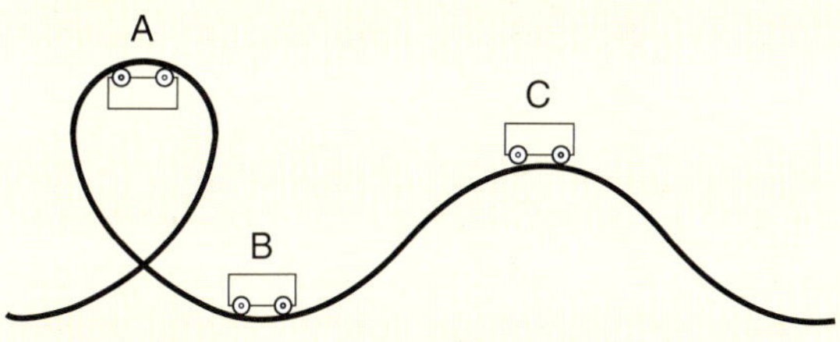

A Top of loop B Bottom of hill C Top of hill

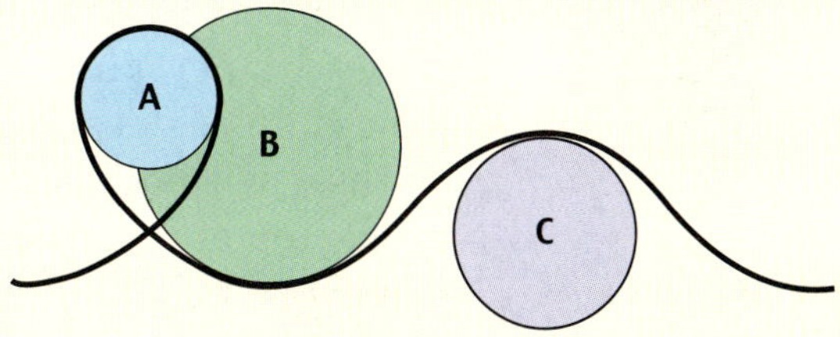

Each of these three situations is basically circular motion.

The type of loop the coaster designers use is called a clothoid. It has a large radius at the bottom,and a smaller radius at the top.

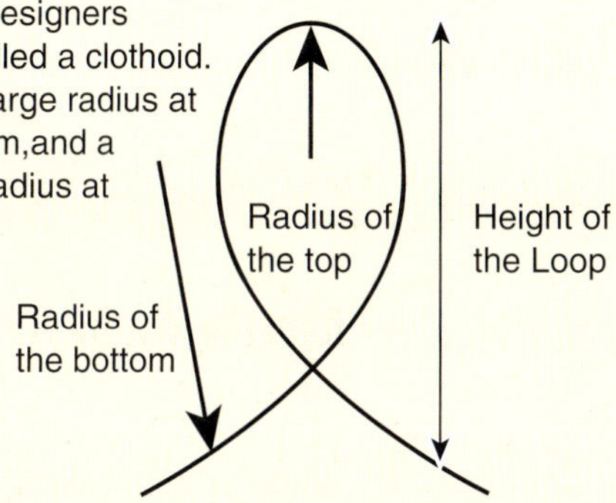

Safety on the Roller Coaster

Test pilots and astronauts experience lots of accelerations during their job performance. To prepare for this, they all go through physical training to see how much acceleration they can endure without getting sick or becoming unconscious. Experiencing an acceleration of more than nine times gravity for a sustained period will cause unconsciousness in most people. Since the acceleration due to gravity is 9.8 m/s^2 or approximately 10 m/s^2, you can refer to other accelerations in terms of 1 g. An acceleration of 2 g is approximately 20 m/s^2 while an acceleration of 8 g is approximately 80 m/s^2. Astronauts sometimes experience as much as 6 g's during liftoff.

Safety on a roller coaster requires that you stay below 4 g for the entire ride. You must never go beyond 4 g for even a short time. Changes in small accelerations may make a better ride than one big thrill from a single large acceleration.

Reflecting on the Activity and the Challenge

The loop on a roller coaster is one of The Big Thrills. People are always nervous that they will fall out of the roller coaster when it is upside down. This does not happen because you arrive there with a large velocity. The gravitational force (weight) at the top of the roller coaster serves as the centripetal force that moves the roller coaster in a circular path. All objects moving in circles require a centripetal force toward the center. With a toy car attached to a string, the tension in the string is the centripetal force. A roller coaster car rounding a turn has the wheels in the track providing the centripetal force. A roller coaster making a turn on its side has the track's normal force as the centripetal force. The upside-down roller coaster has the gravity as the centripetal force. At the bottom of the loop, the normal force provides the centripetal force. This force must be larger than the gravitational force and passengers feel much heavier at the bottom of the loop. In designing your roller coaster, you will have to ensure that the roller coaster has enough speed to make the full circle. You will also have to ensure that it doesn't have so much speed at the bottom that the apparent weight is too great or the passengers may get injured.

Physics To Go

1. A battery-operated toy car is attached to a string.
   a) If the loose end of the string is held to the ground, draw the path of the car.
   b) If the string were to break, draw the path that the car would follow.

2. Consider a car on a road making a turn.
   a) What force has replaced the string of the toy car in **Question 1(a)**?
   b) If the car were to hit a section of ice, draw the path that the car would follow.

3. A person twirls a key chain in a circle. If she twirls it faster, she finds that she holds the chain tighter. Explain why this is necessary.

4. It's a cold night and the roads are icy. If your car is filled with friends, will it be easier or more difficult to make a turn? Explain why.

5. In the equation for circular motion $F_c = \frac{mv^2}{R}$ explain what each of the terms represents.

6. A roller coaster car is traveling east at 20 m/s. After 2 s, it is traveling north at 20 m/s.
   a) Did the speed of the roller coaster car change?
   b) Did the velocity of the roller coaster car change?
   c) How much did the velocity of the roller coaster car change?

7. A roller coaster car is traveling east at 20 m/s. After 16 s, it is traveling north at 20 m/s. The circular curve had a radius of 200 m. Calculate the acceleration of the car.

8. A roller coaster car is traveling in a loop. Identify the six force vectors in the diagram below.

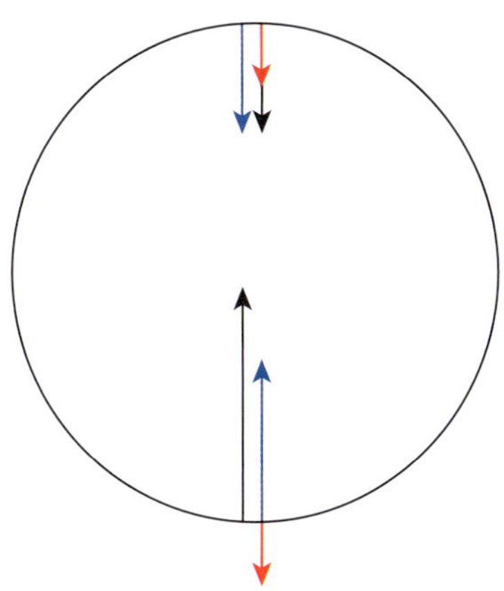

9. In explaining circular motion, someone correctly states that the centripetal force is a name for a force, but it is not an additional force. Explain what this means.

10. Fill in the missing values in the tables that you created in your *Active Physics* log:

Fast-moving roller coaster			
	Required centripetal force	**Force of gravity (weight)**	**Normal force (the force of the track on the car)**
At the top of the loop	4000 N	500 N	
At the bottom of the loop	6000 N	500 N	

Slow-moving roller coaster			
	Required centripetal force	**Force of gravity (weight)**	**Normal force (the force of the track on the car)**
At the top of the loop	800 N	500 N	
At the bottom of the loop	2800 N	500 N	

11. At which section of a loop would the roller coaster passengers feel the heaviest? Why?

12. Safety requires the roller coaster to be able to make the complete loop and to keep the acceleration under 4 g. How can both of these safety features be accomplished at the same time?

13. Design a loop for your roller coaster. Calculate the speeds and accelerations at some key places in the loop.

14. Use the diagram of the Terminator Express roller coaster. Indicate at which of the following points the passengers will feel heavy, where they will feel light, and where it is uncertain.

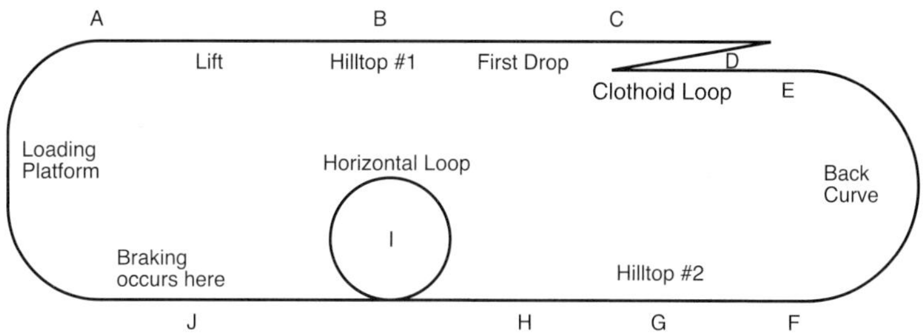

TOP VIEW

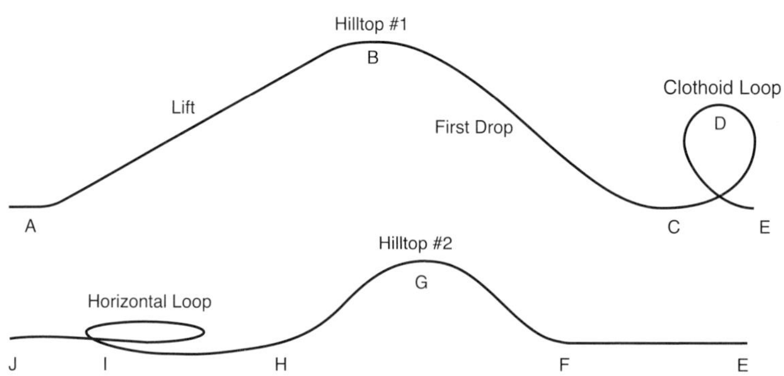

SIDE VIEW

a) C (bottom of hill #1)
b) D (top of the vertical loop)
c) E (bottom of the vertical loop)
d) F (bottom of hill #2)
e) lift hill (going up at constant speed)

15. Using the diagram of the Terminator Express, indicate at which of the following points the centripetal force is up, when it is down, when it is zero, and when it is sideways.

a) C (bottom of hill #1)
b) D (top of the vertical loop)
c) E (bottom of the vertical loop)
d) F (bottom of hill #2)
e) lift hill (going up at constant speed)
f) horizontal loop
g) back curve

Activity 7 Getting Work Done

GOALS

In this activity you will:

- Measure and recognize that the product of force and distance is identical for lifting the object to the same height irrespective of the angle of the ramp.
- Define work where F is the force and d is the displacement and where the dot signifies a special kind of multiplication.
- Explain the relationship between work and gravitational potential energy and spring potential energy.
- Define power as the rate of doing work and the units of power as watts.

What Do You Think?

The greatest drop for a roller coaster is 125 m (400 ft.). The roller coaster must be pulled up to that height to get the ride started.

- **Does it take more energy to slide the roller coaster up a steep incline than a gentle incline?**
- **Why is it more difficult to walk up a steep incline than a gentle incline?**

Record your ideas about these questions in your *Active Physics* log. Be prepared to discuss your response with your small group and the class.

For You To Do

1. From the starting hill, the roller coaster is able to complete its entire trip without any motors, pushes, tires, or chains. The roller coaster car heads down a hill and has enough speed at the bottom to make it up to the next hill.

   a) Why do you think that the roller coaster cannot scale a higher hill than the one from which it began?

2. The roller coaster at the top of the hill is ready to go. It goes up and down the hills and around the curves without any energy input. The roller coaster is a closed system. No energy is added to the system and no energy leaves the system. Nobody adds energy to the roller coaster with motors. No energy is assumed lost by the system to friction or air resistance. The roller coaster as a closed system (an idealized, conceptual roller coaster) will keep on going back and forth forever. The cart will go from point A to B to C to D to E to F to G. It will then reverse and go from G to F to E to D to C to B to A. It will then begin the trip again.

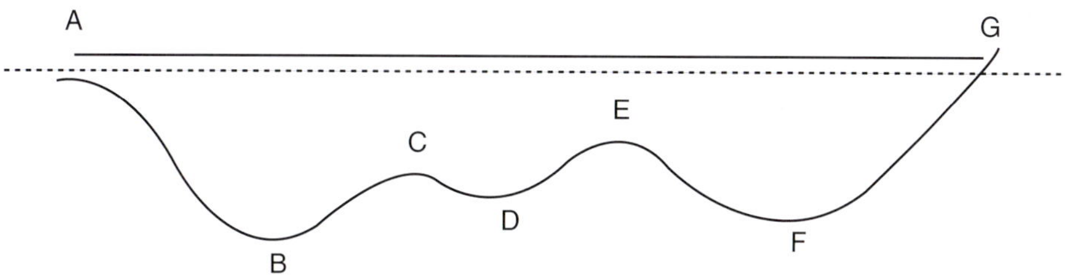

3. You will now investigate the force required to lift a roller coaster car to a certain height. You will use a cart and a track in your classroom. You can pull the cart to the top of the track with the use of a spring scale. The spring scale will indicate the force required to pull the cart. A meter stick can be used to record the distance that the cart moved along the track. You can then vary the length and angle of the track while keeping the height of the track constant.

Pull the cart at a speed that keeps it on the track through the entire run. Before beginning to pull the cart, make sure that the way is clear for the person pulling to walk without obstruction.

a) Create a data table in which you can record the force required to pull the cart up four different tracks. (Reminder: You must always pull the cart to the same height and parallel to the track.)

4. Complete your investigation.

a) What conclusion can you reach about the distance along the track to attain a specific height and the force required to move the cart?

x	y	xy = k
1	12	12
2	6	12
3	4	12
4	3	12
6	2	12

5. When one quantity increases and a second quantity decreases, this is referred to as an inverse relation. If x is one quantity and y is the other quantity, one inverse relation can be described mathematically by the equation $xy = k$ where k is a constant. At left are some x and y values forming an inverse relation where $xy = 12$.

a) Make a graph to show the relationship for the inverse relation $xy = 12$.

6. Create a graph for the data from your experiment.

a) In the equation $xy = 12$, the product of the x and y values always equals 12. Does the product of the force and distance in your experiment always equal a certain value? Make the calculations and record the results on the side of your chart.

b) Why would the values in your experiment not be expected to be *exactly* the same?

7. Any time you take measurements, there is some uncertainty in the measurement. When you weigh yourself on a scale, the weight reading may be off by a little bit. If the scale reads 143 lb., you may actually weigh 143 lb. and a few ounces. The scale does not give you an exact measurement. No measurement is ever exact.

a) What are the uncertainties in your measurements of distance? Could your measurement of distance be off by as much as 3 cm? Could your measurement of distance be off by as much as 1 cm? What is the largest amount that your distance measurement may be off? Write down this value with the notation ± to signify that you may have been under or over by that amount. For instance, if you think that your distance measurement could have been off by 2 cm, you would write this as ± 2 cm.

b) Record the uncertainties in your measurements of force by noting the accuracy of your spring-scale reading.

8. Another way that you can get the cart to the top of the incline is to lift it vertically. Use the spring scale to lift the cart vertically.

a) Record the force required to lift cart vertically and the distance (height) that you lifted it.

b) Does this force and distance fit the same inverse relation as the track measurements?

9. There are other means by which the cart could be lifted to the top of the incline. For example, you could have had an electric motor pull the cart to the top. Brainstorm a list of at least three ways in which the cart could be brought to the top of the incline. (Brainstorming allows for all ideas to be included, even those that appear silly or impractical.)

a) Record your ideas in your *Active Physics* log.

FOR YOU TO READ

Work

The roller coaster must get to the top of the first hill to begin the ride. In the activity, you moved a cart to the top of an inclined ramp by applying a force with the spring scale over a certain distance. In physics, this is called **work.** In physics work has a specific definition. Work is defined through the following equation:

$$W = F \cdot d$$

where F is the force and

d is the displacement.

The dot signifies a special kind of multiplication. The force and displacement can be multiplied if and only if they are partially in the same direction.

In the activity, the spring scale pulled the cart up the incline and the force was in the same direction as the displacement. You found that the work done was the same irrespective of the angle of the incline. The force was larger for a steeper incline, but the distance along the incline was smaller. The product of the force and distance was always the same. That quantity was the work that was done by the spring scale on the cart.

Sample Problem 1

A cart that weighs 300 N is lifted to the top of an incline 2 m above the ground.

a) What is the work done?

b) How much force would be required to lift the same cart to the same height using a 10-m track?

a) Strategy: The minimum force required to lift the cart is equal to its weight. The displacement is the height that the cart was lifted. The force and the displacement are both in the vertical direction.

Givens:

$F = 300$ N

$d = 2$ m

Solution:

$W = F \cdot d$

$W = (300 \text{ N})(2 \text{ m})$

$= 600$ Nm

$= 600$ J

b) Strategy: The work required to lift the cart would be identical since the cart began at the same height and ended at the same height. Since you know the new displacement, you can find the new force.

Givens:

$W = 600$ J

$d = 10$ m

Solution:

$W = F \cdot d$

$600 \text{ J} = F(10 \text{ m})$

$F = 60$ N

By using the track (ramp) only 60 N of force are required to slide the cart up the track instead of the 300 N to lift it. This is why the track is considered a simple machine. The same work is done, but with much less force. Of course, the force must be applied over a longer distance.

Direction of Force in Work Done

It may seem that the force would always be in the same direction as the displacement. This is not always the case. Consider a push lawn mower. The push lawn mower has no motor. It moves because someone pushes it.

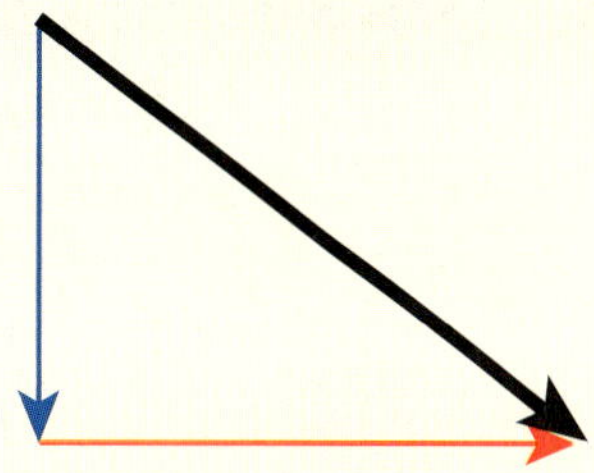

The force is applied along the handle of the lawn mower. The displacement of the lawn mower is the distance along the ground. They are not in the same direction, but there is some work done.

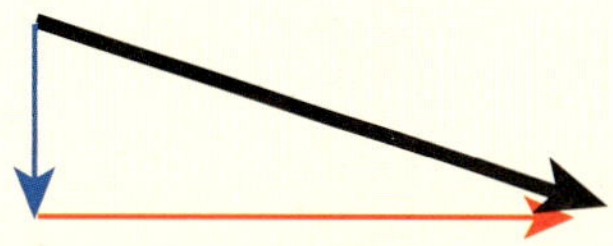

Although the entire force is not in the same direction as the displacement, some of the force is in the same direction of the displacement.

The force along the handle can be broken into its two vector components by finding the horizontal and vertical forces that, when added together, would be identical to the original vector. In the diagram at the top, it appears that the horizontal and vertical component forces are approximately equal in size. In the next diagram, the horizontal vector is much larger than the vertical component. Most of the force applied to the handle is now in the same direction as the displacement. Even though the total force is identical (note that the length of the force vector is the same in each diagram), the horizontal component is larger as the angle gets smaller. The same total force and the same displacement, but more work is done when the horizontal component is greater.

Why then don't you push the lawn mower with a small angle? Although more work would be done, it would hurt your back. Therefore, you sacrifice some work in order to make mowing more comfortable.

Closed and Open Systems

A **closed system** has no external forces on it. With no external forces, a closed system also has no external work done. The ideal roller coaster is a closed system once the roller coaster is on top of the first big hill ready to go. It remains a closed system if friction or air-resistance forces do no work.

To start the roller coaster, work must be applied to the system. The work will increase the energy of the roller coaster system. The work to lift the roller coaster up the track is identical to the work to lift it vertically. The force required is equal to the weight of the car. The vertical displacement is the height that it must be lifted.

$$W = F \cdot d$$

$$W = \text{weight} \cdot \text{height}$$

$$W = mgh$$

The work done on the roller coaster is mgh. This is equal to the change in gravitational potential energy GPE of the roller coaster.

It would be difficult to raise a real roller coaster car by hand as you raised the cart in the activity. You could try to lift it with a large spring. The force of a spring that obeys Hooke's Law is $F = kx$. The force is not constant but changes as the stretch of the spring changes. If you stretch the spring a distance x, then the average force will be $\frac{1}{2}kx$. It is zero when the spring is not stretched at all and a maximum value of kx when the spring is stretched a distance x. The work done by a spring is

$$W = F \cdot d$$

$$W = (\tfrac{1}{2}kx)(x)$$

$$= \tfrac{1}{2}kx^2$$

The work done by a spring is equal to the potential energy of the spring SPE.

The roller coaster car is usually raised with electrical energy supplied by a motor. Electrical energy can be calculated by measuring the voltage, current, and time. Creating steam to push it up the incline could also have raised the roller coaster car. In this method, the heat energy can also be calculated. In all of these methods, work is done by the spring, by the electricity, or by the heat. The roller coaster system gains that amount of energy. The roller coaster has increased its GPE by exactly that amount.

In any closed system, the total energy remains the same. This is an organizing principle of physics and is referred to as the conservation of energy.

Although you treat the roller coaster as a closed system in your simplified analysis, there is work done by friction and work done by air resistance. This work removes energy from the roller coaster. The work done by friction, for instance, becomes heat energy that is dissipated to the air surrounding the roller coaster.

In an **open system**, you can add energy. You do work on the car to bring it to the top of the hill and in this way add gravitational potential energy GPE.

In an open system, you can lose energy. Friction can do work and the car will lose some of its kinetic energy and corresponding GPE and will not get up to the same height that it began.

In a closed system, no external forces act on the system and no energy is able to enter or leave the system. The total energy of the system will remain the same.

Power

A related concept to work is how fast the work is done. In the activity, you pulled the cart up the incline. You could have pulled it up with a variety of speeds. The rate of doing work is **power**.

$$P = \frac{W}{t}$$

Sample Problem

Tomas runs up the stairs in 24 s. His weight is 700 N and the height of the stairs is 10 m.

a) What is the work done by Tomas?

b) How much power must Tomas supply?

Givens:

$F = 700$ N

$d = 10$ m

Solution:

a) $W = F \cdot d$

$W = (700 \text{ N})(10 \text{ m})$

$= 7000$ J

b) $P = \frac{W}{t}$

$= \frac{7000 \text{ J}}{24 \text{ s}}$

$= 290$ W (watts)

Notice that the unit for power is joules per second, or watts. You are familiar with the power ratings of light bulbs in watts. You may have heard of horsepower as another unit for power. One horsepower is the power output of a horse over a specific time. One horsepower is approximately 750 W.

Physics Words

work: the product of the displacement and the force in the direction of the displacement; work is a scalar quantity.

closed system: a physical system on which no outside influences act; closed so that nothing gets in or out of the system and nothing from outside can influence the system's observable behavior or properties.

open system: a physical system on which outside influences are able to act; open so that energy can be added and/or lost from the system.

power: the time rate at which work is done and energy is transformed.

Reflecting on the Activity and the Challenge

A roller coaster ride always begins with a slow, suspenseful ride to the top of the first hill. On the way up, the roller coaster is designed to shake a bit and to make a few extra noises in order to add to the drama. The roller coaster is gaining gravitational potential energy GPE on the way up. The motor is performing work on the roller coaster. Work is a precisely defined term in physics: $W = F \bullet d$. The work supplied by the motor increases the energy of the roller coaster. At the top of the incline, the motor is disengaged and the roller coaster is on its own. There is some work by friction with the air and track that removes energy from the roller coaster. In designing your roller coaster, you will have to include a motor to lift the roller coaster. You will have to decide on the slope of the track going up and the time you want the ride to take. Work and energy will be useful ways of describing what is needed in your design. You will also want to know how fast this work is done. For that you will use the concept of power where $P = W/t$.

Physics To Go

1. A student is asked to use the window pole to slide the window up. If the window moves the same distance up, is the work applied equal in the two cases shown? Is the force applied equal in the two cases?

2. A cart starts at the top of the incline. It slides down the incline a distance l and comes to rest after compressing a spring a distance x.
   a) Compare the GPE of the cart at the top of the incline and at the bottom.
   b) How much work was done on the cart by the force of gravity (the cart's weight)?
   c) How much work was done on the cart by the spring?
   d) What is the spring's SPE when it is compressed?
   e) Before hitting the spring, describe the total energy of the cart.
   f) At which point does the cart begin to slow down?

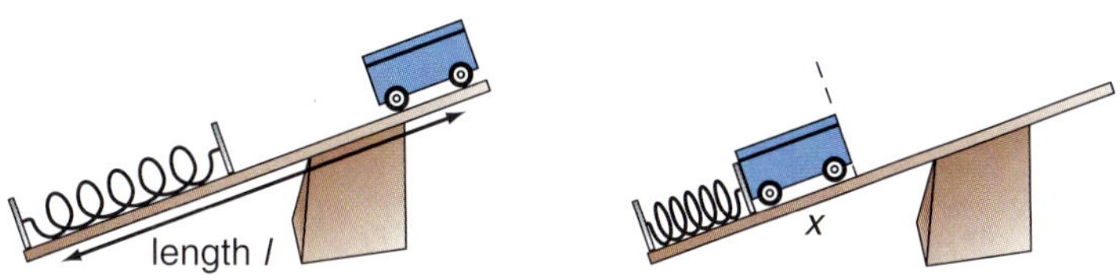

3. Calculate the work done in the following situations:
   a) A waiter applies a force of 150 N to hold a tray filled with plates on his shoulder. He then moves 7 m toward the kitchen door.
   b) A bowler lifts a 60 N bowling ball from the rack to his chest, a vertical distance of 0.5 m.
   c) A girl pulls her sled up a hill. The length of the hill is 40 m and the pulling force required was 75 N.
   d) The weight of an object is 500 N. It is lifted over a body-builder's head, a distance of 0.7 m.

4. Why are you told to conserve energy if the conservation of energy tells you that energy is always conserved? Create a better way of saying "conserve energy."

5. What is the difference between an open and closed system?

6. If you were to fill the cart you used in the activity with clay to represent the people in the roller coaster, what would have changed in the experiment?

7. A roller coaster car that weighs 10,000 N is lifted to the top of the first hill that is 20 m above the ground. To add suspense, the ride up takes 150 s.

   a) Calculate the work done by the motor.

   b) Calculate the power of the motor.

8. In the Terminator Express roller coaster, describe one trip in terms of work and energy.

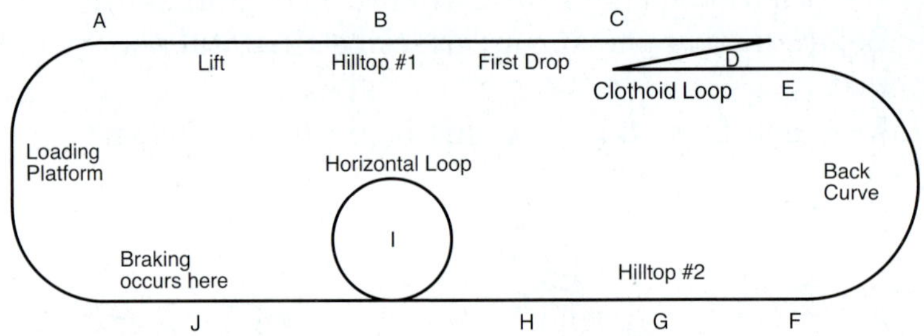

TOP VIEW

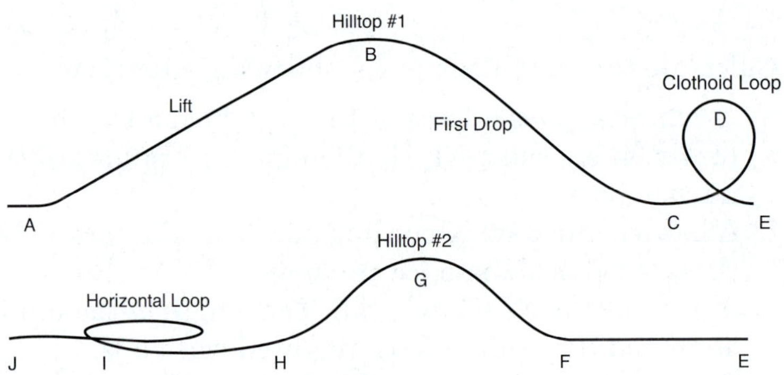

SIDE VIEW

Stretching Exercise

Have the cart pulled up to the top of the incline with a motor. Measure the energy of the motor using voltage, current, and time. Compare the energy from the motor with the increase in GPE of the cart.

Activity 8 Vectors and Scalars

GOALS

In this activity you will:

- Describe instances in which two cars will have the same speed but require different times to reach those speeds.
- Add vectors that are perpendicular to each other.
- Recognize that forces are vectors and energies are scalars.
- Explain how forces and energy considerations provide different insights into roller coaster rides.
- Choose whether energy or force considerations are more appropriate for analyzing aspects of roller coaster rides.

What Do You Think?

"The Snake" roller coaster stays at ground level throughout the ride. The passengers move left, then right, then left again.

- **Which parts of The Snake will be the most thrilling?**
- **If the speed of The Snake always remains the same, why will it still be fun?**

Record your ideas about these questions in your *Active Physics* log. Be prepared to discuss your response with your small group and the class.

For You To Do

Part A: Energy and Forces in a Roller Coaster

1. Your study of roller coasters has actually taken two turns. You have investigated energy changes in roller coasters. You have also investigated forces and accelerations in roller coasters.

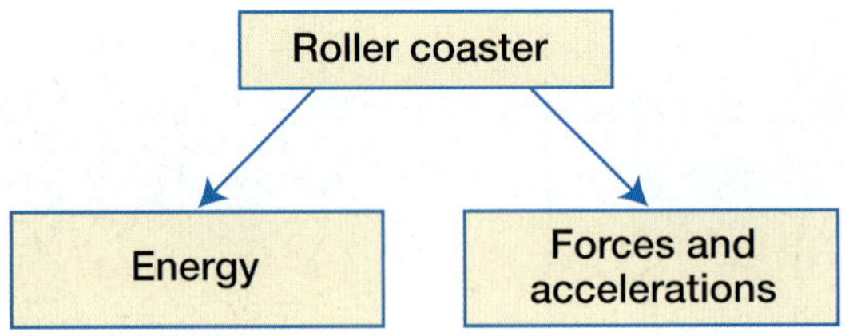

a) Copy this beginning of a concept map into your log. On a set of note-sized pieces of paper, write down some things you know about energy and how it relates to roller coasters. Each note should have one concept only.

2. Sort the concepts into a map.

a) Add these concepts to your log.

3. On a new set of notepaper, write down some things you know about forces and accelerations and how they relate to roller coasters. Each note should have one concept only.

4. Sort the concepts into a map.

a) Add these concepts to your log.

5. The left half of your map reminds you of the relationships between energy concepts. The right half of your map reminds you of the relationships between force and acceleration concepts.

a) Is there a bridge between these two sides of the map? Describe how energy is related to forces and accelerations.

6. You use both energy and force approaches to understand roller coasters because they both provide you with valuable information. Sometimes it is easier to look at a roller coaster as an energy ride, while other times it is best to look at a roller coaster as a force ride. As you become more comfortable with physics, you become better at matching what you want to know with the energy or force approach. Sometimes you need both and sometimes they are redundant.

7. In the roller coaster below, the initial height of the roller coaster is given.

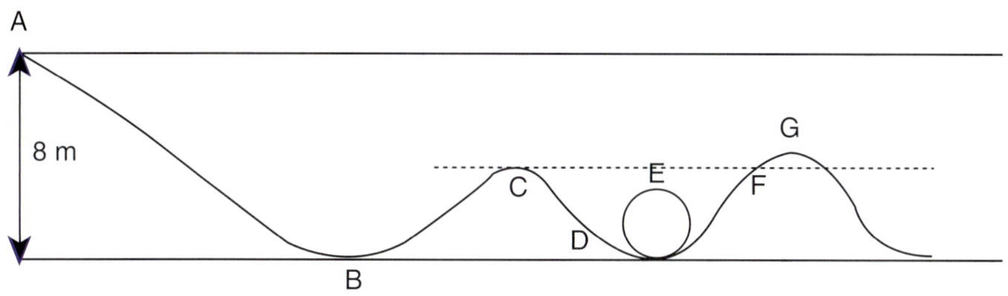

a) At which two points does the roller coaster have the same speed? (Neglect friction.)

b) How did you determine your answer? Write down your approach in your log.

8. Describe how the new roller coaster shown below is different from the roller coaster in **Step 7**.

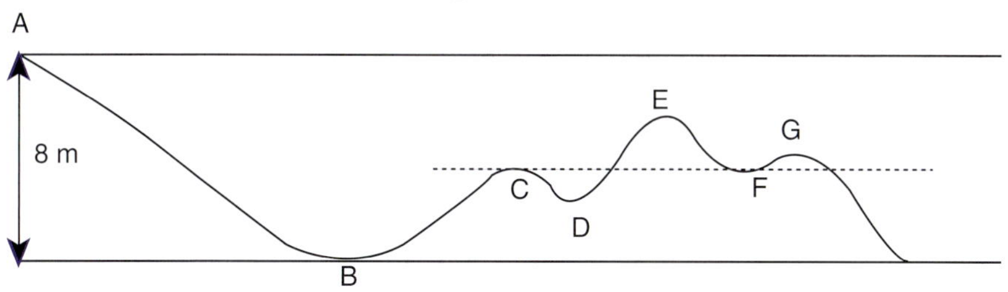

a) In this roller coaster, at which two points does the roller coaster have the same speed?

b) How did you determine your answer? Write down your approach in your log.

9. In either roller coaster on the previous page, the track could have been replaced with the dotted line.

a) Why would the flat track be less fun than the roller coaster track?

10. Look at the following diagram.

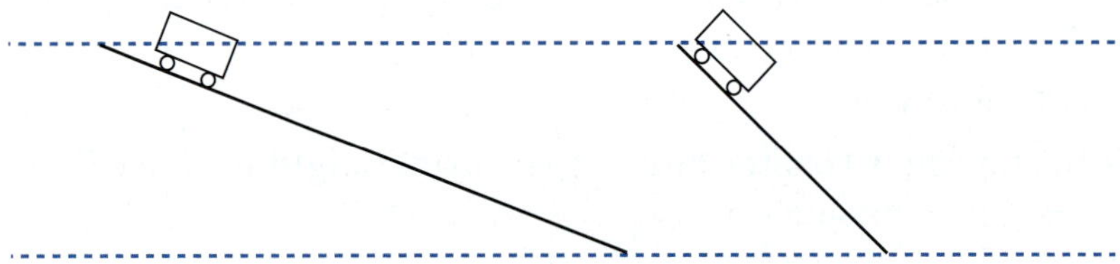

a) Using energy principles, predict which cart would have the greatest speed when it reaches the bottom.

b) Predict which cart will get to the bottom in the least time. On what did you base your response? Write down your explanation in your log.

Part B: Using Vectors to Describe a Path

1. Your teacher will give you a penny.

a) Record the date of the penny. Hide the penny somewhere in the room.

b) Provide a set of detailed instructions to allow another student to find your penny if they start at your desk.

2. Exchange directions and try to find your partner's penny.

a) Did your instructions include how far they have to walk?

b) Did your instructions include any changes in direction (left turns or right turns)?

c) Did your instructions include reaching up or down?

d) Rewrite the instructions so that each instruction describes how far the person should move in meters and in which direction.

e) Compare this new set of directions with your first set. What advantages and disadvantages does each set have?

FOR YOU TO READ

Adding Scalars and Vectors

You can walk 30 m east. You can ride at 60 mph toward Mexico. Both descriptions include a number and a direction. Both are vectors. There are some descriptions that include a number, but no direction. There are 26 students in the classroom. The temperature is 18° C. Physicists have found that whether a number has a direction or not is an extremely important distinction. You can understand the world better if you recognize which quantities can have directions and deal with them accordingly.

It's fairly obvious that some quantities, like force always have directions. Some quantities, like your age, never have direction. There are some quantities, like how fast you are traveling, that can include direction. Your car can be traveling at 30 mph or you can describe the car traveling at 30 mph north.

A quantity with both a number (often referred to as magnitude) and a direction is called a vector. A quantity with a number and no direction is a scalar.

Scalars are easy to add, subtract, multiple, and divide. If you walk 15 km and then walk another 20 km, the total distance traveled is 35 km. After walking 35 km, you know how tired you will be and how worn your shoes will be. This scalar quantity is called distance. Traveling from New York to Florida, your average speed might be 50 mph. This takes into account the total distance traveled and the total time, but does not take into account any turns you made. Speed is also a scalar.

Displacement is a vector. You may walk 15 km north and then walk another 20 km east. Your total displacement is only 25 km. To add vectors, you must draw them and use vector addition. In this case, it is an application of the Pythagorean Theorem.

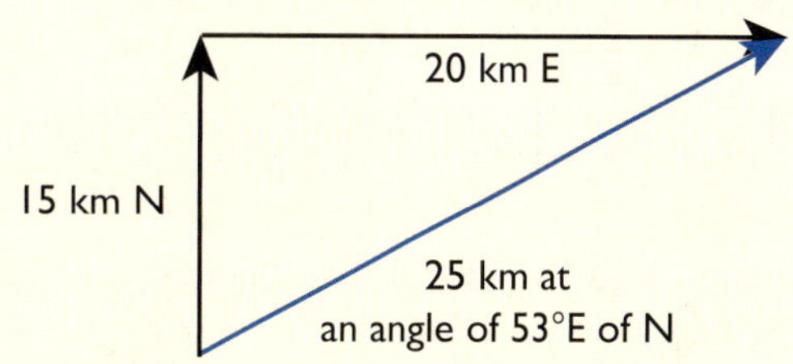

To add these two displacement vectors, you first drew the 15-km vector north. The vector was drawn to scale. To scale means that the length of the vector may be 15 cm or 1.5 cm or 0.15 cm. When you then draw the 20-km east vector you begin at the tip of the 15-km vector and then draw this vector using the same scale. If the 15-km

→

vector was 1.5 cm, then the 20-km vector will be 2.0 cm. You find the sum by drawing the resultant vector from the tail (or beginning) of the first vector to the end (or tip) of the second vector. This completes a triangle. You can then measure the length of the resultant with a ruler. In this case, the vector would be 2.5 cm that corresponds to a displacement of 25 km. You must measure the angle because all vectors have a direction. Using a protractor, you find that the angle is close to 53° east of north (east of the north direction).

Since the vectors are perpendicular, you can also use the Pythagorean Theorem. In any right triangle $a^2 + b^2 = c^2$ where c is the hypotenuse and a and b are the sides. The value c = 25 km is the solution from the Pythagorean Theorem that agrees with the vector addition diagram as it should. Just as you found the length, mathematically, you can also find the angle. The tangent button on the calculator, often labeled "tan" will tell you the angle if you know the lengths of the sides. Divide the side opposite the angle (= 20) by the side adjacent to the angle (= 15). By tapping "inverse tan" the calculator will provide the angle of 53°. Forces are vectors and add the same way as displacements.

Energy—A Scalar Quantity

Energy is a scalar and addition of scalars is simple. As you explored in earlier activities, the roller coaster ride may have GPE (gravitational potential energy) and KE (kinetic energy). It may have used electrical energy to lift the roller coaster to the top of the first hill. All energies can be calculated, but they are all measured in the same units, joules. And to find the total energy at any place or at any time, you just add up all the energies. This is what makes the roller coaster analysis using energies so powerful. After the roller coaster begins moving downhill, the total energy remains the same. The roller coaster begins with GPE and as the cart moves, the GPE converts to KE as the roller coaster picks up speed and then converts back to GPE as the cart goes higher and loses speed. Whatever the energy of the roller coaster is at the beginning of the ride, that is the energy at all times without friction. If two points on the roller coaster have the same height, then they must have the same GPE. If they have the same GPE, then they also have identical KE. It doesn't matter what the car did between the two points. It may have gone up, down, or in a loop-the-loop, but the KE will be the same at all points a specified distance above the ground.

In this activity, you looked at a roller coaster in **Step 7**. The speeds of the roller coasters are the same at points C and F. Both points C and F have the same height and therefore have the same GPE. Since *all* points on the roller coaster have the same total energy (GPE + KE) then both points must have the same KE. The same KE implies the same speed. (KE = $\frac{1}{2}mv^2$).

In the roller coaster in **Step 8** of this activity, the speeds were still the same at points C and F even though the track changed between C and F.

In roller coaster, energy considerations tell you three things:

- **The total energy (GPE + KE) is the same at every point (without friction or motors).**
- **The GPE depends only on the height (GPE = *mgh*) since the mass and the gravitational force remain the same.**
- **If two points on a roller coaster have the same height, the roller coaster is moving at the same speed at those two points.**

Energy considerations are path independent. You can look at the energy at one point and compare it to the energy at a later point. The energy will remain the same. It doesn't matter what happens between the places that are of interest.

In these four roller coaster sections, the cars begin at the top with zero KE and 20 J of GPE. When they reach the bottom, all will have the same KE (kinetic energy). This means that they will all have the same speed. To find this KE or speed, you had to only look at the beginning point and the final point. The path does not affect the final speed.

Force—A Vector Quantity

Although the roller coaster cars all get to the bottom with the same speed, they do not get there in the same time. To find the time, you would have to look at the forces and this becomes a vector problem. In all tracks, the force of gravity is always down. The normal force is always perpendicular to the track.

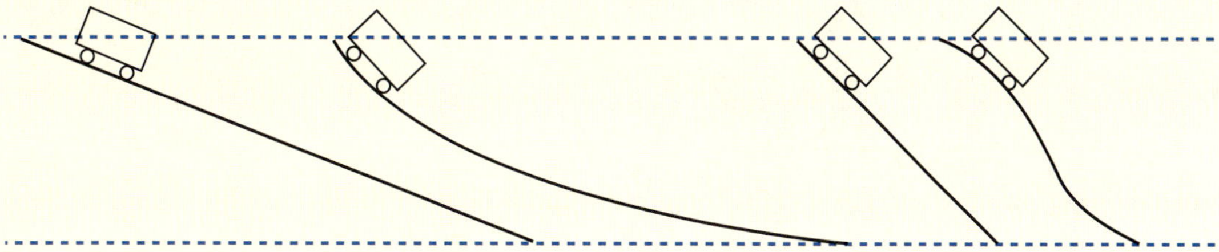

The straight tracks are the easiest to analyze. The force of gravity and the normal force remain in fixed directions. You move down the incline and go faster and faster. The steeper the slope, the larger the gravitational force down the incline and the quicker you get to the bottom. It is a big acceleration for a short time and you reach the maximum speed. On a small incline, there is a small resultant force down the incline. It is a small acceleration for a long time and you reach the same maximum speed.

The inclines with shifting directions add to the thrill. Your speed changes as you move to different heights. As you move closer to the ground, your speed increases. The normal force (the force of the track on you) is always changing direction. This causes you to accelerate in lots of different directions. The changes in the acceleration (both in size and in direction) give you that bouncy feeling and the thrill of the roller coaster. The diagram below shows the gravitational force and normal forces at different points on a roller coaster.

- **On the straight incline, the gravitational force and the normal force remain in fixed directions. The car has a constant acceleration in magnitude and direction.**
- **On the curved incline, the normal force changes direction (it must be perpendicular to the incline) and changes in magnitude. The car has a changing acceleration in magnitude and direction. This provides big thrills.**

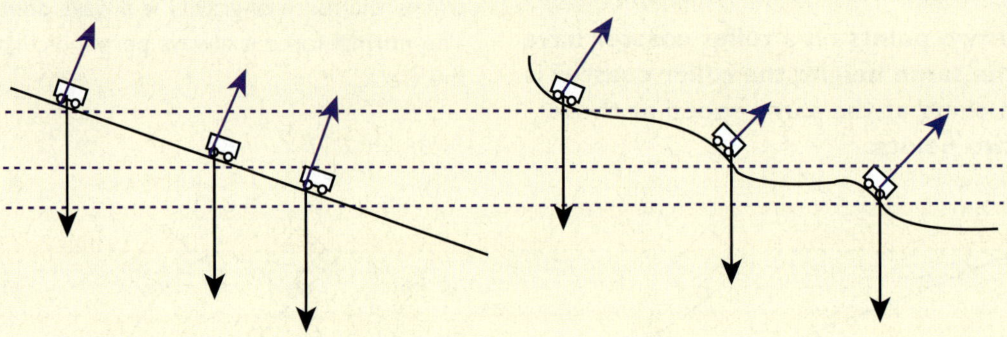

- **The speeds of the cars are identical on the two inclines. When the heights above the ground are the same, the GPE is the same. If the GPE is the same and the total energy is the same, the KE is the same. If the KE is the same, then the speed is the same.**

When to Consider Force or Energy

The mathematics of energy conservation requires simple addition. The mathematics of forces and accelerations requires vector addition. When the roller coaster looks complex, with lots of curves, physicists think of energy first because of the ease of using simple addition rather than vector addition.

When asked about how much time something will take, physicists think about forces and accelerations. Acceleration is the change in speed with respect to time.

Force and energy are related. The force of gravity does work on the roller coaster and increases its KE. Changes in energy always require work by a force. Work is a force applied over a distance ($W = F \cdot d$). The only external force doing work is gravity. When the roller coaster is moving down, the force is down and the displacement is down. There is positive work on the roller coaster and it increases its KE. The normal force does no work since it is always perpendicular to the displacement. No part of the normal force is ever in the direction the roller coaster car is moving.

$W = F \cdot d$ where F is the force of gravity ($F = mg$) and d is the displacement and the dot tells you that the vector multiplication requires you to use only the distance in the same direction of gravity. This is the change in height (Δh).

The work done by gravity is $W = mg\Delta h$. The change in KE is $mg\Delta h$. This is equal to the loss in GPE. GPE = $mg\Delta h$.

Reflecting on the Activity and the Challenge

The thrill of the roller coaster comes from the changing velocities. You can analyze the changes in speed using energy considerations. Energy is a scalar. GPE can be easily calculated at every point on the roller coaster. Once you know the GPE, you can find the KE and then determine how fast the roller coaster moves. Understanding the mathematics of energy is as simple as $2 + 3 = 5$. Energies add with simple arithmetic just like all scalars.

You can also analyze the thrills of changing velocities by noting the forces acting on the roller coaster. Forces are vectors. They have both magnitude and direction. When more than one force acts on a roller coaster (e.g., the gravitational force and the normal force), you have to add them using vector arithmetic. You can always do this with a vector diagram. When the forces are perpendicular, you can also do this mathematically using the Pythagorean Theorem.

Designing a roller coaster requires you to know how fast it will be going at each point. You can use energy considerations to determine this.

You will also have to know how large the forces are because you will need to figure out the strength of the materials needed to provide the forces by the track. If too large a force is applied, the track may break. Adding the forces can provide you with this information.

You will also have to know accelerations of the passengers. Too large an acceleration or a change in acceleration and the riders may get sick or become unconscious. Newton's Second Law relating forces and accelerations ($F = ma$) can help you with this.

Making an exciting roller coaster requires changes in forces. The whips and turns and the ups and downs will change the speeds, the energy, and the forces on the passengers.

Physics To Go

1. A roller coaster makes a sharp right turn. The velocity of the roller coaster car is 5.0 m/s south before the turn and 5.0 m/s west after the turn.
   a) Determine the change in velocity of the roller coaster car using a vector diagram.
   b) Determine the change in velocity of the roller coaster car using the Pythagorean Theorem and the tangent button on your calculator.

2. A roller coaster makes a sharp right turn as it descends a hill. The velocity of the roller coaster is 5.0 m/s south before the turn. After the turn, the velocity of the roller coaster is 12.0 m/s west but it is also pointing downward at an angle of 25°.

   Ignore the downward angle.
   a) Determine the change in velocity of the roller coaster using a vector diagram.
   b) Determine the change in velocity of the roller coaster using the Pythagorean Theorem and the tangent button on your calculator.
   c) How would your answer change if you took into account the downward angle?

3. A roller coaster wall allows the riders to be on their side as they whip around a curve. The normal force (perpendicular to the track) is equal to 3000 N. The weight of the car with riders is 5000 N. What is the total force acting on the riders? (Be sure to include both magnitude and direction.)

4. All roller coasters that begin at the same height have the same speeds at the bottom.

   Explain why these two roller tracks provide the same change in speed.

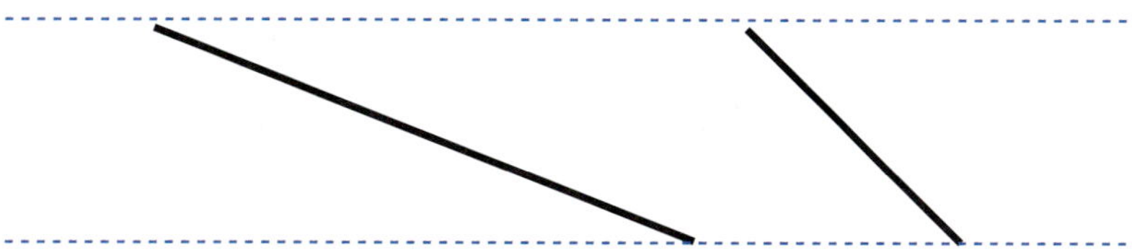

5. Identify the following as vectors or scalars:
   a) distance
   b) displacement
   c) speed
   d) velocity
   e) acceleration
   f) force
   g) kinetic energy
   h) potential energy
   i) work

6. Which of the following are vectors and which are scalars?
   a) Mark traveled 30 km.
   b) Maia's weight (the force of gravity on her) is 600 N.
   c) The roller coaster car had a kinetic energy of 1200 J.
   d) The car was traveling at 30 m/s toward the center of town.

Activity 9 Safety is Required but Thrills are Desired

GOALS

In this activity you will:

- Calculate the speed of the roller coaster at different positions using conservation of energy.
- Calculate the acceleration of the roller coaster at turns.
- Determine if the acceleration is below 4 g for safety.
- Determine if the speed at the top of a loop is sufficient for safety concerns.
- Create sounds and scenery to enhance the thrills of a roller coaster ride.

What Do You Think?

In 2003 a person died on the roller coaster in Disneyland. They closed the roller coaster immediately. Accidents occur very rarely on roller coasters.

- **Does the knowledge that people can get hurt or die on a roller coaster change the thrill of the ride?**
- **Would your answer change if you found out that one-half of all roller coaster rides ended in the death of its passengers?**

Record your ideas about these questions in your *Active Physics* log. Be prepared to discuss your response with your small group and the class.

For You To Do

1. Safety is one of the criteria that you must meet in designing your roller coaster.

   a) List three reasons why safety is a major concern for roller coaster designers.

   b) How safe is safe? Your answer may depend on what injuries you describe for the roller coaster. Nausea and vomiting are one type of injury, broken bones are a second type of injury, becoming unconscious is a third type of injury, and death is the greatest injury. For the four types of injury listed, make an estimate of how many people could get injured on a roller coaster ride before it would be closed to the public. Be sure to include whether these are injuries in a day, a month, or a year.

2. Astronauts going into space had to withstand very large accelerations during rocket launch. After many tests of test pilots and racecar drivers, it was determined that people will become unconscious if the acceleration is greater than 9 g (or 9 times the acceleration due to gravity, that is, 9 x 9.8 m/s^2). Some people black out at 5 g or 6 g. This unconsciousness results from the blood leaving the brain during the high acceleration.

   The roller coaster manufacturer has indicated that the maximum acceleration at any place on the roller coaster can not exceed 4 g (or 4 times the acceleration due to gravity, that is, 4 x 9.8 m/s^2).

   a) At what locations on the roller coaster are there accelerations?

   b) If the roller coaster were to fall straight down, what would be the acceleration?

   c) Is this a safety concern?

3. A roller coaster car is traveling at 30.0 m/s at the bottom of a loop. The radius of the loop is 9.0 m.

a) Using the conservation of energy ($mgh + \frac{1}{2}mv^2$ = constant), calculate the initial height of a roller coaster to give it a speed of 30.0 m/s at the bottom of a loop at ground level. At the top of the roller coaster, the velocity is 0 m/s. At the bottom of the roller coaster, the height h equals 0 m.

b) Using the equation $a = v^2/R$, calculate the acceleration at the bottom of the loop.

c) Is this a safety concern?

d) At what speed would this loop with a radius of 9.0 m begin to be a safety concern?

e) At what speed would a loop with a smaller radius of 7.0 m begin to be a safety concern?

f) How fast would the roller coaster car be traveling at the top of the loop? (The top of the loop is 18.0 m above the ground since the loop has a radius of 9.0 m, and thus a diameter of 18.0 m.) You must use the initial height of the roller coaster that you calculated above to solve this problem.

g) Using the equation $a = v^2/R$, calculate the acceleration at the top of the loop.

h) Is this a safety concern?

i) There are two safety concerns regarding accelerations in a loop. The acceleration cannot exceed 4 g. This excessive acceleration would occur at the bottom of the loop, if at all. The acceleration at the top must be greater than 1 g (9.8 m/s^2). If the acceleration required for circular motion at the top of the loop is less than 1 g, the roller coaster car will leave the track and plummet to the ground. The speed at the top of the roller coaster must be large enough to require acceleration at least as great at 9.8 m/s^2.

4. The track must be strong enough to hold the roller coaster car without breaking. You can calculate the minimum strength of a track by assuming that the roller coaster car is filled with big football players or small Sumo wrestlers.

a) What force would a roller coaster track have to supply to hold up a car filled with passengers if the total mass were 1000.0 kg on a horizontal section of track. (Remember the equation for weight is $W = mg$ where $g = 9.8 \text{ m/s}^2$.)

5. When the roller coaster makes turns, there must be a centripetal force pushing the roller coaster car toward the center of the circle. There are three types of turns that can be analyzed:

   - **a sharp left turn on a flat track**
   - **a sharp left turn where the car is turned on its side and the track is vertical**
   - **a loop where the car is moving in a vertical circle**

   a) For each of the three turns, identify the direction of the centripetal force (the force keeping the car moving in a circle).

   b) For each of the three turns, identify the centripetal force. (The centripetal force could be the force of friction, the gravitational force, the normal force, the force of tension on a rope, the force of the wheels in the track or any combination of these forces.)

6. A roller coaster car has a maximum mass of 1000.0 kg and a sharp left turn where the car is turned on its side and the track is vertical with a radius of 12.0 m and the car is moving at 15.0 m/s.

   a) Calculate the centripetal force required to keep the car moving in the curve.

7. A roller coaster car has a maximum mass of 1000.0 kg and a sharp left turn on a flat track with a radius of 12.0 m and the cart is moving at 15.0 m/s.

   a) Calculate the centripetal force required to keep the car moving in the curve.

8. A roller coaster car has a maximum mass of 1000.0 kg and is about to enter a loop that has a radius of 12.0 m. The car is moving at 15.0 m/s.

a) Calculate the centripetal force required to keep the car moving in the curve.

b) This centripetal force that you calculated is the sum of the normal force of the track up toward the center of the roller coaster car and the force of gravity down to the ground. Copy the following diagram into your log:

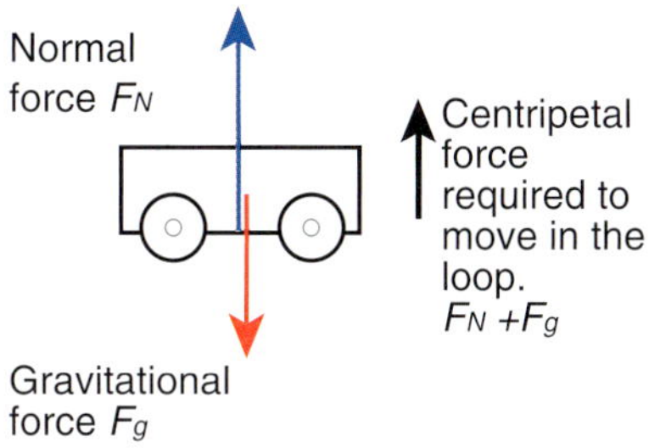

c) Calculate the force of gravity on the car.

d) Calculate the normal force on the car. This is the strength required of the steel that makes up the roller coaster.

9. The roller coaster designer can play all sorts of tricks to produce extra thrills on the same roller coaster.

a) Describe how you can add suspense to the trip up to the top of the first hill.

b) Describe how you can use sounds during the roller coaster ride to add to the thrills.

10. The choice of scenery surrounding the roller coaster can also add to the thrill. The roller coaster can look like it will dive into water when it is descending. The roller coaster can look like it will hit a building when it rounds a curve.

a) Describe three visual effects through the use of scenery that will add to the thrill of your roller coaster.

11. Thrills can come from big speeds. Thrills can also come from acceleration (changing velocity with respect to time). Change in velocity can be a change in speed or direction.

a) Describe three ways in which you can add thrills to a roller coaster design by having changes in velocity.

b) Describe three ways in which you can add thrills to a roller coaster design by having changes in acceleration.

FOR YOU TO READ

Roller Coaster Safety

The roller coaster has to be safe in order to be fun. Analysis of the safety requirements of the roller coaster is a valuable way of reviewing some of the physics in earlier activities.

You know from research with test pilots that people will not be safe if their acceleration is greater than 4 g. A free fall provides an acceleration of 1 g. Roller coasters may have steep inclines but they are generally less than free fall and therefore have an acceleration less than 1 g.

When the roller coaster rips around a corner or moves through the bottom of the loop, the acceleration can be much more than 1 g. Analyze the acceleration at the bottom of a loop. The acceleration can be computed by recognizing that the roller coaster at this location is moving in an arc of a circle. The centripetal acceleration must be toward the center of the circle and can be calculated by using the equation $a_c = v^2/R$. By varying the speed or the radius of the circle in the roller coaster design, you can limit the acceleration to less than 4 g.

Sample Problem 1

A roller coaster car with a mass of 800 kg is traveling at 15.0 m/s at the bottom of a loop. The loop has a radius of 5.0 m.

a) What is the required centripetal acceleration to keep the car moving in a circle?

a) Strategy: Use the equation for centripetal acceleration.

Givens:

$v = 15.0$ m/s

$R = 5.0$ m

Solution:

$$a = \frac{v^2}{R}$$

$$= \frac{(15 \text{ m/s})^2}{5.0 \text{ m}}$$

$$= 45 \text{ m/s}^2$$

This acceleration is greater than 4 g (4×9.8 m/s^2 = 39.2 m/s^2) and is therefore unsafe.

b) One way to lower this acceleration would be to lower the velocity. Assume that the new design gives the car a velocity of 12.0 m/s. Calculate the required centripetal acceleration to keep the car moving in a circle.

b) Strategy: Use the equation for centripetal acceleration, again.

Givens:

$v = 12.0$ m/s

$R = 5.0$ m

$$a = \frac{v^2}{R}$$

$$= \frac{(12.0 \text{ m/s})^2}{5.0 \text{ m}}$$

$$= 29 \text{ m/s}^2$$

This acceleration is now less than 4 g (4 x 9.8 m/s^2 = 39.2 m/s^2) and is therefore safe.

c) Another way to lower the acceleration is to make the loop larger. Using the original speed of 15.0 m/s, calculate the required centripetal acceleration if the radius of the loop were 7.0 m.

c) Strategy: Use the equation for centripetal acceleration, again.

Givens:

$v = 15.0$ m/s

$R = 7.0$ m

$$a = \frac{v^2}{R}$$

$$= \frac{(15.0 \text{ m/s})^2}{7.0 \text{ m}}$$

$$= 32 \text{ m/s}^2$$

→

This acceleration is now less than 4 g ($4 \times 9.8\ m/s^2 = 39.2\ m/s^2$) and is therefore safe. The largest centripetal acceleration (at the bottom of the loop) also requires the largest centripetal force. This maximum force will inform you as roller coaster designer of the strength of materials required to build this part of the roller coaster. The force moving the car in a circle is the normal force. The normal force required to move the car in a circle is much greater than the normal force that would support the car at rest at the bottom of the incline. The "at rest" car requires no net force. The normal force up (provided by the track) must equal the gravitational force down. This is shown with the vector diagram below. (Notice that the gravitational force acts on the center of the car while the normal force acts on the bottom of the car where it touches the surface.)

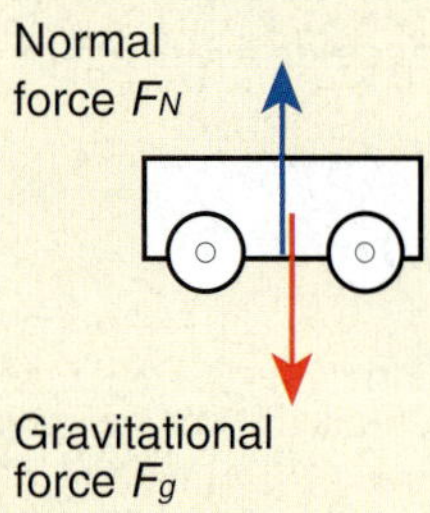

To move the car in a vertical circle, a centripetal force is required. The sum of the normal force and the gravitational force must equal the centripetal force required. Since the gravitational force is down, the normal force must be greater to provide the additional upward force needed. This is shown in the vector diagram below.

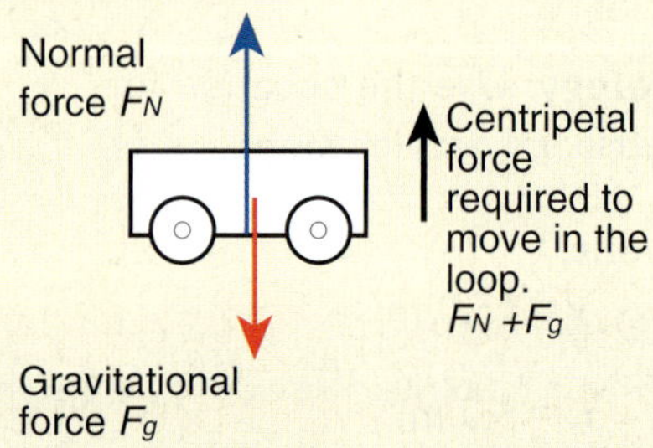

The centripetal force required can be calculated using Newton's Second Law: $F = ma$. In this case $F_c = \frac{mv^2}{R}$.

Sample Problem 2

A roller coaster car with a mass of 800.0 kg is traveling at 15.0 m/s at the bottom of a loop. The loop has a radius of 5.0 m. (**Part (a)** is a repetition of the calculation above.)

a) What is the required centripetal acceleration to keep the car moving in a circle?

b) What is the required centripetal force to keep the car moving in the circle?

c) What is the normal force by the track on the car?

a) Strategy: Use the equation for centripetal acceleration.

Givens:

$v = 15.0\ m/s$

$R = 5.0\ m$

Solution:

$$a = \frac{v^2}{R} = \frac{(15.0 \text{ m/s})^2}{5.0 \text{ m}} = 45 \text{ m/s}^2$$

This acceleration is greater than 4 g ($4 \times 9.8 \text{ m/s}^2 = 39.2 \text{ m/s}^2$) and is therefore unsafe.

b) *Strategy:* Use the equation for centripetal force.

Givens:

$v = 15.0$ m/s

$R = 5.0$ m

$m = 800.0$ kg

Solution:

$$F = \frac{mv^2}{R}$$

$$= \frac{(800.0 \text{ kg})(15.0 \text{ m/s})^2}{5.0 \text{ m}}$$

$$= 36{,}000 \text{ N}$$

This net force of 36,000 N up will allow the cart to move in the vertical circle.

c) *Strategy:* The normal force must be 36,000 N greater than the gravitational force to provide a net force of 36,000 N as required.

Solution: The gravitational force (weight) is:

$$w = mg = (800.0 \text{ kg})(9.8 \text{ m/s}^2)$$

$$= 7840 \text{ N (about 7800 N)}$$

Therefore, the normal force must equal 36,000 N + 7800 N = 43,800 N or about 44,000 N.

This indicates that the strength of the metal of the roller coaster must be at least 44,000 N or the track will break.

Similar calculations can be completed with the force required to make a turn on a horizontal part of the roller coaster and a turn where the roller coaster banks on its side as it whips around a turn.

Another safety feature requires that the speed at the top of the loop is great enough to complete the loop. A car that has too little speed will not make it to the top of the roller coaster and will not be able to move in the circle. It will find itself falling to the ground as the following diagram illustrates.

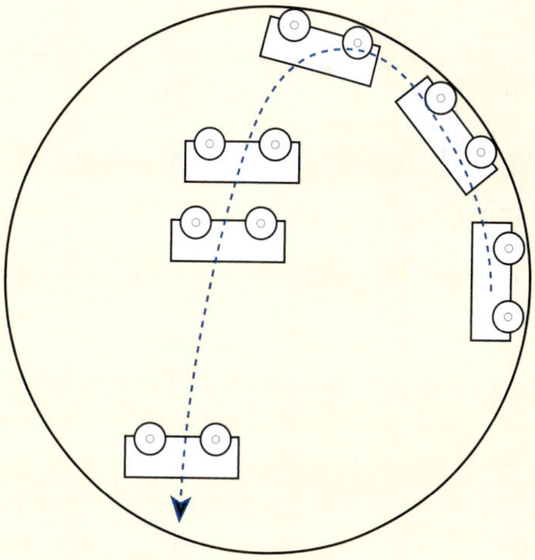

If gravity were the only force acting at the top of the roller coaster, then the car must require a centripetal acceleration equal to gravity.

Sample Problem 3

What is the minimum speed required at the top of the loop to ensure that the car does not leave the track? The car has a mass of 800.0 kg. The loop has a radius of 5.0 m.

Strategy: The minimum speed pertains to the minimum centripetal acceleration of 9.8 m/s². Using the equation for centripetal acceleration, you can find the required speed.

Givens:

$a = 9.8 \text{ m/s}^2$

$R = 5.0 \text{ m}$

Solution:

$$a = \frac{v^2}{R}$$

$$v = \sqrt{aR}$$

$$= \sqrt{(9.8 \text{ m/s}^2)(5.0 \text{ m})}$$

$$= 7.0 \text{ m/s}$$

A speed of 7.0 m/s will be able to complete the upper part of the loop. A speed greater than 7.0 m/s will also be able to make the loop. The greater speed will require a larger centripetal force. The additional force will be provided by the track pushing down on the car.

Reflecting on the Activity and the Challenge

There is lots of creativity in designing a roller coaster. There is lots of creativity in designing a bridge, a building, and a table. All designs are constrained by the physics of the world. A beautiful bridge must also be a bridge that does not collapse. In this activity you learned about the safety features that you must take into account in your design for the roller coaster. You have to ensure that the accelerations are never above 4 g. This will require you to design the curves and loops with radii that limit the accelerations. You must also make sure that if your roller coaster does have a loop, the car will be able to complete the loop. You can vary the radius of any part of the track in the design. You can vary the velocity of the car by changing the launch height for the roller coaster. The higher the first hill, the more speed the coaster will have at the bottom. Safety is required, but thrills are desired. The activity also discussed ways in which you can use sound and scenery to improve the thrills of your design.

Physics To Go

1. An engineering company submits a plan for a roller coaster. What factors will you check to ensure that the roller coaster is safe?

2. A roller coaster car is traveling at 20.0 m/s at the bottom of a loop. The radius of the loop is 12.0 m.
   a) Using the conservation of energy ($mgh + \frac{1}{2}mv^2$ = constant), calculate the initial height of a roller coaster to give it a speed of 20.0 m/s at the bottom of a loop at ground level.
   b) Using the equation $a = v^2/R$, calculate the acceleration at the bottom of the loop.
   c) Is this a safety concern?
   d) At what speed would this loop with radius of 12.0 m begin to be a safety concern?
   e) At what speed would a loop with a smaller radius of 7.0 m begin to be a safety concern?

3. A roller coaster car is traveling at 25.0 m/s at the bottom of a loop. The radius of the loop is 10.0 m.
   a) Calculate the acceleration at the bottom of the loop.
   b) Is this a safety concern?

4. A roller coaster has an initial height of 50.0 m.
   a) What will be the speed of the roller coaster car at the bottom of the hill?
   b) The roller coaster car goes into a loop with a radius of 10.0 m. What is the acceleration required to move in the loop?
   c) What will be the speed of the roller coaster car at the top of the loop?
   d) What will be the required acceleration at the top of the loop to keep the car moving in a circle?
   e) Explain whether the roller coaster is safe at the bottom and the top of the loop.

5. A roller coaster has a loop with a radius of 8.0 m (diameter = 16.0 m).
   a) What speed must the roller coaster car have at the top of the loop if the only force at the top is the force of gravity and the centripetal acceleration is therefore 9.8 m/s^2?
   b) How high must the first hill be to provide this speed at the top of the loop?

6. A roller coaster car, when filled with people, has a mass of 900.0 kg. The roller coaster car rounds a curve on the ground with a radius of 18.0 m at a speed of 12.0 m/s.
   a) What is the centripetal acceleration of the car?
   b) What is the centripetal force on the car?
   c) What will provide this centripetal force?

7. A roller coaster car, when filled with people, has a mass of 900.0 kg. The roller coaster car rounds a curve on the ground with a radius of 15.0 m at a speed of 20.0 m/s.
   a) What is the centripetal acceleration of the car?
   b) What is the centripetal force on the car?
   c) The wheels in the tracks can provide a force of 25,000 N. Is the roller coaster safe?

8. A roller coaster is able to complete a loop when the car has two passengers. The car is loaded with six people.
   a) Will the centripetal acceleration change as a result of the change in mass?
   b) Will the roller coaster be going faster, slower, or the same speed at the bottom of the loop with the extra passengers?
   c) Will the roller coaster require a stronger material because of the increased number of riders?

PHYSICS AT WORK

Mark Rosenzweig

Not a job for the Queasy.

If you're scared of doing loop-the-loops at 100 mph or taking hairpin turns at 4 *g*'s, this is not the profession for you. But if you're one of those people who enjoys extreme excitement and quick thrills, Mark may be living your professional dream.

Mark works for Zamperla, one of the largest designers of roller coasters, spinners, and other stomach-turning amusement park rides in the world. "A lot of my time is spent at places like Disney World, Universal Studios, and other amusement parks like Legoland. I believe that visiting with customers and watching the reaction of park guests as they exit a ride is one of the best ways to judge its success."

Mark is not a scientist by trade. He did not go to graduate school or study engineering in college. After growing up in Long Island and earning an undergraduate degree in psychology at the State University of New York at Oneonta, he took jobs at small amusement parks in Long Island and then Michigan. Then Zamperla, an Italian company with an office in New Jersey, came calling, and the rest is hair-raising history.

As marketing and sales director, Mark is responsible for communicating the wishes and desires of potential riders to Zamperla's team of designers and engineers. And that means understanding many general principles of physics, engineering, and the human body. So, even though he is not a trained physicist, engineer, or human physiologist, he needs knowledge in all these areas to do his job effectively.

"One of our newest rides, called the Volari, will debut in several parks around the world this year," he says. "It's a flying roller coaster—something like hang-gliding. Riders are suspended in the air on their stomachs from an overhead track. Mark's team will work with the parks' staff to balance issues such as *g*-forces with turning angles and the strength of the track for a particular ride. "We can custom make a ride to fit in almost any space," says Mark. "If a customer wants their roller coaster to be indoors, weaving around columns—we can do that. We can do that because we combine an understanding of the physics with an appreciation for the consumers' need to be entertained."

Chapter 4 Assessment

Now that you have finished this chapter, it is time to complete your **Chapter Challenge**. Go back to the beginning of the chapter and read the challenge again. You may have already decided for whom you will be designing the roller coaster. Will it be for a young child, an adult, a physically challenged person, someone who is visually impaired, or a thrill-seeking daredevil?

Review the criteria. All designs must have certain components including:

- **at least two hills**
- **one horizontal curve**

In addition, you will have to provide evidence that the ride is safe. Safety data must include the height, speed, and acceleration of the roller coaster at five designated locations. Finally, you will have to calculate the energy required to get the roller coaster rolling.

Next, review the grading system that you and your classmates established before you started this chapter. Perhaps you may wish to change some of the criteria and the marking scheme now that you have more information about the topic. The more you know about what is expected of you for the **Chapter Challenge**, the better you will be at completing your assignment.

Physics You Learned

Velocity

Acceleration

Gravitational potential energy

Kinetic energy

Spring potential energy

Conservation of energy

Mass and weight

Hooke's Law

Force vectors

Weight changes during acceleration

Newton's First Law

Newton's Second Law

Circular motion

Centripetal acceleration

Centripetal forces

Normal forces

Scalars and vectors

Chapter 5

LET US ENTERTAIN YOU

Scenario

Most entertainment today comes from the communication of sound and light signals. You look forward to television shows, movies, and rock concerts. The sound signals that entertain you come from voices or musical instruments. Light signals make the images you see on TV or in the movies. Specially designed light patterns add to the effect of a rock concert.

Challenge

You have been made part of a committee to design a two- to four-minute sound and light show to entertain other students your age. But unlike the professionals, you have neither the funds nor the technology available to rock stars or MTV™ productions. All the sounds you use must come from musical instruments or sound makers that you build yourself, or from human voices. Some of these sounds may be prerecorded and then played back during your show. If your teacher has a laser and is willing to allow you to use it, you may do so. All other light must come from conventional household lamps.

Criteria

Work with your classmates to agree on the relative importance of the following assessment criteria. Each item in the list has a point value given after it, but your class must decide what kind of grading system to use.

1. The variety and number of physics concepts used to produce the light and sound effects: 30 points

   four or more concepts: 30 points

   three concepts: 25 points

   two concepts: 20 points

   one concept: 10 points

2. Your understanding of the physics concepts: 40 points

   Following your production, you will be asked to:

   a) Name the physics concepts that you used. 10 points
   b) Explain each concept. 10 points
   c) Give an example of something that each concept explains or an example of how each concept is used. 10 points
   d) Explain why each concept is important. 10 points

   As a class, you will have to decide if your answers will be in an oral report or a written report.

3. Entertainment value: 30 points

   Your class will need to decide on a way to assign points for creativity. Note that an entertaining and interesting show need not be loud or bright.

You will have a chance later in the chapter to again discuss these criteria. At that time, you may have more information on the concepts and how you might produce your show. You may want to then propose changes in the criteria and the point values.

Activity 1 Making Waves

What Do You Think?

One of the largest tsunamis (tidal waves) grew from about 0.7 m high in the open ocean to 35 m high when it reached shore.

- **How does water move to make a wave?**
- **How does a wave travel?**

Record your ideas about these questions in your *Active Physics* log. Be prepared to discuss your responses with your small group and with your class.

GOALS

In this activity you will:

- Observe the motion of a pulse.
- Measure the speed of a wave.
- Observe standing waves.
- Investigate the relationship among wave speed, wavelength, and frequency.
- Make a model of wave motion.

For You To Do

1. In an area free of obstacles, stretch out a Slinky® so the turns are a few centimeters apart. Mark the positions of the end of the Slinky by sticking pieces of tape on the floor. Measure the distance between the pieces of tape.
   a) Record the distance between the pieces of tape in your log.

2. With the Slinky stretched out to the tape, grab the spring near one end, as shown in the drawing, and pull sideways 20 cm and back. To move it correctly, move your wrist as if snapping a whip. Observe what happens. You have made a transverse pulse.
   a) In what direction does the spring move as the pulse goes by?
   b) A dictionary definition of *transverse* is: "Situated or lying across." Why is *transverse* a good name for the wave you observed?

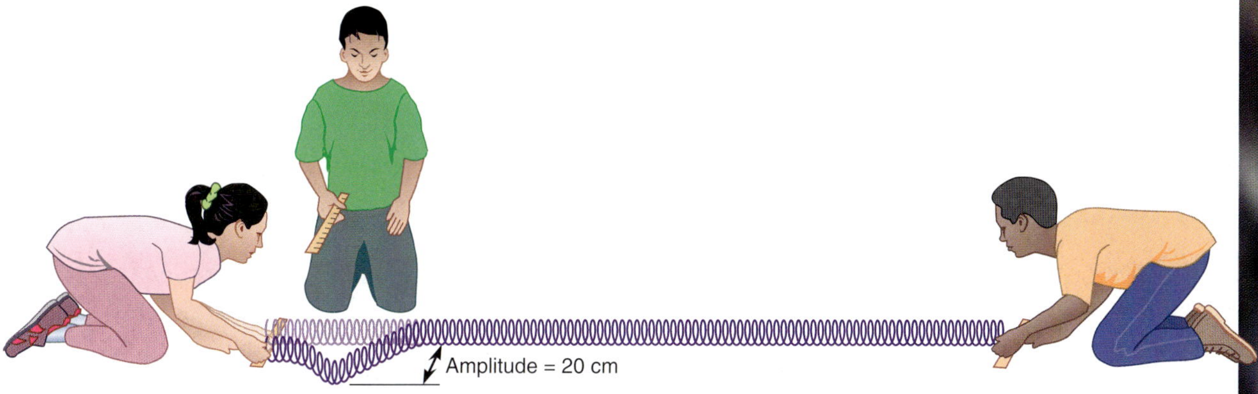

   c) Measure and record the amplitude of the wave. The distance you disturbed the spring is called the amplitude. The amplitude tells how much the spring is displaced.

3. After you have experimented with making pulses, measure the speed of the pulse. You will need to measure the time it takes the pulse to go the length of the spring. Take several measurements and then average the values.
   a) Record your data in the second and third rows of a table like the one on the following page.

Amplitude	Time for pulse to travel from one end to the other	Average time	Speed = $\frac{\text{length of spring}}{\text{average time}}$

4. Measure the speed of the pulses for two other amplitudes, one larger and one smaller than the value used in **Step 3**.
   a) Record the results in the table in your log.
   b) How does the speed of the pulse depend on the amplitude?

5. Now make waves! Swing one end back and forth over and over again along the floor. The result is called a periodic wave.
   a) Describe the appearance of the periodic wave you created.

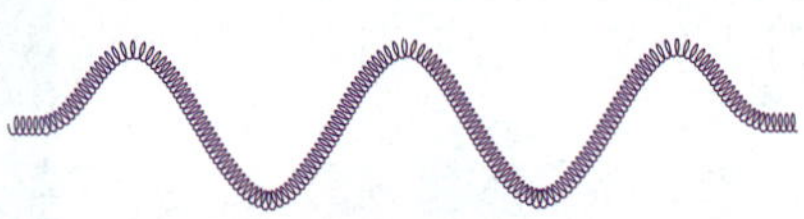

6. To make these waves look very simple, change the way you swing the end until you see large waves that do not move along the spring. You will also see points where the spring does not move at all. These waves are called standing waves.

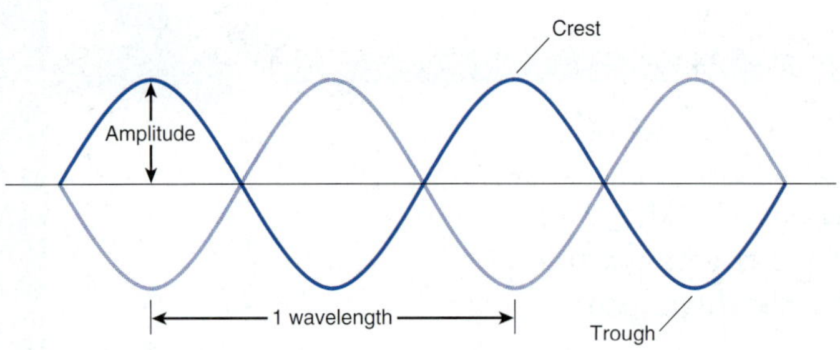

Standing wave

7. The distance from one crest (peak) of a wave to the next is called the wavelength. Notice that you can find the wavelength by looking at the points where the spring does not move. The wavelength is twice the distance between these points. Measure the wavelength of your standing wave.
   a) Record the wavelength of your standing wave in your log.

8. You can also measure the wave frequency. The frequency is the number of times the wave moves up and down each second. Measure the frequency of your standing wave. (Hint: Watch the hands of the person shaking the spring. Time a certain number of back-and-forth motions. The frequency is the number of back-and-forth motions of the hand in one second.)

a) Record the wave frequency in your log. The unit of frequency is the hertz (Hz).

9. Make several different standing waves by changing the wave frequency. Try to make each standing wave shown in the drawings (at right). Measure the wavelength. Measure the frequency.

a) Record both in a table like the one below.

Wavelength (m/cycle)	Frequency (cycles/s or Hz)	Speed (m/s) wavelength × frequency

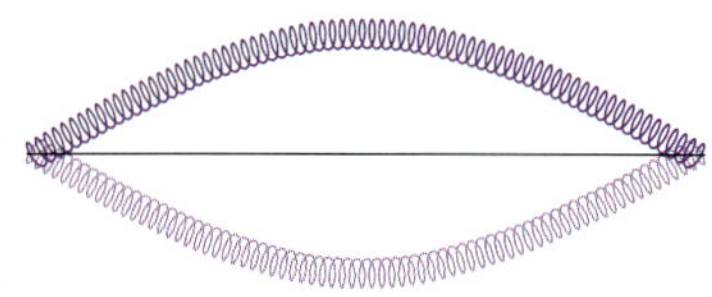

Wavelength = twice Slinky length

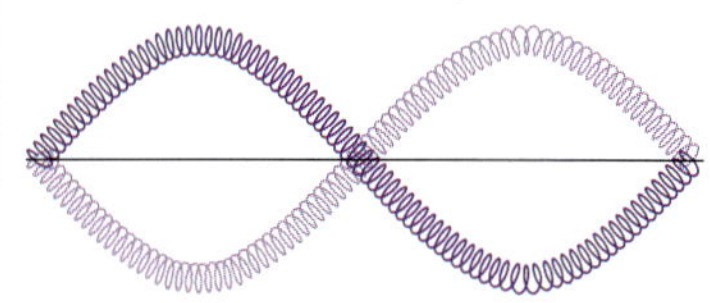

Wavelength = Slinky length

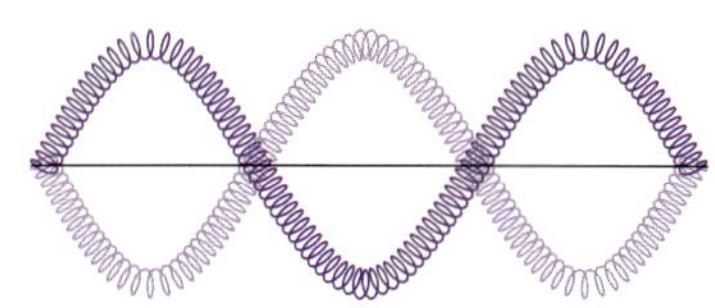

Wavelength = 2/3 Slinky length

b) For each wave, calculate the product of the wavelength and the frequency. Compare these values with the average speed of the pulse that you found in **Steps 3** and **4** above.

10. All the waves you have made so far are transverse waves. A different kind of wave is the compressional (or longitudinal) wave. Have the members of your group stretch out the Slinky between the pieces of tape and hold the ends firmly. To make a compressional wave, squeeze part of the spring and let it go. Measure the speed of the compressional wave and compare it with the speed of the transverse wave.

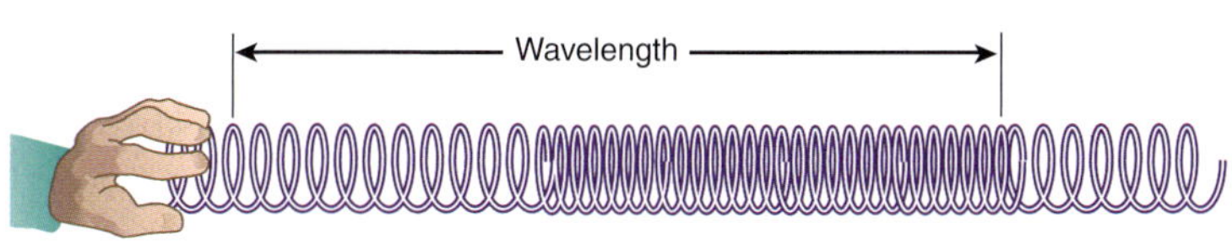

a) Record your results in a table partly like the one after **Step 3**.

b) In what direction does the Slinky move as the wave goes by?

c) A dictionary definition of *compressional* is: "*a*. The act or process of compressing. *b*. The state of being compressed." A dictionary definition of *longitudinal* is: "Placed or running lengthwise." Explain why *compressional* or *longitudinal wave* is a suitable name for this type of wave.

11. To help you understand waves better, construct a wave viewer by cutting a slit in a file card and labeling it as shown.

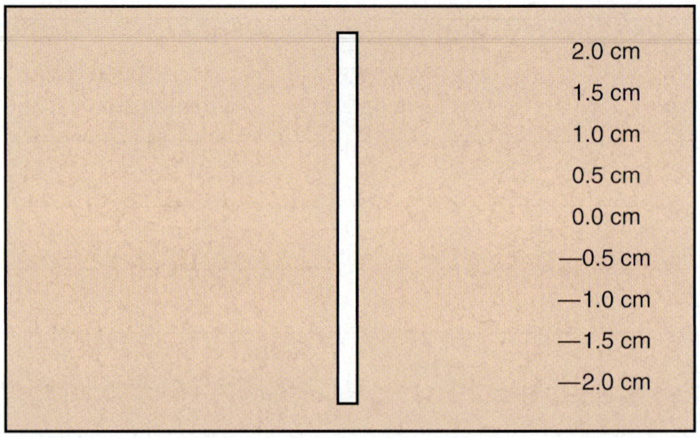

12. Make a drawing of a transverse wave on a strip of adding machine tape. Place this strip under the wave viewer so you can see one part of the wave through the slit.

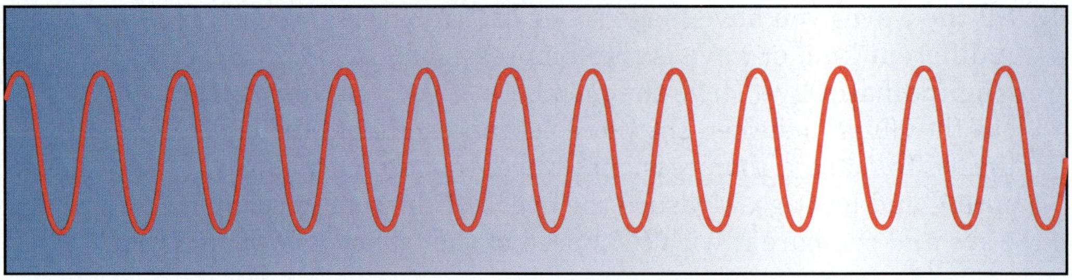

13. With the slit over the tape, pull the tape so that the wave moves. You will see a part of the wave (through the slit) going up and down.

14. Draw waves with different wavelengths on other pieces of adding machine tape. Put these under the slit and pull the adding machine tape at the same speed.

a) Describe what you see.

Physics Words

periodic wave: a repetitive series of pulses; a wave train in which the particles of the medium undergo periodic motion (after a set amount of time the medium returns to its starting point and begins to repeat its motion).

crest: the highest point of displacement of a wave.

trough: the lowest point on a wave.

amplitude: the maximum displacement of a particle as a wave passes; the height of a wave crest; it is related to a wave's energy.

FOR YOU TO READ

Wave Vocabulary

In this activity, you were able to send energy from one end of the Slinky to the other. You used chemical energy in your muscles to create mechanical energy in your arms that you then imparted to the Slinky. The Slinky had energy. A card at the other end of the Slinky would have moved once the wave arrived there. The ability to move the card is an indication that energy is present. The total energy is transferred but it is always conserved.

Of course, you could have used that same mechanical energy in your arm to throw a ball across the room. That would also have transferred the energy from one side of the room to the other. It would have also moved the card.

There is a difference between the Slinky transferring the energy as a wave and the ball transferring the energy. The Slinky wave transferred the energy, but the Slinky basically stayed in the same place. If the part of the Slinky close to one end were painted red, the red part of the Slinky would not move across the room. The Slinky wave moves, but the parts of the Slinky remain in the same place as the wave passes by. A wave can be defined as a transfer of energy with no net transfer of mass.

Leonardo da Vinci stated that "the wave flees the place of creation, while the water does not." The water moves up and down, but the wave moves out from its center.

In discussing waves, a common vocabulary helps to communicate effectively. You observed waves in the lab activity. We will summarize some of the observations here and you can become more familiar with the terminology.

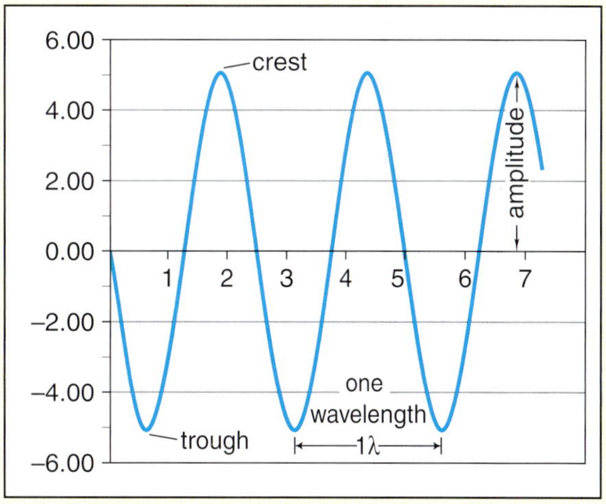

A **periodic wave** is a repetitive series of pulses. In the periodic wave shown in the diagram above, the highest point is called the **crest.** The lowest point is called the **trough**. The maximum disturbance, the **amplitude**, is 5.00 cm. Notice that this is the height of the crest or the height of the trough. It is *not* the distance from the crest to the trough.

The **wavelength** of a periodic wave is the distance between two consecutive points in phase. The distance between two crests is one wavelength or 1 λ. (The Greek letter lambda is used to signify wavelength.) The wavelength of the wave in the diagram is 2.5 cm.

The amplitude of a periodic wave is the maximum disturbance. A large amplitude corresponds to a large energy. In sound, the large amplitude is a loud sound. In light, the large amplitude is a bright light. In Slinkies, the large amplitude is a large disturbance.

The wavelength of the wave in the diagram is 2.5 cm. It is the distance between two crests or the distance between two troughs.

The **frequency** is the number of vibrations occurring per unit time. A frequency of 10 waves per second may also be referred to as 10 vibrations per second, 10 cycles per second, 10 per second, 10 s^{-1}, 10 Hz (hertz). The human ear can hear very low sounds (20 Hz) or very high sounds (20,000 Hz). You can't tell the frequency by examining the wave in the diagram. The "snapshot" of the wave is at an instant of time. To find the frequency, you have to know how many crests pass by a point in a given time.

The **period**, T, of a wave is the time it takes to complete one cycle. It is the time required for one crest to pass a given point. The period and the frequency are related to one another. If three waves pass a point every second, the frequency is three waves per second. The period would be the time for one wave to pass the point, which equals $\frac{1}{3}$ s. If 10 waves pass a point every second, the frequency is 10 waves per second. The period would be the time for one wave to pass the point, which equals $\frac{1}{10}$ second. Mathematically, this relationship can be represented as:

$$T = \frac{1}{f} \text{ or } f = \frac{1}{T}$$

Points in a periodic wave can be "in phase" if they have the same displacement and are moving in the same direction. All crests of the wave shown below are "in phase."

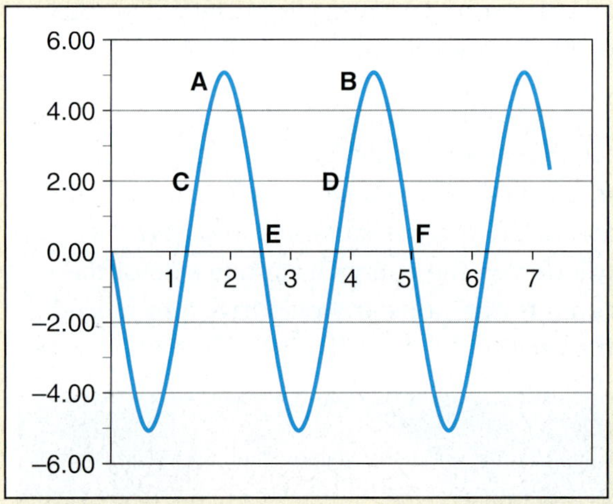

In the wave shown, the following pairs of points are in phase A and B, C and D, E and F.

A **node** is a spot on a standing wave where the medium is motionless. There are places along the medium that do not move as the standing wave moves up and down. The locations of these nodes do not change as the standing wave vibrates. A **transverse wave** is a wave in which the motion of the medium is perpendicular to the motion of the wave. A **longitudinal wave** is a wave in which the motion of the medium is parallel to the direction of the motion of the wave.

PHYSICS TALK

Calculating the Speed of Waves

You can find the speed of a wave by measuring the distance the crest moves during a certain change in time.

$$\text{speed} = \frac{\text{change in distance}}{\text{change in time}}$$

In mathematical language:

$$v = \frac{\Delta d}{\Delta t}$$

where v = speed

d = distance

t = time

Suppose the distance the crest moves is 2 m in 0.2 s. The speed can be calculated as follows:

$$v = \frac{\Delta d}{\Delta t} = \frac{2\text{ m}}{0.2\text{ s}} = 10\text{ m/s}$$

The distance from one crest of a wave to the next is the wavelength. The number of crests that go by in one second is the frequency. Imagine you saw five crests go by in one second. You measure the wavelength to be 2 m. The frequency is 5 crests/second, so the speed is $(5 \times 2) = 10$ m/s. Thus, the speed can also be found by multiplying the wavelength and the frequency.

$$\text{speed} = \text{frequency} \times \text{wavelength}$$

In mathematical language:

$$v = f\lambda$$

where v = speed

f = frequency

λ = wavelength

Physics Words

wavelength: the distance between two identical points in consecutive cycles of a wave.

frequency: the number of waves produced per unit time; the frequency is the reciprocal of the amount of time it takes for a single wavelength to pass a point.

period: the time required to complete one cycle of a wave.

node: a point on a standing wave where the medium is motionless.

transverse pulse or wave: a pulse or wave in which the motion of the medium is perpendicular to the motion of the wave.

longitudinal pulse or wave: a pulse or wave in which the motion of the medium is parallel to the direction of the motion of the wave.

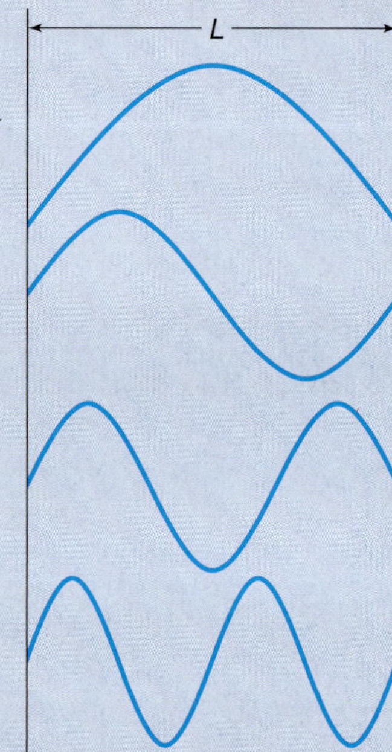

Standing waves happen anywhere that the length of the Slinky and the wavelength have a particular mathematical relationship. The length of the Slinky must equal $\frac{1}{2}$ wavelength, 1 wavelength, $1\frac{1}{2}$ wavelengths, 2 wavelengths, etc. Mathematically, this can be stated as:

$$L = \frac{n\lambda}{2}$$

where L is the length of the Slinky,
λ is the wavelength
n is a number (1, 2, 3...)

Sample Problem 1

You and your partner sit on the floor and stretch out a Slinky to a length of 3.5 m. You shake the Slinky so that it forms one loop between the two of you. Your partner times 10 vibrations and finds that it takes 24.0 s for the Slinky to make these vibrations.

a) How much of a wave have you generated and what is the wavelength of this wave?

Strategy: Draw a sketch of the wave you have made and you will notice that it looks like one-half of a total wave. It is! This is the maximum wavelength that you can produce on this length of Slinky. You can use the equation that shows the relationship between the length of the Slinky and the wavelength.

Givens:

$L = 3.5$ m

$n = 1$

Solution:

$$L = \frac{n\lambda}{2}$$

Rearrange the equation to solve for λ.

$$\lambda = \frac{2\,L}{n}$$

$$= \frac{2\,(3.5 \text{ m})}{1}$$

$$= 7.0 \text{ m}$$

b) What is the period of vibration of the wave?

Strategy: The period is the amount of time for one vibration. You have the amount of time for 10 vibrations.

Solution:

$$T = \frac{\text{time for 10 vibrations}}{10} = \frac{24.0 \text{ s}}{10} = 2.4 \text{ s}$$

c) Calculate the wave frequency.

Strategy: The frequency represents the number of vibrations per second. It is the reciprocal of the period.

Given:

$$T = 2.4 \text{ s}$$

Solution:

$$f = \frac{\text{number of vibrations}}{\text{time}} \text{ or } f = \frac{1}{T}$$

$$= \frac{1}{2.4 \text{ s}}$$

$$= 0.42 \text{ vibrations per second}$$

$$= 0.42 \text{ s}^{-1} \text{ or } 0.42 \text{ Hz}$$

→

d) Determine the speed of the wave you have generated on the Slinky.

Strategy: The speed of the wave may be found by multiplying the frequency times the wavelength.

Givens:

$f = 0.42$ Hz

$\lambda = 7.0$ m

Solution:

$$\begin{aligned} v &= f\lambda \\ &= 0.42 \text{ Hz} \times 7.0 \text{ m} \\ &= 29 \text{ m/s} \end{aligned}$$

Remember that Hz may also be written as 1/s so the unit of speed is m/s.

Sample Problem 2

You stretch out a Slinky to a length of 4.0 m, and your partner generates a pulse that takes 1.2 s to go from one end of the Slinky to the other. What is the speed of the wave on the Slinky?

Strategy: Use your kinematics equation to determine the speed.

Givens:

$d = 4.0$ m

$t = 1.2$ s

Solution:

$$\begin{aligned} v &= \frac{d}{t} \\ &= \frac{4.0 \text{ m}}{1.2 \text{ s}} \\ &= 3.3 \text{ m/s} \end{aligned}$$

Reflecting on the Activity and the Challenge

Slinky waves are easy to observe. You have created transverse and compressional Slinky waves and have measured their speed, wavelength, and frequency. For the **Chapter Challenge**, you may want to create musical instruments. You will receive more guidance in doing this in the next activities. Your instruments will probably not be made of Slinkies. You may, however, use strings that behave just like Slinkies. When you have to explain how your instrument works, you can relate its production of sound in terms of the Slinky waves that you observed in this activity.

Physics To Go

1. a) Four characteristics of waves are amplitude, wavelength, frequency, and speed. For each characteristic, tell how you measured it when you worked with the Slinky.
   b) For each characteristic, give the units you used in your measurement.
   c) Which wave characteristics are related to each other? Tell how they are related.

2. a) Suppose you shake a long Slinky slowly back and forth. Then you shake it rapidly. Describe how the waves change when you shake the Slinky more rapidly.
   b) What wave properties change?
   c) What wave properties do not change?

3. Suppose you took a photograph of a wave on a Slinky. How can you measure wavelength by looking at the photograph?

4. Suppose you mount a video camera on a tripod and aim the camera at one point on a Slinky. You also place a clock next to the Slinky, so the video camera records the time. When you look at the video of a wave going by on the Slinky, how could you measure the frequency?

5. a) What are the units of wavelength?
   b) What are the units of frequency?
   c) What are the units of speed?
   d) Tell how you find the wave speed from the frequency and the wavelength.

e) Using your answer to **Part (d)**, show how the units of speed are related to the units of wavelength and frequency.

6. a) What is a standing wave?
   b) Draw a standing wave.
   c) Add labels to your drawing to show how the Slinky moves.
   d) Tell how to find the wavelength by observing a standing wave.

7. a) Explain the difference between transverse waves and compressional waves.
   b) Slinky waves can be either transverse or compressional. Describe how the Slinky moves in each case.

8. a) When you made standing waves, how did you shake the spring (change the frequency) to make the wavelength shorter?
   b) When you made standing waves, how did you shake the spring (change the frequency) to make the wavelength longer?

9. Use the wave viewer and adding machine tape to investigate what happens if the speed of the wave increases. Pull the tape at different speeds and report your results.

10. A Slinky is stretched out to 5.0 m in length between you and your partner. By shaking the Slinky at different frequencies, you are able to produce waves with one loop, two loops, three loops, four loops, and even five loops.
    a) What are the wavelengths of each of the wave patterns you have produced?
    b) How will the frequencies of the wave patterns be related to each other?

11. A tightrope walker stands in the middle of a high wire that is stretched 10 m between the two platforms at the ends of the wire. He is bouncing up and down, creating a standing wave with a single loop and a period of 2.0 s.

a) What is the wavelength of the wave he is producing?
b) What is the frequency of this wave?
c) What is the speed of the wave?

12. A clothesline is stretched 9 m between two trees. Clothes are hung on the line as shown in the diagram. When a particular standing wave is created in the line, the clothes remain stationary.

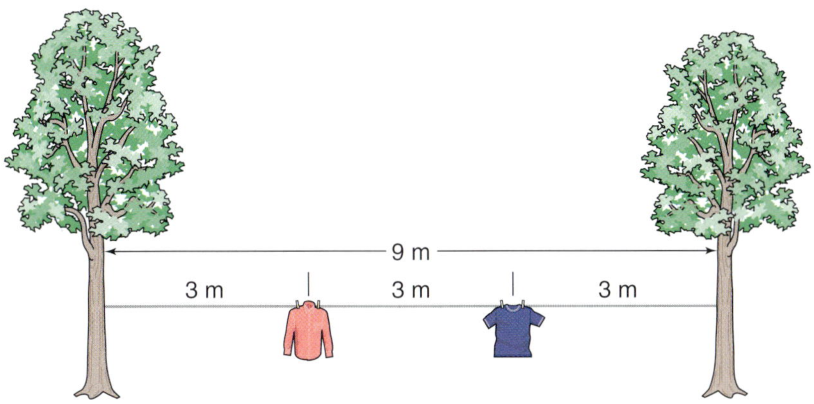

a) What is the term for the positions occupied by the clothes?
b) What is the wavelength of this standing wave?
c) What additional wavelengths could exist in the line such that the clothes remain stationary?

13. During the Slinky lab, your partner generates a wave pulse that takes 2.64 s to go back and forth along the Slinky. The Slinky stretches 4.5 m along the floor. What is the speed of the wave pulse on the Slinky?

14. A drum corps can be heard practicing at a distance of 1.6 km from the field. What is the time delay between the sound the drummer hears (d = 0 m) and the sound heard by an individual 1.6 km away? (Assume the speed of sound in air to be 340.0 m/s.)

Activity 2 Sounds in Strings

GOALS

In this activity you will:

- Observe the effect of string length and tension upon pitch produced.
- Control the variables of tension and length.
- Summarize experimental results.
- Calculate wavelength of a standing wave.
- Organize data in a table.

What Do You Think?

When the ancient Greeks made stringed musical instruments, they discovered that cutting the length of the string by half or two-thirds produced other pleasing sounds.

- **How do guitarists or violinists today make different sounds?**

Record your ideas about this question in your *Active Physics* log. Be prepared to discuss your responses with your small group and with your class.

For You To Do

1. Carefully mount a pulley over one end of a table. Securely clamp one end of a string to the other end of the table.
2. Tie the other end of the string around a mass hanger. Lay the string over the pulley. Place a pencil under the

string near the clamp, so the string can vibrate without hitting the table, as shown in the drawing.

3. Hang one 500-g mass on the mass hanger. Pluck the string, listen to the sound, and observe the string vibrate.

a) Record your observations in your log in a table similar to the following:

Length of vibrating string	Load on mass hanger	Pitch (high, medium, low)

Make sure the area under the hanging mass is clear (no feet, legs). Also monitor the string for fraying.

4. Use a key or some other small metal object. Press this object down on the string right in the middle, to hold the string firmly against the table. Pluck each half of the string.

a) Record the result in your table.

5. To change the string length, press down with the key at the different places shown in the diagrams on the next page. Pluck each part of the string.

a) Record the results in your table.

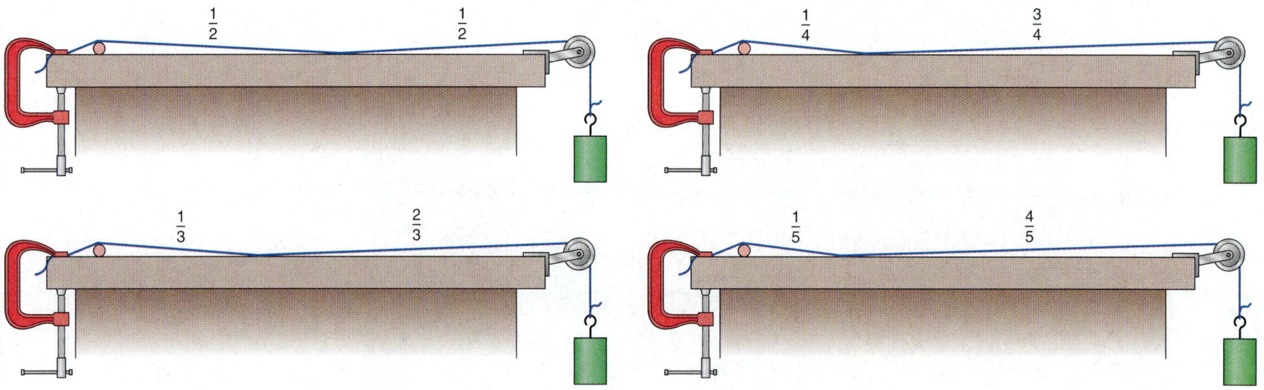

6. When you pluck the string, it does not move at the ends. Look at the drawing under **Step 9** of the **For You To Do** section in **Activity 1**. Measure the length of your string, and find the wavelength of the vibration for each string length.
   a) Record the wavelength in your table.
   b) Look over the data in your table. Make a general statement about what happens to the pitch you hear as you change the length of the string.

Make sure the string is capable of holding 2 kg.

7. Remove the key, so the string is its original length. Pluck the string. To investigate the effect of tightening the string, add a second 500-g mass to the mass hanger. Pluck the string again, observe the vibration, and listen to the pitch of the sound.
   a) Make up a table to record your data in your log.
   b) Add a description of the pitch of the sound to your table. Continue adding weights and observing the sound until the total mass is 2000 g.
   c) Look over your data. As the mass increases, the string becomes tighter, and its tension increases. Make a general statement about what happens to the pitch you hear as you change the tension on the string.

FOR YOU TO READ

Changing the Pitch

Sound comes from vibration. You observed the vibration of the string as it produced sound. You investigated two of the variables that affect the sound of a vibrating string.

When you pushed the vibrating string down against the table, the length of the string that was vibrating became shorter. Shortening the string increased the **pitch** (resulted in a higher pitch). In the same way, a guitarist or violinist pushes the string against the instrument to shorten the length that vibrates and increases the pitch.

When you hung weights on the end of the string, that increased the pitch too. These weights tightened the string, so they created more tension in it. As the string tension increased, the pitch of the sound also increased. In tuning a guitar or violin, the performer changes the string tension by turning a peg attached to one end of a string. As the peg pulls the string tighter, the pitch goes up.

Combining these two results into one expression, you can say that increasing the tension or decreasing the length of the string will increase the pitch.

The string producing the pitch is actually setting up a standing wave between its endpoints. The length of the string determines the wavelength of this standing wave. Twice the distance between the endpoints is the wavelength of the sound. The pitch that you hear is related to the frequency of the wave. The higher the pitch, the higher the frequency. The speed of the wave is equal to its frequency multiplied by its wavelength.

$$v = f\lambda$$

where v = speed

f = frequency

λ = wavelength

If the speed of a wave is constant, a decrease in the wavelength will result in an increase in the frequency or a higher pitch. A shortened string produces a higher pitch.

Reflecting on the Activity and the Challenge

Part of the **Chapter Challenge** is to create a sound show. In this activity you investigated the relationship of pitch to length of the string and tension of the string: the shorter the string, the higher the pitch; the greater the tension, the higher the pitch. You also learned that the string is setting up a standing wave between its two ends, just like the standing wave that you created in the Slinky in **Activity 1**. That's the physics of stringed instruments! If you wanted to create a stringed or multi-string instrument for your show, you would now know how to adjust the length and tension to produce the notes you want. If you were to make such a stringed instrument, you could explain how you change the pitch by referring to the results of this activity.

Physics Words

pitch: the quality of a sound dependent primarily on the frequency of the sound waves produced by its source.

Physics To Go

1. a) Explain how you can change the tension of a vibrating string.
   b) Tell how changing the tension changes the pitch.

2. a) Explain how you can change the length of a vibrating string.
   b) Tell how changing the length changes the sound produced by the string.

3. How would you change both the tension and the length and keep the pitch the same?

4. Suppose you changed both the length and the tension of the string at the same time. What would happen to the sound?

5. a) For the guitar and the piano, tell how a performer plays different notes.
   b) For the guitar and the piano, tell how a performer (or tuner) changes the pitch of the strings to tune the instrument.

6. a) Look at a guitar. Find the tuners (at the end of the neck). Why does a guitar need tuners?
   b) What is the purpose of the frets on a guitar?
   c) Does a violin or a cello have frets?
   d) Why do a violinist and a cellist require more accuracy in playing than a guitarist?

7. a) Using what you have learned in this activity, design a simple two-stringed instrument.
   b) Include references to wavelength, frequency, pitch, and standing waves in your description.
   c) Use the vocabulary of wavelength, frequency, and standing waves from **Activity 1** to describe how the instrument works.

Stretching Exercises

1. Set up the vibrating string as you did in the preceding **For You To Do**. This time, you will measure the frequency of the sound. Set up a frequency meter on your computer. Pick up the sound with a microphone. Investigate how changing the length of the string changes the frequency of the sound. Create a graph to describe the relationship.

2. Set up the vibrating string, computer, and microphone as you did in **Stretching Exercise 1**. This time, investigate how changing the string tension changes the frequency of the sound. Create a graph to describe the relationship.

3. Design an investigation to find how the diameter (thickness) of the string or the type of material the string is made of affects the pitch you hear. Submit your design to your teacher for approval before proceeding to carry out your experiments.

Activity 3 Sounds from Vibrating Air

GOALS

In this activity you will:

- Identify resonance in different kinds of tubes.
- Observe how resonance pitch changes with length of tube.
- Observe the effect of closing one end of the tube.
- Summarize experimental results.
- Relate pitch observations to drawings of standing waves.
- Organize observations to find a pattern.

What Do You Think?

The longest organ pipes are about 11 m long. A flute, about 0.5 m long, makes musical sound in the same way.

- **How do a flute and organ pipes produce sound?**

Record your ideas about this question in your *Active Physics* log. Be prepared to discuss your responses with your small group and with your class.

For You To Do

1. Carefully cut a drinking straw in half. Cut one of the halves into two quarters. Cut one of the quarters into two eighths. Pass one part of the straw out to one member of your group.

2. Gently blow into the top of the piece of straw.
   a) Describe what you hear.
   b) Listen as the members of your group blow into their straw pieces one at a time. Describe what you hear.
   c) Write a general statement about how changing the length of the straw changes the pitch you hear.

3. Now cover the bottom of your straw piece and blow into it again. Uncover the bottom and blow again.
   a) Compare the sound the straw makes when the bottom is covered and then uncovered.
   b) Listen as the members of your group blow into their straw pieces, with the bottom covered and then uncovered. Write a general statement about how changing the length of the straw changes the pitch you hear when one end is covered.

4. Obtain a set of four test tubes. Leave one empty. Fill the next halfway with water. Fill the next three-quarters of the way. Fill the last one seven-eighths of the way.

Make sure the outsides of the tubes are dry.

5. Give each test tube to one member of your group. Blow across your test tube.
   a) Describe what you hear.
   b) Listen as the members of your group blow, one at a time, across their test tubes. Record what you hear.
   c) What pattern do you find in your observations?
   d) Compare the results of blowing across the straws with blowing across the test tubes. How are the results consistent?

PHYSICS TALK

Vibrating Columns of Air

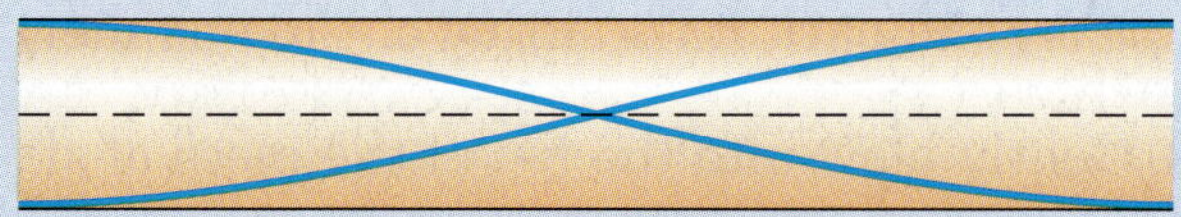

Tube is open at both ends.
1/2 wavelength fits in straw.

The sound you heard when you blew into the straw and test tube was produced by a standing wave. If both ends of the straw are open, the air at both ends moves back and forth. The above drawing shows the movement of the air as a standing wave.

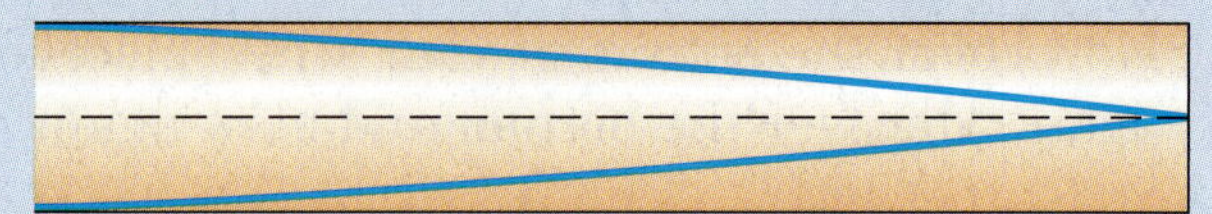

Tube is closed at one end.
1/4 wavelength fits in straw.

When you covered the other end of the straw, you prevented the air from moving at the covered end. This drawing shows the movement of the air as a standing wave.

The velocity of a wave is equal to the frequency multiplied by the wavelength. Therefore,

$$\text{frequency} = \frac{\text{wave speed}}{\text{wavelength}}$$

Using mathematical symbols,

$$f = \frac{v}{\lambda}$$

As the wavelength increases, the frequency decreases. The wavelength in the open straw is half the wavelength in the straw closed at one end. This equation predicts that the frequency of the standing wave in the open straw is twice the frequency of the standing wave in the straw closed at one end.

Physics Words

diffraction: the ability of a sound wave to spread out as it emerges from an opening or moves beyond an obstruction.

FOR YOU TO READ

Compressing Air to Make Sound

Sound is a compression wave. The molecules of air bunch up or spread apart as the sound wave passes by.

At the end where the tube is closed, the air cannot go back and forth, because its motion is blocked by the end of the tube. That's why the wave's amplitude goes to zero at the closed end. At the open end, the amplitude is as large as it can possibly be. This back-and-forth motion of air at the open end makes a sound wave that moves from the tube to your ear.

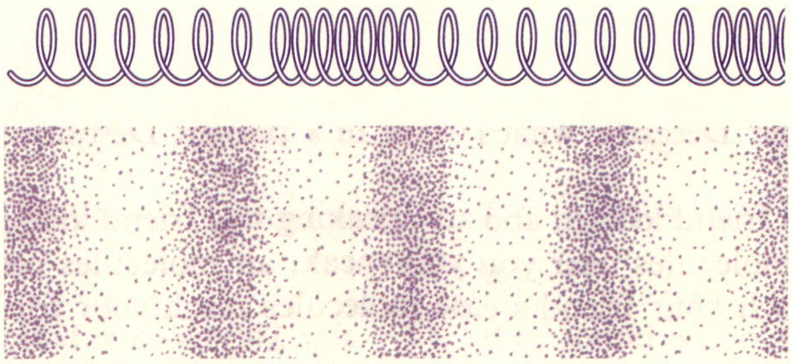

In the compressional Slinky wave, the coils of the Slinky bunched up in a similar fashion when the Slinky wave passed by.

Wave Diffraction

As the sound wave leaves the test tube in this activity, it spreads out. In the same way, when you speak to a friend, the sound waves leave your mouth and spread out. You can speak to a group of friends because the sound leaves your mouth and moves out to the front and to the sides.

This ability of the sound wave to spread out as it emerges from an opening is called **diffraction**. The smaller the opening, the more spreading of the sound. The spreading of the wave as it emerges from two holes can be shown with a diagram.

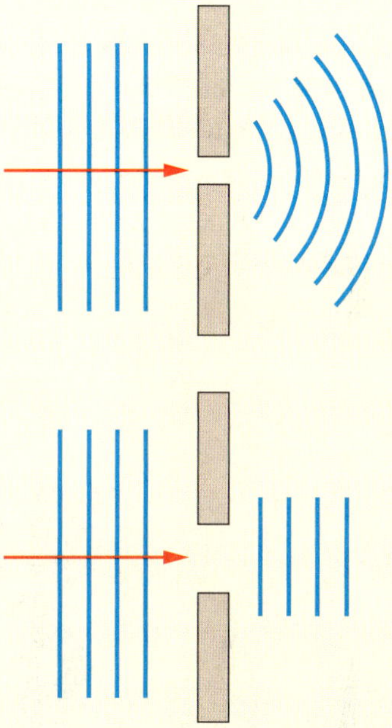

The wave on the top is going through a small opening (in comparison to its wavelength) and diffracts a great deal. The wave on the bottom is going through a large opening (in comparison to its wavelength) and shows little diffraction.

Cheerleaders use a megaphone to limit the diffraction. With a megaphone, the mouth opening becomes larger. The sound wave spreads out less, and the cheering crowd in front of the cheerleader hears a louder sound.

→

A new musical instrument that you can make uses a straw instead of a test tube.

Take a straw and cut the ends to form a V as show below.

Flatten the V end of the straw and blow this "trumpet." You can shorten the trumpet, decrease the wavelength of the standing wave, and increase the frequency of the sound. Try making a sound. As you emit the sound, use scissors to cut ends off the straw. Listen to the different tones.

You can probably make a trombone by inserting one straw within another.

The sound diffracts from the small opening. You can add a horn to one end of the straw and limit the diffraction. The effect will be that the sound appears louder because it doesn't spread out. You can make a horn out of a piece of paper, as shown.

You may want to adapt this idea of a trumpet and a megaphone and make diffraction a part of your light and sound show.

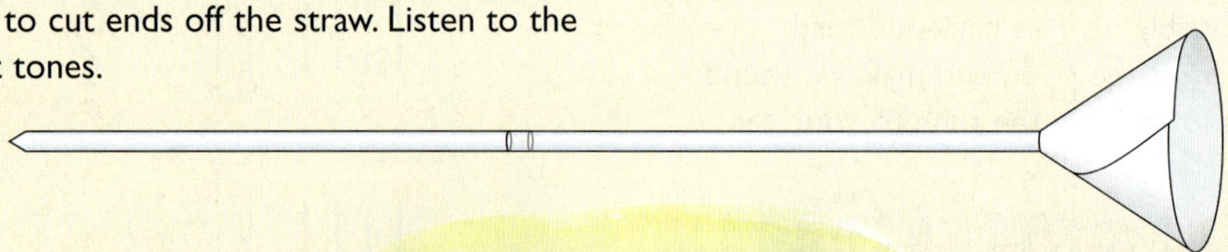

Reflecting on the Activity and the Challenge

In this activity you have observed the sounds produced by different kinds of pipes. If the pipe is cut to a shorter length, the pitch of the sound increases. Also, when the pipe is open at both ends, the pitch is much higher than if the pipe were open at only one end. You have seen how simple drawings of standing waves in these tubes help you find the wavelength of the sound. If the tube is closed at one end, the air has zero displacement at that end. If the tube is open at one end, the air has maximum displacement there.

For your sound show, you may decide to create some "wind" instruments using test tubes or straws, or other materials approved by your teacher. When it comes time to explain how these work, you can refer to this activity to get the physics right.

Physics To Go

1. a) You can produce a sound by plucking a string or by blowing into a pipe. How are these two ways of producing sound similar?
   b) How are these two ways different?

2. a) For each piece of straw your group used, make a full-sized drawing to show the standing wave inside. Show both the straw closed at one end and open at both ends.
   b) Next to each drawing of the standing waves, make a drawing, at the same scale, of one full wavelength. You may need to tape together several pieces of paper for this drawing.
   c) Frequency times the wavelength is the wave speed. The speed is the same for all frequencies. From your answer to **Part (b)**, what can you predict about the frequencies of the standing waves in the straw pieces?
   d) How well do your predictions from **Part (c)** agree with your observations in this activity?

3. a) What is the length, in meters, of the longest organ pipe?
   b) Assume this pipe is closed at one end. Draw the standing wave pattern.
   c) For this pipe, how long is the wavelength of this standing wave?
   d) Why does a long wavelength indicate that the frequency will be low? Give a reason for your answer.

4. a) Suppose you are listening to the sound of an organ pipe that is closed at one end. The pipe is 3 m long. What is the wavelength of the sound in the pipe?
   b) The speed of sound in air is about 340 m/s. What is the frequency of the sound wave?
   c) Now suppose you are listening to the sound of an organ pipe that is open at both ends. As before, the pipe is 3 m long. What is the wavelength of the sound in the pipe?
   d) What is the frequency of the sound wave?

5. Suppose you listen to the sound of an organ pipe that is closed at one end. This pipe is only 1 m long. How does its frequency compare with the frequency you found in **Question 4, Part (b)**?

6. Waves can spread into a region behind an obstruction.
   a) What is this wave phenomenon called?
   b) Draw a diagram to illustrate this phenomenon.

Stretching Exercises

1. If you have a good musical ear, add water to eight test tubes to make a scale. Play a simple piece for the class.

2. Obtain a 2- to 3- meter-long piece of a 7- to 10-centimeter-diameter plastic pipe, like that used to filter water in small swimming pools. In an area free of obstructions, twirl the pipe overhead. What can you say about how the sound is formed? Place some small bits of paper on a stool. Twirl the pipe and keep one end right over the stool. What happens to the paper? What does that tell you about the air flowing through the pipe? Try to play a simple tune by changing the speed of the pipe as you twirl it.

3. Carefully cut new straw pieces, as you did in **For You To Do**, **Step 1**. This time, you will measure the frequency of the sound. Set up a frequency meter on your computer. Place the microphone near an open end of the straw.
   As before, each person blows into only one piece of straw. Make the sound and record the frequency. Now cover the end of the straw and predict what frequency you will measure. Make the measurement and compare it with your prediction. Repeat the measurements for all of the lengths of straw. Record your results, and tell what patterns you find.

Activity 4 Reflected Light

GOALS

In this activity you will:

- Identify the normal of a mirror.
- Measure angles of incidence and reflection.
- Observe the relationship between the angle of incidence and the angle of reflection.
- Observe changes in the reflections of letters.
- Identify patterns in multiple reflections.

What Do You Think?

Astronauts placed a mirror on the Moon in 1969 so that a light beam sent from Earth could be reflected back to Earth. By timing the return of the beam, scientists found the distance between the Earth and the Moon. They measured this distance to within 30 cm.

- **How are you able to see yourself in a mirror?**
- **If you want to see more of yourself, what can you do?**

Record your ideas about these questions in your *Active Physics* log. Be prepared to discuss your responses with your small group and with your class.

For You To Do

1. Place a piece of paper on your desk. Carefully aim the laser pointer, or the light from a ray box, so the light beam moves horizontally, as shown on the opposite page.

2. Place a glass rod in the light beam so that the beam spreads up and down. Shine the beam on the piece of paper to be sure the beam passes through the glass rod.

3. Carefully stand the plane mirror on your desk in the middle of the piece of paper. Draw a line on the paper along the

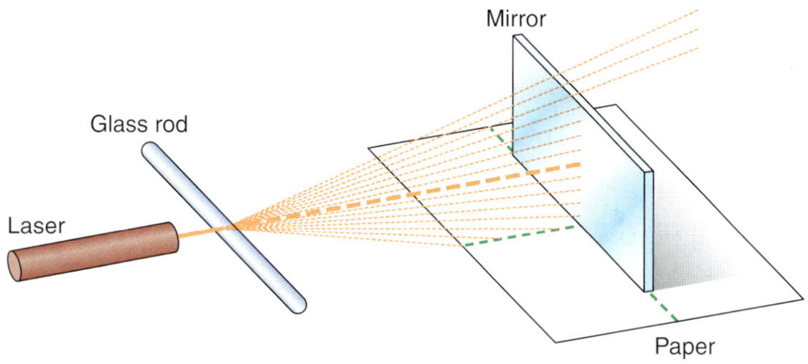

front edge of the mirror. Now remove the mirror and draw a dotted line perpendicular to the first line, as shown. This dotted line is called the **normal**.

Do not use mirrors with chipped edges. Make sure the ends of the glass rod are polished.

Never look directly at a laser beam or shine a laser beam into someone's eyes. Always work above the plane of the beam and beware of reflections from shiny surfaces.

Physics Words

normal: at right angles or perpendicular to.

4. Aim the light source so the beam approaches the mirror along the normal. Be sure the glass rod is in place to spread out the beam.

   a) What happens to the light after it hits the mirror?

5. Make the light hit the mirror at a different angle.

   a) What happens now?

   b) On the paper, mark three or more dots under the beam to show the direction of the beam as it travels to the mirror. The line you traced shows the incident ray. Also make dots to show the light going away from the mirror. This line shows the reflected ray. Label this pair of rays to show they go together.

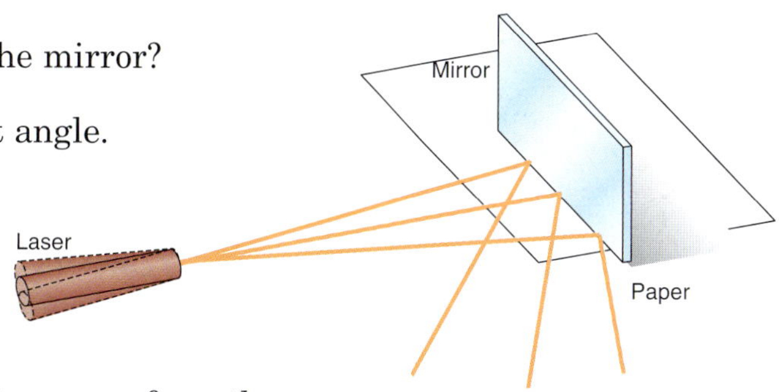

6. Turn the light source so it starts from the same point but strikes the mirror at different angles. For each angle, mark dots on the paper to show the direction of the incident and reflected rays. Also, label each pair of rays.

7. Most lab mirrors have the reflecting surface on the back. In addition, the light bends as it enters and leaves the glass part of the mirror. In your drawing, the rays may not meet at the mirror surface. Extend the rays until they do meet.
   a) Measure these angles for one pair of your rays.

8. Turn off the light source and remove the paper. Look at one pair of rays. The diagram shows a top view of the mirror, the normal, and an incident and reflected ray. Notice the angle of incidence and the angle of reflection in the drawing. Using a protractor, measure these angles for one pair of rays.
   a) Record your data in a table.
   b) Measure and record the angles of incidence and reflection for all of your pairs of rays.

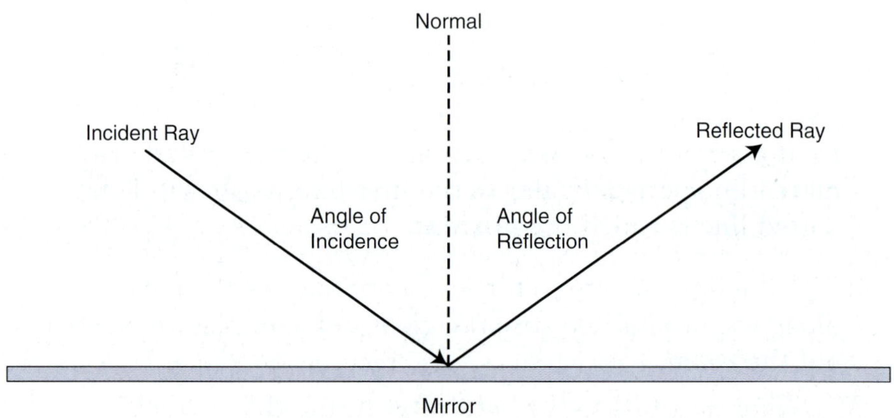

   c) What is the relationship between the angles of incidence and reflection?
   d) Look at the reflected rays in your drawing. Extend each ray back behind the mirror. What do you notice when you have extended all the rays? The position where the rays meet is the location of the image of the light source. All of the light rays leave one point in front of the mirror. The reflected rays all seem to emerge from one point behind the mirror. Wherever you observed the reflection, you would see the source at this point behind the mirror.
   e) Tape a copy of your diagram in your log.

9. Hold the light source, or any object, near the mirror and look at the reflection. Now hold the object far away and again look at the reflection.

a) How is the position of the reflection related to the position of the object?

10. Set up a mirror on another piece of paper, and draw the normal on the paper. Write your name in block capital letters along the normal (a line perpendicular to the mirror). Observe the reflection of your name in the mirror.

    a) How can you explain the reflection you see?

    b) Which letters in the reflection are closest to the mirror? Which are farthest away?

    c) In your log, make a sketch of your name and its reflection.

11. Carefully stand up two mirrors so they meet at a right angle. Be sure they touch each other, as shown in the drawing.

12. Place an object in front of the mirrors.

    a) How many images do you see?

    b) Slowly change the angle between the mirrors. Make a general statement about how the number of images you see changes as the angle between the mirrors changes.

Physics Words

angle of incidence: the angle a ray of light makes with the normal to the surface at the point of incidence.

angle of reflection: the angle a reflected ray makes with the normal to the surface at the point of reflection.

ray: the path followed by a very thin beam of light.

FOR YOU TO READ

Images in a Plane Mirror

An object like the tip of a nose reflects light in all directions. That is why everybody in a room can see the tip of a nose. Light reflects off a mirror in such a way that the **angle of incidence** is equal to the **angle of reflection**. You can look at the light leaving the tip of a nose and hitting a mirror to see how an image is produced and where it is located. Each **ray** of light leaves the nose at a different angle. Once it hits the mirror, the angle of incidence must equal the angle of reflection. There are now a set of rays diverging from the mirror. If you assume that the light always travels in straight lines, you can extend these rays behind the mirror and find where they "seem" to emerge from. That is the location of the image.

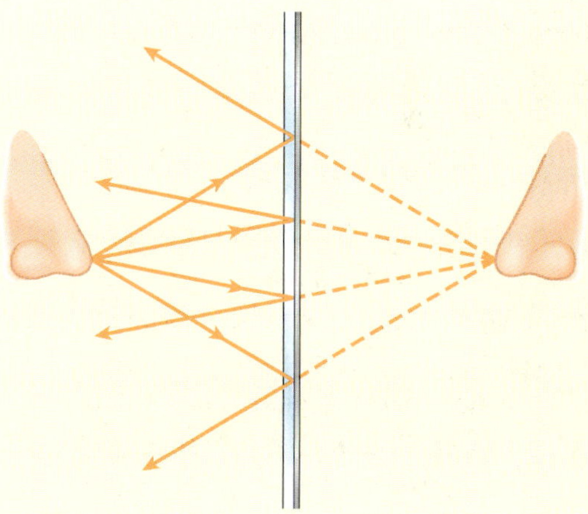

The mirror does such a good job of reflecting that it looks as if there is a tip of a nose (and all other parts of the face) behind the mirror. If you measure the distance of the image behind the mirror, you will find that it is equal to the distance of the nose (object) in front of the mirror. This can also be proved using geometry.

Diffraction of Light

As you begin to study the reflection of light rays, it is worthwhile to recognize that light is a wave and has properties similar to sound waves.

In studying sound waves, you learned that sound waves are compressional or longitudinal. The disturbance is parallel to the direction of motion of the wave. In sound waves, the compression of the air is left and right as the wave travels to the right. You saw a similar compressional wave using the compressed Slinky.

Light waves are transverse waves. They are similar to the transverse waves of the Slinky. In a transverse wave, the disturbance is perpendicular to the direction of the wave. In the Slinky, the disturbance was up and down as the wave traveled to the right. In light, the fields (the disturbance) are perpendicular to the direction of motion of the waves.

You also read that sound waves diffract—they spread out as they emerge from small openings. You can find out if light waves spread out as they emerge from a small opening. Try this: Take a piece of aluminum foil. Pierce the foil with a pin to create a succession of holes, one smaller than the next. Shine the laser beam through each hole and observe its appearance on a distant wall. You will be able to observe the diffraction of light.

Sample Problem

Light is incident upon the surface of a mirror at an angle of 40°.

a) Sketch the reflection of the ray.

Strategy: The angles of incidence and reflection are always measured from the normal. The Law of Reflection states that the angle of incidence is equal to the angle of reflection. Since the angle of incidence is equal to 40°, the angle of reflection is also 40°.

Given:

$$\theta_i = 40°$$

Solution:

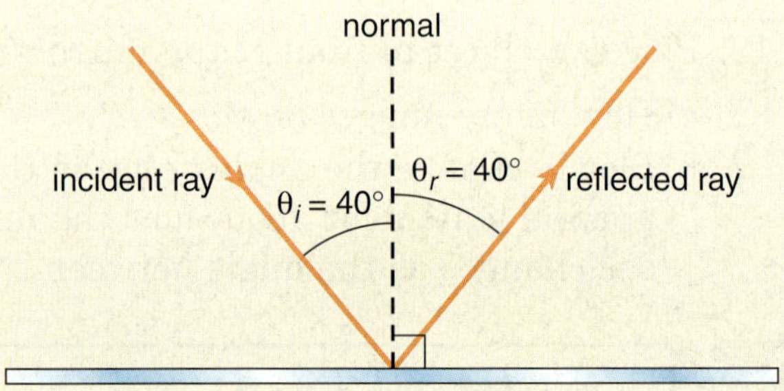

b) At what angle, as measured from the surface of the mirror, did the beam strike the mirror?

Strategy: The angle of incidence is measured from the normal. The question is asking for the complementary angle.

Solution:

$$\theta_i = \theta_r = 40°$$

$$90° - 40° = 50°$$

The angle between the light beam and the mirror is 50°.

Reflecting on the Activity and the Challenge

In this activity you aimed light rays at mirrors and observed the reflections. From the experiment you discovered that the angle of incidence is equal to the angle of reflection. Therefore, you can now predict the path of a reflected light beam. You also experimented with reflections from two mirrors. When you observed the reflection in two mirrors, you found many images that made interesting patterns.

This activity has given you experience with many interesting effects that you can use in your sound and light show. For instance, you may want to show the audience a reflection in one mirror or two mirrors placed at angles. You can probably create a kaleidoscope. You will also be able to explain the physics concept you use in terms of reflected light.

Physics To Go

1. How is the way light reflects from a mirror similar to the way a tennis ball bounces off a wall?

2. a) What is the normal to a plane mirror?
   b) When a light beam reflects from a plane mirror, how do you measure the angle of incidence?
   c) How do you measure the angle of reflection?
   d) What is the relationship between the angle of incidence and the angle of reflection?

3. Make a top-view drawing to show the relationships among the normal, the angle of incidence, and the angle of reflection.

4. a) Suppose you are experimenting with a mirror mounted vertically on a table, like the one you used in this activity. Make a top-view drawing, with a heavy line to represent the mirror and a dotted line to represent the normal.
   b) Show light beams that make angles of incidence of 0°, 30°, 45°, and 60° to the normal.
   c) For each of the above beams, draw the reflected ray. Add a label if necessary to show where the rays are.

5. a) Stand in front of a mirror.
   b) Move your hand toward the mirror. Which way does the reflection move?
   c) Move your hand away from the mirror. Which way does the reflection move?
   d) Use what you learned about the position of the mirror image to explain your answers to **Parts (b)** and **(c)**.

6. Suppose you wrote the whole alphabet along the normal to a mirror in the way you wrote your name in **Step 10** of **For You To Do**.
   a) Which letters would look just like their reflections?
   b) Write three words that would look just like their reflections.
   c) Write three letters that would look different than their reflections.
   d) Draw the reflection of each letter you gave in **Part (c)**.

7. Why is the word *Ambulance* written in an unusual way on the front of an ambulance?

8. Use a ruler and protractor and a ray diagram to locate the image of an object placed in front of a plane mirror. Be careful! You must measure as carefully as you can to obtain the most accurate answer.

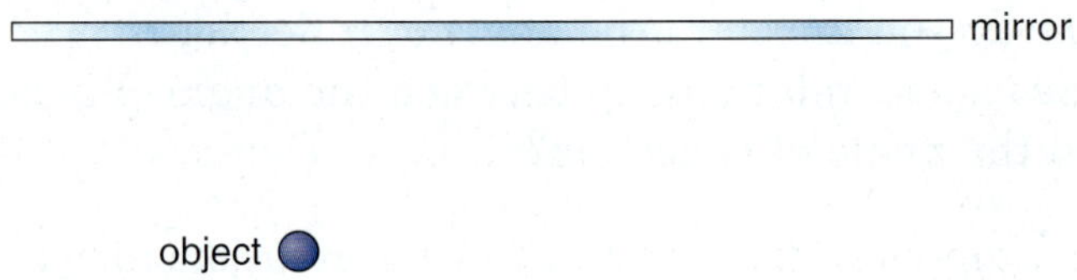

9. Locate the image of the lamp shown in the diagram.

10. After reflecting off mirrors A, B, and C, which target will the ray of light hit?

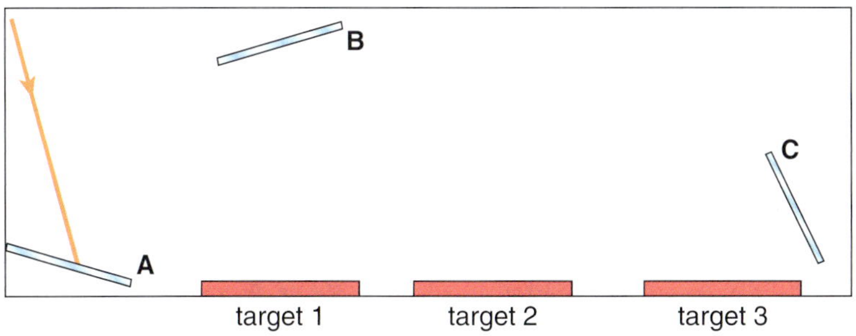

Stretching Exercises

1. Carefully tape together three small mirrors to make a corner reflector. Shine a flashlight down into the corner. Where does the reflected beam go?

2. Build a kaleidoscope by *carefully* inserting two mirrors inside a paper towel holder. You can also use three identical mirrors. Do not force the mirrors into the tube. Tape the edges of the mirrors together, with the mirrored surfaces inside. Describe what you see through your kaleidoscope.

3. Carefully tape together one edge of two mirrors so they can move like a hinge, with the mirrored surfaces facing each other. Place a small object between the mirrors. Investigate how the number of images you see depends on the angle between the mirrors. You will need a protractor to measure this angle. Plot a graph of the results. What mathematical relationship can you find between the angle and the number of images?

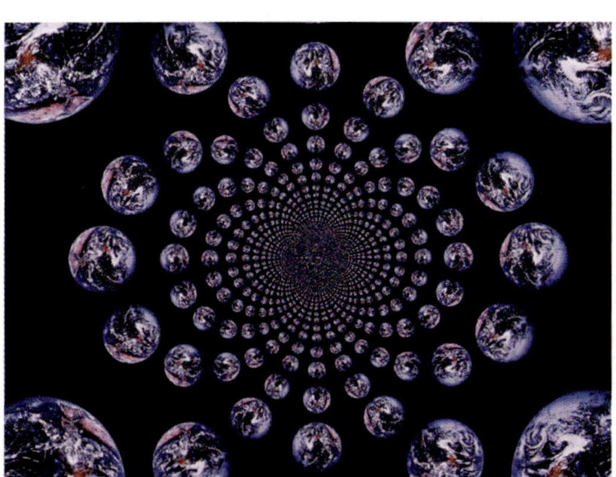

Activity 5 Curved Mirrors

GOALS

In this activity you will:

- Identify the focus and focal length of a curved mirror.
- Observe virtual images in a convex mirror.
- Observe real and virtual images in a concave mirror.
- Measure and graph image distance versus object distance for a convex mirror.
- Summarize observations in a sentence.

What Do You Think?

The curved mirror of the Palomar telescope is five meters across. Mirrors with varying curvatures are used in amusement parks as fun-house mirrors. Store mirrors and car side-view mirrors are also curved.

- **How is what you see in curved mirrors different from what you see in ordinary flat mirrors?**

Record your ideas about this question in your *Active Physics* log. Be prepared to discuss your responses with your small group and with your class.

For You To Do

1. Carefully aim a laser pointer, or the light from a ray box, so the light beam moves horizontally, as you did in the previous activity. Place a glass rod in the light beam so that the beam spreads up and down.

2. Place a convex mirror in the light beam, as shown in the diagram.

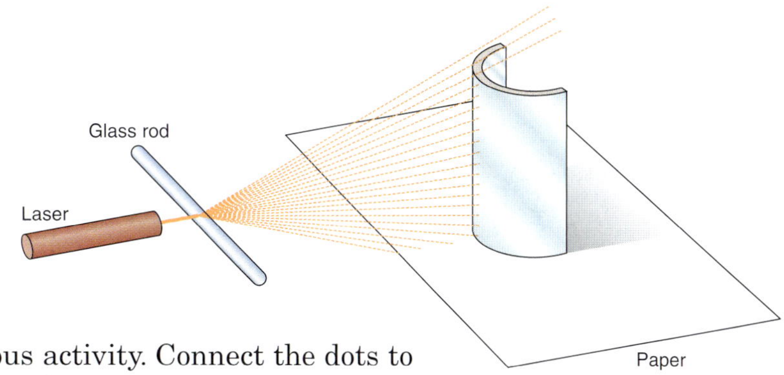

3. Shine a beam directly at the center of the mirror. This is the incident beam. Show its path by placing three or more dots on the paper, as you did in the previous activity. Connect the dots to make a straight line. Find the reflected ray and mark its path in a similar way. Label the two lines so you will know they go together.

Never look directly at a laser beam or shine a laser beam into someone's eyes. Always work above the plane of the beam and beware of reflections from shiny surfaces.

4. You will move the light source sideways to make a series of parallel beams. To make sure the incident beams are parallel, line up each one with the dots you made to show the incoming beam in **Step 3**. Mark the path of the incoming ray with three dots.

5. Each parallel beam makes a reflected beam. Show the path of each of these reflected rays. Label each incident and reflected beam so you will know that they go together.
   a) Write a sentence to tell what happens to the parallel beams after they are reflected.
   b) Make a drawing in your *Active Physics* log to record the path of the light.

Physics Words

focus: the place at which light rays converge or from which they appear to diverge after refraction or reflection; also called focal point.

focal length: the distance between the center of a lens and either focal point.

6. Remove the mirror. With a ruler, extend each reflected ray backwards to the part of the paper that was behind the mirror.
   a) You probably noticed that all the lines converge in a single point. The place where the extended rays meet is called the **focus** of the mirror. The distance from this point to the mirror is called the **focal length**. Measure and record this focal length.

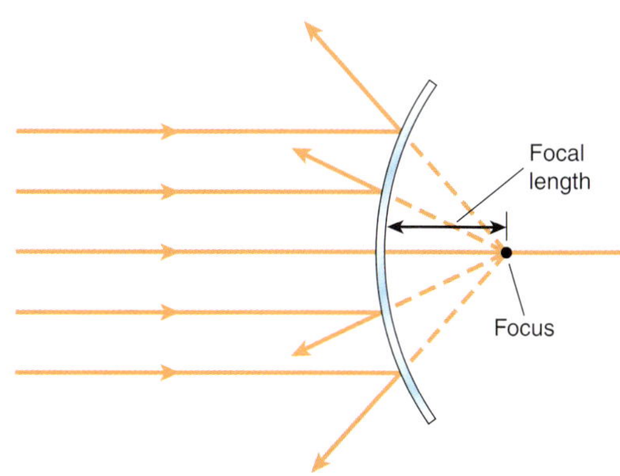

7. Place the concave side of the mirror in the light beam. To help you remember the name *concave*, think of the

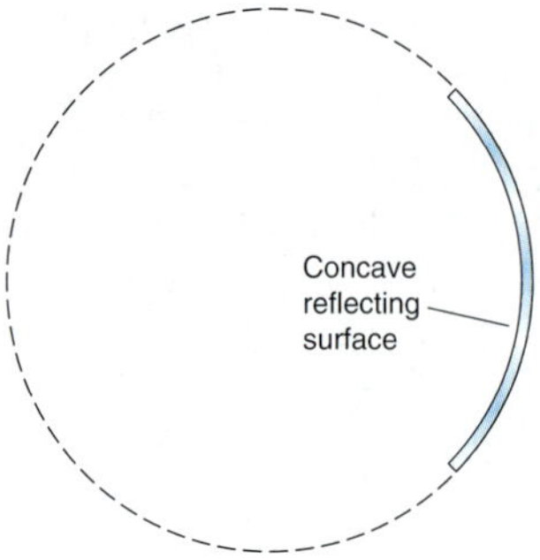

concave mirror as "caving in." Repeat **Steps 3** through **5** for this mirror.

a) Write a sentence to tell what happens to the parallel beams after they are reflected from the concave mirror.

b) Make a drawing in your *Active Physics* log to record the path of the light. The place where the beams cross is called the focus. The distance from the focus to the mirror is the focal length.

c) Measure and record the focal length.

d) How do concave and convex mirrors reflect light differently? Record your answer in your log.

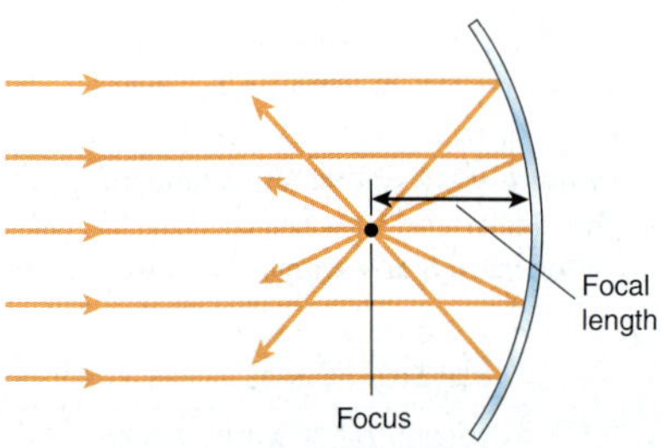

8. Use the concave mirror. Use a 40-W light bulb or a candle as a light source, which will be called the "object." Carefully mount your mirror so it is at the same height as the light source. Place a light bulb about a meter away from the mirror. Put the bulb slightly off the center line, as shown, so that an index card will not block the light from hitting the mirror.

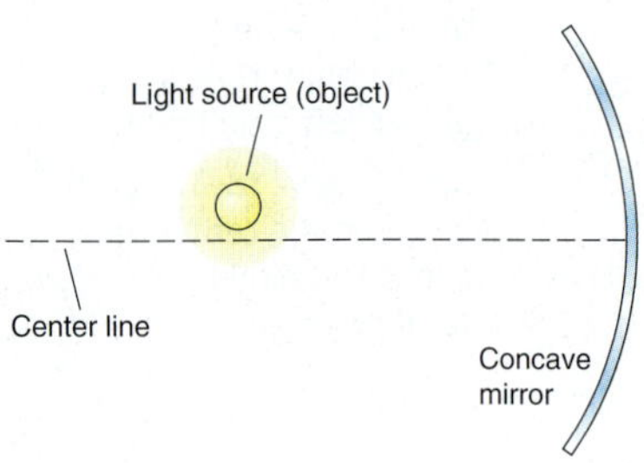

9. Try to find the image of the object on an index card. Move the card back and forth until the image is sharp. The image you found is called a real image because you are able to project it on a card.

a) Record the distance of the bulb from the mirror and of the image on the file card from the mirror. Put your results in the first line of a table like the one below.

Distance of bulb from mirror	Distance of image from mirror

10. Carefully move the mirror closer to the object. Find the sharp image, as before, by moving the index card back and forth.
    - a) Record the image and object distances in your table.
    - b) Repeat the measurement for at least six object locations.
    - c) Draw a graph of the image distance (y-axis) versus the object distance (x-axis).
    - d) Write a sentence that describes the relationship between the image distance and the object distance.

11. A mathematical relation that describes concave mirrors is

$$\frac{1}{f} = \frac{1}{D_o} + \frac{1}{D_i}$$

where

f is the focal length of that particular mirror

D_o is the object distance

D_i is the image distance

You have measured D_o and D_i. Calculate $\frac{1}{D_o}$ and $\frac{1}{D_i}$. Find their sum for each pair of data.

- a) Record your calculations in your log.
- b) Are your sums approximately equal? If so, you have mathematically found the value of $\frac{1}{f}$ for the mirror you used.

12. A convex mirror cannot form a real image that can be projected onto a screen. It can form an image behind the mirror, like a plane mirror.
    - a) Record in your log descriptions of the image in a convex mirror when the mirror is held close and when the mirror is held far from the object.

PHYSICS TALK

Making Real Images

To find how a concave mirror makes a **real image**, you can view a few rays of light. Each ray of light obeys the relation you found for plane mirrors (angle of incidence = angle of reflection). In this case, you choose two easily drawn rays.

Physics Words

real image: an image that will project on a screen or on the film of a camera; the rays of light actually pass through the image.

Look at the drawing. It shows rays coming into a concave mirror from a point on a light bulb. One ray comes in parallel to the dotted line, which is the axis of the mirror. This ray reflects through the focus. The other ray hits the center of the mirror. This ray reflects and makes the same angle with the mirror axis going out as it did coming in. Where these rays meet is the image of the top of the light bulb.

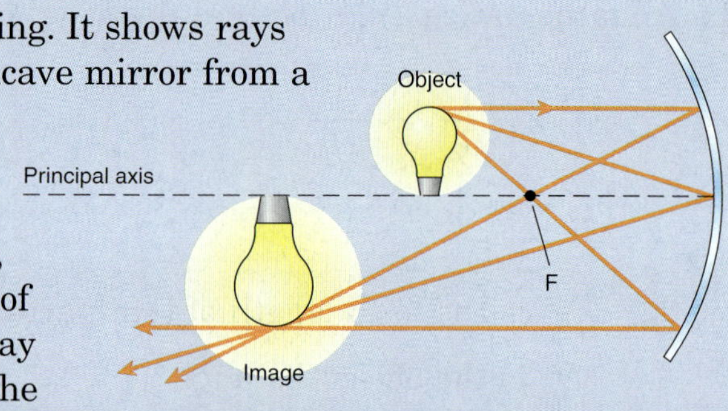

The next drawing shows the same mirror, but with the object much further from the mirror. Notice how the image in this second drawing is much smaller and much closer to the focus.

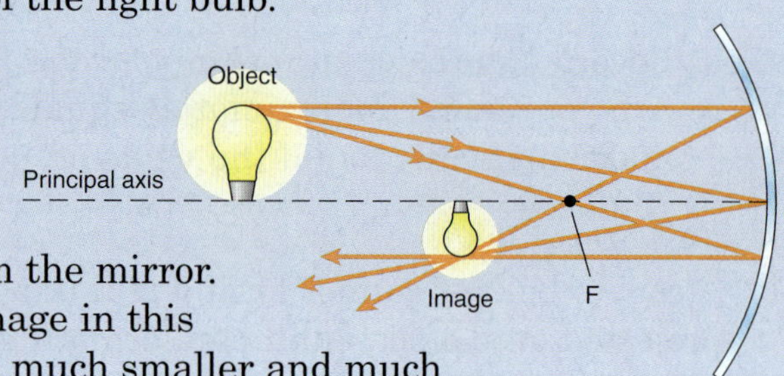

As you have seen, the position of the object and image are described by the equation below.

$$\frac{1}{f} = \frac{1}{D_o} + \frac{1}{D_i}$$

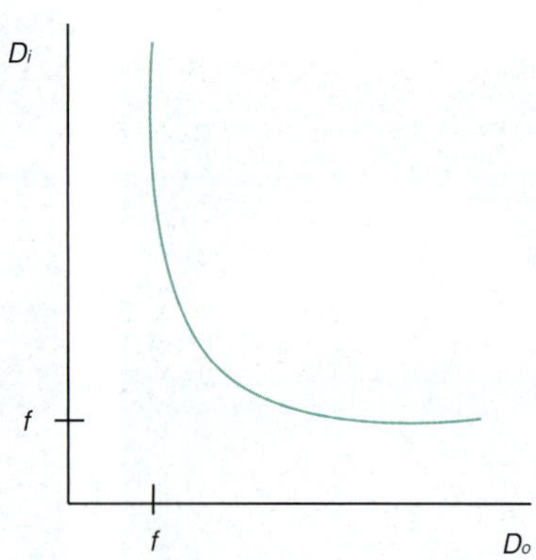

Look at the graph of this equation at left. Notice that as the object distance decreases, the image distance becomes very large. As the object distance increases, the image distance moves towards the focal length (*f*). Also notice that neither the object distance nor the image distance can be less than the focal length.

Reflecting on the Activity and the Challenge

You have observed how rays of light are reflected by a curved mirror. You have seen that a concave mirror can make an upside-down real image (an image on a screen). You have also seen that the image and object distances are described by a simple mathematical relationship. In addition, you have seen that there is no real image in a convex mirror, and the image is always smaller than the object.

You may want to use a curved mirror in your sound and light show. You may want to project an image on a screen or produce a reflection that the audience can see in the mirror. What you have learned will help you explain how these images are made.

Since the image changes with distance, you may try to find a way to have a moving object so that the image will automatically move and change size. A ball suspended by a string in front of a mirror may produce an interesting effect. You may also wish to combine convex and concave mirrors so that some parts of the object are larger and others are smaller. Convex and concave mirrors could be shaped to make some kind of fun-house mirror.

Remember that your light show will be judged partly on creativity and partly on the application of physics principles. This activity has provided you with some useful principles that can help with both criteria.

Physics To Go

1. a) Make a drawing of parallel laser beams aimed at a convex mirror.
   b) Draw lines to show how the beams reflect from the mirror.

2. a) Make a drawing of parallel laser beams aimed at a concave mirror.
   b) Draw lines to show how the beams reflect from the mirror.

3. a) Look at the back of a spoon. What do you see?
   b) Look at the inside of a spoon. What do you see?

4. a) If you were designing a shaving mirror, would you make it concave or convex? Explain your answer.
   b) Why do some makeup mirrors have two sides? What do the different sides do? How does each side produce its own special view?
   c) How does a curved side mirror on a car produce a useful view? How can this view sometimes be dangerous?
   d) Why does a dentist use a curved mirror?

D_i (cm)	D_o (cm)
549	15
56	25
20	50
18	91
14	142

5. a) A student found the real image of a light bulb in a concave mirror. The student moved the light bulb to different positions. At each position, the student measured the position of the image and the light bulb. The results are shown in the table on the left. Draw a graph of this data.
   b) Make a general statement to summarize how the image distance changes as the object distance changes.
   c) If the object were twice as far away as the greatest object distance in the data, estimate where the image would be.
   d) If the object were only half as far from the mirror as the smallest object distance in the data, estimate what would happen to the image.

6. A ball is hung on a string in front of a flat mirror. The ball swings toward the mirror and back. How would the image of the ball in the mirror change as the ball swings back and forth?

7. a) A ball is hung on a string in front of a concave mirror. The ball swings toward the mirror and back. How would the image of the ball in the mirror change as the ball swings back and forth?
   b) How could you use this swinging ball in your light show?

8. Outdoors at night, you use a large concave mirror to make an image on a card of distant auto headlights. You make the image on a card. What happens to the image as the car gradually comes closer?

9. The diagram shows a light ray R parallel to the principal axis of a spherical concave (converging) mirror. Point F is the focal point and C is the center of curvature. Draw the reflected light ray.

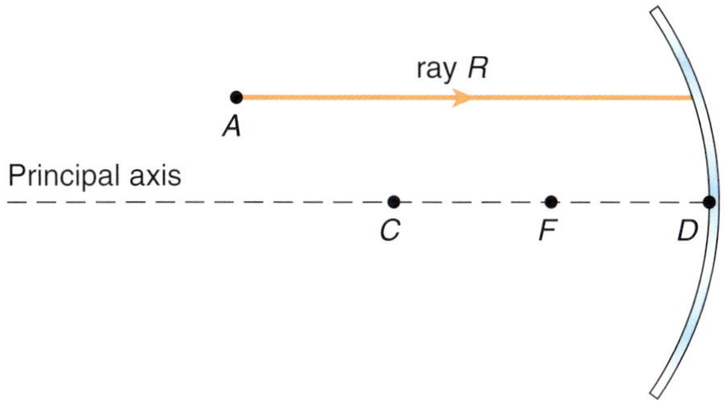

10. The diagram shows a curved mirror surface and a light bulb and its image. In relation to the focal point of the mirror, where is the light bulb (object) most likely located?

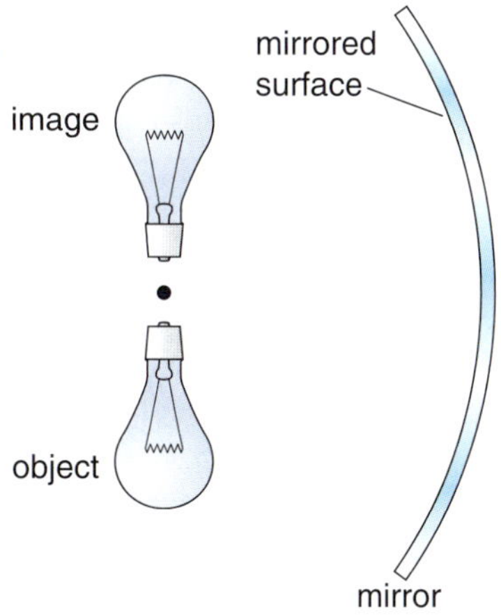

11. A candle is located beyond the center of curvature, C, of a concave spherical mirror having a principal focus, F, as shown in the diagram. Sketch the image of the candle.

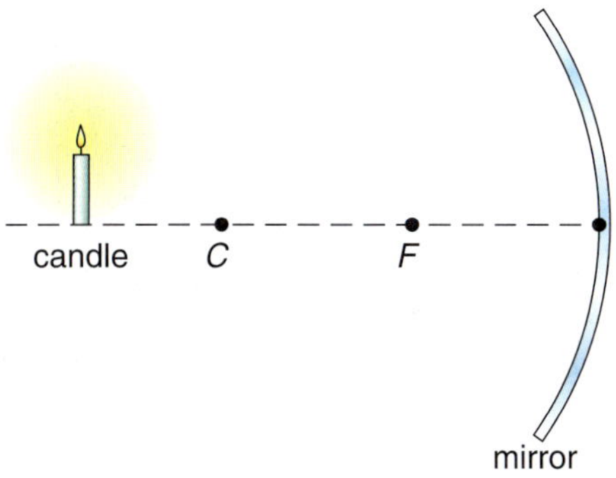

12. The diagram shows four rays of light from object AB incident on a spherical mirror with a focal length of 0.04 m. Point F is the principal focus of the mirror, point C is the center of curvature, and point O is located on the principal axis.

    a) Which ray of light will pass through F after it is reflected from the mirror?

    b) As object AB is moved from its position toward the left, what will happen to the size of the image produced?

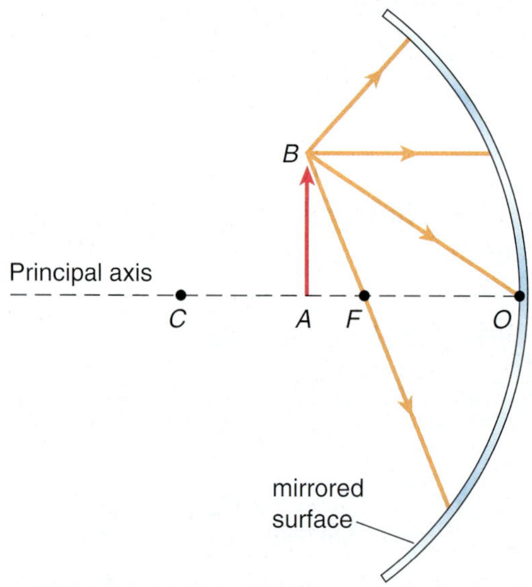

Activity 6 Refraction of Light

GOALS

In this activity you will:

- Observe refraction.
- Measure angles of incidence and refraction.
- Measure the critical angle.
- Observe total internal reflection.

What Do You Think?

The Hope Diamond is valued at about 100 million dollars. A piece of cut glass of about the same size is worth only a few dollars.

- **How can a jeweler tell the difference between a diamond and cut glass?**

Record your ideas about this question in your *Active Physics* log. Be prepared to discuss your responses with your small group and with your class.

For You To Do

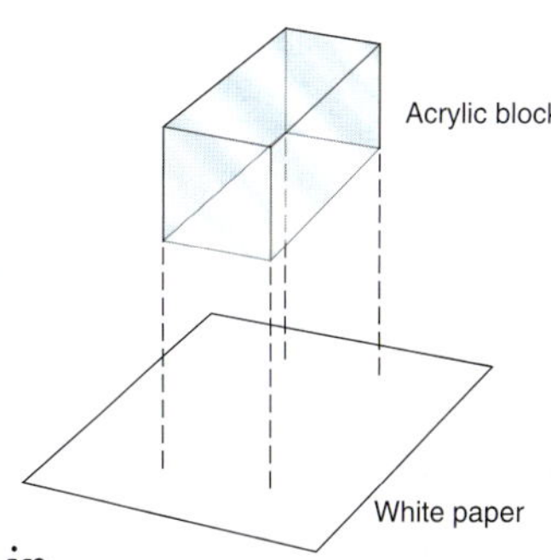

1. Place an acrylic block on a piece of white paper on your desk.
2. Carefully aim a laser pointer, or the light from a ray box, so the light beam moves horizontally, as you did in

Never look directly at a laser beam or shine a laser beam into someone's eyes. Always work above the plane of the beam and beware of reflections from shiny surfaces.

previous activities. Place a glass rod in the light beam so that the beam spreads up and down.

3. Shine the laser pointer or light from the ray box through the acrylic block. Be sure the beam leaves the acrylic block on the side opposite the side the beam enters. Mark the path of each beam. You may wish to use a series of dots as you did before. Label each path on both sides of the acrylic block so you will know that they go together.

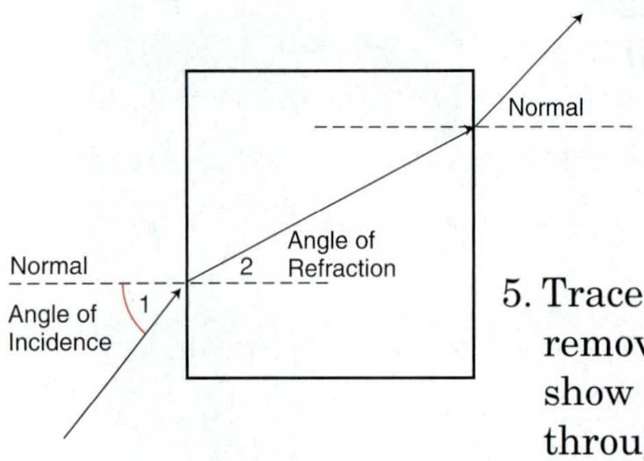

4. The angle of incidence is the angle between the incident laser beam and the normal, as shown in the diagram. Choose two other angles of incidence and again mark the path of the light, as you did in **Step 3**. As before, label each pair of paths.

5. Trace the outline of the acrylic block on the paper and remove the acrylic block. Connect the paths you traced to show the light beam entering the acrylic block, traveling through the acrylic block, and emerging from the acrylic block. Draw a perpendicular line at the point where a ray enters or leaves the acrylic block. Label this line the normal.

6. Measure the angles of incidence (the angle in the air) and refraction (the angle in the acrylic block).

a) Record your measurements in tables like the one shown.

Angle of incidence	Angle of refraction	Sine of angle of incidence	Sine of angle of refraction	$\frac{\text{Sin } \angle i}{\text{Sin } \angle R}$

b) Use a calculator to complete the chart by finding the sines of the angles (sin button on calculator).

c) Is the value of $\frac{\sin \angle i}{\sin \angle R}$ a constant? This value is called the index of refraction for the acrylic block.

7. Set up the acrylic block on a clean sheet of white paper. This time, as shown in the drawing (next page), aim the beam so it leaves the acrylic block on the side, rather than at the back.

Physics Words

critical angle: the angle of incidence for which a light ray passing from one medium to another has an angle of refraction of 90°.

index of refraction: a property of a medium that is related to the speed of light through it; it is calculated by dividing the speed of light in vacuum by the speed of light in the medium.

Snell's Law: describes the relationship between the index of refraction and the ratio of the sine of the angle of incidence and the sine of the angle of refraction.

8. Make the first angle of incidence (angle 1) as small as possible, so the second angle of incidence (angle 2) will be as large as possible. Adjust angle 1 so that the beam leaves the acrylic block parallel to the side of the acrylic block, as shown. Measure the value of angle 2.

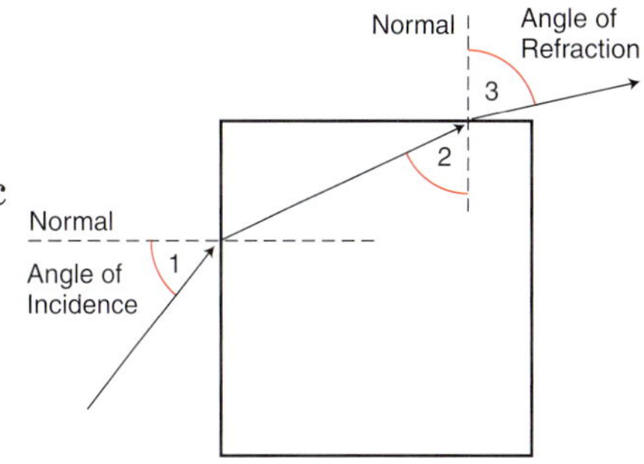

a) Record the value of angle 2. It is called the **critical angle**.

b) What happens to the beam if you make angle 2 greater than the critical angle?

c) What you observed in **(b)** is called "total internal reflection." What is reflected totally, and where?

9. It is possible to bend a long, rectangular acrylic block so the light enters the narrow end of the acrylic block, reflects off one side of the acrylic block, then reflects off the other and back again to finally emerge from the other narrow end. Try to bend an acrylic block rectangle so that the light is reflected as described.

FOR YOU TO READ

Snell's Law

Light refracts (bends) when it goes from air into another substance. This is true whether the other substance is gelatin, glass, water, or diamond. The amount of bending is dependent on the material that the light enters. Each material has a specific **index of refraction**, *n*. This index of refraction is a property of the material and is one way in which a diamond (very high index of refraction—lots of bending) can be distinguished from glass (lower index of refraction—less bending). The index of refraction is a ratio of the sine of the angle of incidence and the sine of the angle of refraction.

Index of refraction: $n = \dfrac{\sin \angle i}{\sin \angle R}$

This equation is referred to as **Snell's Law**.

As light enters a substance from air, the light bends toward the normal. When light leaves a substance and enters the air, it bends away from the normal. If the light is entering the air from a substance, the angle in that substance may be such that the angle of refraction is 90°. In this special case, the angle in the substance is called the critical angle. If the angle in the substance is greater than this critical angle, then the light does not enter the air but reflects back into the substance as if the surface were a perfect mirror. This is the basis for light fibers where laser light reflects off the inner walls of glass and travels down the fiber, regardless of the bend in the fiber.

Reflecting on the Activity and the Challenge

The bending of light as it goes from air into a substance or from a substance into air is called refraction. It is mathematically expressed by Snell's Law. When light enters the substance at an angle, it bends towards the normal. When light leaves the substance at an angle, it bends away from the normal. As you create your light show for the **Chapter Challenge**, you may find creative uses of refraction. You may decide to have light bending in such a way that it spells out a letter or word or creates a picture. You may wish to have the light travel from air into glass to change its direction. You may have it bend by different amounts by replacing one material with another. Regardless of how you use refraction effects, you can now explain the physics principles behind them.

Physics To Go

1. A light ray goes from the air into an acrylic block. In general, which is larger, the angle of incidence or the angle of refraction?

2. a) Make a sketch of a ray of light as it enters a piece of acrylic block and is refracted.

   b) Now turn the ray around so it goes backward. What was the angle of refraction is now the angle of incidence. Does the turned-around ray follow the path of the original ray?

3. A light ray enters an acrylic block from the air. Make a diagram to show the angle of incidence, the angle of refraction, and the normal at the edge of the acrylic block.

4. Light rays enter an acrylic block from the air. Make drawings to show rays with angles of incidence of 30° and 60°. For each incident ray, sketch the refracted ray that passes through the acrylic block.

5. a) Light is passing from the air into an acrylic block. What is the maximum possible angle of incidence that will permit light to pass into the acrylic block?

   b) Make a sketch to show your answer for **Part (a)**. Include the refracted ray (inside the acrylic block) in your sketch.

6. a) A ray of light is already inside an acrylic block and is heading out. What is the name of the maximum possible angle of incidence that will permit the light to pass out of the acrylic block?

   b) If you make the angle of incidence in **Part (a)** greater than this special angle, what happens to the light?

   c) Make a sketch to show your answer for **Part (b)**. Be sure to show what happened to the light.

7. a) Make a drawing of a light ray that enters the front side of a rectangular piece of acrylic block and leaves through the back side.

   b) What is the relationship between the direction of the ray that enters the acrylic block and the direction of the ray that leaves the acrylic block?

   c) Use geometry and your answer to **Question 2 (b)**, to prove your answer to **Question 7 (b)**.

8. You have seen the colored bands that a prism or cut glass or water produce from sunlight. Light that you see as different colors has different wavelengths. Since refraction makes these bands, what can you say about the way light of different wavelengths refracts?

Stretching Exercises

1. Cover the acrylic block with a red filter. Shine a red laser beam into the acrylic block, as you did in **For You To Do, Steps 1** through **3**. What happens? How can you explain what happens?

2. Find some $\frac{1}{2}$" diameter clear tubing, about 2 m long. Plug one end. Pour clear gelatin in the other end, through a funnel, before the gelatin has had time to set. Arrange the tubing into an interesting shape and let the gelatin set. You may wish to mount your tube on a support or a sturdy piece of cardboard, which can be covered with interesting reflective material. Fasten one end of the tube so laser light can easily shine straight into it. When the gelatin has set, turn on the laser. What do you see? This phenomena is called total internal reflection.

3. Place a penny in the bottom of a dish or glass. Position your eye so you can just see the penny over the rim of the glass. Predict what will happen when you fill the glass with water. Then try it and see what happens. How can you explain the results?

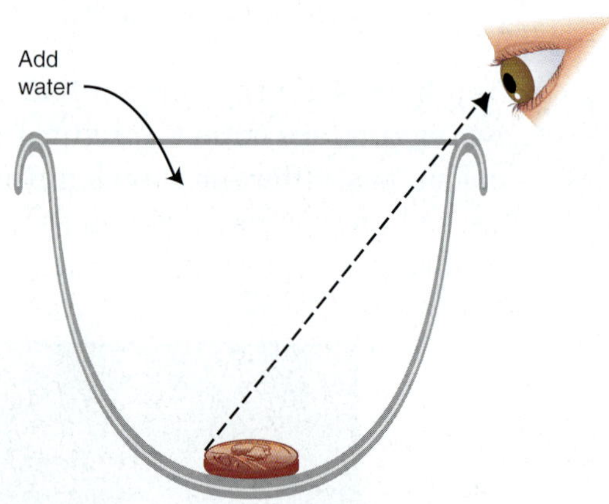

4. Place an empty, clear drinking glass over a piece of a newspaper. When you look through the side of the glass near the bottom, you can see the printing on the newspaper. What do you think will happen if you fill the glass with water? Try it and see. How can you explain the result? Does it help to hold your fingers over the back of the glass?

Activity 7 Effect of Lenses on Light

GOALS

In this activity you will:

- Observe real images.
- Project a slide.
- Relate image size and position.

What Do You Think?

Engineers have created special lenses that can photograph movie scenes lit only by candlelight.

- **How is a lens able to project movies, take photographs, or help people with vision problems?**

Record your ideas about this question in your *Active Physics* log. Be prepared to discuss your responses with your small group and with your class.

For You To Do

1. Look at the lens your teacher has given you.

   a) Make a side-view drawing of this lens in your log. This is a convex lens.

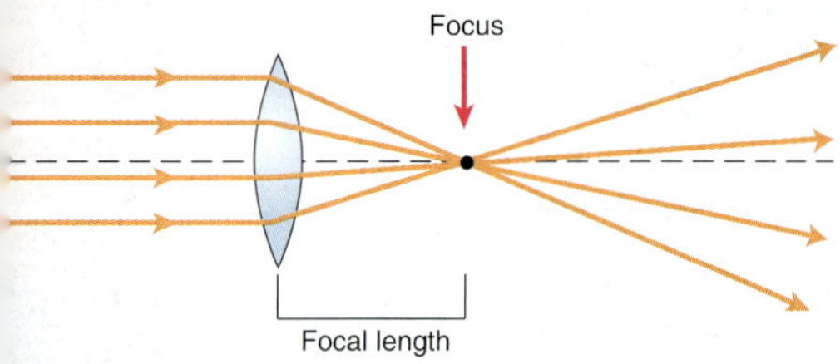

2. Point the lens at a window or at something distant outside. Use a file card as a screen. Look for the image on the file card. Move it back and forth until you see a sharp image of the distant object.
   a) Describe what you see. Is the image large or small? Is it right side up or upside down? Is it reversed left to right? This image is called "real" because you can project it on the screen.

3. Measure the distance between the image and the lens. If the object is very far away, this distance is the focal length of the lens. The position of this image is the focus of the lens. It is the same location at which parallel rays of light would converge.
   a) Approximate the object distance.
   b) Measure the image distance.
   c) Record your object and image distance. Note that the image distance is also the focal length of the lens.

Do not use lenses with chipped edges. Mount lenses securely in a holder. Use only light sources with enclosed or covered electrical contacts. Keep flammables/combustibles away from the candleholder.

4. Set up a 40-W light bulb or a candle to be a light source. Mount the lens at the same height as the light source. If you are using a light bulb, point it right at the lens, as shown.

5. Place the light bulb about a meter away from the lens. Try to find the image of the light bulb on a screen. The screen can be a file card or a sheet of paper.
   a) Record your results in a table, including the distance and appearance of the image.

6. Adjust the position of the object to create a larger image.
   a) Describe how the position of the object, the image, and the size of the image have changed. Record the results in a table.

7. Create an object by carefully cutting a hole in the shape of an arrow in an index card. Have someone in your group hold the card close to the light bulb.
   - a) Can you see the object on the screen? Describe what you see.
   - b) Have the person holding the object move it around between the light bulb and the convex lens. What happens?

8. Project the object onto the wall. Can you make what you project larger or smaller?
   - a) In your log indicate what you did to change the size of the image.

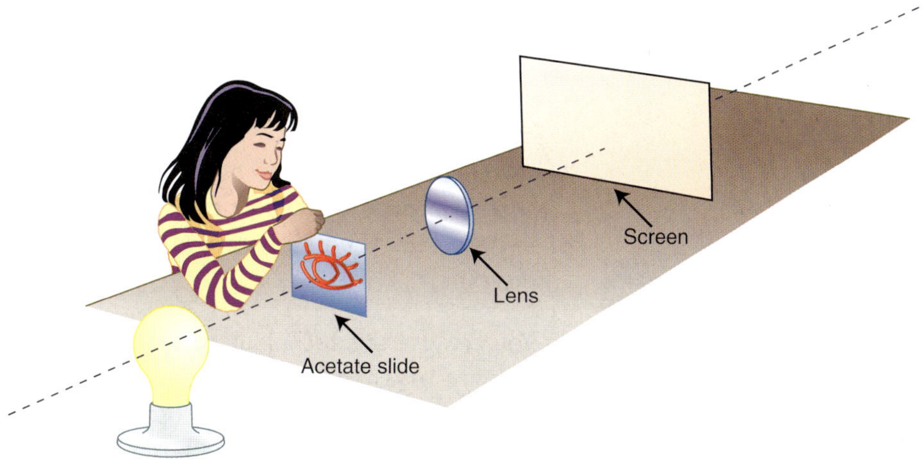

Caution: Lamps get very hot. Be careful not to touch the bulb or housing surrounding the bulb.

9. Create a slide by drawing with a marking pen on clear acetate. Try placing a 100-W light bulb and the slide in different positions.
   - a) Describe how you can project a real, enlarged image of your slide onto a screen or wall.
   - b) How can you use the lens to change the size of the image?
   - c) Explore the effect of different lenses. In your log, record how you think this effect might be part of your light and sound show.

FOR YOU TO READ

Lens Ray Diagrams

You are probably more familiar with images produced by lenses than you are with images from curved mirrors. The lens is responsible for images of slides, overhead projectors, cameras, microscopes, and binoculars.

Light bends as it enters glass and bends again when it leaves the glass. The **convex converging lens** is constructed so that all parallel rays of light will bend in such a way that they meet at a location past the lens. This place is the focal point.

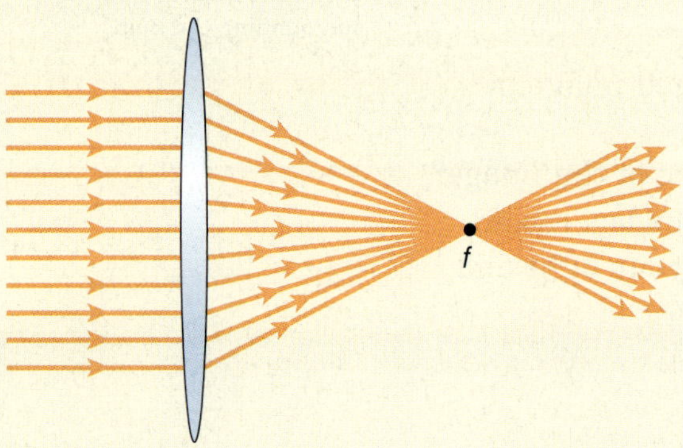

If an object is illuminated, it reflects light in all directions. If these rays of light pass through a lens, an image is formed.

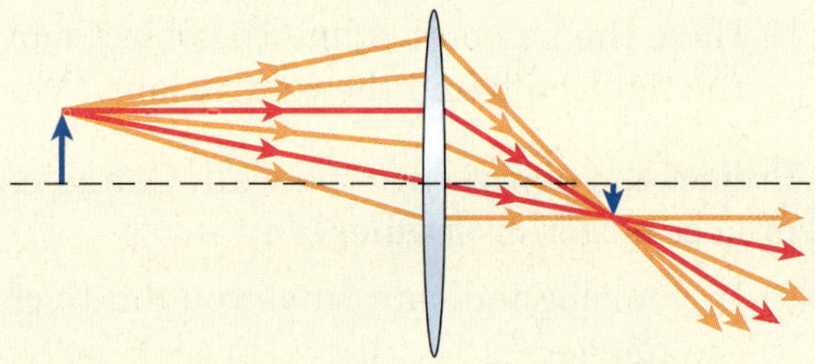

Although all of the light rays from the object help to form the image, you can locate an image by looking at two easy rays to draw—the ray that is parallel to the principal axis and travels through the focal point and the ray that travels through the center of the lens undeflected. (These rays are in red in the diagram.)

You can use this technique to see how images that are larger (movie projector), smaller (camera), and the same size (copy machine) as the object can be created with the same lens.

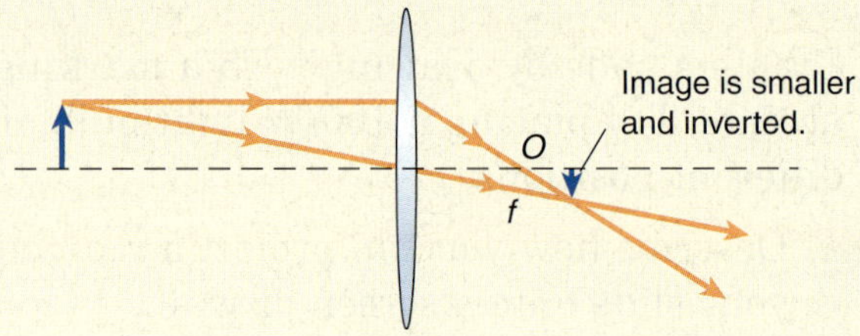

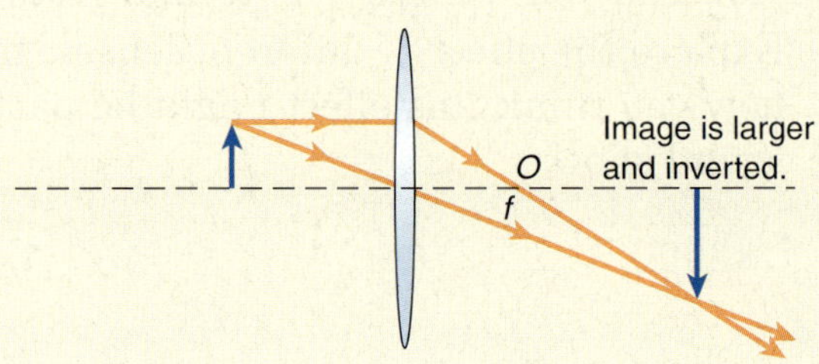

If the object is close to the lens (an object distance smaller than the focal distance), then an image is not formed. However, if you were to view the rays emerging, they would appear to have come from a place on the same side of the lens as the object. To view this image, you put your eye on the side of the lens opposite the object and peer through it—it's a magnifying glass!

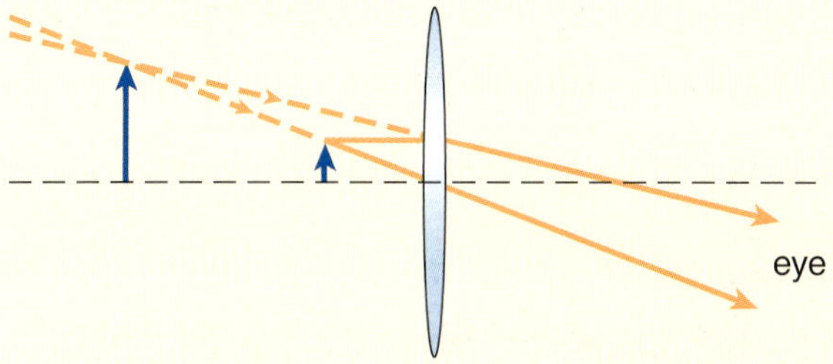

Sample Problem

The diagram shows a lens and an object.

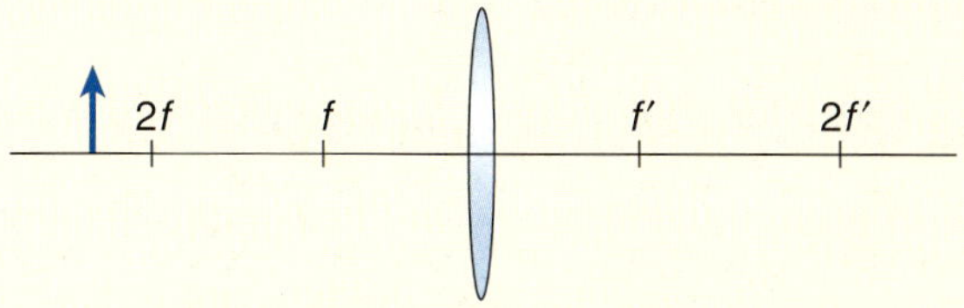

a) Using a ray diagram, locate the image of the object shown.

b) Describe the image completely.

Strategy: Choose a location on the object to be the origin of the rays. A simple choice would be the tip of the arrow. At least two rays must be drawn to locate the image.

Givens:

See the diagram.

Solution:

a)

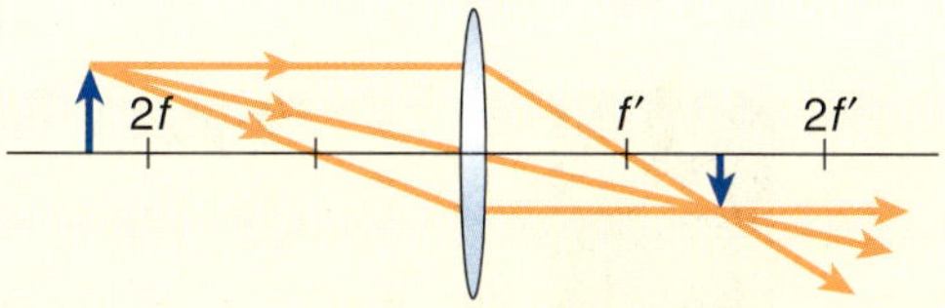

b) The image is real, reduced, and inverted.

As the object moves closer to the lens, its size will increase. At $d_o = f$ there will be no image and at $d_o < f$ the image will be virtual and upright.

Physics Words

converging lens: parallel beams of light passing through the lens are brought to a real point or focus (if the outside index of refraction is less than that of the lens material); also called a convex lens.

Reflecting on the Activity and the Challenge

You have explored how convex lenses make real images. You have found these images on a screen by moving a card back and forth until the image was sharp and clear, so you know that they occur at a particular place. Bringing the object near the lens moves the image away from the lens and enlarges the image, but if the object is too close to the lens, there is no real image. These images are also reversed left to right and are upside down. You may be able to use this kind of image in your sound and light show. You have also projected images of slides on a wall. You may be able to add interest by moving the lens and screen to change the size of these images.

Physics To Go

1. a) What is the focus of a lens?
   b) If the image of an object is at the focus on a lens, where is the object located?
   c) What is the focal length of a lens?
   d) How can you measure the focal length of a lens?

2. a) You set up a lens and screen to make an image of a distant light. Is the image in color?
   b) Is the image right side up or upside down?
   c) Did the lens bend light to make this image? How can you tell?
   d) A distant light source begins moving toward a lens. What must you do to keep the image sharp?

3. a) You make an image of a light bulb. What can you do to make the image smaller than the light bulb?
   b) What can you do to make the image larger than the light bulb?

4. a) You have two lights, a lens, and a screen, as shown on opposite page. One light is at a great distance from the lens. The other light is much closer. If you see a sharp image of the distant light, describe the image of the closer light.
   b) If you see a sharp image of the closer light, describe the image of the more distant light.

c) Could you see a sharp image of both lights at the same time? Explain how you found your answer.

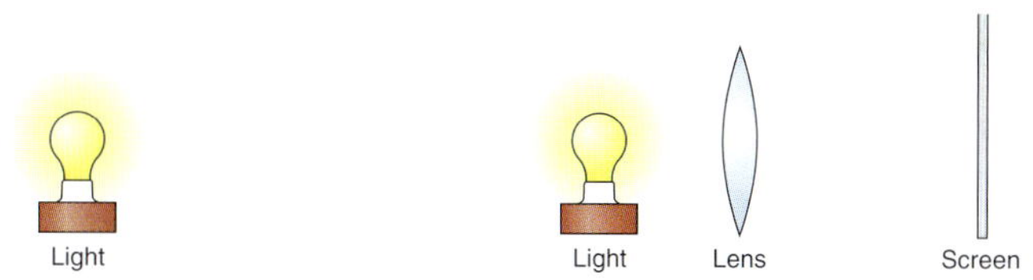

5. Research how a camera works. Find out where the image is located. Also find out how the lens changes so that you can photograph a distant landscape and also photograph people close up.

6. Using a ray diagram, locate the image formed by the lens below.

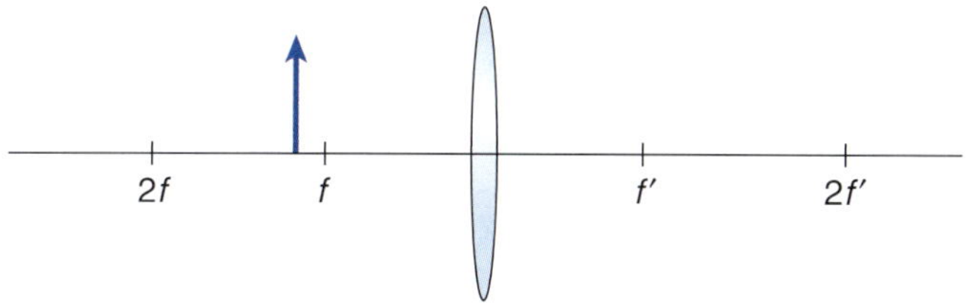

7. An object 1.5 cm tall is placed 5.0 cm in front of a converging lens of focal length 8.0 cm.
   a) Determine the location of the image.
   b) Completely describe the image.

8. A relative wants to show you slides from her wedding in 1972. She brings out her slide projector and screen.
   a) If she puts the screen 2.8 m from the projector and the lens has a focal length of 10.0 cm, how far from the lens will the slide be so that her pictures are in focus?
   b) If each slide is 3.0 cm tall, how big will the image be on the screen?

9. The diagram shows an object 0.030 m high placed at point X, 0.60 m from the center of the lens. An image is formed at point Y, 0.30 m from the center of the lens. Completely describe the image.

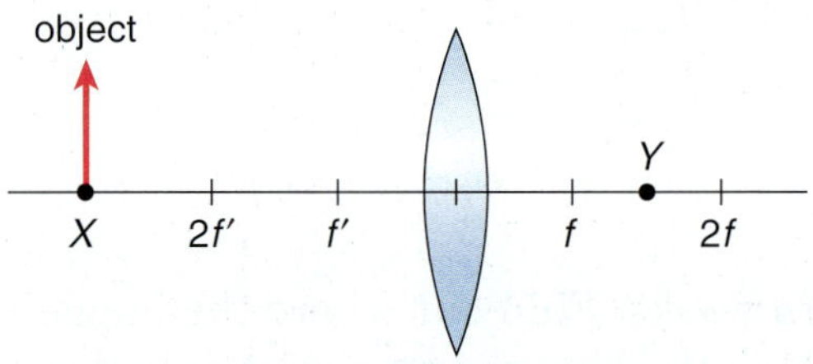

10. The diagram represents an object placed two focal lengths from a converging lens. At which point will the image be located?

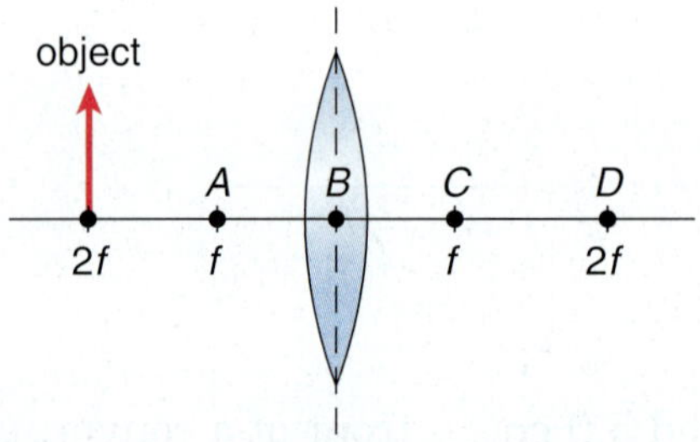

11. The diagram shows a lens with an object located at position A. Describe what will happen to the image formed as the object is moved from position A to position B.

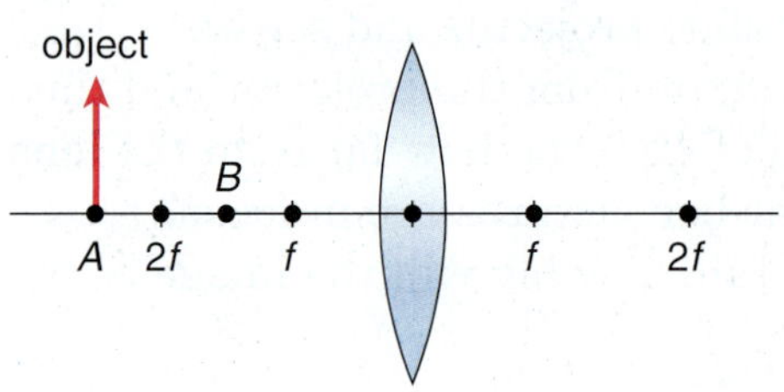

Stretching Exercises

1. To investigate how the image position depends on the object position, find a convex lens, a white card, and a light source. Find the image of the light source, and measure the image and object distance from the lens. Make these measurements for as wide a range of object distances as you can. In addition, make an image of an object outside, such as a tree. Estimate the distance to the tree. The image of a distant object, like the tree, is located very near the focus of the lens. Draw a graph of the results. Compare the graph with the equation

$$\frac{1}{f} = \frac{1}{D_o} + \frac{1}{D_i}$$

2. Find a camera with a shutter that you can keep open (with a bulb or time setting). Place a piece of waxed paper or a piece of a plastic bag behind the lens, where the film would be if you took a picture. Find the image and compare it to the images you made in this activity. Focus the lens for objects at different distances. Investigate how well the object and image location fit the lens equation $\frac{1}{f} = \frac{1}{D_o} + \frac{1}{D_i}$.

   Remember that the focal length of the lens is typically printed on the lens.

3. Research how the concept of "depth of field" is important in photography. Report to the class on what you learn.

Activity 8 Color

GOALS

In this activity you will:

- Analyze shadow patterns.
- Explain the size of shadows.
- Predict pattern of colored shadows.
- Observe combinations of colored lights.

What Do You Think?

When a painter mixes red and green paint, the result is a dull brown. But when a lighting designer in a theater shines a red and a green light on an actress, the actress's skin looks bright yellow.

- **How could these two results be so different?**
- **How are the colors you see produced?**

Record your ideas about these questions in your *Active Physics* log. Be prepared to discuss your responses with your small group and with your class.

For You To Do

1. Carefully cut out a cardboard puppet that you will use to make shadows.

2. Turn on a white light bulb only. Move the puppet around and observe the shadow.
   a) Describe the shadow you see.
   b) What happens to the shadow if you move the puppet sideways or up and down?
   c) What happens to the shadow if you move the puppet close to the screen?
   d) What happens to the shadow if you move the puppet close to the bulb?

Caution: Lamps get very hot. Be careful not to touch the bulb or housing surrounding the bulb.

3. Look at the drawing. It shows a top view of a puppet halfway between the light and the screen.

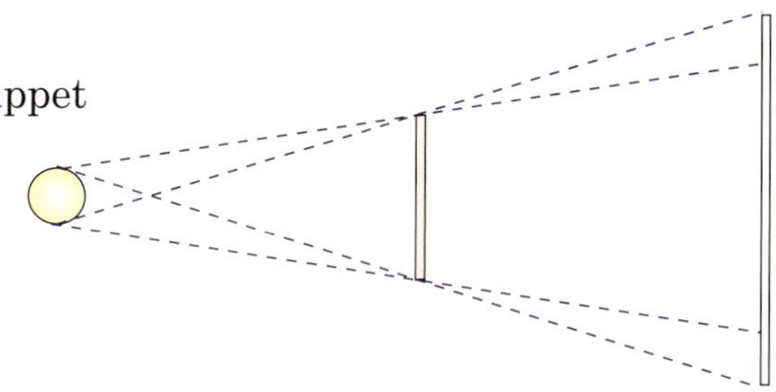

   a) Make a copy of this drawing in your log. Draw light rays going from the light to the screen. (Light rays will go in straight lines in all directions from all parts of the light.)
   b) Use the top-view drawing you drew to answer these questions: Which part of the screen receives light? Which part receives no light? Which part receives some light?
   c) Is the shadow larger or smaller than the puppet? Explain how you found your answer.
   d) Now copy the other two top-view drawings and show the path of the light rays.

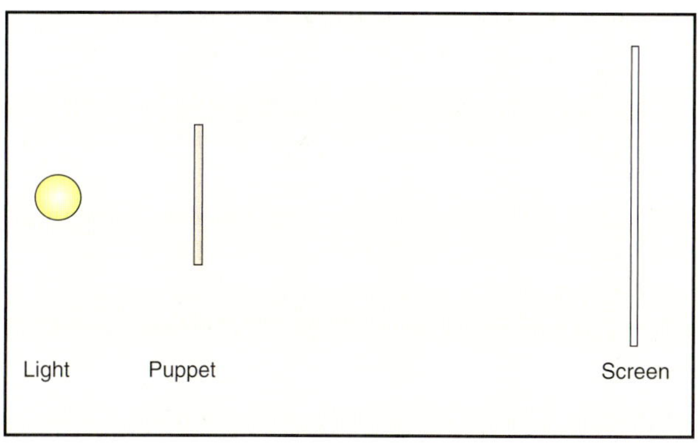

   e) On your drawings, show which part of the screen does receive light and which part does not receive light and which part receives some light.
   f) For each of these two drawings, tell whether the shadow is larger or smaller than the puppet. For each one, explain how you found your answer.

4. Turn off the white bulb. Turn on red and green bulbs. They should be aimed directly at the center of the screen.
   a) What color do you see on the screen?
   b) Predict what color the shadows will be if you bring your puppet between the bulbs and the screen. Record your prediction, and give a reason for it.
   c) Make a top-view drawing to show the path of the light rays from the red and green bulbs.
   d) On your drawing, label the color you will see on each part of the screen.

5. Turn off the green bulb and turn on a blue one. Repeat what you did in **Step 4**, but with the blue and red bulbs lit.

6. Turn off the red bulb and turn on the green one. Repeat what you did in **Step 4**, but with the blue and green bulbs lit.

7. Turn on the red bulb so all three—red, blue, and green—are lit. Repeat what you did in **Steps 5** and **6**.

Reflecting on the Activity and the Challenge

Different colored lights can combine to make white light. When an object blocks all light, it creates a dark shadow. Since some light comes from all parts of the bulb, there are places where the shadow is black (no light) and places where the shadow is gray (some light reaches this area). An object illuminated by different colored lights can create shadows that prevent certain colors from reaching the wall and allowing other colors to pass by.

In your light show creation, you may choose to use the ideas of colored shadows to show how lights can be added to produce interesting combinations of colors. By moving the object or the lights during the show, you may be able to produce some interesting effects. Lighting design is used in all theater productions. It requires a knowledge and understanding of how lights work, as well as an aesthetic sense of what creates an enjoyable display.

Physics To Go

1. Show how a shadow is created.
2. How can moving the light, the object, and the screen all produce the effect of enlarging the shadow?
3. Explain why a gray halo surrounds a dark shadow made by a light bulb and an object.
4. a) Why is your shadow different at different times of the day?
   b) What is the position of the Sun when your shadow is the longest? The shortest?
5. Why is the gray halo about your shadow so thin when you are illuminated by the Sun?
6. a) Suppose you shine a red light on a screen in a dark room. The result is a disk of red light. Now you turn on a green light and a blue light. The three disks of light overlap as shown. Copy the diagram into your journal. Label the color you will see in each part of the diagram.

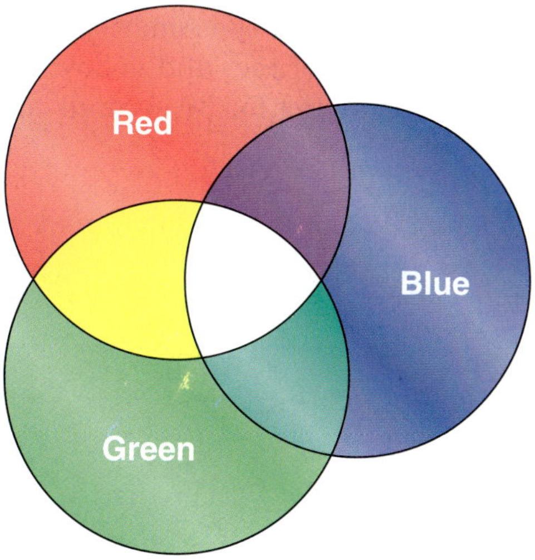

   b) Add the labels "bright," "brighter," and "brightest" to describe what you would see in each part of your diagram.

7. a) Make a drawing of an object in red light. The object casts a shadow on the screen. Label the color of the shadow and the rest of the screen.
   b) Repeat **Part (a)** for an object in green light.
   c) Now make a copy of your drawing for **Part (b)**. Add a red light, as in **Part (a)**. Label the color of all the shadows.

8. List some imaginative ways that you can add colors to your light show.

Stretching Exercises

1. With the room completely dark, shine a red light on various colored objects. Compare the way they look in red light with the way they look in ordinary room light.

2. View 3-D pictures with red and blue glasses. Explain how each eye sees a different picture.

3. Shine a white light and a red light on a small object in front of a screen. What colors are the shadows? How is this surprising?

4. Prepare a large drawing of the American flag but with blue-green in the place of red, and yellow in the place of blue. Stare at the drawing for 30 seconds and then look at a white surface.

PHYSICS AT WORK

Alicja and Dennis Phipps

Alicja Phipps has always been interested in electronics. As a child she wanted to be a television repair person. She now works with her husband, Dennis Phipps, in their company, Light & Sound Entertainment, which designs original content programming in a variety of areas—from rock concerts to the Olympics.

Light & Sound Entertainment got its name because Dennis believes strongly that the two are and should be linked. "It's terrible when the music of a production does not match what you are seeing," he says. "More and more theaters have lighting specialists come in and set up the theater with everything preset on a computerized lighting board. There will be a setting for 'outdoor lighting' and 'nighttime' or 'sunset'. The problem with that is that every production will look the same, when in reality nighttime in Canada looks very different from nighttime in Florida. There are also different lighting considerations depending on your audience. A production in front of a live audience needs different lighting than one being recorded on video. For instance, the human eye picks up shadows a lot better than a camera will. If a production is intended for both a live audience and video, lighting needs to be arranged accordingly."

Dennis continues, "The sound of a production is only as good as its setup, and nothing can replace the actual setting. The Red Rocks Theater in Colorado is terrific, for example, because stone has a very high reverberation rate, which is great for guitars. A huge wooden room like Carnegie Hall also provides a unique sound. However, these spaces and materials are not readily available." The hardest projects, Dennis says, are those in which you cannot control the elements. "Sound elements include the size and shape of the space, reverberation, feedback, and temperature."

"New media has its own set of challenges," explains Alicja who oversees the conversion of live events into various other formats, such as CD-ROMs, virtual reality, and Web sites. "We have to think about how much information (sound and image) we will be able to fit on a disc or on to a Web page and how long it will take to load. If it takes too long, no one will ever see or hear what we've done."

"We enjoy the creative process of every production," claim Alicja and Dennis. "Each one is a unique challenge."

Chapter 5 Assessment

With what you learned about sound and light in this chapter, you are now ready to dazzle the world. However, you have neither the funds nor the technology available to professionals. All sounds you use to capture the interest of the class must come from musical instruments that you build yourself, or from human voices. Some of these sounds may be prerecorded and then played in your show. If your teacher has a laser and is willing to allow you to use it, you may do so. All other light must come from conventional household lamps. Gather with your committee to design a two- to four-minute sound and light show to entertain other students your age.

Review the criteria by which you decided that your show will be evaluated. The following suggestions were provided at the beginning of the chapter:

1. The variety and number of physics concepts used to produce the light and sound effects.
2. Your understanding of the physics concepts:
   a) Name the physics concepts that you used.
   b) Explain each concept.
   c) Give an example of something that each concept explains or an example of how each concept is used.
   d) Explain why each concept is important.
3. Entertainment value

At this time you may wish to propose changes in the criteria. Also decide as a class if you wish to modify or keep the point value you established at the beginning of the chapter.

Enjoy the sound and light productions!

Physics You Learned

Compressional and transverse waves

Wave speed = wavelength × frequency

Standing waves

Pitch and frequency

Sound production in pipes and vibrating strings

Controlling frequency of sounds produced electronically

Angle of incidence and angle of reflection

Location of image in plane and curved mirrors

$\frac{1}{f} = \frac{1}{D_o} + \frac{1}{D_i}$ in curved mirrors

Real images

Angle of incidence and angle of refraction

Lenses and image formation

$\frac{1}{f} = \frac{1}{D_o} + \frac{1}{D_i}$ in lenses

Color addition

$$n = \frac{\sin\angle i}{\sin\angle R}$$

Chapter 6

DESIGNING THE UNIVERSAL DWELLING

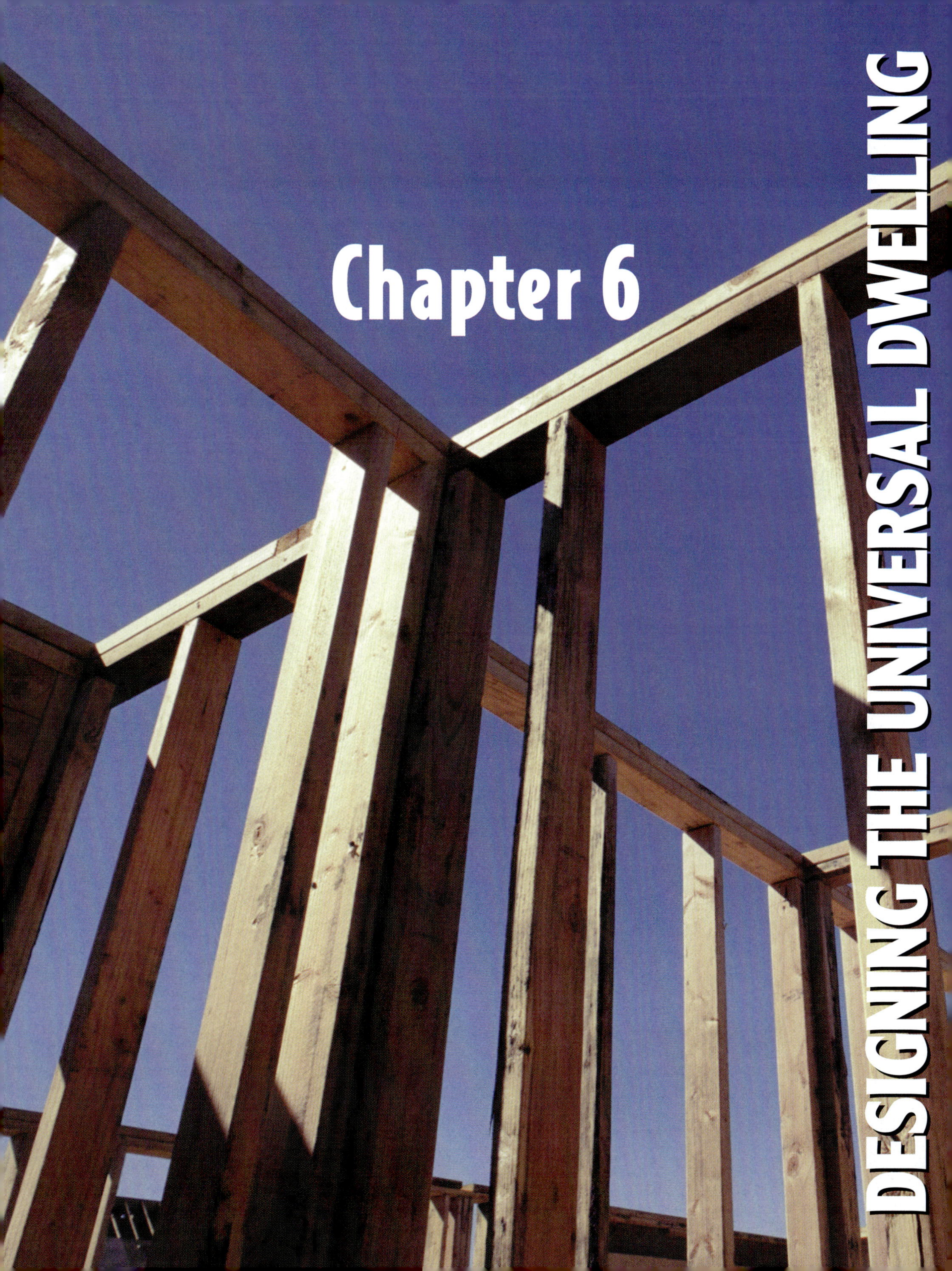

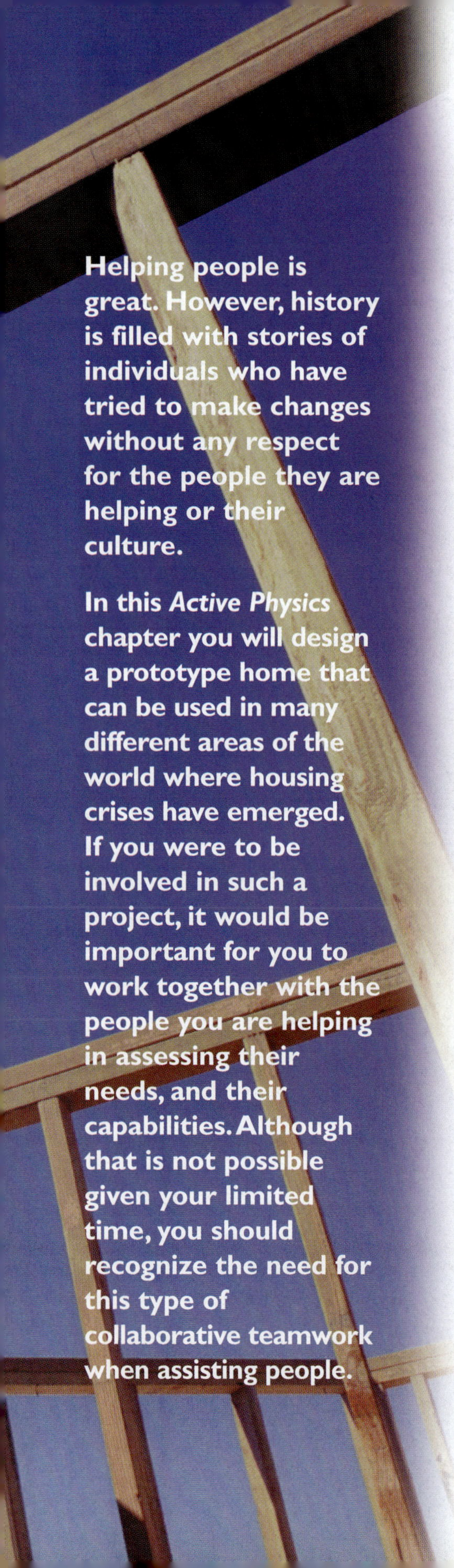

Helping people is great. However, history is filled with stories of individuals who have tried to make changes without any respect for the people they are helping or their culture.

In this *Active Physics* chapter you will design a prototype home that can be used in many different areas of the world where housing crises have emerged. If you were to be involved in such a project, it would be important for you to work together with the people you are helping in assessing their needs, and their capabilities. Although that is not possible given your limited time, you should recognize the need for this type of collaborative teamwork when assisting people.

Scenario

Imagine you and your team members are part of an international group called Homes For Everyone (HFE). The purpose of your organization is to address the growing housing shortage in many areas throughout the world. You have recently been sent to work with a self-help community group in a faraway area. Here is a letter you might write home.

Dear Mom and Dad,

Greetings! I've finally settled into my new home. Sometimes I think that I'm not in another country but on another planet. Everything here is so different. Sorry about the water splotches on this letter. It's pouring rain outside, and the roof is leaking. I have to remind myself that at least I have a roof over my head. Many people here have lost their homes from the last hurricane and are crowded into their relatives' already crowded spaces.

We're trying to figure out what we can do to help with the housing situation here. We're still not sure what the solution is. Right now I'm trying to learn more about these people who have been so generous and gracious to me. When we arrived the whole village got together to greet our group. I felt so honored! I expect the same kind of welcome from my family when I return home!

Tomorrow we are going to start looking around. I'll keep you posted on what we discover and on our progress. Signing off for now. Miss you all!

Love,

Kim

HFE plans to design a "universal dwelling" to meet the need for homes in diverse environments. The group needs a design that can be constructed quickly and simply at the building site. The design should also use the least amount of materials to create the most living space. To make mass production possible, the dwelling design should be uniform, but you should be able to make simple changes to meet local conditions. It should be energy efficient in any climate.

Challenge

After completing the nine activities in this chapter you will be challenged to do the following to present to the HFE Architectural Committee:

1. Develop scale drawings of the floor plan and all side views of the universal dwelling. Following are specifications for the drawings:
   a) The drawings should be done on a scale of 1 inch = 4 feet (1:48).
   b) The scale drawings should include all sides of the house, showing roof lines, roof overhangs, and placement and dimensions of all windows for a selected climatic region and culture.
   c) The scale drawings should show the floor plans of the living spaces and include specifications for the thickness of the wall, types of insulation and their thickness, and kinds and types of windows or ventilation openings.
   d) The plans should show the geographic orientation of the house (in which direction the house should face).

2. Write a two-page explanation which gives your reasons for the following:
   a) your choices for the shape and dimensions of the dwelling
   b) the energy consideration that went into the design of the dwelling
   c) the changes that could be made to the basic design to take into account different climates and cultures
   d) the things you included to make your dwelling attractive.

Criteria

The HFE Architectural Committee will use criteria similiar to the ones below in evaluating your drawings and your written presentation. Discuss and decide as a class the exact criteria the committee should use.

- **(20%) The drawings and written presentation should meet all specifications listed in the Chapter Challenge.**
- **(10%) The drawings and written presentation should explain how you considered surface-to-volume ratio in your design.**
- **(20%) The written presentation should explain how the house will accommodate differing seasonal and climatic conditions in terms of heating and cooling. It should also show you have a basis for scaling the dwelling appropriate to the family and culture for which it is designed.**
- **(20%) The design of the roof and the placement and dimensions of windows should account for climatic, solar, and latitude considerations, and should show you have accounted for the advantages and disadvantages of windows.**
- **(20%) The placement and kind of insulation materials and thickness you choose should show you know the principles which make insulation effective.**
- **(10%) The appearance of your house and the interior layout should show consideration for fundamental human needs.**

Activity 1 Factors in Designing the "Universal Dwelling"

GOALS

In this activity you will:

- Identify essential characteristics that all human dwellings have in common.
- Identify characteristics of human dwellings that vary with environment.
- Measure the dimensions of living spaces in your home.
- Calculate the floor areas of living spaces in your home.

What Do You Think?

All humans have a basic need for shelter. Examine the pictures of dwellings shown on this page.

- **What are some common characteristics of all these dwellings?**
- **How are they alike? How are they different?**

Record your ideas about these questions in your *Active Physics* log. Be prepared to discuss your responses with your small group and the class.

For You To Do

1. "Brainstorming" is a process in which you simply generate a large number of ideas. The rule of brainstorming is that all ideas should be accepted and no idea should be evaluated or thrown out. With your group, brainstorm a list of the characteristics of a "universal dwelling" for people anywhere in the world. In this case, set your team's goal at 100 different characteristics. Use the following questions to guide your brainstorming:
   - a) What are the functions of a home?
   - b) What features are essential to any home?
   - c) What are some of the factors that determine the size and shape of a home?

2. After brainstorming, narrow the lists down and develop two lists of ten or fewer items each. The first list should identify essential characteristics that should be present in all homes. The second list should identify special characteristics that are absolutely necessary for some environments, but that are non-essential in other environments.
   - a) Record the lists in your log using these two headings:
     - Essential characteristics of the universal dwelling.
     - Characteristics that must be modified for various climatic and cultural conditions.

3. Return to your group and answer the following questions in your log:
   - a) What things do you need to know about people, family sizes, and lifestyles in order to design a universal dwelling?
   - b) How large should a universal dwelling be? How could you make an educated guess?
   - c) How many of your "essential characteristics of the universal dwelling" require an energy input of some kind?

Reflecting on the Activity and the Challenge

You may think that this activity didn't move you very far toward meeting the **Chapter Challenge**, but it really has gotten you started. Through interaction with others, you have identified and shared ideas about the two basic aspects of dwellings for humans: function and form. Function involves the many things that a dwelling must do for people, and you have identified basic functions. Form involves the physical characteristics that a dwelling must have to support necessary functions, and you have started thinking about the size of a dwelling as perhaps the most basic part of form. Congratulations! You're on the way to meeting the **Chapter Challenge**.

As you progress through this chapter, you a1nd your group may find that you will need to make modifications to your original plans. Don't be concerned about this, as this is typical of the planning and research process in which you are involved.

Physics To Go

1. a) Describe the weather in the area of the country where you live.
   b) Describe some features of homes in your area that provide protection against the weather.

2. Choose another area of the country, or in the world, which has weather different from yours. Describe the weather in that area. Describe some features of homes in that area that provide protection against the weather.

3. A square foot (ft.^{2}) is a unit of measurement of surface area. Using a ruler, draw an accurate diagram of a square foot. (You may have to tape two pieces of paper together.) Label the dimensions of width and length.

4. Use your diagram of a square foot. Estimate the number of square feet in the room in which you are presently located.

5. a) Measure the length and width (in feet) of the room in which you are presently located.
   b) To calculate the area of a rectangle you multiply the length by the width.
   $A = l \times w$
   Calculate the area of the room you are in.

c) How close was your estimate in **Question 4** to the area you calculated?

6. A family has decided to install a new floor in the kitchen of their home. The kitchen floor is a rectangle 8 feet wide and 12 feet long. The new floor will be covered with tiles which are square pieces measuring 1 foot on each side. How many tiles will be needed?

7. The family described in **Question 6** has changed its mind and has decided to use a type of flooring material that is available in square pieces measuring 2 feet on each side. How many pieces will be needed?

8. To prepare for **Activity 2**, determine the total floor area of your own or someone else's living spaces. Measure the amount of floor space in the bedrooms, the living areas, the kitchen, bathroom(s), and any other areas. Include any closets in your measurement. Place the data you collect on a 3×5 card, using the format shown here. All the information will be kept anonymous.

The Total Floor Area of a Home	
Combined area of bedrooms in square feet:	______
Combined area of kitchens/eating areas in square feet:	______
Combined area of living room, den, etc., in square feet:	______
Total area of all the above spaces in square feet:	______
Combined area of bathroom(s) in your home in square feet:	______
Total areas of all other living spaces in square feet:	______
GRAND TOTAL of all living spaces in square feet:	______
The number of people occupying the space:	______

Stretching Exercises

Do a survey in your neighborhood. Make a list of, or draw features of homes that are well-suited to the local climate.

Activity 2

What is the "Right Size" for a Universal Dwelling?

GOALS

In this activity you will:

- Calculate the average number of square feet per person for a sample of homes represented in your class.
- Recognize ranges in family size and the number of square feet per person in the class's sample of homes.
- Use data on the class's sample of homes and family sizes to decide upon the size needed for a "basic" universal dwelling, and decide how the size of the dwelling will vary for large and small families.

What Do You Think?

In some prisons in the United States, two inmates occupy an 8-foot by 10-foot cell. On the other hand, some people live in mansions of 10,000 to 20,000 square feet!

- **How much space do people need in their shelters to have a decent, humane life?**

Record your ideas about this question in your *Active Physics* log. Be prepared to discuss your responses with your small group and the class.

For You To Do

1. Give the "Total Floor Area of a Home" index cards that you completed for today's class to your teacher. He or she will then read aloud the Grand Totals of the living spaces in square feet and the number of people who live in some of the homes.

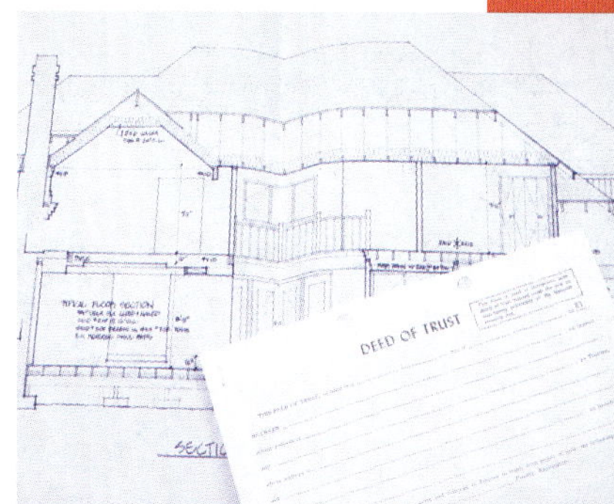

a) Enter these numbers in your log.

b) Calculate the average number of square feet per person for at least five of the homes in your community.

2. Discuss the answers to the questions below with your group and then record the answers in your log.

a) How much total living space for all functions should your universal dwelling have per person?

b) Could you design a universal dwelling that would combine some living space functions? If so, how might you do this?

c) Will bathroom space be necessary for your universal dwelling? If not, explain how you might modify your dwelling if no indoor plumbing will be available. (Indoor plumbing is not available to many people in many countries.)

d) What size family (number of people) and what total square footage will you use for designing your "basic" universal dwelling? Explain your group's reasoning for choosing these family and dwelling sizes.

e) Decide on two additional sizes for your universal dwelling, the first to accommodate a larger family and the second to accommodate a smaller family. What will the square footage and family sizes of these versions be? What is your group's reasoning for deciding upon these specific larger and smaller sizes?

Reflecting on the Activity and the Challenge

Based upon the data that your class has collected, and your group discussions about what is needed in terms of living space for a universal dwelling, your group and other groups in the class have probably decided on universal dwellings of various sizes. This is another step toward completing the **Chapter Challenge**. You didn't just draw a size out of a hat. Your decision on size can be defended in terms of the data that you have collected about the sizes of homes of class members. You have also included necessary functions and various family sizes as factors in your decision.

There are probably a range of appropriate answers, but there is also opportunity to make changes and improvements as you proceed.

Physics To Go

1. Make a floor plan drawing of a room in your home, or someone else's home, or in your school. A floor plan drawing shows what the room looks like when viewed from above. It should show the shape of the room and the positions and sizes of the doors and windows. You will need a scale for your drawing, for example, one square on a sheet of graph paper could equal one square foot.

2. Make a vertical cross-section drawing of the same room you used in **Question 1**. A vertical cross-section drawing should show what the room looks like when viewed from the side, including how high it is, and the shape and size of the windows and doors.

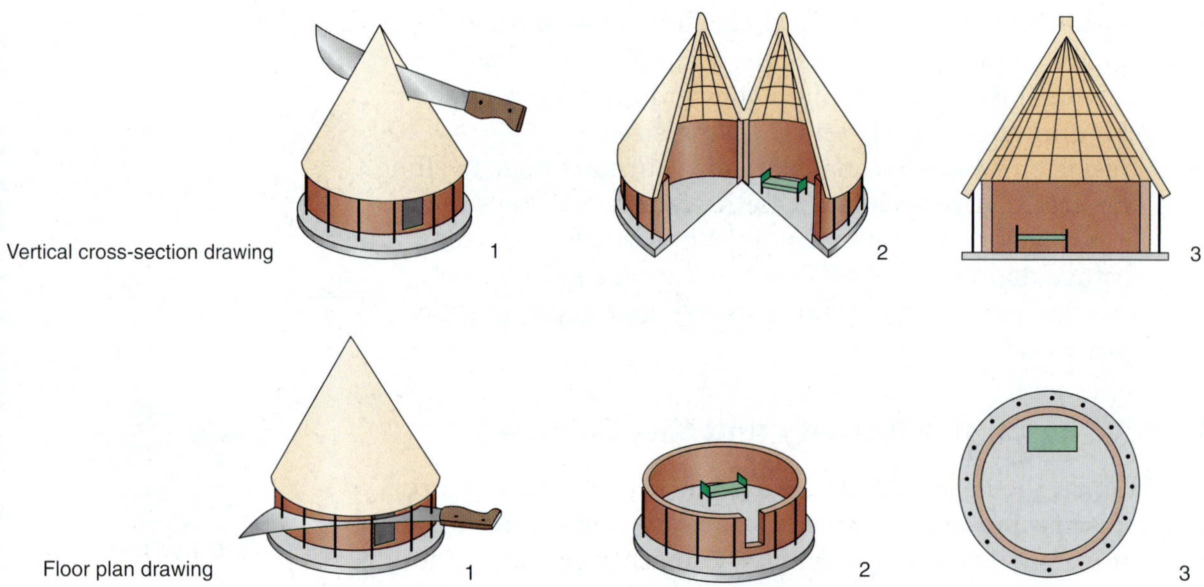

3. A house lived in by an elderly man and woman has 1500 square feet of floor area. Calculate the individual living space in square feet per person.

4. The house described in **Question 3** is sold to a young man and woman who have three children. Calculate the individual living space for the new family.

5. Here is a floor plan for houses being built for some people in southern India.

   a) Calculate the number of square feet of living space the house has.

   b) If the average family size is 6 to 8 persons, what is the average living space per person?

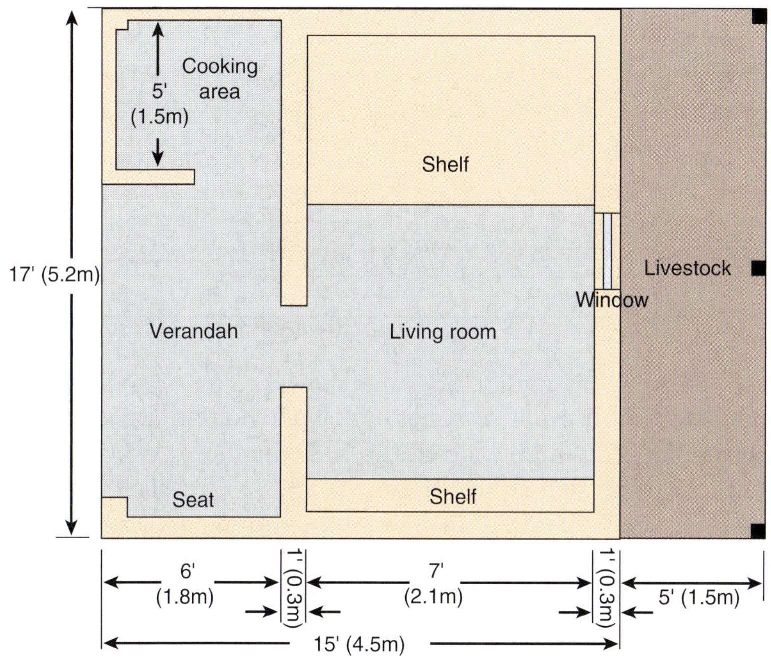

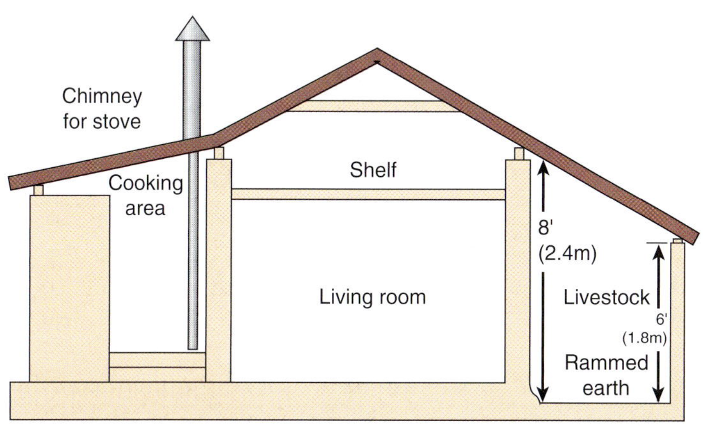

Typical family house—vertical cross-section drawing.

6. Look at the floor plan from Malawi of "a house that grows." It allows the homeowner to add rooms to the house as the household income or the family grows.

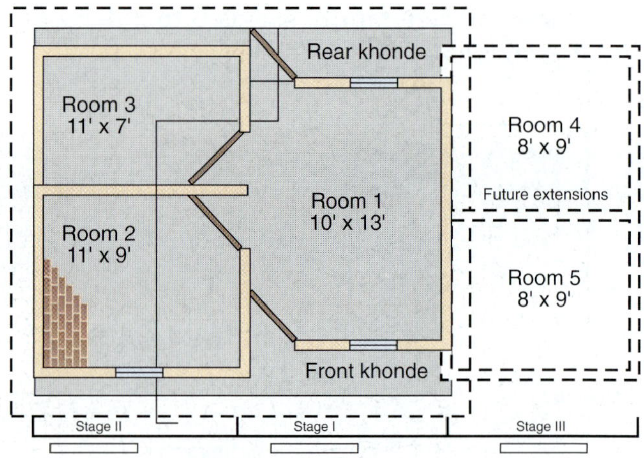

a) Calculate the number of square feet per person for a newly married couple who builds Stage I (Room 1).
b) The couple has two children and decides to build Stage II (Rooms 2 and 3). Calculate how many square feet there are per person for this family of four.
c) The couple has two more children and then builds Stage III (Rooms 4 and 5). Calculate the number of square feet per person for this family of six.

7. Make a floor plan drawing and a vertical cross-section drawing for your universal dwelling. Show the total square footage of each room and the total square footage of the dwelling. Include a scale for your drawing.

Stretching Exercises

1. Do research in your school library or on the Internet to find information on the size of the "average American home."

2. Locate information on the amount of living space used by an average person in another country.

Activity 3 The Shape of the "Universal Dwelling"

GOALS

In this activity you will:

- Explain how the surface area of a two-dimensional figure is related to its shape when the perimeter is held constant.
- Calculate the surface-to-volume ratio of simple three-dimensional shapes.
- Relate the shape of a dwelling to optimization of living space for a limited amount of building materials.

What Do You Think?

Throughout the world, homes take on many different shapes.

- **Why are most homes in the United States built in rectangular shapes?**

Record your ideas about this question in your *Active Physics* log. Be prepared to discuss your responses with your small group and the class.

For You To Do

1. Make a closed 20-inch loop of string. You will also need pins and 1-inch square grid graph paper attached to a sheet of cardboard as shown on the next page.

2. Carefully place four pins at the intersections of grid lines. Run the closed loop of string around the pins. Adjust the

Perimeter	Length of Long Sides	Length of Short Sides	Area Inside Perimeter

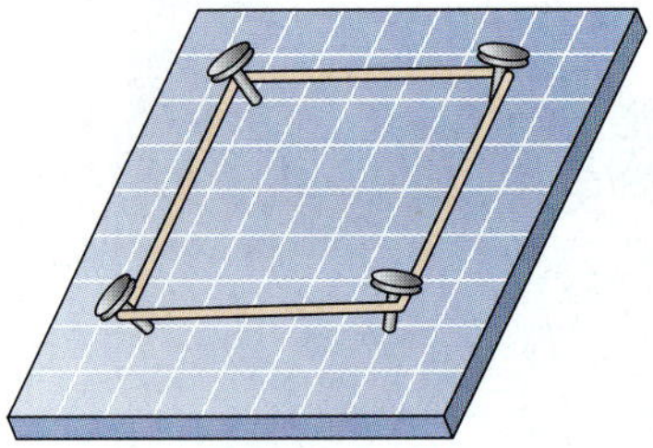

pins until you have constructed a rectangle on your grid paper. The loop should form a tight fit around the perimeter.

a) Count and record the number of 1-inch squares that lie within this perimeter. Record your data in a table similar to the one shown on the following page.

3. By carefully moving the pins, construct at least five more rectangles.

a) Record the results in the table in your log.

4. Use the data you collected to help you answer the following questions:

a) Which rectangle has the greatest perimeter, or do they all have equal perimeters?

b) What happens to the area of the rectangle as one pair of sides gets longer while the perimeter stays the same?

c) Which rectangle seems to have the largest internal area?

5. With your partner, use your loop of string and graph paper to determine how triangular (3-sided), pentagonal (5-sided), hexagonal (6-sided), octagonal (8-sided), and other shapes compare to that of a square of the same perimeter.

a) Record your results in a table. (You can determine the area of the shapes by counting the number of boxes in the graph paper enclosed by the string. You will probably need to count half-boxes as well.)

b) What appears to happen to the enclosed area as the number of sides increase?

c) What shape do you think will generate the largest area for a given perimeter?

6. Extend your two-dimensional knowledge to three dimensions. Since homes are surrounded by walls and not string, you must consider how the information about enclosing the largest possible area with a loop of string can be used in designing your home.

a) If you only had a limited quantity of material for walls to enclose a home, which shape would give the most living area? Why?

7. Meet with the other members of your group and, using the information you have gathered from these activities, choose a design shape for your universal dwelling. Your group will also have to determine the ceiling height, number of stories, and surface area-to-volume ratio for your basic dwelling design.

a) Record your group's decisions regarding these specifications in your log.

The shape of the universal dwelling will be:

The ceiling height of the rooms will be:

The living space of the dwelling in square feet will be:

The number of stories in the dwelling will be:

The volume of the dwelling will be:

Reflecting on the Activity and the Challenge

Your group has reached some tentative decisions about how many square feet of living space per person your universal dwelling should have. You have to be concerned about making sure that you provide enough living space. You also need to make sure that the building is not too expensive to produce. This means that a good design will not require a large building to contain the floor space.

In this activity, you explored how you may construct a dwelling with the required floor space, while keeping the dimensions of the building small and the amount of materials needed to construct the building as small as possible.

Now you have addressed the two most basic questions about the form of the universal dwelling: size and shape. You can defend your choice of shape in terms of keeping the cost of materials low. You are on your way to being able to present your plan to the Architectural Committee with confidence.

Physics To Go

1. The total area of all the faces of an object is called the surface area. To calculate the total surface area, find the area of each of the faces, and then add these areas.

   To find the surface area of the following cube, you can imagine unfolding the cube to form a flat pattern.

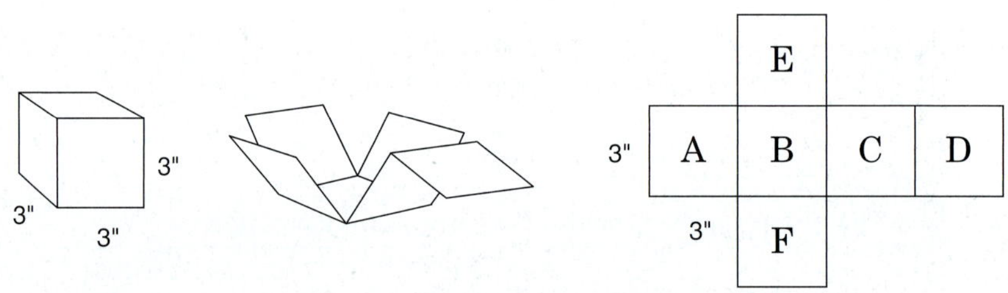

   a) What is the area of face A?
   b) What is the total surface area of the cube?

2. To find the surface area of the shape shown in the diagram below (a rectangular prism) you can once again imagine unfolding it to form a flat pattern.

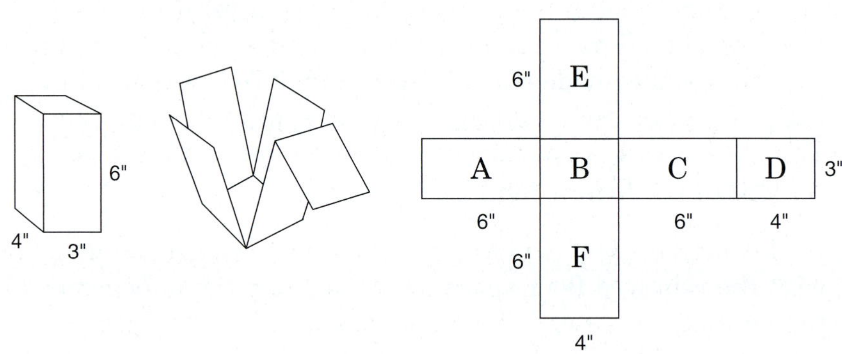

a) Calculate the areas of A, B, C, D, E, and F.
b) What is the surface area of the shape?

3. Find the surface areas of the following:

a) 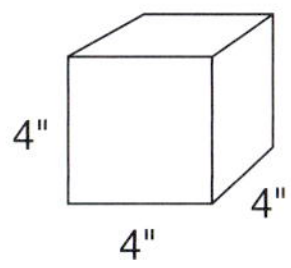

b) 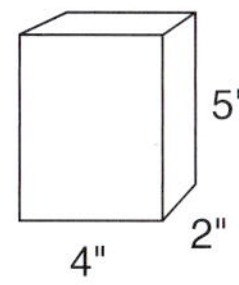

c) 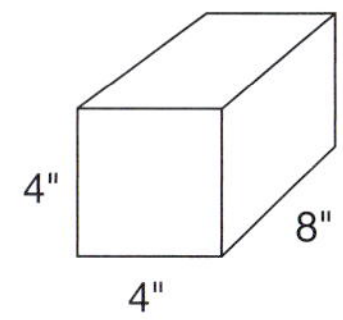

4. To calculate the volume of a rectangular shape you can use the following relationship:

   Volume = length × width × height

   Calculate the volume of each figure in **Question 3**.

5. Find the surface area-to-volume ratio of each figure in **Question 3**. (Divide the surface area by its volume.) Which figure has the greatest surface area-to-volume ratio? The lowest?

6. A house was designed with outside walls that form a square base 27 feet on each side.

   a) Calculate the perimeter (distance around) of the house in feet.
   b) Calculate the total floor area of the house in square feet.

7. A house was designed with an outside wall that forms a circle that has a radius of 17.2 feet.

   a) Calculate the circumference (distance around) of the house in feet. The equation for the circumference of a circle is:
   Circumference = 2 × π × radius.
   You can use a value of 3.14 for π.
   b) Calculate the total floor area of the house in square feet. The equation for the area of a circle is
   Area = π × radius × radius.

8. Use numbers to compare the houses in **Questions 6** and **7**.

   a) If the height of the outside walls of the two houses were equal, how would the amount of paint needed to cover the walls compare? How significant is the difference?

b) How do the total floor areas of the two houses compare? How significant is the difference?
c) If living space of at least 180 square feet per person is needed, how many persons could live in each house? How significant is this difference?

9. A research outpost for a scientist in a remote part of Alaska has a hut shelter which is a cube eight feet on each side. The scientist sometimes visits a nearby family who live in a two-story cabin also shaped as a cube. The cabin's dimensions are double the dimensions of the hut, measuring 16 feet on each side. Use the numbers above to make the following comparisons:
   a) Compare the total floor areas of the two dwellings as a ratio. (The cabin has an upstairs level.)
   b) Compare the outside surface areas of the two dwellings including the bottom floor of each. Express the comparison as a ratio.
   c) Compare the volumes of the two dwellings as a ratio.
   d) Calculate the surface area-to-volume ratio for each dwelling. (Divide the outside surface area of each dwelling by its volume. The answers will be in units of square feet per cubic foot.) Which dwelling has the greatest possibility for heat loss through its surfaces per cubic foot of inside volume?

10. Using the information you gathered in this activity, modify your floor plan drawing and section drawing for your universal dwelling. Show the total square footage of each room and the total square footage of the dwelling. Include a scale for your drawing.

Activity 4 Solar Heat Flow in the "Universal Dwelling"

GOALS

In this activity you will:

- Construct a three-dimensional scale model of a universal dwelling to meet the design specifications of your group.
- Conduct a heat transfer experiment on the model.
- Interpret heat transfer data collected from the experiment.

What Do You Think?

Throughout one cloudless day in June, at 45° north latitude, over 300 million joules of solar radiation strike each square meter of area at the Earth's surface.

- **What is the relationship between how fast a building heats and how fast it cools? (Do buildings that heat quickly, cool slowly; or do buildings that heat quickly, cool quickly?)**

Record your ideas about these questions in your *Active Physics* log. Be prepared to discuss your responses with your small group and the class.

For You To Do

1. Using the resources that have been provided, construct a scale model of your universal dwelling using the design specifications for the basic unit. The model home should be built on a scale of 1:48. That means that one inch on the model will equal four feet, or 48", on the real home. In order to build the model quickly, build only the walls and roof of your home. Don't worry about putting a bottom (floor) on the model or placing windows and doors at this time.

2. Using the figure below, assemble a heat lamp, ring stand support and clamp, timer, measuring tools and thermometers. Conduct a timed heat-transfer experiment of 20 min. Place the nearest edge of the model 20 cm from the bulb of the heat lamp, with the lamp at approximately 45° from the center of the model home. (A 45° angle is halfway between the vertical and the horizontal.)

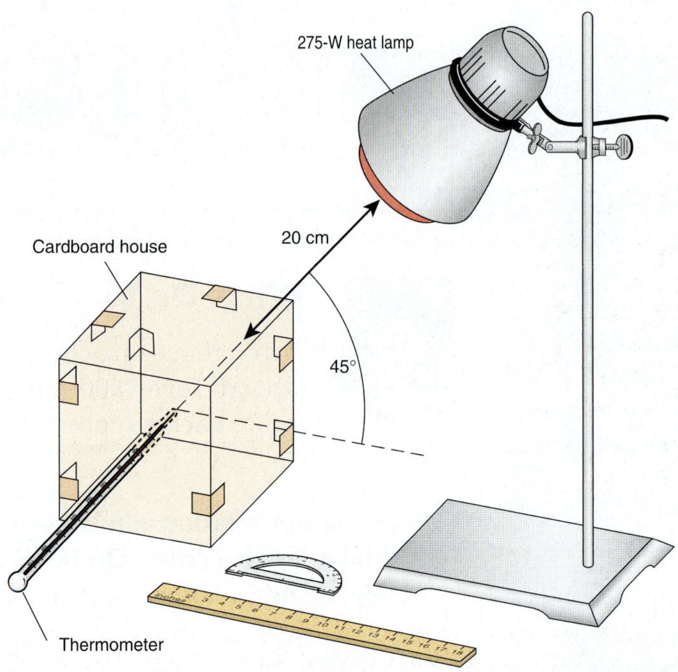

3. Carefully make a small hole in one wall. After making the hole, place the thermometer (or temperature probe) into one side of your model so that the bulb or probe is located in the dead center of the model. Be sure you can read the thermometer once it is in place!

If the thermometer should break, *immediately* notify your instructor.

4. When you have your model, the thermometer, and the heat lamp placed properly, have your teacher check it before you begin heating.
   a) Record the temperature of the thermometer before heating and record this as the temperature at 0 time.
   b) Turn on the heat lamp and, after 30 s, record the temperature.
   c) Continue this procedure for 10 min, recording temperature at each 30-s interval.

Caution: Heat lamps get very hot. Be careful not to touch the bulb or housing surrounding the bulb.

5. Turn off the lamp at the end of the 10-min heating phase.
   a) Continue to record the temperature every minute after the lamp is turned off for another 10 min.
   b) Calculate and record the change in temperature during each timed interval of heating and cooling.
   c) Graph your recorded data (temperatures against time) on the graph paper provided by your teacher. Your time axis should extend from 0 min (when you began) to 20 min (after the experiment was completed), and be placed on the *x*-axis.

6. Use the graph and the data you recorded to answer the following questions:
   a) What was the total temperature increase in your model home during the heating phase?
   b) What was the total temperature decrease in your model home during the cooling phase?
   c) Was the rate of temperature increase constant? How do you know?
   d) Was the rate of temperature decrease constant? How do you know?
   e) What time of day did the heating phase represent?
   f) What time of day did the cooling phase represent?
   g) What does your experiment suggest in regard to the potential of your model home to be heated by solar energy?

Reflecting on the Activity and the Challenge

Over the past three activities you have made good progress in deciding upon the design specifications for your universal dwelling. In the last activity, you explored finding the "best fit" between living space and minimum surface area for your universal dwelling.

There are several reasons why you would want to keep the surface area of the dwelling small, while keeping the interior as large as possible. One reason is that the materials for construction of the surface area of a building are expensive and a smaller surface area requires fewer materials. Another reason has to do with controlling the inside temperature of the building. In this activity, you constructed a model of your universal dwelling and investigated its heating and cooling properties.

The Architectural Committee probably will have some hard questions for you about your plans for heating and cooling the dwelling you are designing. That's because the greatest cost in operating most homes is for heating and cooling the home. This activity has given you some baseline information about how your dwelling design responds to the best kind of energy: free energy from the Sun. It will be important for you to learn how to use solar energy to best advantage in the next activities. You do not want to design a building that is expensive to heat or cool.

Physics To Go

1. If four feet of length on a real home is represented by one inch on a scale model of the home, what length on the scale model would represent one foot on the real home?

2. A 1:48 scale drawing of the floor plan of a home must fit one $8\frac{1}{2} \times 11$-inch sheet of paper. What maximum length and width in feet can the home have?

3. Home designers sometimes use a 1:96 scale ($\frac{1}{8}$ inch represents 1 foot) when making drawings of home plans. What advantages would a 1:96 scale have over a 1:48 scale ($\frac{1}{4}$ inch represents 1 foot)?

4. If the heat lamp represented the Sun at its noon position during the heating phase of the activity, which side of the model dwelling—north, south, east, or west—must have been facing the Sun?

5. Describe how you would need to change the position of the heat lamp relative to the model dwelling to have the lamp represent the Sun during the morning or afternoon.

6. Do you think that all of the radiation from the heat lamp that hit the model dwelling was absorbed by the dwelling? Give evidence to support your answer.

7. Explain how you think heat travelled from the inside surface of the model dwelling to the thermometer at the center of the model during the heating phase. Obviously, the heat had to travel through the air inside the model. How did it do that?

Stretching Exercises

1. Write down three questions that you would ask someone who uses solar heating or solar cooling for their home.

2. If you know of any people who use solar energy to heat their homes, or have designed ways to keep solar energy out to keep their homes cool, you may wish to talk with them about how they have done this. Ask them about the special things that they have done to control the effects of solar energy on their homes.

Activity 5

The Role of Insulation: Investigating Insulation Types

GOALS

In this activity you will:

- Explain how insulation retards heat flow in terms of conduction.
- Explain how insulation retards heat flow in terms of convection.
- Conduct an experiment to test how the thickness and kind of insulating material influence the rate of heat transfer.

What Do You Think?

Insulation used in homes is usually a lightweight material that is designed to reduce the flow of thermal energy through the walls or ceiling.

- **Is insulation in a home more important in the heat of summer or the cool of winter?**

Record your ideas about this question in your *Active Physics* log. Be prepared to discuss your responses with your small group and the class.

For You To Do

1. Place the insulating material supplied to your team in each of the three cans. Insert a thermometer in the absolute center of each material.
   a) Measure the thickness of the insulation surrounding the thermometer in each can.
   b) Record the temperature shown by each thermometer.

2. Place the three cans in a hot water bath.
   a) Record the temperature of each can every minute for 10 min.

Do not change the temperature setting on the hot water bath.

3. Carefully check that the cans are not too hot to touch. Carefully remove the cans from the bath.
   a) Record the temperature each minute for another 10 min.

Use a cloth or paper towel to grasp the cans when removing them from the baths. Moisture may form on the outside—be careful that the cans do not slip from your grasp.

4. Now place the three cans in an ice water bath and repeat the steps above.

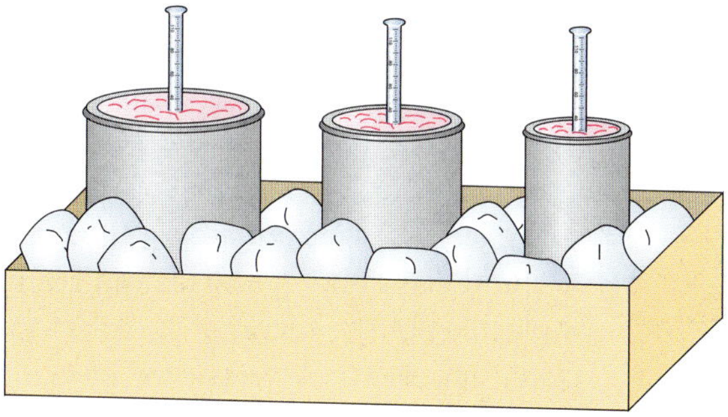

   a) Record the temperature in each can every minute for 10 min of cooling in the bath and for another 10 min after removing them from the bath.

If the thermometer should break, *immediately* notify your instructor.

5. Display the data on a graph.
   a) Graph your heating and cooling data for each of the three cans on a single graph, labeling each appropriately.

6. Use your graph to answer the following questions:
   a) Compare the three graphs that you made. What differences do you note among the graphs?
   b) What do you conclude about the effectiveness of insulation in heating as opposed to cooling situations?
   c) What do you conclude about the effectiveness of using varying thickness of insulating material?

FOR YOU TO READ

Transmission of Thermal Energy

You have most likely studied in prior science classes the three means by which heat is transmitted: **conduction**, **convection**, and **radiation**. As you read the descriptions of these means of transmission, try to recall experiments or investigations that you pursued if you studied this earlier.

In the case of conduction, the thermal energy possessed by a material is transmitted to another material by direct contact of the materials with one another. The rapidly moving (and hence high energy) atoms in a red hot iron bar come in direct contact with, let us say, cold water. The thermal energy of the iron atoms is conducted by physical contact directly to the water molecules, which, in turn, heat other adjacent molecules and so on, causing the heat to dissipate throughout the water, warming it in the process. In convection, the molecules or atoms of a fluid (a liquid or a gas) that have more thermal energy will move faster than the surrounding molecules of that material. As a result, they will move further apart and, therefore, become pushed by the colder, more closely packed molecules or atoms. As the hotter molecules spread out through the colder surrounding material, they carry their thermal energy (motion) with them, losing it along their path to other colder molecules by conduction. Convection then, is the transmission of heat from one place to another due to physical movement of a warmer fluid through a colder fluid (typically in response to the force of gravity).

Radiation is the third form of energy transmission. In this case, the electromagnetic radiation given off by high energy (hot) materials can be transmitted through objects, or even a vacuum, at the speed of light. When this energy strikes a material that can absorb it, the radiant energy causes the atoms or molecules of the material it strikes to move faster and the molecules now have more heat energy.

Reflecting on the Activity and the Challenge

Homes For Everyone (HFE) realizes that it is important to control the interior temperatures of their universal dwellings regardless of climate and location. It is important because of energy, comfort, and health concerns.

One of the criteria that the Architectural Committee will use to evaluate your plan is "the kind of insulation materials and thickness you choose should show you know the principles which make the insulation effective." That's exactly what this activity was about, so it will be very helpful in meeting the **Chapter Challenge**.

Physics To Go

1. Explain how conduction, convection, and radiation were involved in both the heating and cooling phases of the model dwelling in **Activity 4**.

2. Which of the three ways of transferring thermal energy was most directly involved in the investigation in **Activity 4**? Explain your answer.

3. In this activity you tested the effects of two variables on the transfer of thermal energy. Identify the variables and explain what you learned about the effect of each on thermal energy transfer.

4. If you wanted to reduce the amount of thermal energy conducted through a particular kind of material to one-half the amount, what change in the thickness of the material would be needed? Also provide answers for reducing the transfer of thermal energy to one-third, and one-fourth the amount.

5. Would insulating a home be more, less, or about equally important for keeping a home warm in a cold climate or keeping a home cool in a hot climate? Use evidence from the data for your answer.

6. Years ago, the answer to keeping warm during cold weather was to put on more clothes. Today, the answer can be to put on different clothes. Comment on these two ways of keeping warm in terms of the variables tested in this activity, and the evolution of clothing available for winter.

7. Find out what kinds and thickness of insulating materials are used in the walls and ceiling of your home. Your parents or building supervisor may have this information.

8. Go to a building supply store and examine various insulating materials. The insulating properties of these materials are given in "R values." Find out what an R value is.

Physics Words

conduction: (of heat) the energy transfer from one material or particle to another when the materials or particles are in direct contact.

convection: the heat transfer resulting from the movement of the heated substance, like air or water currents.

radiation: (heat transfer) electromagnetic radiation strikes a material that can absorb it, causing the particles in the material to have more energy often resulting in a higher temperature.

9. Some rigid insulating material has an aluminum foil surface bonded to the sheet. What purpose does this aluminum foil surface serve? How does it influence the R value of the material?

10. How much insulation is recommended for a home in your area of the country?

11. Is the amount of insulation recommended the same for floors, walls, and ceilings? Why?

Stretching Exercises

1. Design a beverage container to keep a cold drink cold, and a hot drink hot.

2. Design an experiment to test the insulating effectiveness of natural material such as grass, dry mud, adobe brick, or stone. After your teacher approves your design, conduct the experiment.

Activity 6 Investigating Insulation Placement in Your Universal Dwelling

GOALS

In this activity you will:

- Identify and explain which surface in a house is vulnerable to the greatest heat loss in terms of conduction and convection.
- Infer an optimum thickness for home insulation.
- Identify limitations of using scale models in conducting experiments.

What Do You Think?

The density (mass per unit volume) of air decreases with temperature, leading to the common knowledge expression, "heat rises."

- **Why are attics of homes so hot on a summer day?**
- **What are these same attics like on a cold winter day?**

Record your ideas about these questions in your *Active Physics* log. Be prepared to discuss your responses with your small group and the class.

For You To Do

1. Cut the cardboard as insulation so that it will fit snugly within the ceiling of your model home. Use masking tape to fix it in place.

2. Set up your experiment in the same manner as you did for **Activity 4**. Be sure to keep the lamp at the same distance (20 cm) and at the same 45° angle as in the first experiment.
   a) Construct a temperature vs. time graph of your data as you did in **Activity 4**.
   b) Compare this heat transfer experiment data with data from **Activity 4** and interpret the results.
   c) How did insulating the ceiling affect the heating and cooling of your model?

3. Insulate the walls of your model in the same way as you did the ceiling. Repeat the heat transfer experiment.
   a) Compare and interpret these results with the first two experiments.

4. Use the results of **Activities 4** and **5** to answer the following questions:
   a) Compare the results of your team's findings with those of the other teams.
   b) How do you explain the similarities and differences?
   c) Given the scale of your model home, to what thickness of insulation in a real home does your $\frac{1}{4}$ inch of cardboard insulation correspond?
   d) What effect do you predict would occur if you used twice the thickness of cardboard in your model? Three times the thickness? Defend your responses with data from **Activity 5**.

Reflecting on the Activity and the Challenge

In this activity, you insulated your model home and then tested it to see what effects insulating will have on controlling the internal temperature of the home. The Architectural Committee will expect you to be able to show that you know where insulation is most effective in trapping energy. You now have gathered scientific data which you can use to show that you know where to place insulation most effectively in a home.

Physics To Go

1. In insulating a real home you probably wouldn't use cardboard as the insulating material as you did for the model dwelling. What materials might you use, and why?

2. What considerations, other than the effect of the insulation, might guide your choices of insulating materials for a home? Think about fire prevention.

3. Homes are usually designed to have about twice as much insulating effect in ceilings than in exterior walls. Why?

4. Could a vacuum be a good insulator? Explain why or why not, and under what conditions.

5. Compare and contrast the purpose of insulation in hot weather versus cold weather. Should placement of insulation be different for these different purposes?

6. Insulating materials for homes are rated on a relative scale of R values. Sheets or rolls of different kinds of materials in various thickness are available, and each has an assigned R value. The R values of two or more thickness can be added together to find the total insulating value of combined layers. What combinations of the following materials could be used to provide R-19 insulation for a wall and R-38 insulation for a ceiling?
   - R-8 expanded polystyrene board 2 inches thick
   - R-11 fiberglass wool $3\frac{1}{2}$ inches thick
   - R-30 fiberglass wool $9\frac{1}{2}$ inches thick

7. In preparation for tomorrow's class work, collect the following information tonight at your home:
   a) Measure all the windows in your home and calculate the total number of square feet of window area of your home. Record your calculation.
   b) Measure all the exterior wall surface areas of your home, and calculate the total number of square feet (including the window areas) of exterior walls of your home. Record your calculation.
   c) Divide the total window area by the total wall surface area. Record your calculation.

Stretching Exercises

1. In constructing homes today, high, open (cathedral) ceilings are often part of the design plans for the living space. What advantages and disadvantages do high ceilings have with respect to heating and cooling considerations for the home?

2. Contemporary homes in America today are built with full basements, half-buried basements, or are constructed on top of concrete slabs cast on the surface. What heating advantages and disadvantages do you believe would exist for each of these house foundation types? Explain your responses for each type of foundation.

Activity 7 The Role of Windows... Placing Windows in Your Universal Dwelling

GOALS

In this activity you will:

- Calculate the ratio of window-to-wall areas for typical homes.
- Experimentally compare the heat transfer properties of a scale model of a home with and without windows.
- Explain where windows should be placed in a home to optimize solar heating effects for various seasons and climates.

What Do You Think?

It is often said, "A window hasn't been made that is as good as a well-insulated wall."

- **Should all sides of the model dwelling have the same area of windows? Why or why not?**

Record your ideas about these questions in your *Active Physics* log. Be prepared to discuss your responses with your small group and the class.

For You To Do

1. Compile a chart in your group to summarize your calculations of window area to wall area ratios in your homes.
   a) What is the average window to wall area ratio for your group?

2. Use all the information and resources available to you. Decide upon the number, sizes, and placement of the windows for your model home. Remember, you want to provide for optimal lighting and ventilation and you also want to allow solar energy to enter the home and heat it when needed. However, you do not want to transfer large amounts of energy from the home to the outside when direct sunlight does not enter the window.
   a) Approximate the window to wall area ratio you plan to use in your model home.
   b) How does the window to wall area ratio for your model home compare to the average window to wall area ratio you calculated for your group? Explain why you chose the ratio you did.

Caution: The cutting tools are very sharp. Be careful and get help from your teacher as needed.

3. Draw outlines of the windows to scale on the walls of your model and carefully cut out the windows with a cutting tool.

4. Use masking tape to tape transparent food wrap over the window cutouts.

5. Repeat the heat transfer experiment with the heat lamp on your model home. Place the heat lamp and model home so that the lamp is shining on the home at the same angle and at the same distance as in previous activities.
   a) Record and graph your findings as you did in previous activities.
   b) Compare your heating and cooling graphs for this activity with those in **Activities 4** and **6**. How do the graphs differ? Explain your results.

Reflecting on the Activity and the Challenge

Virtually all dwellings have windows of one type or another to provide interior light and to provide for ventilation and temperature control. The windows in a home built today are extremely well-designed and engineered in comparison to those of 30 years ago. The problem your group faces is a complicated one. You wish to design and place windows in your Homes For Everyone (HFE) universal dwelling to provide for optimal lighting and ventilation, but you also know that windows have energy advantages and disadvantages. Windows, which can allow solar energy to enter the home and heat it when needed, lose heat through conduction, convection, and radiation to a much greater degree than the insulated walls.

You will need to convince the Architectural Committee that you understand that windows offer both advantages and disadvantages. You have gathered evidence by comparing the heating and cooling curves from this activity to the corresponding curves from **Activity 6**.

Physics To Go

1. Heat can enter and leave a home through windows. How are conduction, convection, and radiation involved in:

   a) The transfer of heat into a home through its windows?

   b) The transfer of heat out of a home through its windows?

2. The amount of heat gained or lost through a window is directly proportional to the surface area of the window.

   a) Compare the amount of heat expected to be transferred through a window 2 feet wide by 2 feet high to the amount expected for a window 4 feet wide by 4 feet high.

   b) Glass doors behave about the same as windows regarding heat transfer. Compare the amount of heat expected to be transferred through a glass patio door 8 feet wide by 6 feet high to the amount expected for a 2-feet-wide by 2-feet-high window.

3. Visit a building supply store in your community. Examine the windows available. Find out about the following:
   a) The R value of the cheapest and most expensive windows.
   b) The difference between thermopane and single pane windows.
   c) Why thermopane windows are built the way they are.
   d) What "low E" windows are, and how they function.

4. Even the highest quality windows do not have as much insulating effect as a well-insulated wall. Explain why.

5. Window drapes and shades can provide privacy and beauty in a home. Explain how they may also be used to control heat transfer through windows.

6. Designs for energy-efficient homes in parts of the United States that have cold winters have very few, if any, windows on the north side of the home. Why?

7. People who live in warm climates often use air conditioners to cool their homes. What, if any, problems do windows present related to air conditioners?

Stretching Exercises

Find out about passive solar heating for homes. Answer the following:

a) How can adding more windows to a house serve to warm the house during the day?

b) How can you prevent heat from escaping from these extra windows during the night?

c) What can be used to store the heat that is collected during the day?

d) Where should windows for passive solar heating be placed to be most effective?

Activity 8 Investigating Overhangs and Awnings

GOALS

In this activity you will:

- Explain how the altitude and heating effect of the Sun varies with seasons.
- Design overhangs and awnings to control solar heating through the windows of a home during summer and winter.
- Conduct a heat transfer experiment on a scale model of a home to test the effectiveness of overhangs and awnings for controlling solar heating during summer and winter.

What Do You Think?

Heating and air conditioning are expensive items for operating the average American home.

- **What can be done so that windows will let Sun enter the dwelling when it is cold outside, and block sunlight from entering when it is hot outside?**

Record your ideas about these questions in your *Active Physics* log. Be prepared to discuss your responses with your small group and the class.

For You To Do

1. Find the latitude of the location for which you are designing your model home.
2. In the temperate regions, the Sun is low in the sky at noon in winter and high in the sky at noon

in summer. In fact, it is this changing elevation of the Sun which causes the winter and summer temperature differences.

a) Use the example below and the figure to the left to help you calculate the angle of the Sun at noon at the winter and summer solstices in your area.

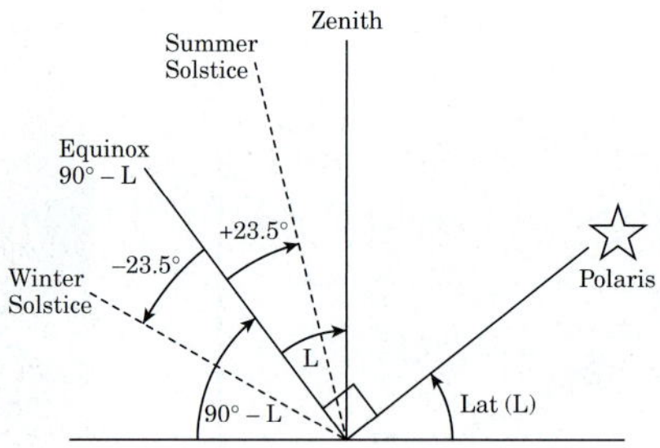

Finding Sun angles for a given latitude

Example:
The latitude of St. Louis, Missouri, is 38° north. Calculate the noon altitude of the Sun as viewed from St. Louis on the equinoxes and solstices.

Sun angle on equinoxes = (90°) minus (latitude angle)
= 90° –38°
= 52°

Sun angle on summer solstice = (Sun angle on equinoxes) plus (23.5°)
= 52° + 23.5°
= 75.5°

Sun angle on winter solstice = (Sun angle on equinoxes) minus (23.5°)
= 52° –23.5°
= 28.5°

b) Use the same figure and example to help you calculate the angle of the Sun at noon at the winter and summer solstices in the area for which you are designing a model home.

3. Set up your model home and the heat lamp as in previous activities.

4. Use a protractor and the Sun angle data to adjust the heat lamp so that it matches the angle position of the Sun at noon at the summer solstice.

a) Measure, record, and graph the heating and cooling curves for your model home.

5. With the lamp turned on, place a small piece of tag board (or other lightweight cardboard) over one of the windows on the side of the model facing the heat lamp. Try different widths of tag board and the angles of attachment until you find the width and/or angle that will block the lamp's light entirely but still provide someone inside with a view. Use masking tape to attach awnings and/or roof overhangs to your model.

6. Repeat the heat transfer experiment on your model home that is now fitted with awnings or overhangs for summer solstice conditions.
   a) Record and graph your data.

7. Do not change the awnings or overhangs on your model, but adjust the lamp so its position matches the angle of the Sun at the winter solstice. Repeat the heat transfer experiment.
   a) Record and graph this data on the same graph.
   b) What differences do you note in the summer and winter solstice heating graphs? How do you explain the differences?

Reflecting on the Activity and the Challenge

One of the main requirements is that the universal dwelling must be able to be used in a wide variety of locations on Earth. Since the dwelling also depends on energy from the Sun, you must include in your design the knowledge about how solar energy varies with location on the Earth. This activity has prepared you to answer questions that the Architectural Committee will ask about how your design allows for solar energy at different locations.

Physics To Go

1. Imagine yourself standing outdoors at noon at your home latitude. You are facing south straight toward the horizon. You are doing this on four special days during the year, whose titles are listed on the next page. For each special day give its date, explain its meaning in terms of seasons, and then predict the angle of the Sun above the horizon.

a) Vernal equinox
b) Summer solstice
c) Autumnal equinox
d) Winter solstice

2. Latitudes in the continental United States vary from about 25° north to 50° north. Calculate the angle of the Sun above the south horizon on the summer solstice for the latitudes listed below:

   a) 25° north
   b) 30° north
   c) 35° north
   d) 40° north
   e) 45° north

3. One town in the United States is 2000 miles west of another town. At noon on the winter solstice at each town, the Sun is observed to have an altitude of 26°. What is the latitude of each town?

4. The latitude of Mexico City, Mexico is 19° north. Do a calculation to show yourself that the Sun will shine on the north side of a home in Mexico City at noon on the summer solstice. State a rule for locations in Earth's Northern Hemisphere where this can happen.

5. Sketch a design for a roof overhang or window awning that will always shade a south-facing window from the noontime Sun during the six months between the vernal and autumnal equinoxes, but will allow the noon Sun to shine on the entire window during the other six months of the year.

6. The Sun shines on windows on the east and west sides of a home, too, but for only part of the day. What could be done to control heat gains and losses through those windows?

7. Do you think the added cost of building a roof overhang or using awnings on your model dwelling would be worth the additional expense? Why or why not?

8. Shades are more effective than awnings for blocking sunlight from entering windows on the east and west sides of buildings. Why is this so?

Activity 9 Too Hot, Too Cold, Just Right

GOALS

In this activity you will:

- Experimentally determine the final temperature when two liquids of different temperatures are mixed.
- Experimentally determine the final temperature when a hot metal is added to cold water.
- Calculate the heat lost and the heat gained of two objects after they are placed in thermal contact.
- Determine if energy is conserved when two objects are placed in thermal contact and reach an equilibrium temperature.
- Explain the concept of entropy as it relates to objects placed in thermal contact.

What Do You Think?

As you add cold milk to hot coffee, you expect that the milk will get a bit warmer and the coffee will get a bit colder.

- **What determines the final temperature of the coffee and milk?**

Record your ideas about this question in your *Active Physics* log. Be prepared to discuss your responses with your small group and the class.

For You To Do

1. In this activity, you will determine the final temperature of a cold water and hot water mixture. Styrofoam® cups work well as containers in this activity. The insulation "protects" the experiment from the environment.

   Pour 100 mL of hot water into a Styrofoam cup. Measure its temperature.

   Use a heat-proof holder while pouring, such as a glove or tongs.

   a) Record the temperature of the hot water. Measure the temperature of 100 mL of cold water.
   b) Record the temperature of the cold water.
   c) Predict the final temperature of the mixture of hot and cold water. Add the cold water to the hot water. Measure the final temperature.
   d) Record the final temperature.

2. Vary the experiment by changing the amount of cold water. Mix 100 mL of hot water with 50 mL of cold water; 75 mL of cold water; 125 mL of cold water; and 150 mL of cold water.

   a) Create a data table. Record your observations.
   b) Construct a graph of the results. Plot the final temperature on the y-axis and the amount of cold water added on the y-axis.
   c) Use your graph to predict the final temperature when 108 mL of cold water is added to 100 mL of hot water.

   Clean up any spilled water immediately, especially off the floor so that no one slips on it.

3. The Law of Conservation of Energy informs you that if the cold water gained heat energy (as indicated by its rise in temperature), then the hot water must have lost an equal amount of heat energy.

   The amount of energy change can be expressed using the following equation:

$$\Delta Q = mc\Delta t$$

where ΔQ is a measure of heat energy in joules
m is the mass of the substance in grams
c is the specific heat of the substance (the specific heat of water is 4.18 J/g°C)
Δt is the change in temperature

The equation requires you to use the mass of the water. For water, a volume of 50 mL has a mass of 50 g. Converting from volume to mass is easy for water, since the density of water is 1 g/mL.

$$D = \frac{m}{V} \text{ or } m = DV$$

Heat lost by hot water = Heat gained by cold water

$$(mc\Delta t)_{\text{hot water}} = (mc\Delta t)_{\text{cold water}}$$

Sample Problem

100.0 g of hot water at 80.0°C is mixed with 60.0 g of cold water at 20.0°C. The final temperature is 57.5°C. Show that energy was conserved.

Strategy: You can use the Law of Conservation of Energy to solve this problem.

Givens:

$m_h = 100.0$ g
$m_c = 80.0$ g
$c = 4.18$ J/g°C
$t_h = 80.0$°C
$t_c = 20.0$°C
$t_f = 57.5$°C

Solution:

Heat lost by hot water = Heat gained by cold water

$$(mc\Delta t)_{\text{hot water}} = (mc\Delta t)_{\text{cold water}}$$

$$(100.0 \text{ g})(4.18 \text{ J/g°C})(80.0\text{°C} - 57.5\text{°C}) = (60.0\text{g})(4.18 \text{ J/g°C})(57.5\text{°C} - 20.0\text{°C})$$

$$9405 \text{ J} = 9405 \text{ J}$$

The energy lost by the hot water is gained by the cold water.

4. Cool water can be heated with the addition of hot water. How well would a piece of hot metal heat the cool water?

   Design an experiment to compare the effect of adding 100 g of hot water and 100 g of heated metal to separate Styrofoam cups of cool water.

One way to heat the metal is to place it in a bath of hot water for three to five minutes. The length of time you need to keep the metal in the water bath depends upon the size of the metal. You can then use tongs to gently lift the metal from the hot water. As soon as the metal is out of the water, you will need to hold it over several pieces of paper towel folded to make a small mat and shake off drops of the hot water so that none of the hot water enters the beaker with the cold water. You want to try to place only the metal gently into the cold water.

Physics Words

temperature: a measure of the average kinetic energy of the molecules of a material.

a) Predict whether the individual cups of cool water will reach the same final temperature.

b) Do you think it matters what kind of metal you use? For example, will copper or zinc produce the same final temperature? Explain your answer.

5. Ask your teacher to comment on your experimental design. Make sure that your design is approved before you continue with the experiment.

6. Conduct the experiment.

a) Record your results.

b) Energy is conserved whether the cold water is mixed with hot water or hot metal. Look at this equation again:

Heat lost by metal = Heat gained by water

$$(mc\Delta t)_{\text{metal}} = (mc\Delta t)_{\text{water}}$$

The value of c for the metal may be different from the value of c for the water. Determine the value of the specific heat, c, for the metal that you used to heat the water.

c) Predict what would happen to the temperature of the cold water if the same amount of a different hot material with a lower specific heat than the metal you used were placed in the same volume of cold water.

d) Explain in your own words what the specific heat allows you to predict.

FOR YOU TO READ

Conservation of Energy

Energy is conserved. Kinetic energy can become gravitational potential energy and vice versa. When an object like a book is dropped to the ground, the gravitational

potential energy becomes kinetic energy as the book gains speed. As the book hits the ground and stops, some of the kinetic energy is converted to sound, as you hear a "thump." The rest of the kinetic energy of the book becomes heat energy and the temperatures of the book and of the ground both rise a bit.

You can calculate gravitational potential energy and kinetic energy. You also now know how to calculate heat energy. Heat energy is part of the total energy picture. Conservation of energy demands that the sum of all of the energies must remain constant.

Temperature and Heat

Temperature and heat energy are not the same. Temperature is a measure of the kinetic energy of the molecules of the material. You can measure this energy with a thermometer. The temperature and the kinetic energy of the molecules change when the object touches a material of a higher or lower temperature.

Heat energy has to do with the nature of the material, the mass of the material, and the temperature of the material. For example, 100 g of hot water has more energy than 100 g of cold water because of a difference in temperature. A swimming pool of 10,000 kg of cold water will have more energy than 1 kg of hot water, mainly because of a difference in mass. If the 1 kg of hot water is poured into the swimming pool of 10,000 kg of cold water, the temperature of the pool water will rise a tiny amount. The temperature of the hot water will drop considerably.

When you mix cold milk with hot coffee, you expect the cold milk to warm a bit and the coffee to cool a bit. They will soon arrive at the same temperature. This can be explained using the conservation of energy. The milk gained some energy and the hot coffee lost some energy. It might be clearer if you look at some numbers. If the cold milk is at 5°C and the hot coffee is at 90°C, the final temperature of the milk-coffee

mixture could be 88°C. In this case the temperature of the milk rose 83°C and the temperature of the coffee fell 2°C. Energy was conserved. If you knew the mass of the coffee and the milk, you could compute the gain and loss of the energy by each substance using the relation $Q = mc\Delta t$. The change in energy of both the milk and the coffee would be the same.

This situation is quite common. If you put a piece of cold metal into the coffee, there would be a similar effect. The coffee could cool from 90°C to 88°C and the metal could warm from 5°C to 88°C. If you knew the masses of the metal and the coffee and the specific heat of the metal, you could again compute the gain and loss of the energy with the relation $Q = mc\Delta t$. Again, the change in energy would be the same for the metal and for the coffee.

When a piece of cold metal is placed in hot coffee, it never happens that the metal gets even colder and the coffee gets even hotter. It never happens that the coffee heats up from 90°C to 92°C and the metal cools from 5°C to 3°C. The conservation of energy would be satisfied if the cold metal lost energy and the coffee gained energy.

If something never happens, you must assume that nature has placed a restriction on it. This restriction is called **entropy.** It informs you that the two materials in contact will reach a common equilibrium temperature. The transfer of heat energy can only take place in one direction. A cooler metal will heat up when placed in contact with the hot coffee, but the cooler metal will never become cooler when placed in contact with the hot coffee.

This irreversibility of heat and the related concept of entropy help to distinguish the past from the future. If you watch a movie of a pendulum moving back and forth, you may not be able to tell whether the film is being played forward or backward. If you watch someone break an egg and fry it, the film would look quite silly when played backward. It doesn't make sense that the egg could get un-fried and then return to its shell.

Reflecting on the Activity and the Challenge

As you design your universal dwelling, you must be very aware of the temperature. You have investigated what happens when you change the shape of the dwelling or add windows. You also know that a cold drink will warm up if it sits on the table and that a hot drink will cool down if it sits on the same table. All objects in the dwelling will reach the same final temperature – the equilibrium temperature. Since water requires lots of energy to heat up or cool down, some people have used water or ice to provide comfort in their dwellings. You can now calculate the energy changes when cold objects and warm objects are put in contact. You may want to consider cold water or ice as a type of air conditioning unit. Similarly, you may want to consider the idea of heating water by solar energy and then moving that hot water through the house to heat up the air.

Physics Words

entropy: a measure of the degree of disorder in a system or a substance.

Physics To Go

1. A hot cup of coffee at 90°C is mixed with an equal amount of milk at 80°C. What would be the final temperature if you assume that coffee and milk have identical specific heats?

2. Explain why heating up a pot of water when you only need enough for one cup of tea is wasteful of time and wasteful of energy consumption.

3. A container of water can be heated with the addition of hot water or the addition of a piece of hot metal. If the mass of the water is equal to the mass of the metal, which material will have the greatest effect on the water temperature? Explain your answer.

4. Suppose 200 g of water at 50°C is mixed with 200 g of water at 30°C.

   a) What will be the final temperature?
   b) Calculate the energy gained by the cold water.
   c) Calculate the energy lost by the hot water.

5. Suppose 200 g of water at 50°C is placed in contact with 200 g of iron at 30°C. The final temperature is 48°C.

   a) Calculate the energy gained by the iron. The specific heat c of iron is 0.45 J/g°C.
   b) Calculate the energy lost by the hot water.
   c) If the final temperature could have been measured more accurately, would you expect that it would have been a bit more or less than 48°C? Why?

6. Suppose 100 g of water at 50°C is placed in contact with 200 g of iron at 30°C. The final temperature is 46.5°C.

   a) Calculate the energy gained by the iron. The specific heat c of iron is 0.45 J/g°C.
   b) Calculate the energy lost by the hot water.
   c) If the final temperature could have been measured more accurately, would you expect that it would have been a bit more or less than 46.5°C? Why?

7. Suppose 300 g of water at 50°C must be cooled to 40°C by adding cold water. The temperature of the cold water is 10°C. How much of the cold water must be added to the hot water to bring the temperature down to 40°C?

PHYSICS AT WORK

Ray Aguilera

HABITAT FOR HUMANITY

Ray is the construction manager at the Valley in the Sun, Arizona, Habitat for Humanity branch. Habitat for Humanity International is a non-profit organization that seeks to eliminate poverty, housing, and homelessness from the world. They do this by building and selling homes for no profit, to families who cannot get conventional financing. Homeowners also become partners in the process by contributing 500 hours of "sweat equity" toward the construction of their own home.

Our former president, Jimmy Carter, has been deeply committed to Habitat since 1984. Each year former President Carter and his wife, Rosalynn, join Habitat volunteers to build homes and raise awareness of the critical need for affordable housing.

Ray Aguilera was on his way to becoming a lawyer, when one summer he got a job building homes, and he has never stopped. Ray gets a great deal of satisfaction from his work with Habitat. "I enjoy going from an empty lot and watching something magical grow out of it," he says.

Arizona has a different climate than many other places in the country and takes special considerations when planning homes. In this southwestern desert, the temperatures are often in the 100s and rarely very cold. For example, the foundations and footings for houses do not need as much concrete as other places because the soil conditions are so different. And, in Arizona, you never have to worry about winter frosts. "The most essential characteristic of building a house in Arizona," states Ray, "is to keep it energy efficient. We also strive to design houses that will blend in with the existing environment. In Arizona, we make more of an effort to keep the hot out and the cold in. To do this we use double-paned windows, and as much insulation as we can fit between the walls and in the attic."

Chapter 6 Assessment

You and your group have been investigating the design of your model universal dwelling for this entire chapter. You have learned a great deal about this problem and have explored possible solutions. You now have reached the point where you must finalize your plans and prepare your presentation for the hearing before the Homes for Everyone (HFE) Architectural Committee. Good luck!

Using the resources that you have been provided, develop scale drawings of your model universal dwelling. Refer to the **Chapter Challenge** given at the beginning of this chapter to guide your group's activities. Your presentations will be strictly limited to five minutes, so make every drawing and sentence count!

Review the criteria that you decided the HFE Architectural Committee should use in evaluating your drawings and your written presentation.

Physics You Learned

Surface area-to-volume ratio

Brief statements of conduction, convection, radiation

Heating curves and Cooling curves
- With opaque backgrounds
- With windows
- With insulation
- With awnings

Properties of insulation

Effect of thickness on insulation effectiveness

Brief review of latitude and solar position at equinox

Chapter 7

ELECTRICITY FOR EVERYONE

Throughout history people have tried to help other people. However, changes often have been made without any respect for the personal and cultural needs of those who are being assisted.

If you ever become involved in a self-help community group, it would be important for you to work together with the people you are helping in assessing their needs and their capabilities. Although that is not possible given your limited time in class, you should recognize the need for this type of collaborative teamwork when assisting people.

Scenario

The Homes For Everyone (HFE) Architectural Committee has just accepted your design for a "universal dwelling" to meet the growing housing shortage in many diverse areas of the world. The organization would now like you to develop an appliance package that would help meet the basic needs for healthy, enjoyable living for the families who will reside in the universal dwellings.

The source of electrical energy chosen for this particular project is a wind generator. The following is a description of the wind-generator system chosen for HFE. Try to get a sense of the meaning of unfamiliar words. When the chapter is completed, you will understand these terms.

The wind-generator system chosen for HFE is a highly reliable, mass-produced model that has an output of 2400 W (2.4 kW). Experience has shown that in areas having only moderate average wind speed (6 to 8 km/h) the generator system will deliver a monthly energy output of about 90 kWh (kiloWatt-hours) to the home, or about 3 kWh per day.

Direct current (DC) from the wind-driven generator is stored in batteries that allow storage of electrical energy to keep the home going for four windless days. The batteries deliver DC electricity, but most home appliances are designed to use alternating current (AC). An inverter changes the DC from the batteries into AC before it enters the home. A circuit breaker rated at 2400 W protects the batteries from overheating if too much energy is asked for at any single time. Finally, a kiloWatt-hour meter is provided to keep track of the amount of electrical energy that has been used. The result is that the dwelling will have the same kind of electricity delivered to it as do most homes in the U.S., but less electrical power and energy will be available than for the average homes in the U.S.

Challenge

You will use your experience with electricity in your home and what you learn in this chapter to decide which electrical appliances, powered by a wind generator, can and should be provided for the HFE dwellings.

1. Your first task is to decide what electrical appliances can and should be used to meet the basic needs of the people whose HFE dwelling will be served by a wind generator.

   - **Use the list of appliances on pages 496-497, any additional information that you can gather about appliances, and the characteristics of the wind-generator system to decide what appliances to include in an "appliance package" for HFE.**
   - **As part of your decision-making process, determine if it seems best to provide a basic appliance package that would be the same for all dwellings, or if packages should be adapted with "options" to allow for factors such as different family sizes, climates, or other local conditions.**
   - **Describe how each appliance in your package will contribute to the well-being of the people who live in the dwelling.**

2. Your second task is educational. The people will need to be instructed how to stay within the power and energy limits of their electrical system as they use their appliances.

   - **You must outline a training manual for volunteers who will be going into the field to teach the inhabitants about the HFE wind-generator system and the appliances. The volunteers have no special knowledge of electricity. Therefore, the volunteers need a "crash course" that will prepare them to teach the people to use their electrical system with success.**
   - **Two factors will be especially important to teach: the power demand of the combination of appliances being used at any one time may not exceed 2400 W, and the average daily total consumption of electrical energy should not exceed 3 kWh.**

Criteria

The criteria for this challenge will be judged on the basis of 100 points. Discuss the criteria below, add details to the criteria if it would be helpful, and agree as a class on the point allocation.

- **The list of appliances to be included in the HFE appliance package must be as comprehensive as possible, and it must be clear how each appliance will enhance the health or well-being of the people who live in the dwelling.**
- **The outline of the training manual for HFE volunteers must explain the difference between 2400 W and 3 kWh. It must also give clear examples of how use of the appliances in the package can be scheduled to stay within the power and energy limits of the electrical system on both a daily and a long-term basis.**

Activity 1 Generate

GOALS

In this activity you will:

- Trace energy transformations.
- Begin developing a personal model for electricity.

What Do You Think?

Electricity affects most parts of your life. You pay for it, over and over, in the form of electric bills and batteries. Also, most products that you purchase are manufactured by processes that use electricity, so you pay for electricity in indirect ways, too.

- **Is there any "free" electricity available and, if so, why pay for it?**

Record your ideas about this question in your *Active Physics* log. Be prepared to discuss your responses with your small group and the class.

For You To Do

1. You will be provided with a bulb, bulb base, connecting wires and a generator. Assemble the bulb, bulb base, connecting wires, and hand generator, and turn the crank of the generator to make the bulb light. Never turn the crank too fast. You can strip the gears!

   a) Draw a diagram of how you assembled the equipment for the bulb to light.

   b) Under what conditions will the bulb not light? Use words and a diagram in your answer.

   c) What are the effects of changing the speed or direction of cranking the generator?

   d) What are the effects of reversing the connections of the wires to the bulb or to the generator?

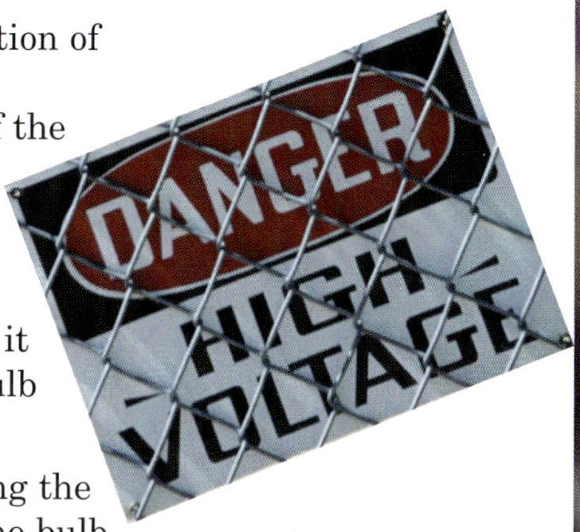

2. Replace the bulb that you have been using with a blinking bulb, the kind used in some toys, flashlights, and decorations. As before, use the generator to make it light, and keep cranking the generator to make the bulb go through several on and off cycles.

   a) Describe any difference that you can feel in cranking the generator when the bulb is on compared to when the bulb is off.

   b) How do you think that the blinking bulb works? What makes it go on and off?

3. Replace the blinking bulb with a strand of steel wool. Connect the ends of the steel wool to alligator clips on the generator. Crank the generator and observe what happens to the steel wool. Be careful not to touch the hot steel wool! You may push the steel wool with the point of a pencil to provide a better contact with the socket.

   a) Describe the appearance of the steel wool.

   b) What factors affect whether or not the steel wool glows, how much it glows, and for how long?

   c) What were the similarities and differences between the steel wool and the light bulbs as used above?

The steel wool will get very hot. Do not touch it while conducting the experiment. Allow it to cool before removing it.

4. Was the electrical energy that you used to "light things up" in this activity "free"? Did you get something for nothing? Using your observations in this activity, write a short paragraph to answer each of the following questions:
   - a) What was the energy source for each part of the activity (bulb, blinking bulb, steel wool)? Was it free energy, at no cost?
   - b) What forms of energy were involved in the activity, and in what order did the forms appear?
   - c) How is the energy source used in this activity different from the source used to light a bulb in a flashlight, or in a house lamp?

Reflecting on the Activity and the Challenge

This activity has given you some experience with a process that is involved in the electrical system you will use for the HFE dwelling: using a generator to provide energy for electric light bulbs. The generator and the light bulbs used in this activity are scaled-down versions of the ones to be used for the dwelling, but they work in the same way. One additional feature will exist in the electrical system for the dwelling: the electrical energy from the generator will be able to be stored in batteries until it is needed to operate lights and other appliances.

Part of your challenge is to write a training manual to help instructors teach the inhabitants about their wind-generator system. You will probably want to include what you learned in this activity in your manual.

Physics To Go

1. Make a chart with two columns, the first one labelled "Word" and the other labelled "Meaning."
   a) In the first column make a list of "electricity words"—words that you have heard used in connection with electrical units of measurement, parts of electrical systems, or how electricity behaves.
   b) In the second column write what *you* think each word means, or describes.

2. You know that electricity comes "out of the wall." You also know that it "starts" in a power plant. Draw a picture that shows how *you* think the electricity is "created" and how it gets to your home.

3. Explain what *you* think electricity is, how it behaves, and how it does what it does.

4. A variety of energy sources are used to operate light bulbs. Identify as many energy sources as you can that are used to power light sources.

5. The kind of light bulb you used in this activity is called "incandescent." Another kind of light bulb often used is called "fluorescent." Look up the meaning of the two words and explain how they are related to what glows to cause each kind of bulb to give off light.

6. "You don't get something for nothing." Explain how this expression applies to using a hand-operated generator to light a bulb.

Stretching Exercises

Some light bulbs are frosted on the inside, while others are clear. Some light bulbs have built-in reflectors to make them serve as a spotlight or a floodlight. Research different types of light bulbs. On a poster, describe how each is different, how each works, and where and why each is useful.

Activity 2 Lighten Up

GOALS

In this activity you will:

- Qualitatively describe current, resistance, voltage.
- Define: coulomb, ampere, volt.
- Compare series and parallel circuits.
- Recognize generator output limit.
- Extend a personal model of electricity.

What Do You Think?

Lights were the first electric appliances for homes.

- **How do light bulbs, and the electricity that makes them glow, work?**

Record your ideas about this question in your *Active Physics* log. Be prepared to discuss your responses with your small group and the class.

For You To Do

1. Before you begin this activity, remind yourself about the "feel" of the generator and the brightness of the bulb when the generator was used to energize one bulb in **Activity 1**. Connect a single bulb to the generator and crank. Use your observations of a single bulb as a basis for comparison when you use two or more bulbs during this activity.

2. There are two distinct ways to connect more than one light bulb to the generator. Look at the two diagrams showing three bulbs connected in series and in parallel.

   a) Describe in your log how the two circuits are different.

   b) Make predictions about how each circuit operates.

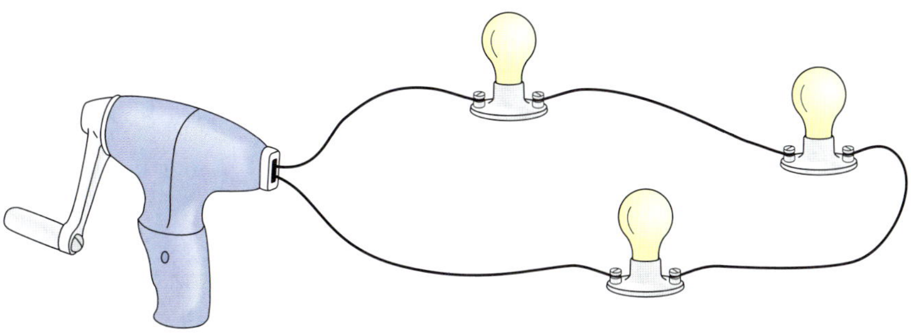

A series circuit.

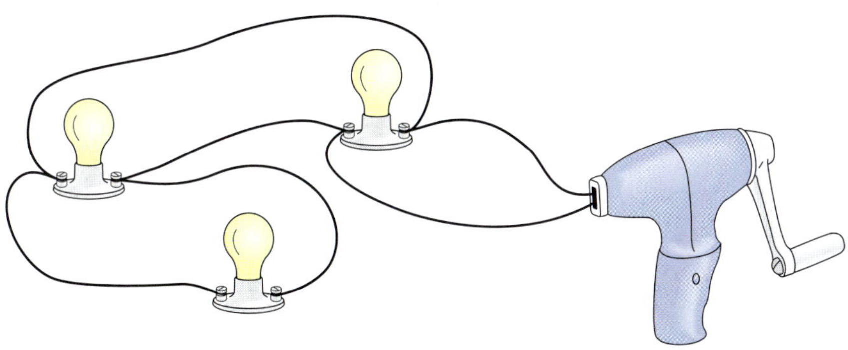

A parallel circuit.

3. Connect *two* bulbs in series with the generator. Use the diagram showing three bulbs connected in series to help you. Crank the generator, and notice the "feel" of the generator and the brightness of the bulbs. Repeat this for three bulbs, and four bulbs.

   a) Describe what happens and try to explain why it happens.

4. What would happen if, in a series circuit of several bulbs, one bulb were to be disconnected, or burn out? Try it by unscrewing one bulb from its base while the circuit is operating.

   a) Describe what happens, and try to explain why it happens.

Physics Words

electric charge: a fundamental property of matter; charge is either positive or negative.

proton: a subatomic particle that is part of the structure of the atomic nucleus; a proton is positively charged.

electron: a negatively charged particle with a charge of 1.6×10^{-19} coulombs and a mass of 9.1×10^{-31} kg.

coulomb: the SI unit for electric charge; one coulomb (1 C) is equal to the charge of 6.25×10^{18} electrons.

5. Connect *two* bulbs in parallel with the generator and, again, observe the "feel" of the generator and the brightness of the bulbs. Repeat this for three bulbs, and four bulbs.

   a) Describe your observations and compare them to your predictions for a parallel circuit.

6. What would happen if one bulb were to fail in a parallel circuit? Try it by unscrewing one bulb.

   a) Describe what happens, and try to explain why it happens.

FOR YOU TO READ

The Language of Electricity

Now you are ready to become acquainted with some of the basic language of electricity. You are ready to learn the meanings of, and use of, some of the "electricity words" that you identified in **Activity 1**. Here are some theories and definitions about electricity to help you:

- There are two kinds of electric charges, positive and negative, named by Benjamin Franklin. Protons have positive charge, and electrons have negative charge. Like kinds of electric charges repel, and unlike kinds attract.

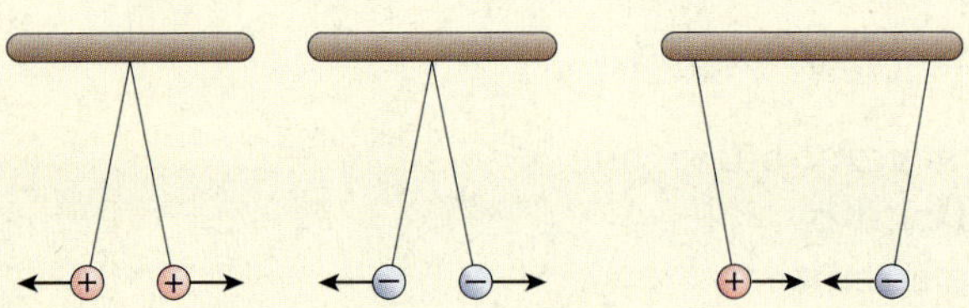

Like charges repel, and opposite charges attract.

- There is a smallest amount of the property called **electric charge**, the amount possessed by one proton and one electron. While **protons** and **electrons** differ in several ways (such as mass), an electron and a proton have an identical amount of charge.
- It is electrons that move in electric circuits of the kind you have been exploring. They flow through the circuit path, delivering energy, often in the form of heat and light, as they go. Protons, although present in the materials from which circuits are made, do not flow because they are locked within atoms. Positive particles sometimes do move in circuits where part of the path is liquid or gas.
- Scientists have agreed upon a standard "package" of electric charge, called the coulomb. Just as 12 eggs are called a dozen, 6.25 billion basic charges are called a **coulomb**. To provide a sense of how big this number is, one coulomb is approximately the charge transferred during a lightning bolt!
- Scientists have agreed upon a standard rate of flow of the **electric current** in circuits. When one coulomb of charge passes through a point in a circuit during each second of time, the current is said to be one **ampere**, symbol A, or often abbreviated to amp.

- Different materials offer different resistance, or opposition, to the flow of electric charge through them. Tungsten, from which light bulb filaments are made, has high electrical resistance. When electricity flows through it, the tungsten metal "robs" energy from the moving electrons and gets hot. Copper, by contrast, has low resistance; electrons transfer very little energy when flowing through copper. **Electrical resistance** is measured in **ohms** (Ω).
- Batteries or the hand generator provide energy to the electrons. These electrons are then able to light bulbs, heat wires, or make motors turn. The energy given to each coulomb of charge is measured in **volts** (V).

Reflecting on the Activity and the Challenge

In this activity you were introduced to the parallel and series circuits and to the electrical terms that you will need to know and be able to use for planning electric circuits to be used in the universal dwelling.

It is a fact that homes are "wired" using parallel circuits. Individual houses, apartments, mobile homes, or any other dwellings that receive electricity from a power company, have circuits of the parallel kind. Some older homes have as few as four circuits, and newer homes usually have many more. Each circuit in a home may have several light bulbs and other electrical appliances "plugged in," all in parallel. When electrical appliances are hooked up in parallel, if one is off or diconnected, the others can still be on. In a series circuit, if any appliance is disconnected, the other appliances cannot work. In your training manual, you will need to explain why the circuits in the universal dwelling are wired in parallel.

Physics Words

electric current: the flow of electric charges through a conductor; electric current is measured in amperes.

ampere: the SI unit for electric current; one ampere (1 A) is the flow of one coulomb of charge every second.

electrical resistance: opposition of a material to the flow of electrical charge through it: it is measured in ohms (Ω); the ratio of the potential difference to the current.

ohm: the SI unit of electrical resistance; the symbol for ohm is Ω.

volt: the SI unit of electric potential; one volt (1 V) is equal to one joule per coulomb (J/C).

Physics To Go

1. What kind of circuit, series or parallel, would you choose for household wiring, and why? Write a short paragraph to explain your choice.

2. Did the generator used in this activity seem to have an "output limit"? In other words, did you arrive at conditions when the generator could not make the bulbs glow brightly even though you tried to crank the generator? Discuss this in a few sentences.

3. There is a great big generator at the power plant that sends electricity to your home. The wind generator chosen for HFE is much smaller than the generators used at power plants, but much larger than the one used for this activity. What implications might the output limit of the HFE electrical system have for the number of light bulbs and other electrical appliances that can be used in the HFE appliance package that you will recommend? Discuss this in a short paragraph.

Stretching Exercises

Thomas Edison is arguably one of the greatest inventors in world history. When we think of Edison, we think of the light bulb and the changes that this invention has made on the world. Edison dreamed of a world where we could read at night, where we could stroll down a lit street and where we could enjoy daytime all the time. Electricity and the light bulb have made that dream a reality. We live in Edison's dream!

Edison once said that genius is 1% inspiration and 99% perspiration. Explain the meaning of this phrase.

Construct a list of Edison's major inventions. (Edison had 1903 patents in his name!)

Activity 3 Ohm's Law

GOALS

In this activity you will:

- Calculate the resistance of an unknown resistor given the potential drop and current.
- Construct a series circuit.
- Properly use a voltmeter and ammeter in a series circuit.
- Graph the relationship between voltage and current for a resistor that obeys Ohm's Law.

What Do You Think?

Lighting makes some rooms conducive to work and other rooms more relaxing.

- **What determines the brightness of a light bulb?**
- **How can a dimmer switch make the same light bulb appear dim or bright?**

Record your ideas about these questions in your *Active Physics* log. Be prepared to discuss your responses with your small group and the class.

For You To Do

1. Imagine a closed box with something inside it. If you wanted to discover the contents, you could shake the box, weigh it, roll it, or twirl it. The sounds that you hear could help you decide what is inside.

   In this activity, you will be given a "black box." Inside the box is an electrical resistor. Outside the box are two terminals. The box will be set up as a part of an electrical circuit.

   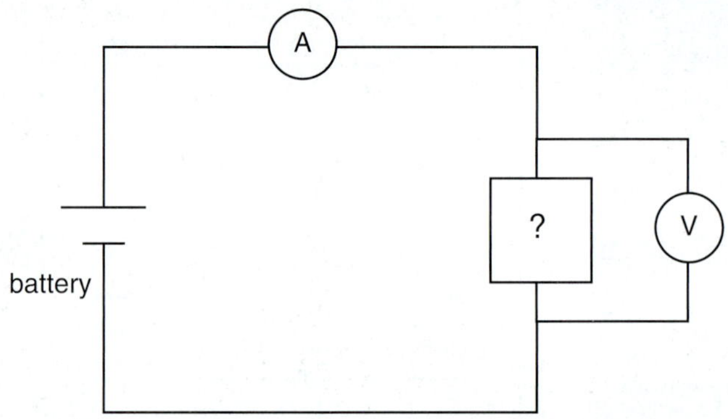

   You will be told the voltage across the black box and the current in the ammeter.

2. The box contains either a 3-Ω (ohm), a 10-Ω, or a 15-Ω resistor. By varying the voltage of the battery, you can learn how the current of the circuit and the voltage across a 3-Ω resistor changes. You can then conduct similar studies of the 10-Ω and 15-Ω resistor.

3. After you gather your data, your teacher will give you the values of the ammeter and voltmeter of the circuit with the black box. From the information, you will have to determine the contents of the black box.

4. Before beginning your study, please note the following circuit rules:

   - The ammeter is always placed in series in the circuit.
   - The positive terminal of the ammeter is always closest to the positive terminal of the battery.

- The voltmeter is always placed in parallel in the circuit. The simplest way to measure the voltage drop is to take the two leads from the voltmeter and touch the two ends of the resistor simultaneously.
- The positive terminal of the voltmeter is always closest to the positive terminal of the battery.
- Keep the connection in the circuit on for as long as it takes to read the ammeter and voltmeter and no longer.
- You can vary the voltage with a variable voltage supply or by adding additional batteries into the circuit.

5. Plan your experiment.

a) Write down your plan in your *Active Physics* log.

b) Create a table that will summarize the data you intend to collect.

c) Describe how you will predict the contents of the black box from your data and the values given by the teacher.

6. Share your plan with your teacher.

7. After your teacher has approved your plan, conduct your experiment.

a) Record your data in your *Active Physics* log.

b) Record in your log the values given by your teacher for the black box.

c) Which resistor is inside the black box?

d) Explain how you know.

e) How confident are you about the contents of the black box? (1 = I have no idea; 10 = I am completely sure; nothing else is possible.)

Physics Words

Ohm's Law: the relationship among resistance, voltage, and current.

PHYSICS TALK

Ohm's Law Relates Resistance to Voltage and Current

The measurements of voltage and current showed that increasing the voltage increased the current in the circuit. This was true for all three resistors. These resistors also obey **Ohm's Law**, which states that as the voltage increases in a fixed rate, the current increases at the same rate. The ratio of the voltage to the current is a constant for any one of these resistors. This is expressed mathematically:

$$R = \frac{V}{I}$$

where R is the resistance in ohms (Ω),

V is the voltage in volts (V), and

I is the current in amperes (A).

The equation is known as Ohm's law. This equation may be rearranged to calculate the value of any of the terms.

$$V = IR \qquad I = \frac{V}{R}$$

Some resistors obey Ohm's Law over a wide range of voltages. For these resistors, the value of R always remains the same.

Sample Problem

A 2-Ω resistor is placed in a circuit. Record the currents corresponding to voltage measurements of 10 V and 30 V.

Strategy: You are asked to calculate current for a known resistor. You can use Ohm's Law, but you will need to rearrange the equation to solve for current.

Givens:

$$R = 2\ \Omega$$

$$V = 10\ \text{V and } 30\ \text{V}$$

Solution:

$$I = \frac{V}{R}$$

$$I = \frac{10\ \text{V}}{2\ \Omega} = 5\ \text{A} \qquad I = \frac{30\ \text{V}}{2\ \Omega} = 15\ \text{A}$$

FOR YOU TO READ

Graphing Ohm's Law

Ohm's Law expresses the relationship between the voltage and current ($R=V/I$). For certain resistors, the ratio of the voltage and current is a constant. Recording the varying voltage and the corresponding current would allow you to make a graph. If the resistor obeys Ohm's Law, the ratio of the voltage and current remains the same and the graph would be a straight line. If the current is on the *x*-axis and the voltage on the *y*-axis, the slope of the line is equal to the resistance. (If the current is on the *y*-axis and the voltage on the *x*-axis, the slope of the line is equal to $1/R$.)

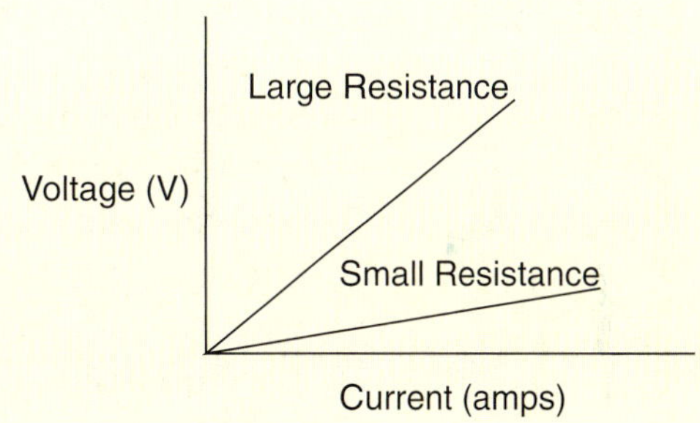

Many electrical devices do not obey Ohm's Law. A graph of the voltage versus current for a neon gas lamp is shown below.

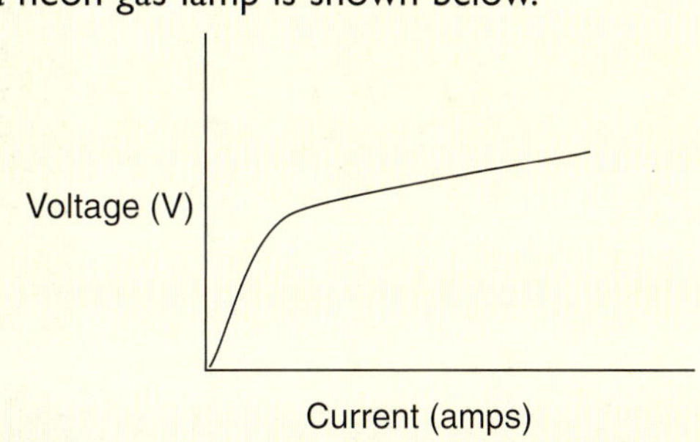

The resistance has a high value (large slope) for small currents and a low value (small slope) for large currents. The neon gas lamp does not obey Ohm's Law because the resistance is not constant.

The brightness of a bulb is dependent on the power (the rate of energy) supplied to the bulb. The power, as you will learn in the next activity, is equal to the product of the voltage and the current ($P = VI$). You have learned in this activity that increasing the voltage increases the current. This, in turn, increases the power to the light bulb. More power is more energy every second and that will make the bulb brighter.

The dimmer switch is a variable resistor. It controls the voltage and the current available to the light bulb and affects its brightness.

Reflecting on the Activity and the Challenge

Ohm's Law expresses the relationship between voltage and current for many resistors. Your appliances are all electrical resistors. These resistors can get hot (in toasters) or help a motor turn (in fans) or create light (in bulbs). Knowing the resistance allows you to use Ohm's Law to compute the voltage and current required in a circuit. As you develop your appliance package, you will need to consider the voltage and currents used in the circuit. You may wish to choose appliances that have different resistances to limit the power required in the circuit.

Physics To Go

1. A resistor is placed in a circuit. Calculate the resistance in each of the following cases.
   a) The current in the circuit is 3 A and the voltage drop across the resistor is 12 V.
   b) The current in the circuit is 2 A and the voltage drop across the resistor is 6 V.

2. A resistor of 5 Ω is placed in a circuit. The voltage drop across the resistor is 12 V. What is the current through the resistor?

3. A resistor is placed in a circuit. The current in the circuit is 2 A and the voltage drop across the resistor is 8 V. The voltage is then increased to 12 V. What will be the new current?

4. Using the information in the data table, construct a graph following the directions below.
   - Mark an appropriate scale on the *x*-axis labeled "Current (A)."
   - Mark an appropriate scale on the *y*-axis labeled "Voltage (V)."
   - Plot the data points for voltage versus current.
   - Draw the best-fit line.

Current (A)	Voltage (V)
0.010	2.3
0.020	5.2
0.030	7.4
0.040	9.9
0.050	12.7

   a) Using your graph, find the slope of the best-fit line.
   b) What physical quantity does the slope of the graph represent?

5. Your hair dryer has a resistance of 9.6 Ω and you plug it into the bathroom outlet. Assume household voltage to be 120 V.
   a) What current will it draw?
   b) Suppose your brother has an identical hair dryer and plugs it into the same part of the circuit. What current will the two hair dryers draw?

c) If the maximum current the circuit breaker in the system can handle is 20 A, what do you think will happen?

6. Two wires are tested in a lab setting. Current was measured as the voltage across the wire was varied. The results of the experiment are shown in the graph below. Both wire A and B obey Ohm's Law. Which wire has the greatest resistance?

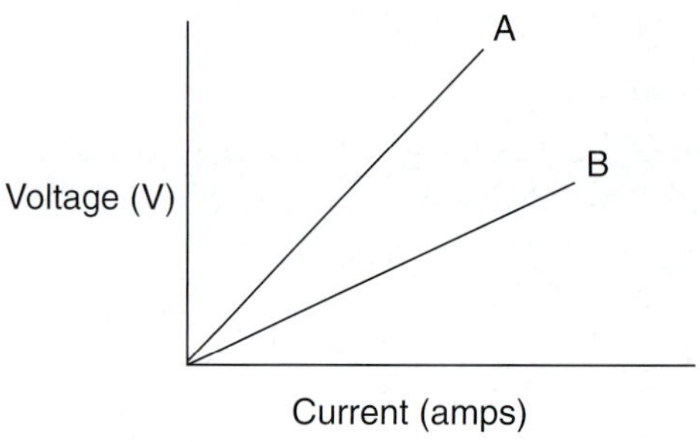

Activity 4 Load Limit

GOALS

In this activity you will:

- Define: watt, insulator, conductor.
- Apply $P = IV$.
- Measure the power limit of a 120-V household circuit.
- Differentiate between a fuse and a circuit breaker.
- Understand the need for circuit breakers and fuses in a home.

What Do You Think?

Everybody has at one time "blown a fuse" or "tripped" a circuit breaker.

- **What is a fuse or circuit breaker?**
- **Exactly what conditions do you think make a fuse "blow" or a circuit breaker "trip"?**

Record your ideas about these questions in your *Active Physics* log. Be prepared to discuss your responses with your small group and the class.

For You To Do

1. Your teacher will do this part of the activity as a demonstration for the entire class. You will intentionally exceed the load limit of the circuits that supply electricity to your classroom. Observe the number of light bulbs needed to reach the load limit and the manufacturer's ratings of the bulbs.

   a) Calculate the load limit of the circuit, in watts. This will apply to the load limit of the circuits for your Homes For Everyone (HFE) dwelling.

2. Knowing that the energy per unit of electric charge supplied to the circuit is 120 V (volts), use the answer to the previous calculation and the equation:

   $$\text{Power} = \text{current} \times \text{voltage } (P = IV)$$

   to calculate the current, in amperes, that must have been flowing in the circuit when the load limit was reached.

   a) Show your calculation in your log. Assume a voltage of 120 V.

   b) How much current, in amps, must be flowing through the filament of a 60-W light bulb when it is operating in a 120-V household circuit? A 100-W light bulb? Show your calculations in your log.

   c) Fuses and circuit breakers are rated in amperes, usually 15 or 20 A for most household circuits. In your log use the equation $P = IV$ to show how you can calculate the load limit, in watts, of a 120-V household circuit protected by a 15-A circuit breaker.

PHYSICS TALK

Did you ever wonder why the lights go out in some rooms but not others when a fuse blows? Why does turning on the hair dryer blow a fuse but turning on the radio does not?

In the appliance package that you will be creating, there is a load limit of 2400 W. You are not restricted to a load limit in your own home. If you have the money, the electric company will set up more circuits allowing you to use more electricity. The company can then collect more money from you. There may actually be a limit. One person could not be allowed to use all the electricity from a massive power plant—at least, we hope not.

There are some restrictions on each circuit in your home. A fuse or circuit breaker protects each circuit. The protection is needed because too much current can heat or burn the wires in the walls of your home. An electrical fire can be extremely dangerous.

You probably think about the fuse or circuit breaker only when it turns off the electricity to a part of your home. The circuit is shut down when too much current is required. In most household circuits, the maximum current is usually 15 or 20 A (amperes). Each appliance in your home requires a certain amount of current to run properly. The power rating of the appliance, the voltage of the circuit, and the current required are related by the following equation:

$$P = VI$$

where P is the power in watts (W),

I is the current in amperes (A), and

V is the voltage in volts (V).

This equation can be rearranged to calculate the value of any of the terms.

$$I = \frac{P}{V} \qquad V = \frac{P}{I}$$

→

The voltage of almost all of the circuits in your home is 120 V. If the fuse or circuit breaker can handle a maximum of 15 A, then the power limit is (120 V) × (15 A) or 1800 W. If a toaster is 1200 W and a hair dryer is 1000 W, they cannot both be on any one line in your house since they total 2200 W, which is more than the 1800-W limit. If you want to run both appliances at the same time, you must use different circuits, each with its own 1800-W limit.

A second way of viewing the circuit is to look at the current requirements of each appliance. Since the appliances are in parallel, the total current will be the sum of all of the individual currents. In the example above, the current of the 1200-W toaster can be found using the power equation.

$$P = VI$$

$$1200\ W = 120\ V \times I$$

$$I = 10\ \text{A}$$

Similarly, the hair dryer requires about 8 A.

The total current is about 18 A. This is more than the 15-A fuse can tolerate.

Sample Problem 1

A 12-V starter battery in a car supplies 48 A of current to the starter. What is the power output of the battery?

Strategy: You are asked to find the power, so you use the power equation that is specific to electrical circuits. Power is defined as change in energy or work per unit time. However, in electrical circuits it can be found by multiplying the voltage (1 V = 1 J/C) times the current (1 A = 1 C/s). You notice that this results in joules per second, which are watts, the unit for power (1 J/s = 1 W).

Givens:

$$V = 12\ \text{V}$$

$$I = 48\ \text{A}$$

Solution:

$$P = VI$$

$$= (12\ \text{V})\ (48\ \text{A})$$

$$= 576\ \text{W (or about 580 W)}$$

Sample Problem 2

A 75-W study lamp is plugged into the 120-V household outlet in your room. What current does the outlet supply to the light bulb?

Strategy: Again, use the power equation but rearrange the equation to solve for current.

Givens:

$$P = 75\ \text{W}$$

$$V = 120\ \text{V}$$

Solution:

$$I = \frac{P}{V}$$

$$= \frac{75\ \text{W}}{120\ \text{V}}$$

$$= 0.63\ \text{A}$$

Light bulbs do not draw a lot of current.

Sample Problem 3

A light bulb operating at 120 V draws 0.5 A. Determine the bulb's:

a) resistance

b) power

Strategy: Ohm's Law can be used to determine the resistance of the light bulb. The power can be determined using $P = VI$.

→

Givens:

$V = 120\text{ V}$

$I = 0.5\text{ A}$

Solution:

a) $V = IR$

Solving for R,

$$R = \frac{V}{I}$$

$$= \frac{120\text{ V}}{0.50\text{ A}}$$

$$= 240\ \Omega$$

b) $P = VI$

$$= (120\text{ V})(0.50\text{ A})$$

$$= 60\text{ W}$$

The power could also be calculated using either of the following:

$$P = I^2R$$

or $$P = \frac{V^2}{R} \quad \left(\text{Since } I = \frac{V}{R},\ I^2 = \frac{V^2}{R^2}\right)$$

These relationships are derived using Ohm's Law.

Reflecting on the Activity and the Challenge

The load limit of the electrical system for the universal dwelling is set at 2400 W, as outlined in the chapter **Scenario**. It is also established by the design of the windmill power plant that 120 V will be applied to circuits within the dwelling. In this activity you learned what load limit means, and how to relate it to current and voltage. If the people in the dwelling try to run appliances that require more than 2400 W, the fuse will blow. In your home, you can always choose a different line to run the extra appliances. With only one generator, this is not an option in the universal dwelling. This will have direct application soon when you begin selecting appliances to be used in the dwelling.

Physics To Go

1. Explain in detail what load limit means, and include maximum current, in amperes, as part of your explanation.

2. Find out about the power rating, in watts, of at least six electrical appliances. You can do this at home, at a store that sells appliances, or by studying a catalog. Some appliances have the watt rating stamped somewhere on the device itself, but for others you may have to check the instruction book for the appliance or find the power rating on the original package. Also, your local power company probably will provide a free list of appliances and their power ratings on request. Bring your list to class.

   If the appliance lists the current in amps, you can assume a voltage of 120 V and calculate the power (in watts) by using the equation $P = IV$.

Make sure small appliances are unplugged before handling them.

3. List three appliances you would include in the HFE appliance package that will be part of the chapter **Assessment**. For each appliance, list the power demand. For each appliance, describe how it will contribute to the well-being of the people living in the dwelling.

4. An electric hair dryer has a power rating of 1200 W and is designed to be used on a 120-V household circuit. How much current flows in the hair dryer when it is in use?

5. A 120-V circuit for the kitchen of a home is protected by a 20-A circuit breaker. What combinations of the appliances listed below can be used on the circuit at the same time without the circuit breaker shutting off the circuit?
   - 1000-W toaster
   - 1200-W frying pan
   - 300-W blender
   - 600-W coffeemaker

6. How many 60-W incandescent light bulbs can be operated at the same time on a 120-V, 15-A circuit in a home? How many energy-efficient 22-W fluorescent bulbs can operate on a similar circuit?

7. Some electrical appliances are rated in horsepower (HP).

   $$1 \text{ HP} = 746 \text{ W}$$

   What amount of current flows in a 0.8 HP vacuum cleaner operating on a 120-V circuit?

8. Some electrical appliances are rated in amps. What is the power in watts of a 6-A appliance designed to operate on a 120-V circuit?

9. A 1500-W hair dryer is plugged into the outlet in your bathroom. How much current does this hair dryer draw?

10. When you turn on the toaster in the kitchen, it draws 8.0 A of current from the line.

    a) Find the power output of the toaster.

    b) You plug another toaster in the same outlet and the circuit breaker snaps off. What do you think the maximum current is for the type of breaker you are using?

11. A 3-W clock operates at 120 V.

    a) How much current does the clock draw?
    b) Determine the clock's resistance.
    c) If the maximum current that can be drawn from the outlet is 15 A, how many clocks would it take to blow the fuse? (Assume all clocks could be plugged into the outlet.)

12. An iron has a resistance of 13.1 Ω. Could two identical irons operate on the same fuse? (Assume V = 120 V and I_{max} = 15 A.)

13. The load limit for a particular fuse is 2400 W. Which combination of the following devices would blow the fuse: mini-refrigerator (P = 300 W), microwave oven (R = 19 Ω), hair dryer (I = 12 A), coffeemaker (R = 9.2 Ω).

Stretching Exercises

Find out about the electrical system of your home or the home of a friend or acquaintance. **With the approval of the owner or manager, and with adult supervision**, locate the load center, also called the main distribution panel, for the electrical system. Open the panel door and observe whether the system uses circuit breakers or fuses. How many are there, and what is the ampere rating shown on each fuse or circuit breaker? You may find some larger fuses or breakers that control large, 240-V appliances such as a kitchen range (electric stove); if so, how many are there, what are their ampere ratings and, if you can, determine what they control.

Warning: The inside of a load center is a dangerous area. Do not touch anything. Doing so could cause injury or death. Always have a qualified person help you.

In some load centers there is a list of what rooms or electrical devices are controlled by each fuse or breaker, but often the list is missing or incomplete.

With the approval of the owner or manager, and with adult supervision, you can develop a list that indicates what each fuse or breaker controls. To do so, shut off one circuit and go through the entire house to find the lights and outlets that are "dead" (check outlets with a lamp that you can carry around easily). Those items that are "off" are controlled by that fuse or breaker. List them. Then repeat the same process.

Do not touch any exposed electrical connections or wiring harnesses. Do not attempt to look into or insert anything into any wiring entry points on the panel. Do not reset any circuit breakers or attempt to change any fuses.

Report your findings to your teacher in the form of a list or diagram of the house showing what is controlled by each fuse or circuit breaker.

Activity 5 Who's In Control?

GOALS

In this activity you will:

- Explain how a variety of automatic electrical switches work.
- Select switches and control devices to meet particular needs.
- Insert a switch in a parallel circuit to control a particular lamp.

What Do You Think?

Many electrical switches are operated manually (by hand), and many others are automatic, turning appliances on and off in response to a variety of conditions.

- **List as many different kinds of automatic switching devices as you can.**
- **What are the conditions that cause the on/off action of the switch?**

Record your ideas about these questions in your *Active Physics* log. Be prepared to discuss your responses with your small group and the class.

For You To Do

1. Assemble the circuit as shown in the diagram. (Each number corresponds to a different wire.)

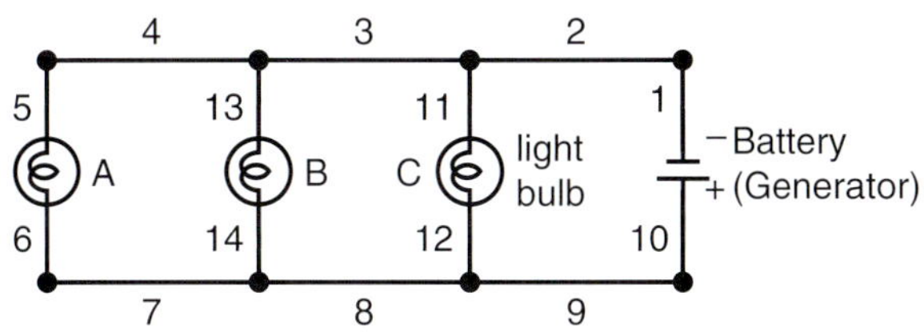

a) Draw a diagram of the circuit. Label the bulbs: A, B, C. (The bulb on the left should be A, the bulb in the middle B, and the bulb on the right C.)

Have your teacher approve your circuit before proceeding.

2. Compare the circuit you assembled with the one in the following diagram that shows the circuit you used in **Activity 2**. This time you used additional wires, 14 wires in total. It is sometimes easier to place switches into circuits if you use a few additional wires in the circuit. Crank the generator to be sure that all bulbs operate in this circuit.

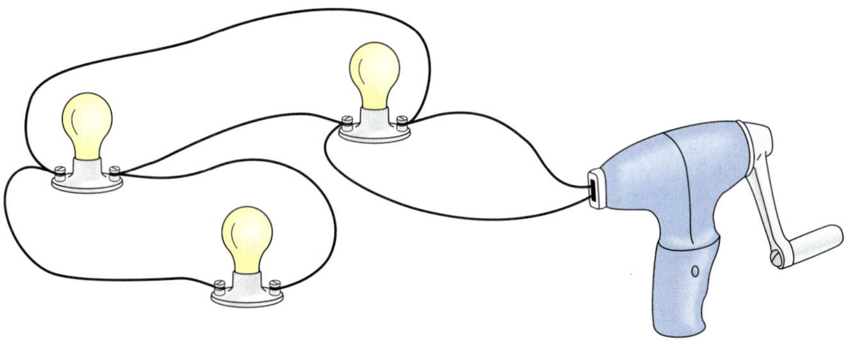

a) Identify the additional new wires on your circuit diagram.

b) What type of circuit (series or parallel) is represented in both diagrams?

3. Predict which wire could be replaced with a switch if you wished to turn all three bulbs on and off.

a) Record your prediction in your log.

b) Replace that wire with a switch. Does it work as predicted?

4. Predict which wire could be replaced with a switch if you wished to turn only bulb A on and off.
   - a) Record your prediction in your log.
   - b) Replace that wire with a switch. Does it work as predicted?
   - c) Mark the location of the switch on the circuit diagram in your log by writing "Switch A" and drawing an arrow from the word "Switch A" to the place where the switch should be placed.

5. Repeat **Step 4** for bulb B and then for bulb C.
   - a) Remember to record your predictions.
   - b) Replace the wire you chose with a switch. Are your predictions beginning to improve?
   - c) Draw two additional diagrams to show the location of switch B and switch C.

FOR YOU TO READ

Switches

Regardless of how an electrical switch may be activated, most switches work in the same basic way. When a switch is "on," a good conductor of electricity, usually copper, is provided as the path for current flow through the switch. Then, the circuit containing the switch is said to be "closed," and the current flows. When a switch is turned "off," the conducting path through the switch is replaced by an air gap. Since air has very high resistance, the current flow through the switch is interrupted, and the circuit is said to be "open."

Circuit Diagrams

An electrical circuit can be represented by a simple line diagram, called a schematic diagram, or wiring diagram. You may have seen wiring diagrams for cars, stereo systems, or homes. They may appear complicated, but if you know what the different symbols represent, they are relatively easy to follow.

The wire that carries the electric current is represented by a straight line. If a wire crosses over another wire, a line crossing a straight line is shown. If the wires join, a heavy black dot is shown. A lamp or light bulb is shown by a loop in the center of a circle. A battery or a generator is represented by a number of unequal lines. A switch is a line shown at an angle to the wire. There are many other symbols that are used by electricians.

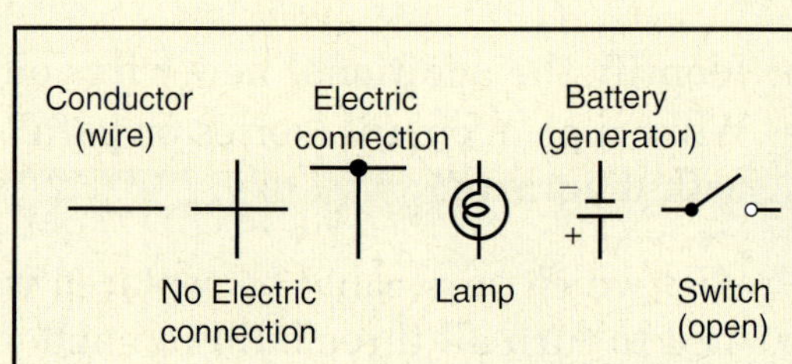

PHYSICS TALK

Multiple resistors, like the light bulbs in **Activity 2**, can be set up in series or parallel. The properties of the series and parallel circuits are quite different. Light bulbs in series will all go out when one bulb is removed. Light bulbs in parallel will remain on when one bulb is removed. In this activity, you learned how switches could control which resistors in a parallel circuit will get current.

Computing the voltage, current, resistance, and power of appliances in your circuit require you to understand Ohm's Law ($V = IR$) and to calculate power ($P = VI$). Series and parallel circuits distribute the current differently and require different analysis.

We will begin with a few general organizing principles for circuits and then show how you can use these principles to determine the voltage, current, power, and resistance for each appliance in a circuit and for the circuit as a whole.

The current, measured in amperes (A), in an electrical circuit is a measure of the flow of charge. One ampere is defined as one coulomb per second. If charge is flowing in the circuit, then the current must remain constant. In a series circuit, the current that goes through the first light bulb must then go through the second light bulb and then through the third light bulb. The equation that describes this is:

$$I_{total} = I_1 = I_2 = I_3$$

In a parallel circuit, the current splits at certain junctions and then joins together at other junctions. The current entering any junction must equal the current leaving any junction if charge is to be conserved. In the part of the circuit shown on the following page, you can see that 1 A and 8 A enter the junction and 2 A, 3 A, and 4 A leave the junction.

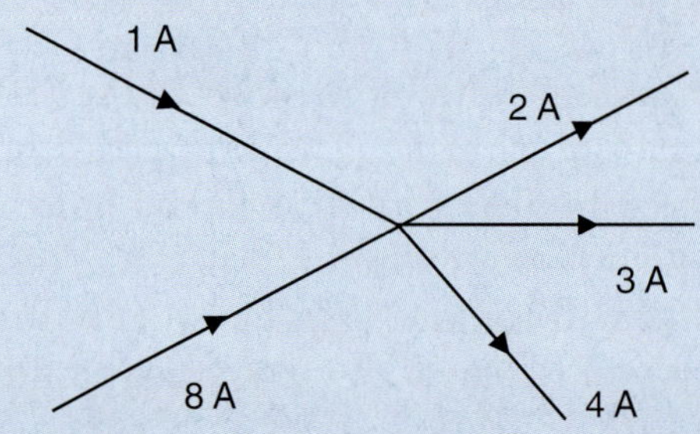

The equation that describes this is:

$$I_{total} = I_1 + I_2 + I_3$$

The voltage (also called the potential difference), measured in volts, in an electrical circuit is a measure of the energy per charge. One volt is defined as one joule per coulomb. The battery provides a certain amount of energy to every coulomb of charge. The energy must be conserved. The energy in the circuit is distributed to the resistors (appliances) in the circuit.

In a series circuit, all of the charges go through all of the resistors. The energy that the charge begins with must be distributed to the resistors. If a battery provides 6 V, 2 V may be transferred to the first resistor, 3 V to the second resistor and 1 V to the third resistor. The equation that describes this is:

$$V_{total} = V_1 + V_2 + V_3$$

In a parallel circuit, the charges that go through one resistor do not go through any other resistor. All of the energy of that charge must be transferred to that one resistor. If a battery provides 6 V, then 6 V must be transferred to each resistor. In the parallel circuit shown in the diagram on the next page, there are 8 A of current. The charges in these 8 A have been provided with 6 V by the battery (6 V = 6 J/C). The 8 A of current splits and 2 A go through the first resistor, 1 A through the second resistor and 5 A through the third resistor. The three currents are shown in three different colors in the sketch on the next page. The 2 A of "red" current drops all of its

6 V in the first resistor. The 1 A of "blue" current drops all of its 6 V in the second resistor. The 5 A of "green" current drops all of its 6 V in the third resistor.

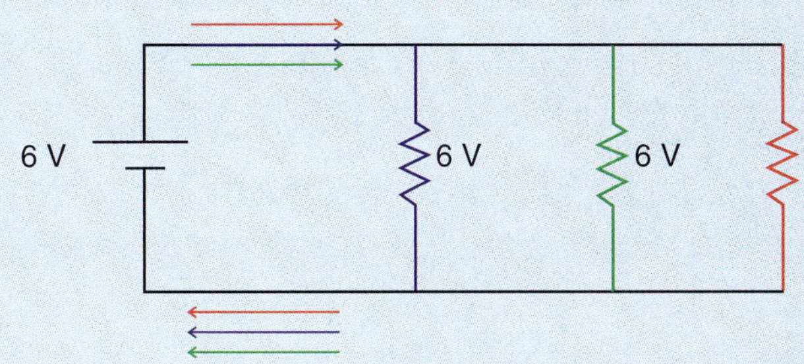

The equation that describes this is:

$$V_{total} = V_1 = V_2 = V_3$$

Having used the conservation of charge and the conservation of energy to determine how currents and voltages are distributed in series and parallel circuits you can now use Ohm's Law to derive equations for the total resistance of multiple resistors.

Series Circuit

Current: $I_{total} = I_1 = I_2 = I_3$

Voltage: $V_{total} = V_1 + V_2 + V_3$

Resistance: $R_{total} = R_1 + R_2 + R_3$

Ohm's Law: Ohm's Law can be applied to the entire circuit or to any resistor in the circuit.

$V_{total} = I_{total}R_{total} \quad V_1 = I_1R_1 \quad V_2 = I_2R_2 \quad V_3 = I_3R_3$

Power: Power can also be applied to the entire circuit or to any resistor in the circuit.

$P_{total} = V_{total}I_{total} \quad P_1 = V_1I_1 \quad P_2 = V_2I_2 \quad P_3 = V_3I_3$

Sample Problem 1

Given the following series circuit, find:

a) the total resistance
b) the total current
c) the current through each resistor
d) the voltage for each resistor
e) the total power
f) the power through each resistor

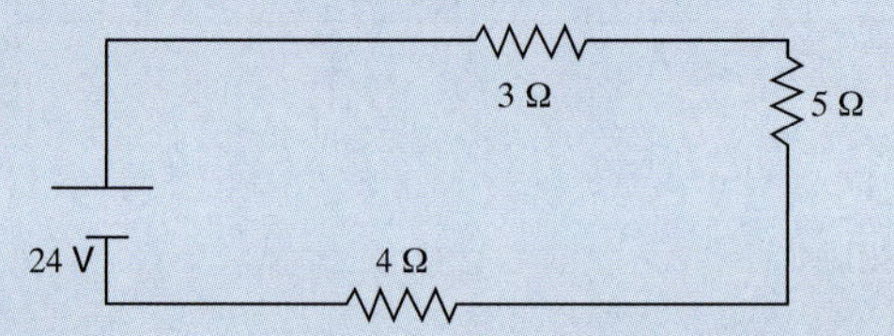

Strategy: This is a series circuit. All the current goes through each of the resistors. First calculate the total resistance. Then using Ohm's Law, calculate the total current. You can also use Ohm's Law to calculate the voltage for each resistor. Then, using the equation for power, calculate the total power and the power through each resistor.

Givens:

$V = 24\text{ V}$

$R_1 = 3\ \Omega$

$R_2 = 5\ \Omega$

$R_3 = 4\ \Omega$

Solution:

a) Total resistance: $R_{total} = R_1 + R_2 + R_3$

$$= 3\ \Omega + 5\ \Omega + 4\ \Omega$$

$$= 12\ \Omega$$

b) Total current: $V_{total} = I_{total} R_{total}$

$$24\text{ V} = I_{total}(12\ \Omega)$$

$$I_{total} = 2\text{ A}$$

c) If you know that the total current is 2 A, then you know that the current through each resistor is 2 A. It is helpful to place this information in the diagram.

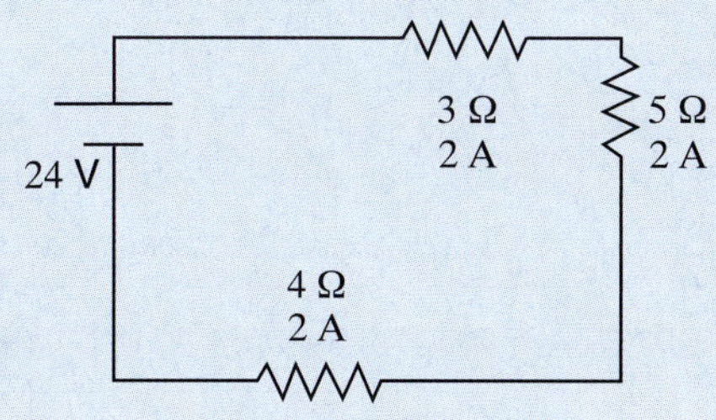

d) Once you know two of the four variables (*V, I, P, R*), you can find the other two variables. In this case, you know *I* and *R* and can find the voltage using Ohm's Law for each resistor.

$V_1 = I_1R_1$	$V_2 = I_2R_2$	$V_3 = I_3R_3$
$V_1 = (2\ \text{A})(3\ \text{W})$	$V_2 = (2\ \text{A})(5\ \text{W})$	$V_3 = (2\ \text{A})(4\ \text{W})$
$V_1 = 6\ \text{V}$	$V_2 = 10\ \text{V}$	$V_3 = 8\ \text{V}$

Notice the sum of the voltage drops:

6 V + 10 V + 8 V = 24 V

This is the voltage supplied by the battery.

e) You can now find the total power.

$P_{total} = V_{total}I_{total}$

$P_{total} = (24\ \text{V})(2\ \text{A})$

$P_{total} = 48\ \text{W}$

f)

$P_1 = V_1I_1$	$P_2 = V_2I_2$	$P_3 = V_3I_3$
$P_1 = (6\ \text{V})(2\ \text{A})$	$P_1 = (10\ \text{V})(2\ \text{A})$	$P_1 = (8\ \text{V})(2\ \text{A})$
$P_1 = 12\ \text{W}$	$P_1 = 20\ \text{W}$	$P_1 = 16\ \text{W}$

Notice the sum of the power in the resistor:

12 W + 20 W + 16 W = 48 W

That is the power supplied by the battery.

Parallel Circuits

Current: $I_{total} = I_1 = I_2 = I_3$

Voltage: $V_{total} = V_1 + V_2 + V_3$

Resistance: $\dfrac{1}{R_{total}} = \dfrac{1}{R_1} + \dfrac{1}{R_2} + \dfrac{1}{R_3}$

Ohm's Law: Ohm's Law can be applied to the entire circuit or to any resistor in the circuit.

$V_{total} = I_{total}R_{total} \quad V_1 = I_1R_1 \quad V_2 = I_2R_2 \quad V_3 = I_3R_3$

Power: Power can also be applied to the entire circuit or to any resistor in the circuit.

$P_{total} = V_{total}I_{total} \quad P_1 = V_1I_1 \quad P_2 = V_2I_2 \quad P_3 = V_3I_3$

Sample Problem 2

Given the following parallel circuit, find:

a) the current through each resistor
b) the total current
c) the total power
d) the power in each resistor
e) the total resistance

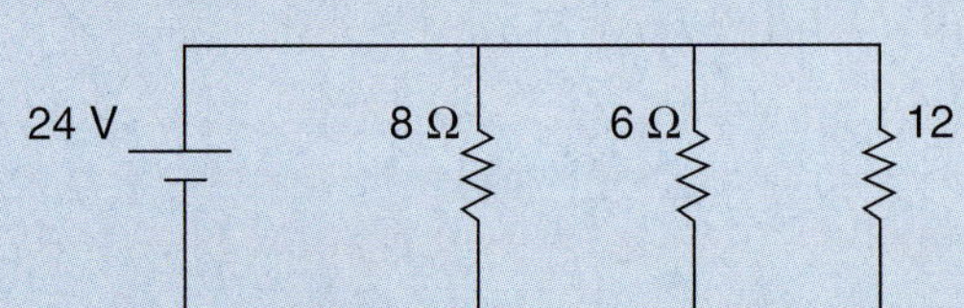

Strategy: This is a parallel circuit. The current follows different paths to each resistor. In a parallel circuit, the voltage drops across each resistor are equal. You can put this information in the diagram immediately.

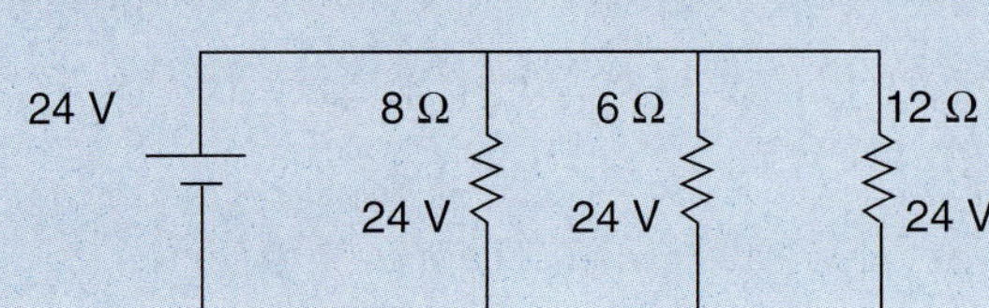

Once you know two of the four variables (V, I, P, R), you can find the other two variables. In this case, you know V and R. You can find the current using Ohm's Law for each resistor. You can find the total current by adding the currents through each resistor.

Use the power equation to calculate the power once you know the current.

a) $V = IR$

$I_1 = \frac{V_1}{R_1}$ $\quad I_2 = \frac{V_2}{R_2}$ $\quad I_3 = \frac{V_3}{R_3}$

$I_1 = \frac{24\text{ V}}{8\ \Omega}$ $\quad I_2 = \frac{24\text{ V}}{6\ \Omega}$ $\quad I_3 = \frac{24\text{ V}}{12\ \Omega}$

$I_1 = 3\text{ A}$ $\quad I_2 = 4\text{ A}$ $\quad I_3 = 2\text{ A}$

b) If the resistors have currents of 3 A, 4 A and 2 A, then the total current must be equal to

3 A + 4 A + 2 A = 9 A

c) $P_{total} = V_{total}I_{total}$

$P_{total} = (24\text{ V})(9\text{ A})$

$P_{total} = 216\text{ W}$

d) $P_1 = V_1I_1$ $\quad P_2 = V_2I_2$ $\quad P_3 = V_3I_3$

$P_1 = (24\text{ V})(3\text{ A})$ $\quad P_1 = (24\text{ V})(4\text{ A})$ $\quad P_1 = (24\text{ V})(2\text{ A})$

$P_1 = 72\text{ W}$ $\quad P_1 = 96\text{ W}$ $\quad P_1 = 48\text{ W}$

Notice the sum of the power in the resistors:

72 W + 96 W + 48 W = 216 W

That is the power supplied by the battery.

e) You can find the total resistance of the circuit easily by using Ohm's Law for the entire circuit.

$V_{total} = I_{total}R_{total}$

$R_{total} = \frac{V_{total}}{I_{total}}$

→

$$= \frac{24\ \text{V}}{9\ \text{A}}$$

$$= 2.67\ \Omega \text{ (or about } 3\ \Omega)$$

You can also find the total resistance by adding the individual resistors in parallel. Notice that in this equation, you are dealing with fractions.

$$\frac{1}{R_{total}} = \frac{1}{R_1} + \frac{1}{R_2} + \frac{1}{R_3}$$

$$= \frac{1}{8\ \Omega} + \frac{1}{6\ \Omega} + \frac{1}{12\ \Omega}$$

Using the least common denominator of 24:

$$\frac{1}{R_{total}} = \frac{3}{24\ \Omega} + \frac{4}{24\ \Omega} + \frac{2}{24\ \Omega}$$

$$= \frac{9}{24\ \Omega}$$

$$R_{total} = \frac{24\ \Omega}{9}$$

$$= 2.67\ \Omega$$

This agrees with the simpler method above, as it must.

Reflecting on the Activity and the Challenge

Part of the problem you are facing with the electrical system for the Homes For Everyone (HFE) dwellings is to assure that the people who live in them will not exceed the 2400 W power limit of the system by having too many appliances in use at any one time. Depending on what you choose to include in the HFE appliance package, it may be necessary to use switching devices—automatic, manual, or both—to assure that the people who live in the homes will stay within the power limit as they use their appliances.

Perhaps you could also use automatic switches as "fail safe" devices to prevent accidentally using up too much electrical energy by, for example, forgetting to shut off lights that are not in use.

Physics To Go

1. In your log, describe several possibilities for using switching devices to address the power limit problem in your universal dwelling. Write your ideas in your log.

2. Electric switches are available which act as timers to turn appliances on and off at chosen times or for chosen intervals. Identify one or more ways a timer switch would be useful in an HFE dwelling.

3. Look at the wiring diagrams shown. Copy each into your log. Position and draw a switch in each circuit which would allow you to turn two lights on, and leave one light off.

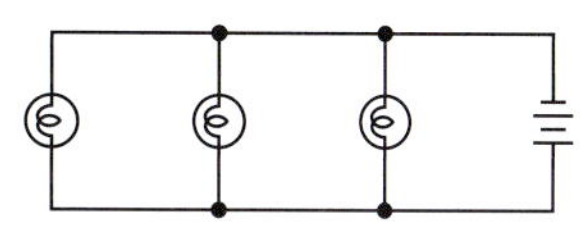

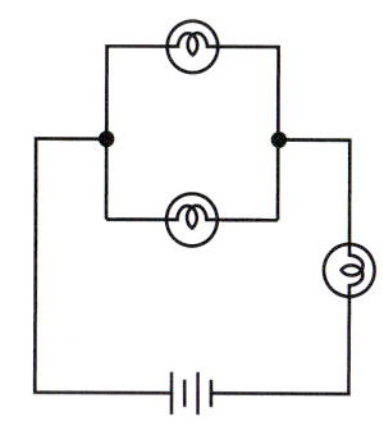

4. A 12-Ω, 14-Ω, and 4-Ω resistor are connected in series to a 12-V power supply. Find:

   a) the total resistance of the circuit
   b) the current flowing through each resistor
   c) the voltage drop across each resistor

5. Three resistors of 8 Ω, 6 Ω, and 4 Ω are connected in series to a battery of six 1.5-V dry cells, which are also connected in series.

   a) Draw a circuit diagram for this situation.
   b) Calculate the total voltage provided by the battery.
   c) Calculate the total resistance.
   d) Find the total current.
   e) What is the voltage drop across each individual resistor?

6. The following chart represents a set of three resistors arranged in series. Draw a diagram of the circuit. Then use the facts that you know about series circuits to fill in the empty spaces.

	V	I	R
Total	120 V		
R_1			10 Ω
R_2			5 Ω
R_3			25 Ω

7. A 6-Ω, 3-Ω, and 18-Ω resistor are connected in parallel to an 18-V power supply. Find:
   a) the total resistance of the circuit
   b) the total current through the circuit
   c) the current flowing through each resistor

8. You are given three 10-Ω resistors by your teacher. You are told to arrange them in the following ways. Sketch a diagram for each arrangement. What values will you have for total resistance in each case?
   a) All three resistors in series.
   b) All three resistors in parallel.
   c) One resistor in series and two in parallel.
   d) Two resistors in series with one in parallel.

9. The following chart represents a set of three resistors arranged in parallel. Draw a diagram of the circuit. Then use the facts that you know about parallel circuits to fill in the empty spaces.

	V	*I*	*R*
Total	24 V		
R_1			18 Ω
R_2			12 Ω
R_3			36 Ω

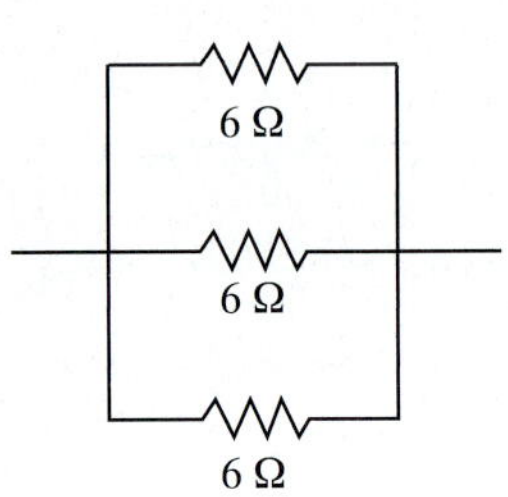

10. Which circuit segment below has the same total resistance as the circuit segment shown in the diagram to the left?

A
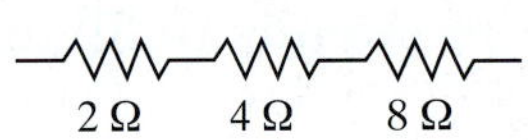

C
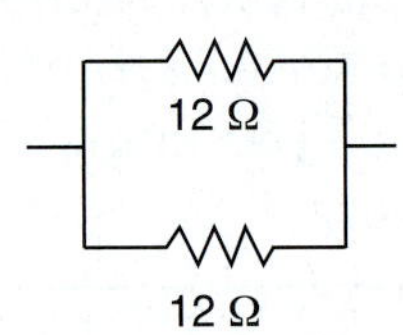

B
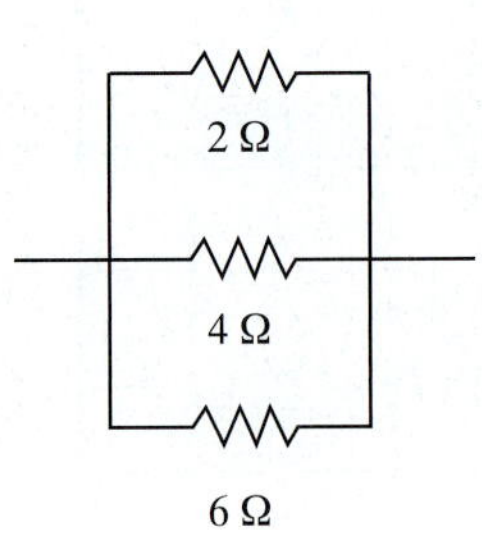

D
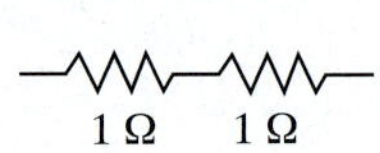

11. The diagram to the right shows the current in three branches of a direct-current electric circuit. What is the value and the direction of the current between junction P and point W?

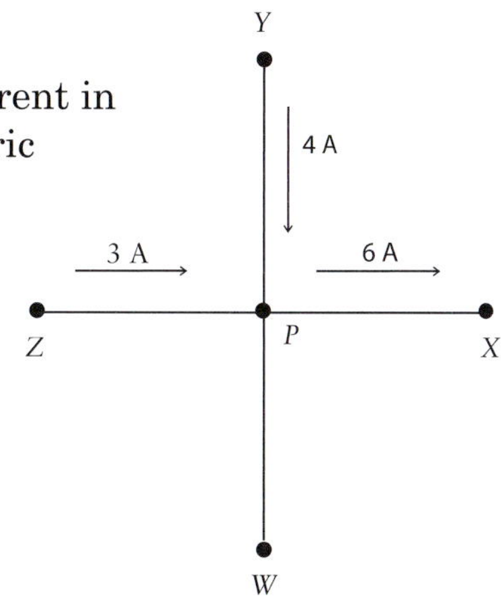

12. What is the current in the circuit represented in the diagram below?

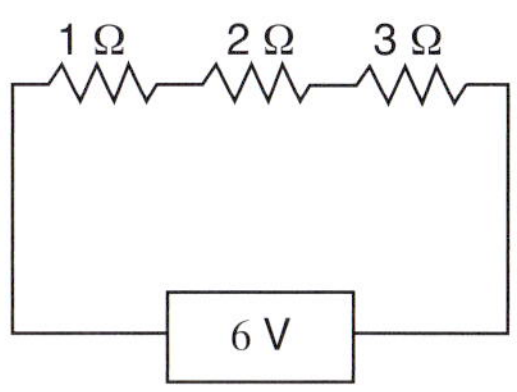

13. What is the current in the ammeter A in the diagram below?

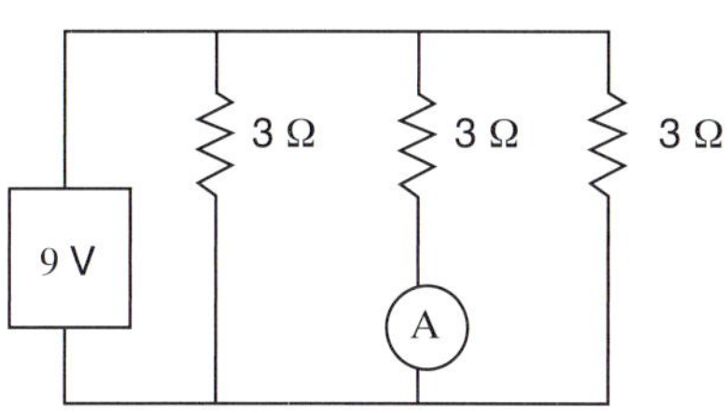

Stretching Exercises

Shop, in a store or catalog, for electrical switching devices controlled by a variety of conditions such as light/dark, high/low temperature, motion, etc. The devices may be either "built-in" to appliances (example: waterbed thermostat) or separate (example: a clock timer designed to control appliances plugged into it). Find as many different kinds of switching devices as you can, and note which ones may be useful for the HFE electrical system, and for what purpose. In your log, write a brief report on your findings.

Activity 6 Cold Shower

GOALS

In this activity you will:

- Calculate the heat gained by a sample of water.
- Calculate the electrical energy dissipated by a resistor.
- Recognize and quantify an ideal mechanical/electrical equivalent of heat.
- Calculate the efficiency of a transformation of electrical energy to heat.
- Explore the power ratings and energy consumption levels of a variety of electrical appliances.

What Do You Think?

The entire daily energy output of a Homes For Everyone (HFE) generator would not be enough to heat water for an average American family for a day.

- **If an electrical heating coil (a type of resistor) were submerged in a container of water, and if a current were to flow through the coil to make it hot, what factors would affect the temperature increase of the water? Identify as many factors as you can, and predict the effect of each on the water temperature.**

Record your ideas about this question in your *Active Physics* log. Be prepared to discuss your responses with your small group and the class.

For You To Do

1. Assemble and use an electric calorimeter according to the directions given by your teacher. Add a measured amount of cold tap water to the calorimeter.

   a) Record the mass of the water. (You will need to find the mass of an empty container as well as the mass of the container and the measured amount of water.)

   b) Measure and record the temperature of the water.

   c) Record the watt rating of the resistor that will be used to heat the water.

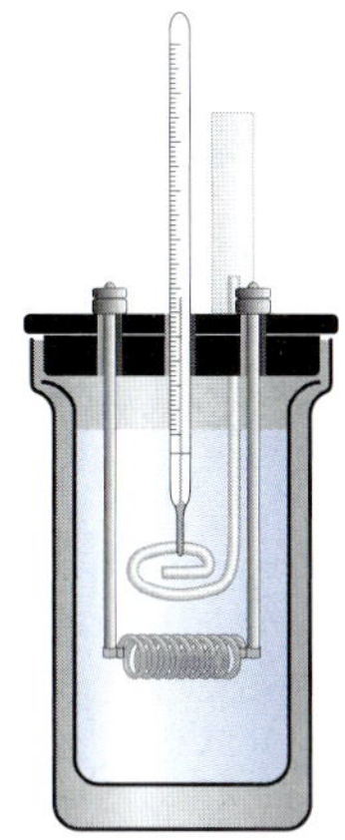

Always make sure the coil (or heater) is completely submerged in the water.

Promptly dry up any spilled water.

2. Mark the time at which you begin sending electric current through the resistor. Keep the electric heater operating for the amount of time recommended by your teacher. When you stop the current, note the time, stir the water, and measure the final, maximum temperature of the water.

Do not try to stir the water with the thermometer. If the thermometer should break, immediately notify your instructor.

Mass of water (grams):

Cold water temperature (°C):

Resistor power (watts):

Heating time (seconds):

Final water temperature (°C):

   a) In your log, record the final water temperature (°C) and the heating time (seconds).

3. The heat energy gained by an object can be found using the equation:

   Heat energy = mass of object × specific heat of material × temperature change

   Use the equation to calculate the heat energy gained by the water in the calorimeter. The specific heat of water is 4180 J/kg°C.

   a) Show your calculation in your log.

4. Power rating, in watts, is expressed as the amount of energy that something consumes per unit of time. Mathematically, power is expressed as

   $$\text{Power} = \frac{\text{energy}}{\text{time}}$$

   Then it is true that: Energy = power × time

   Energy can be expressed in kiloWatt-hours (kWh) if power is in kiloWatts and time is in hours, OR energy can be expressed in joules, if power is expressed in watts, and time in seconds. Use the equation to calculate the electrical energy, in joules, consumed by the resistor in the calorimeter.

   a) Show your calculation in your log.

5. The energy calculated from the temperature change of the water should be equal to the energy calculated from the power rating of the appliance and the time.

a) Compare these two values for the heat energy. If the values are not the same, what must you conclude?

b) Where could the "lost" energy have gone?

6. If all of the electrical energy did not heat up the water, then you know your system for heating water is less than 100% efficient. Calculate the efficiency of your water heating device. (Hint: If the appliance were 100% efficient, all of the electrical energy would have heated the water; if the appliance were 50% efficient, half of the electrical energy would have heated the water.)

a) Record your calculations in your log. How efficient was the calorimeter?

PHYSICS TALK

What is the world's greatest invention? A comedian once replied: "The thermos. When you put a cold drink in it, it stays cold. When you put a hot drink in it, it stays hot. How does the thermos know?"

A cup of hot water left on a table cools down. A cup of cold water left on the same table warms up. How can you change the temperature or keep the temperature constant?

Thermal energy is a form of energy. Energy in one form must come from energy in another form. Energy must be conserved. If you wish to heat up water, you must supply it with a source of energy like a flame, an electrical heater, or a hot metal.

If a system is isolated from outside sources of energy, one part of the system may warm up and another part may cool down, but the total energy must remain constant.

In this activity, you calculated the electrical energy that was used to heat the water. Energy can be calculated using the following equation:

$$E = VIt$$

where E is the energy in joules (J),

V is the voltage in volts (V),

I is the current in amperes (A), and

t is the time in seconds (s).

Since $P = VI$, the equation may also be written as:

$$E = Pt$$

Sample Problem I

A 12-V starter battery in a car supplies 48 A of current to the starter. If the starter draws energy for 15 s, how much energy does the starter use?

Strategy: You can use the equation to calculate energy.

Givens:

$V = 12\text{ V}$

$I = 48\text{ A}$

$t = 15\text{ s}$

Solution:

$$\begin{aligned} E &= VIt \\ &= (12\text{ V})(48\text{ A})(15\text{ s}) \\ &= 8640\text{ J (or about 8600 J)} \end{aligned}$$

Recall that volts are joules per coulomb and amperes are coulombs per second. When you multiply these together you get joules per second, the power unit. Multiplying this by seconds gives you joules, units of energy.

Heat energy is measured in units of joules (J). The heat energy gained by an object can be found using the following equation:

Heat energy = mass of object × **specific heat** of material × temperature change

Physics Words

specific heat: the amount of energy required to raise the temperature of 1 kg of a material by 1°C.

Where heat energy is expressed in joules (J),

mass is in kilograms (kg),

specific heat is in joules per kilogram degree Celsius (J/kg°C), and temperature is in degrees Celsius (°C).

You may also see this equation written as follows:

$$E_H = mc\Delta T$$

The symbol for specific heat is *c*. The symbol Δ represents a change in something, in this case, temperature.

To calculate the heat energy gained by a material, you must know the specific heat of the material. Different materials require different amounts of energy to raise the temperature of a given mass of the substance. Different substances have different capacities to hold heat. Water holds heat very well, whereas silver does not hold heat well. Water is said to have a higher specific heat than silver. The specific heat of a material is the amount of energy required to raise 1 kg of the material by 1°C.

The specific heat of water is 4180 J/kg°C.

Sample Problem 2

You apply 8640 J of energy to 100.0 g of water. How much would it change the temperature of the water?

Strategy: The amount of temperature change depends on several factors. One is the specific heat of the material with which you are working. In this case, the material is water. The other factor is the amount of material you have.

Givens:

$E = 8640$ J

$m = 100.0$ g $= 0.10$ kg

$c = 4180$ J/kg°C

→

Solution:

$$E = mc\Delta T$$

$$\Delta T = \frac{E}{mc}$$

$$\Delta T = \frac{8640 \text{ J}}{(0.10 \text{ kg})(4180 \text{ J/kg°C})}$$

$$= 20.6°\text{C (or about } 21°\text{C)}$$

Reflecting on the Activity and the Challenge

In this activity your knowledge of how to calculate electric power consumption was extended to include how to calculate electric energy consumption. You also learned that heating water electrically requires a lot of energy and can be quite inefficient. All of this knowledge applies directly to the selections you will make for electrical appliances to be used in the universal dwelling.

Physics To Go

1. The calorimeter did not allow the water to trap 100% of the energy delivered to it by the resistor. Some of the heat energy probably escaped from the water. Identify and explain ways in which you think heat energy may have escaped from the water, reducing the efficiency of the calorimeter.

2. The calorimeter used in this activity can be thought of as a scaled-down, crude version of a household hot water heater. The efficiencies of hot water heaters used in homes range from about 80% for older models to as much as 92% for new, energy-efficient models. Identify and explain ways in which you think heat escapes from household hot water heaters, and how some of the heat loss could be prevented.

3. From what you have learned so far, discuss the possibilities for providing electrically heated water for Homes For Everyone (HFE). Is a standard water heater of the kind used in American homes desirable, or possible, for HFE? What other electrical options exist for accomplishing part or all of the task of heating water for HFE?

4. For most Americans, the second biggest energy user in the home, next to the heating/air conditioning system, is the water heater. A family of four that heats water electrically (some use gas or oil to heat water) typically spends about \$35 per month using a 4500-W heater to keep a 160-L (40 gallon) tank of water hot at all times. The water is raised from an average inlet temperature of 10°C (50°F) to a temperature of 60°C (140°F), and the average family uses about 250 L (60 gallons) of hot water per day for bathing and washing clothes and dishes.

   In the above description, explain what each of the following numbers represents: 35, 4500, 160, 40, 10, 60, 50, 140, 250, 60.

5. Make a list with three columns in your log. The first column will be Name of Appliance, the second column will be Power (Watts), the third column will be Form of Energy Delivered.

   Choose 5 appliances from the list of Home Electrical Appliances beginning on page 496 that have a wide range of power. Use information from the list of appliances to fill in the first two columns. In the third column, write the form of energy (heat, light, motion, sound, etc.) that you think each appliance is designed to deliver.

   What pattern, if any, do you think exists between the power rating of an appliance and the form of energy it is designed to provide? Explain your answer.

6. Make a new list with six columns similar to the one shown on the next page.

   Choose 5-10 appliances from the list. Record the name and power rating of that appliance. Record the approximate time that this appliance is used in one day. Calculate the time that this appliance is used in one month (assuming that a month has 30 days). Calculate the electrical energy that the appliance consumes.

Appliance	Power (Watts)	#hours/ day (est.)	#hours/ month	Energy/month (Wh/month)	Energy/month (kWh/month)

The energy in watt-hours (Wh) is found by multiplying the power times the time in hours in one month.

$$\text{Energy} = \text{power} \times \text{time}$$

Calculate the energy in kiloWatt-hours (kWh) by dividing the watt-hours by 1000 since there are 1000 W in 1 kW.

7. You use a 1500-W hair dryer for 5 minutes every morning to dry your hair.
   a) How much electrical energy are you changing to heat every day?
   b) If you could transfer all of that energy to heating up water for coffee, how much water could you heat from 20°C to 90°C?

8. Copper wire has a specific heat of 387 J/kg°C. How many joules of energy would it take to heat 200 g of this copper wire from 20°C to 150°C?

9. If 42 J of energy are supplied to a 20-g sample of mercury (c = 140 J/kg °C), what is the temperature change of the sample?

Stretching Exercises

1. Find out about EnergyGuide® labels. Recently, the United States government established a federal law that requires EnergyGuide labels to be displayed on major appliances such as water heaters, refrigerators, freezers, dishwashers, clothes washers, air conditioners, furnaces, and heat pumps. The bright yellow EnergyGuide label allows consumers to compare the energy costs and efficiencies of appliances. Visit a store where appliances are sold and record the information given on the EnergyGuide labels of competing brands of one kind of appliance, such as water heaters. Prepare a short report on which appliance you would purchase, and why.

2. Research ways to reduce the amount of electrical energy needed to provide hot water for your own home or an HFE dwelling. Some possibilities may include (a) using solar energy and/or a "tempering tank" to heat the water partially, followed by "finishing off" the heating electrically, and (b) tankless "instant" water heaters which use electricity to heat water, but only when hot water is needed. Prepare a report on your findings.

Home Electrical Appliances

Average Power and Average Monthly Energy Use for a Family of Four

Family Data

	Power (watts)	Energy/mo. (kWh/mo.)
Big Appliances		
Air Conditioner		
(Room)	1360	
(Central)	3540	
Clothes Washer	512	
Clothes Dryer	5000	
Dehumidifier	645	
Dishwasher	1200	
Freezer	400	
Humidifier	177	
Pool Filter	1000	
Kitchen Range	12,400	
Refrigerator	795	
Space Heater	1500	
Waterbed	350	
Water Heater	4500	
Small Refrigerator	300	
Lights & Minor Appliances (combined)		
Kitchen		
Baby Food Warmer	165	
Blender	300	
Broiler (portable)	1200	
Can Opener	100	
Coffee Maker		
Drip	1100	
Percolator	600	
Corn Popper		
Oil-type	575	
Hot Air-type	1400	
Deep Fryer	1500	
Food Processor	370	
Frying Pan	1200	
Garbage Disposal	445	
Sandwich Grill	1200	
Hot Plate	1200	
Microwave Oven	750	
Mixer	150	

Home Electrical Appliances

	Power (watts)	Energy/mo. (kWh/mo.)
Roaster	1400	
Rotisserie	1400	
Slow Cooker	200	
Toaster	1100	
Toaster Oven	1500	
Trash Compactor	400	
Waffle Iron	1200	
Entertainment		
Computer	60	
Radio	70	
Television	90	
Stereo	125	
VCR	50	
Personal Care		
Air Cleaner	50	
Curling Iron	40	
Hair Dryer	1200	
Hair Rollers	350	
Heat Blanket	170	
Heat Lamp	250	
Heat Pad	60	
Iron	1100	
Lighted Mirror	20	
Shaver	15	
Sun Lamp	300	
Toothbrush	1	
Miscellaneous		
Auto Engine Heater	850	
Clock	3	
Drill (1/4")	250	
Fan (attic)	375	
Fan (window)	200	
Heat Tape	240	
Sewing Machine	75	
Skill Saw	1000	
Vacuum Cleaner	650	
Water Pump (well)	335	

Please note: Average values of power are shown. The power of a particular appliance may vary considerably from the value in the table. Energy use will vary with family size (a four-member family is assumed for the tabled values), personal preferences and habits, climate, and season. Similar information in greater detail is available free upon request from most electric utilities.

Activity 7 Pay Up

GOALS

In this activity you will:

- Analyze household electric bills, relating energy consumed, billing rate, and cost.
- Compare the costs of operating a variety of electrical appliances in terms of power ratings, amount of time each appliance is used, and billing rate.
- Appreciate energy consumption differences across cultures.

What Do You Think?

Eggs are priced by the dozen, and electricity is priced by the kiloWatt-hour.

- **What factors determine the amount of money the electric company charges you for the use of an appliance?**

Record your ideas about this question in your *Active Physics* log. Be prepared to discuss your responses with your small group and the class.

For You To Do

1. Look at the copies of the electric bills provided. If possible, also obtain a copy of the monthly electric bill for your home or that of an acquaintance. Compare the electric bills provided below, or the ones from the homes of individuals in your group.

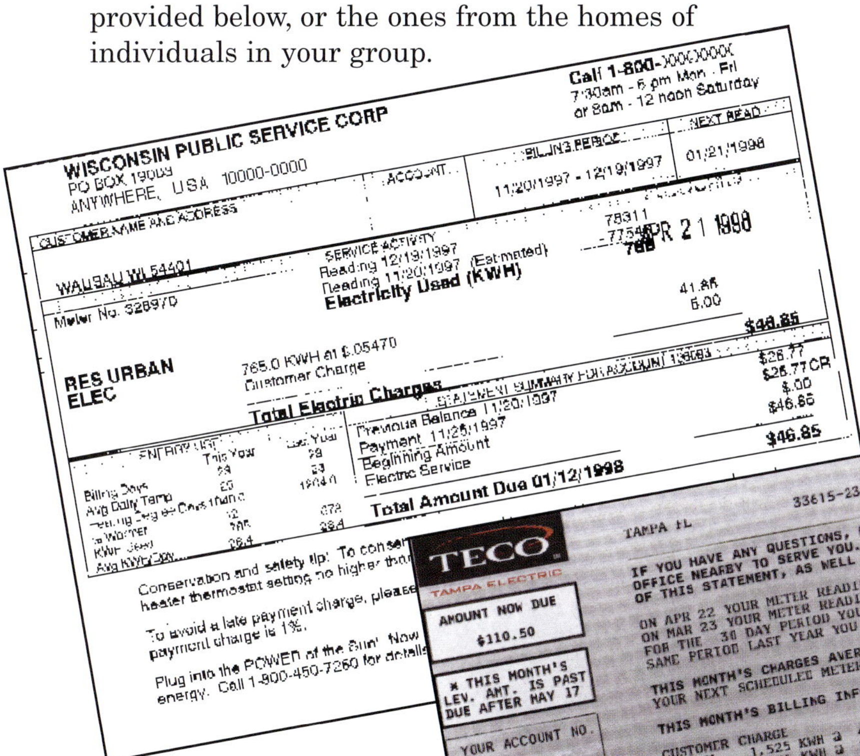

WISCONSIN PUBLIC SERVICE CORP
PO BOX 19003
ANYWHERE, USA 10000-0000

Call 1-800-XXXXXXX
7:30am - 6 pm Mon - Fri
or 8am - 12 noon Saturday

BILLING PERIOD 11/20/1997 - 12/19/1997
NEXT READ 01/21/1998

WAUSAU WI 54401
Meter No. 328970

SERVICE ACTIVITY
Reading 12/19/1997 78311
Reading 11/20/1997 (Estimated) 77546
Electricity Used (KWH) 765

RES URBAN ELEC
765.0 KWH at $.05470 41.85
Customer Charge 5.00
Total Electric Charges $46.85

STATEMENT SUMMARY FOR ACCOUNT
Previous Balance 11/20/1997 $26.77
Payment 11/26/1997 $26.77CR
Beginning Amount $.00
Electric Service $46.85
Total Amount Due 01/12/1998 $46.85

Conservation and safety tip: To conserve ... heater thermostat setting no higher than ...

To avoid a late payment charge, please ... payment charge is 1%.

Plug into the POWER of the Sun! Now ... energy. Call 1-800-450-7260 for details ...

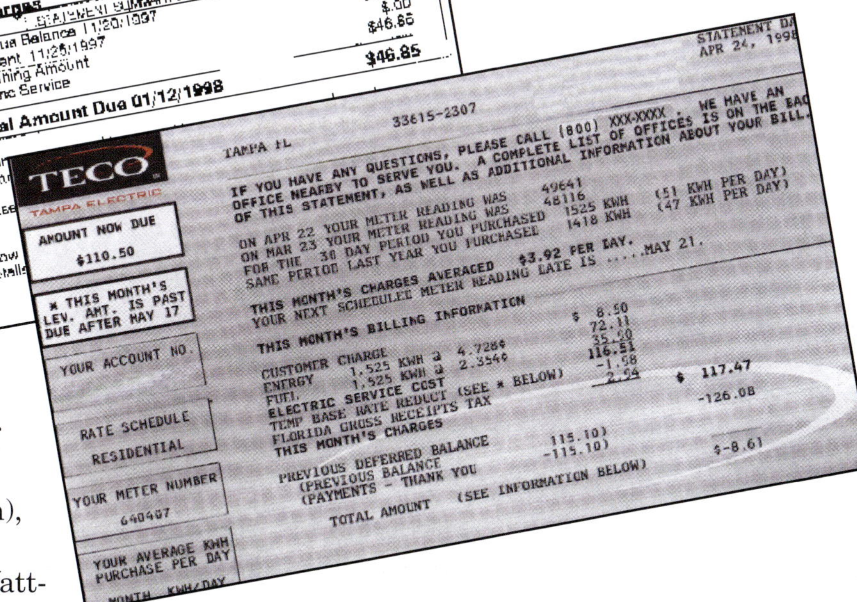

TECO
TAMPA ELECTRIC

STATEMENT DATE APR 24, 1998

TAMPA FL 33615-2307

AMOUNT NOW DUE $110.50

* THIS MONTH'S LEV. AMT. IS PAST DUE AFTER MAY 17

YOUR ACCOUNT NO.

RATE SCHEDULE RESIDENTIAL

YOUR METER NUMBER 640487

YOUR AVERAGE KWH PURCHASE PER DAY
MONTH KWH/DAY

IF YOU HAVE ANY QUESTIONS, PLEASE CALL (800) XXX-XXXX. WE HAVE AN OFFICE NEARBY TO SERVE YOU. A COMPLETE LIST OF OFFICES IS ON THE BACK OF THIS STATEMENT, AS WELL AS ADDITIONAL INFORMATION ABOUT YOUR BILL.

ON APR 22 YOUR METER READING WAS 49641
ON MAR 23 YOUR METER READING WAS 48116
FOR THE 30 DAY PERIOD YOU PURCHASED 1525 KWH (51 KWH PER DAY)
SAME PERIOD LAST YEAR YOU PURCHASED 1418 KWH (47 KWH PER DAY)

THIS MONTH'S CHARGES AVERAGED $3.92 PER DAY.
YOUR NEXT SCHEDULED METER READING DATE IS .....MAY 21.

THIS MONTH'S BILLING INFORMATION

CUSTOMER CHARGE	$ 8.50	
ENERGY 1,525 KWH @ 4.728¢	72.11	
FUEL 1,525 KWH @ 2.354¢	35.90	
ELECTRIC SERVICE COST	116.51	
TEMP BASE RATE REDUCT (SEE * BELOW)	-1.98	
FLORIDA GROSS RECEIPTS TAX	2.94	
THIS MONTH'S CHARGES		$ 117.47
PREVIOUS DEFERRED BALANCE		-126.08
(PREVIOUS BALANCE	115.10)	
(PAYMENTS - THANK YOU	-115.10)	
TOTAL AMOUNT (SEE INFORMATION BELOW)		$-8.61

a) What is the range of electrical energy, in kiloWatt-hours (kWh), used per month?

b) Is the cost per kiloWatt-hour the same on all bills, or are there differences?

c) What factors might account for differences among electric bills within your group? Identify as many factors as you can and explain what you think is the effect of each factor on the bill.

d) What is the average monthly amount of electrical energy used in the homes represented in your group, and how does it compare to the 90 kWh per month available to Homes For Everyone (HFE) dwellings?

2. In a previous activity, you estimated the time that an appliance is used, and the amount of energy required to use it for one month. Now you will calculate the expense of that energy. Combine the lists from the group members and calculate the expense of running the electrical appliances by using the following equation:

   Cost = Energy × Price per unit of energy

   The price for electricity can be taken from the electric bills that were provided.

   a) Record your calculations in your log.

3. Compare the cost of electricity in your town to other places in the United States. The "low" for electrical costs is approximately $0.03 per kWh. The "high" for electrical costs is approximately $0.16 per kWh.

   a) What are the high and low costs of the electrical bills that you examined?

   b) How might your consumption of electricity change if you had to pay for your own electrical bills?

FOR YOU TO READ

Determining Energy Consumption

The electrical energy consumed by appliances can be determined in different ways. If the power rating of an appliance is known, and if the power remains at a steady value while the appliance is in use (example: light bulb), the energy, in kWh, can be calculated by multiplying the power of the appliance, in kiloWatts, by the time, in hours, for which the appliance is used (Energy = Power × Time).

Determining the energy consumed by some appliances, however, is tricky, because the appliances vary in power while they are in use. For example, a refrigerator may cycle on and off under the control of a thermostat. Therefore, it is not operating during all of the time it is "plugged in" and the calculation described above would lead to a misleadingly high value for the heater's energy consumption. For such appliances whose power varies throughout time, a kiloWatt-hour meter is used to measure the total energy consumed, with variations in power throughout time taken into account. The same kind of meter, a kiloWatt-hour meter, is used by the power company to measure the total electrical energy used in your home. The meter is mounted somewhere at your home so that it can be read by the power company's "meter reader" person.

Another way to determine the energy used by electrical appliances is to use the experience of power companies or corporations that sell electrical supplies. Extensive lists of appliances, their power ratings, and each appliance's average energy use per month by a typical family are available free from such sources.

PHYSICS TALK

The cost of operating an electrical appliance can be calculated using the following equation:

Cost = Energy × Price per unit of energy

where energy is in kiloWatt-hours (1000 Wh = 1 kWh).

Sample Problem

An electric coffeemaker uses an average of 22.5 kWh of energy each month. If the family is charged \$0.12/kWh for electricity, what is the average monthly cost of operating the coffeemaker?

Strategy: Use the equation for cost.

Givens:

Energy = 22.5 kWh

Price = \$0.12/kWh

Solution:

Cost = Energy × Price per unit of energy

= 22.5 kWh x \$0.12/kWh

= \$2.70

Reflecting on the Activity and the Challenge

In this activity you learned that the average American family's energy use exceeds the maximum available for our HFE dwellings. Imagine how your own standard of living would change if you were restricted to 90 kWh per month.

The **Chapter Challenge** requires that you need to be thinking about two things at the same time as you select appliances for the universal dwelling: the power consumed by each appliance and the amount of time for which each appliance is used. Both need to be taken into account to stay within the power and energy limits of the HFE electrical system.

Physics To Go

1. A 1200-W hair dryer is used by several members of a family for a total of 30 minutes per day during a 30-day month. How much electrical energy is consumed by the hair dryer during the month? Give your answer in:

   a) watt-hours
   b) kiloWatt-hours

2. If the power company charges $0.15 per kWh for electrical energy, what is the cost of using the hair dryer in **Question 1** during the month? What is the cost for a year?

3. Not enough heat from the furnace reaches one bedroom in a home. The homeowner uses a portable 1350-W electric heater 24 hours per day to keep the bedroom warm during four cold winter months. At $0.12 per kiloWatt-hour, how much does it cost to operate the heater for the four months? (Assume two 30-day and two 31-day months.)

4. Prepare your personal list of electrical appliances to recommend to your group to be included in the HFE appliance package. Remember that you will have to justify why you chose a certain appliance. Be prepared to contribute your ideas to your group's decision-making process when completing the challenge.

5. A portable CD player is rated at approximately 20 W and uses 4 AA batteries.

   a) Estimate the number of hours that you can listen to the music on a CD player before the batteries need replacing.
   b) Calculate the energy requirements of the CD player.
   c) Estimate the cost of 4 AA batteries.
   d) Calculate the cost per kiloWatt-hour of a battery.
   e) Compare battery costs with the cost of electricity from the utilities (use approximately $0.10 per kiloWatt-hour).

Activity 8 More for Your Money

GOALS

In this activity you will:

- Measure and compare the energy consumed by appliances.
- Measure and compare the efficiencies of appliances.
- Choose an appropriate appliance for efficiency, based on tests performed.

What Do You Think?

Some hot water heaters and furnaces for homes are more than 90% efficient.

- **If high-efficiency appliances cost more, are they worth the added cost?**

Record your ideas about this question in your *Active Physics* log. Be prepared to discuss your responses with your small group and the class.

For You To Do

1. Place one liter of cold tap water in a beaker made of Pyrex® glass. Make sure the outside of the beaker is dry so it does not slip from your grasp. Measure the temperature of the water.

a) Record the temperature of the water in your log.

b) Record the quantity of water in milliliters and grams. (1 mL of water has a mass of 1 g.)

2. Place the beaker of water in a microwave oven of known power, in watts. Mark the time at which the oven is turned on at its highest power level. After two minutes, stop the time measurement. Carefully check that the beaker is not too hot to grasp and remove the beaker from the oven, stir the water, and check the water temperature, all as quickly as possible.

microwave oven power = _________ watts

time required to heat water = _________ seconds

final water temperature = _________ °C

a) Why is it important to complete the temperature measurement as quickly as possible? Record your answer in your log.

3. Prepare an identical Pyrex beaker containing the same amount of cold tap water, preferably at the same original temperature as the water used above.

a) Record the mass and temperature of the water in your log.

4. Have ready a hot plate that has not been turned on (a "cold" hot plate) and that has a known power rating, in watts. Also have ready a clock or stopwatch capable of measuring time, in seconds, for an interval of several minutes. Place the beaker of cold water on the hot plate, and mark the time at which the hot plate is turned on at its highest setting. Gently stir the water while it is heating and monitor the temperature of the water. When the temperature of the water has increased to the value of the water from the microwave, mark the time and shut the hot plate off.

a) Record the power of the hot plate, the time required to heat the water, and its final temperature in your log.

5. Repeat the process a third time with the coil.

   a) Record all your observations in your log.

6. Energy (in joules) = Power (in watts) × Time (in seconds)

   Calculate the energy used by each appliance to cause equal temperature increases in equal amounts of water.

   a) Show your calculations in your log.

   b) Which appliance is the "winner"? Choose a way to express a comparison of the performance of the three appliances.

   c) Was the method used to compare the three appliances fair? How could the fairness be improved?

   d) The beaker that served as the container for the water also was heated by the three appliances. Did it affect the outcome of the comparison? Might another kind of container be more or less effective to use with either appliance; for example, might using a metal pan as the container on a hot plate make the hot plate perform more efficiently?

Reflecting on the Activity and the Challenge

You know that the electrical appliances used in the universal dwelling cannot exceed 90 kWh per month of energy consumption. In this activity you discovered that some appliances are more efficient than others. That is, a highly efficient appliance can accomplish a task while using less electrical energy than a low-efficiency appliance for the same task. Obviously you will want to make a careful selection for each kind of appliance based on efficiency so that Homes For Everyone (HFE) inhabitants can have the greatest possible benefit from the electrical system.

Physics To Go

1. Are some cooking utensils (pots, pans, etc.) better than others for certain purposes? Write what you think about the effectiveness of different cooking utensils, and what you could do to find out about their comparative effects on efficiency.

2. Does either the hot plate or the microwave oven seem to be a good choice to include in the HFE appliance package? Why, or why not?

3. You probably have concluded that the most efficient appliance of the three tested is the one that used the least energy, but do you know the actual efficiency of the appliances? Explain how you could calculate the efficiencies, and try it.

4. Calculate the energy used, in joules, by each of the following:
   a) a 1500-W hair dryer operating for 3 minutes
   b) a 1200-W hair dryer operating for 4 minutes

5. If both situations described in **Question 4** result in the same dryness of hair, which hair dryer is more efficient?

6. A 10-A electric pencil sharpener is used 2 minutes every working day.
   a) Calculate the power of the electric pencil sharpener.
   b) How much energy is used by the sharpener in one day?
   c) How much energy could be saved in one year (assume 5 working days per week all year) by using a manual pencil sharpener?

PHYSICS AT WORK

The Schultzes

LIVING CLEAN AND FREE

Bob-o Schultze and his family live "off-the-grid," miles from the nearest electrical power line. Their 2,400 square foot home in northern California includes two computers, a dishwasher, a microwave oven, a washing machine, and a satellite television. It is entirely powered by clean, free, renewable energy.

"Living off-the-grid does not require any major lifestyle changes," explains Mr. Schultze who has been living without traditional energy for over 30 years. "It takes no more technical sophistication or work than the basic upkeep of any house or car."

The Schultzes use three natural power sources: water, wind, and Sun. There is a hydroelectric facility in a creek on the property, a wind machine behind the house, and photovoltaic facilities on tracks following the Sun. All three feed into one large industrial, deep-cycle, 6-V battery and there is never an electric bill. "We are fortunate to have a great site, but every site has at least one source of natural, renewable energy and most have two. The Sun is always an option and if conditions allow for a wind machine, which must be 20-30 feet higher than anything within 500 feet, the power of the wind can also be harnessed."

"I was inspired by the Grateful Dead," explains Bob-o whose name was created from Robert by children many years ago and has stuck. "I moved to a very rural area—30 miles from any electrical power—and I wanted my music. I started using regular D batteries, which drained very quickly, then a car battery, which drained almost as quickly, and then I developed a hydroelectric plant."

All of the appliances in the Schultze's home are the most energy-efficient available. "Compact fluorescent light uses only 25% of the power of standard lighting—and simply turning things off when they are not being used can save 500 watt-hours a day," says Bob-o, emphasizing that energy should be conserved regardless of its source. "We use nine to ten kiloWatt-hours per day, and the average for a family of four is about thirty. As more people become aware of the finite nature of traditional energy sources, the use of renewable energy will increase," claims Bob-o. The Schultzes are just a little ahead of many people in this regard.

Chapter 7 Assessment

You learned a great deal about electricity and electrical terms in this chapter. Read the **Scenario** from the beginning of the chapter once again. Do you now understand all the terms used in the description of the wind-generator system?

Now that you have completed the activities in this chapter, you are ready to complete the **Chapter Challenge**. You will be asked to do the following:

- **List the appliances to be included in the HFE "appliance package." Your list must be as comprehensive as possible, and it must be clear how each appliance will enhance the health or well-being of the people who live in the dwelling.**
- **Develop an outline of the training manual for HFE volunteers. Your manual must explain the difference between 2400 W and 3 kWh. It must also give clear examples of how use of the appliances in the package can be scheduled to stay within the power and energy limits of the electrical system on both a daily and a long-term basis.**

Review the criteria which you and your class established for how your appliance package and training manual will be graded. Your work will be judged on the basis of 100 points. If necessary, discuss the criteria once again, add details to the criteria if it would be helpful, and agree on the finalized point allocation.

Physics You Learned

Simple circuits

Generators

Series circuits

Parallel circuits

Power

$P = IV$ (watts)

Energy = Pt (kiloWatt-hours; joules)

Load limits

Fuses

Switches

Utility bills

Costs for electricity

Electrical efficiency

Chapter 8

TOYS FOR UNDERSTANDING

Scenario

In this *Active Physics* chapter, you will try to help educate children through the use of toys. With your input, the Homes for Everyone (HFE) organization has developed an appliance package that will allow families living in the "universal dwelling" to enjoy a healthy and comfortable lifestyle. The HFE organization would now like to teach the children living in these homes, and elsewhere, more about electricity and the generation of electricity. They hope that this may encourage interest in children to use electricity wisely, as well as encourage development of alternative sources for electrical energy by future generations.

The HFE organization will work with a toy company to provide kits and instructions for children to make toy electric motors and generators. These toys should illustrate how electric motors and generators work and capture the interest of the children.

In an effort to help others, people often make changes or introduce new products without considering the personal and cultural impact on those whom they are trying to assist. If you ever become involved in a self-help community group, such as HFE, it would be important for you to work together with the people you are assisting to assess their needs, both personal and cultural. Although that is not possible given your limited time in class, you should recognize the need for collaborative teamwork in evaluating the impact of any new product on an established community.

Challenge

Your task is to prepare a kit of materials and instructions that the toy company will manufacture. Children will use these kits to make a motor or generator, or a combination electric motor/generator. It will serve both as a toy and to illustrate how the electric motors in home appliances work or how electricity can be produced from an energy source such as wind, moving water, a falling weight, or some other external source.

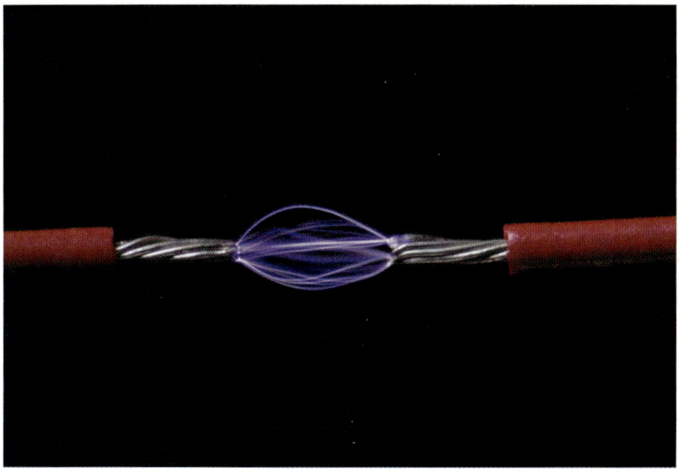

Criteria

Your work will be judged by the following criteria:

- **(30%) The motor/generator is made from inexpensive, common materials, and the working parts are exposed but with due consideration for safety.**
- **(40%) The instructions for the children clearly explain how to assemble and operate the motor/generator device, and explain how and why it works in terms of basic principles of physics.**
    - **(30%) If used as a motor, the device will operate using a maximum of four 1.5-V (volt) batteries (D cells), and will power a toy (such as a car, boat, crane, etc.) that will be fascinating to children.**

    OR

    - **(30%) If used as a generator, the device will demonstrate the production of electricity from an energy source such as wind, moving water, a falling weight, or some other external source and be fascinating to children.**

Activity 1 The Electricity and Magnetism Connection

GOALS

In this activity you will:

- Describe electric motor effect and generator effect in terms of energy transformations.
- Use a magnetic compass to map a magnetic field.
- Describe the magnetic field near a long, straight current-carrying wire.

What Do You Think?

Generators produce electricity. Motors use electricity.

- **What is the significance of motors and generators to your standard of living? That is, how would your life be different if you had no motors or generators?**

Write your answer to these questions in your *Active Physics* log. Be prepared to discuss your ideas with your small group and other members of your class.

For You To Do

1. Set up the equipment as shown in the diagram, or as directed by your teacher.

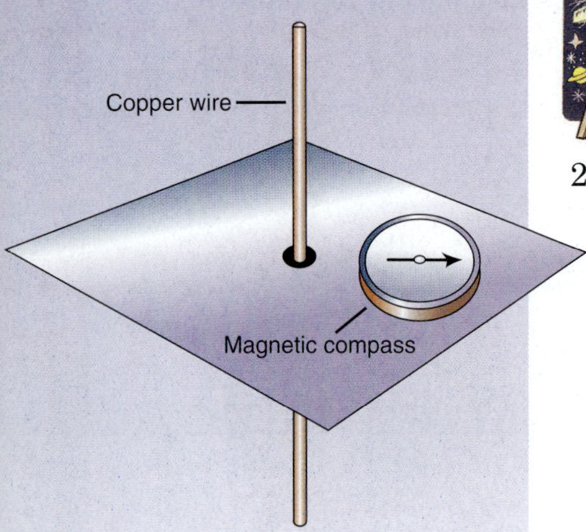

2. The needle of a compass is a balanced magnet. It can be used as a magnetic field detector. If any magnet is present, the compass will respond. It usually aligns itself with Earth's magnetic field. With no current flowing in the wire, verify that the compass always points in the same direction, north, no matter where it is placed on the horizontal surface.

a) Sketch the compass direction at different places on the horizontal surface in your log.

3. Bring another type of magnet, such as a bar magnet, into the area near the compass needle.

a) Describe your observations in your log.

b) What happens to the dependable north-pointing property of a compass when the compass is placed in a region where magnetic effects, in addition to Earth's magnetic field, exist?

4. You will now make a map of the magnetic field of the bar magnet. Place the magnet on another piece of paper and trace its position. Place the compass at one location and note the direction it points. Remove the compass.

a) Put a small arrow at the location from which you removed the compass to signify the way in which it pointed.

b) Place the compass at a second location about at the tip of the first arrow. Remove the compass and place another small arrow in this location to signify the way in which the compass pointed.

c) Repeat the process at an additional 20 locations to get a map of the magnetic field of a bar magnet. Tape the piece of paper of the map in your log.

Do not adjust the power supply settings provided by your teacher.

5. Return the compass to the horizontal surface surrounding the wire. Observe the orientation of the compass. Send a current through the wire. The direction of the flow of electrons which make up the current in the wire is from the negative terminal of the power supply to the positive terminal. Move the compass to different locations on the horizontal surface, observing the direction in which the compass points at each location. Make observations on all sides of the wire, and at different distances from the wire.

a) Record how the compass was oriented when the bar magnet was removed.

b) Describe any pattern that you observe about how the compass behaves when it is near the current-carrying wire. Use a sketch and words to describe your observations in your log.

c) From your observations, what effect does the electric current appear to have on the wire?

6. Reverse the direction of the current in the wire by exchanging the contacts of the power supply. Repeat your observations.

   a) Describe the results.

   b) Make up a rule for remembering the relationship between the direction of the current in a wire and the direction of the magnetism near the wire (i.e., when the current is up, the magnetic field . . .). Anyone told your rule should be able to use it with success. Write your rule in your log. Include a sketch. (Hint: One of the rules that physicists use makes use of your thumb and fingers.)

Reflecting on the Activity and the Challenge

This activity has provided you with knowledge about a critical link between electricity and magnetism, which is deeply involved in your challenge to make a working electric motor or generator. The response of the compass needle to a nearby electric current showed that an electric current itself has a magnetic effect which can cause a magnet, in this case a compass needle, to experience force. You have a way to go to understand and be able to be "in control" of electric motors and generators, but you've started along the path to being in control.

Physics To Go

1. If 100 compasses were available to be placed on the horizontal surface to surround the current-carrying wire in this activity, describe the pattern of directions in which the 100 compasses would point in each of the following situations:

   a) no current is flowing in the wire

   b) a weak current is flowing in the wire

   c) a strong current is flowing in the wire

2. If a vertical wire carrying a strong current penetrated the floor of a room, and if you were using a compass to "navigate" in the room by always walking in the direction indicated by the north-seeking pole of the compass needle, describe the "walk" you would take.

3. Use the rule which you made up for remembering the relationship between the direction of the current flowing in a wire and the direction of the magnetic field near the wire to make a sketch showing the direction of the magnetic field near a wire which has a current flowing:

   a) downward
   b) horizontally

4. Physicists remember the orientation of the magnetic field of a current by placing their left thumb in the direction of the electric current and noting whether the fingers of their left hand curve clockwise or counterclockwise. Copy the following diagrams into your log. Use this rule to sketch the direction of the magnetic field in each case.

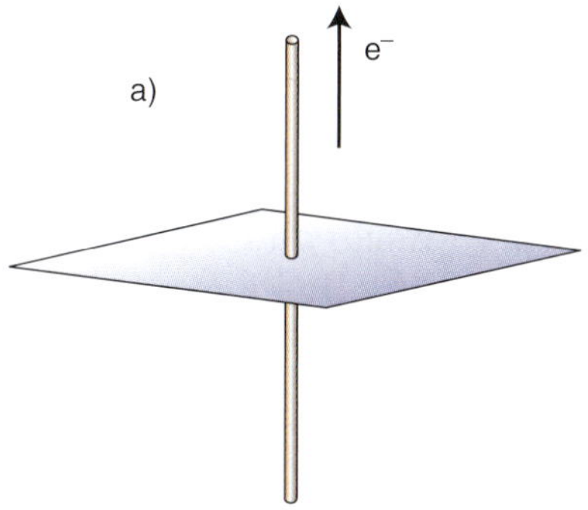

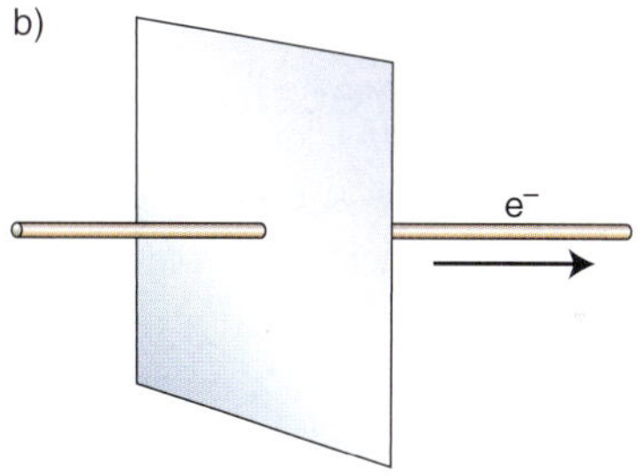

5. Imagine that a second vertical wire is placed in the apparatus used in this activity, but not touching the first wire. There is room to place a magnetic compass between the wires without touching either wire. If a compass were placed between the wires, describe in what direction the compass would point if the wires carry equal currents:

   a) which are in opposite directions
   b) which are in the same direction

6. A hollow, transparent plastic tube is placed on a horizontal surface as shown in the diagram. A wire carrying a current is wound once around the tube to form a circular loop in the wire. In what direction would a compass placed inside the tube point? (Plastic does not affect a compass; only the current in the wire loop will affect the compass.)

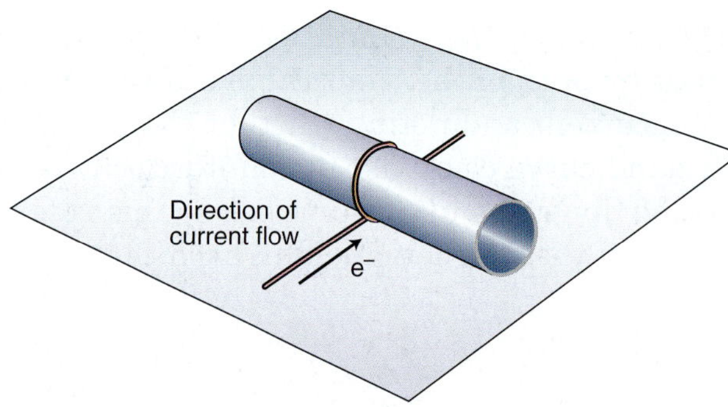

Stretching Exercises

Use a compass to search for magnetic effects and magnetic "stuff." As you know, a compass needle usually aligns in a north-south direction (or nearly so, depending on where you live). If a compass needle does not align north-south, a magnetic effect in addition to that of the Earth is the cause, and the needle is responding to both the Earth's magnetism and some other source of magnetism. Use a compass as a probe for magnetic effects. Try to find magnetic effects in a variety of places and near a variety of things where you suspect magnetism may be present. Try inside a car, bus, or subway. The structural steel in some buildings is magnetized and may cause a compass to give a "wrong" reading. Try near the speaker of a radio, stereo, or TV. Try near electric motors, both operating and not operating.

Do not bring a known strong magnet close to a compass, because the magnet may change the magnetic alignment of the compass needle, destroying the effectiveness of the compass.

Make a list of the magnetic objects and effects that you find in your search.

Activity 2 Electromagnets

GOALS

In this activity you will:

- Describe and explain the magnetic field of a current-carrying solenoid.
- Compare the field of a solenoid to the field of a bar magnet.
- Identify the variables of an electromagnet and explain the effects of each variable.

What Do You Think?

Large electromagnets are used to pick up cars in junkyards.

- **How does an electromagnet work?**
- **How could it be made stronger?**

Write your answer to these questions in your *Active Physics* log. Be prepared to discuss your ideas with your small group and other members of your class.

For You To Do

1. Wind 50 turns of wire on a drinking straw to form a solenoid as shown in the diagram on the next page. Use sandpaper to carefully clean the insulation from a short section of the wire ends to allow electrical connection of the solenoid to the generator.

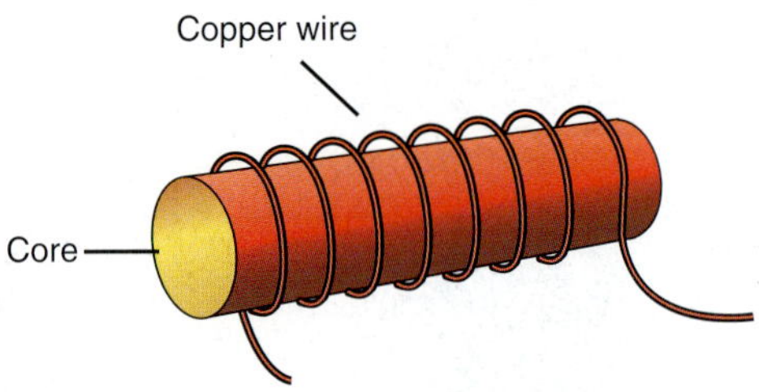

2. Carefully connect the wires from the generator to the wire ends of the solenoid. Bring one end of the solenoid near the magnetic compass and crank the generator to send a current through the solenoid. Observe any effect on the compass needle. Try several orientations of the solenoid to produce effects on the compass needle.

   a) Record your observations in your log.

   b) How can you tell the "polarity" of an electromagnet; that is, how can you tell which end of an electromagnet behaves as a north-seeking pole?

3. Predict what you can do to change the polarity of an electromagnet.

   a) Write your answer in your log.

   b) Test your prediction.

4. Use the solenoid wound on the drinking straw as an electromagnet to pick up paper clips.

   a) Record your observations in your log.

5. Carefully, slip a nail into the drinking straw to serve as a new core. Again, test the effect on a compass needle.

   a) Record your observations in your log.

6. Use the solenoid wound on the nail to pick up paper clips.

   a) Record your observations in your log.

   b) What evidence did you find that the choice of core material for an electromagnet makes a difference?

7. Predict what will happen when you increase the current running through the coiled wire solenoid. This can be done by increasing the speed at which you crank the generator.

   a) Write your answer in your log.

   b) Test your prediction by measuring how many paper clips can be picked up.

8. Predict what will happen when you increase the number of turns of wire forming the solenoid.

   a) Write your answer in your log.

   b) Test your prediction by measuring how many paper clips can be picked up.

Reflecting on the Activity and the Challenge

An electromagnet, often constructed in the shape of a solenoid, and having an iron core, is the basic moving part of many electric motors. In this activity you learned how the amount of current and the number of turns of wire affect the strength of an electromagnet. You will be able to apply this knowledge to affect the speed and strength with which an electric motor of your own design rotates.

Physics To Go

1. Explain the differences between permanent magnets and electromagnets.
2. The diagram below shows an electromagnet with a compass at each end. Copy the diagram and indicate the direction in which the compass needles will be pointing when a current is generated.

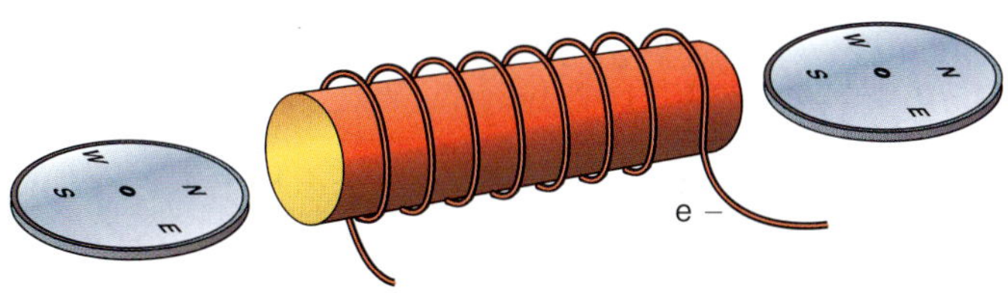

3. Which of the following will pick up more paper clips when an electric current is sent through the wire:
   a) A coil of wire with 20 turns, or a coil of wire with 50 turns?
   b) Wire wound around a cardboard core, or wire wound around a steel core?
4. Explain conditions necessary for two electromagnets to attract or repel one another, as do permanent magnets when they are brought near one another.
5. Explain what you think would happen if, when making an electromagnet, half of the turns of wire on the core were made in one direction, and half in the opposite direction.

Stretching Exercises

1. Find out how both permanent magnets and electromagnets are used. Do some library research to learn how electromagnets are used to lift steel in junkyards, make buzzers, or serve as part of electrical switching devices called "relays." For other possibilities, find out how magnetism is used in microphones and speakers within sound systems, or how "super-strong" permanent magnets made possible the small, high-quality, headset speakers for today's portable radio, tape and CD players. Prepare a brief report on your findings.

2. Do some research to find out about "magnetic levitation." "Maglev" involves using super-conducting electromagnets to levitate, or suspend objects such as subway trains in air, thereby reducing friction and the "bumpiness" of the ride.

   a) What possibilities do "maglev" trains, cars, or other transportation devices have for the future?
   b) What advantages would such devices have?
   c) What problems need to be solved? Prepare a brief report on your research.

3. Identify as many variables as you can that you think will affect the behavior of an electromagnet, and design an experiment to test the effect of each variable. Identify each variable, and describe what you would do to test its effects. After your teacher approves your procedures, do the experiments. Report your findings.

Activity 3 Detect and Induce Currents

What Do You Think?

In 1820, the Danish physicist Hans Christian Oersted placed a long, straight, horizontal wire on top of a magnetic compass. Both the compass and the wire were resting on a horizontal surface, and both the length of the wire and the compass needle were oriented north-south. Next, Oersted sent a current through the wire, and happened upon one of the greatest discoveries in physics.

- **What do you think Oersted saw?**

Write your answer to this question in your *Active Physics* log. Be prepared to discuss your ideas with your small group and other members of your class.

GOALS

In this activity you will:

- Explain how a simple galvanometer works.
- Induce current using a magnet and coil.
- Describe alternating current.
- Recognize the relativity of motion.

For You To Do

1. Wrap 10 turns of wire to form a coil that surrounds a magnetic compass. Wrap the wire on a diameter corresponding to the north-south markings of the compass scale, as shown in the diagram. Hold the turns of wire in place with tape, or use the method recommended by your teacher. Use sandpaper to carefully remove the insulation from a short section of the wire ends to allow electrical connection.

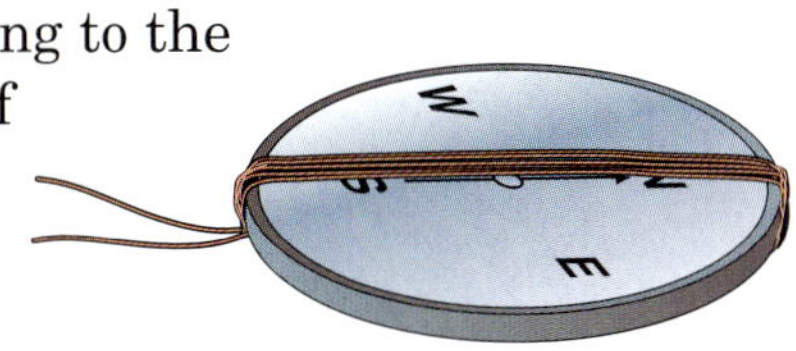

2. In **Step 1**, you constructed a galvanometer, a device to detect and measure small currents. Carefully connect a hand generator, a light bulb, and the galvanometer, as shown on the next page (in a series circuit). Rest the galvanometer so that the compass is horizontal, with the needle balanced, pointing north, and free to rotate. Also, turn the galvanometer, if necessary, so that the compass needle is aligned parallel to the turns of wire which pass over the top of the compass.

3. Crank the generator to establish a current in the circuit. Think of the compass needle as a meter such as the one in the speedometer of a car. The amount it moves corresponds to the amount of current. The glow of the light bulb verifies that current is flowing.

   a) Does the compass-needle galvanometer also indicate that current is flowing? How? In your log, use words and a sketch to indicate your answer.

4. The amount of current flowing in the circuit can be varied by changing the speed at which the generator is cranked, and the amount of current is indicated by the brightness of the light bulb. Vary the speed at which you crank the generator, and observe the galvanometer.

   a) How does the galvanometer indicate changes in the amount of current? Use words and sketches to indicate your answer.

5. Change the direction in which you crank the generator.

   a) What evidence does the galvanometer provide that changing the direction in which the generator is cranked has the effect of changing the direction of current flow in the circuit? Use words and sketches to give your answer.

6. Carefully connect each wire end of a galvanometer to a wire end of a solenoid wound on a hollow core of non-magnetic

material, such as a cardboard tube. Orient the galvanometer so that it is ready to detect current flow.

7. Hold a bar magnet in one hand and the solenoid steady in the other hand. Rapidly plunge one end of the bar magnet into the hollow core of the solenoid, and then stop the motion of the magnet, bringing the end of the magnet to rest inside the solenoid. Another person should hold the galvanometer in a steady position so that it will not be disturbed if the solenoid is moved. Observe the galvanometer during the sequence. You may need to practice this a few times.
   - a) Write your observations in your log.

8. Remove the magnet from the solenoid with a quick motion, and observe the galvanometer during the action.
   - a) Record your observations.
   - b) A current is produced! How does the direction of the current caused, or induced, when the end of the magnet is entering the solenoid, compare to the direction of the current when the magnet is leaving the solenoid?
   - c) How can you detect the direction of the current in each case?

9. Modify and repeat **Steps 7** and **8** to answer the following questions:

   a) What, if anything, about the created or induced current changes if the opposite end of the bar magnet is plunged in and out of the solenoid?

   b) How does the induced current change if the speed at which the magnet is moved in and out of the solenoid is changed?

   c) What is the amount of induced current when the magnet is not moving (stopped)?

   d) What is the effect on the induced current of holding the magnet stationary and moving the solenoid back and forth over either end of the magnet?

   e) What is the effect of moving both the magnet and the solenoid?

Reflecting on the Activity and the Challenge

In this activity you discovered that you can produce electricity. A current is created or induced when a magnet is moved in and out of a solenoid. The current flows back and forth, changing direction with each reversal of the motion of the magnet. Such a current is called an alternating current, and you may recognize that name as the kind of current that flows in household circuits. It is frequently referred to by its abbreviated form, "AC." It is the type of current that is used to run electric motors in home appliances. Part of your **Chapter Challenge** is to explain to the children how a motor operates in terms of basic principles of physics or to show how electricity can be produced from an external energy source. This activity will help you with that part of the challenge.

Physics To Go

1. An electric motor takes electricity and converts it into movement. The movement can be a fan, a washing machine, or a CD player. The galvanometer may be thought of as a crude electric motor. Discuss that statement, using forms of energy as part of your discussion.

2. Explain how the galvanometer works to detect the amount and direction of an electric current.

3. How could the galvanometer be made more sensitive, so that it could detect very weak currents?

4. An electric generator takes motion and turns it into electricity. The electricity can then be used for many purposes. The solenoid and the bar magnet, as used in this activity, could be thought of as a crude electric generator. Explain the truth of that statement, referring to specific forms of energy in your explanation.

5. If the activity were to be repeated so that you would be able to see only the galvanometer and not the solenoid, the magnet, and the person moving the equipment, would you be able to tell from only the response of the galvanometer what was being moved, the magnet or the solenoid or both? Explain your answer.

6. Part of the **Chapter Challenge** is to explain how the motor and generator toy works.
   a) Write a paragraph explaining how a motor works.
   b) Write a paragraph explaining how a generator works.

7. In generating electricity in this activity, you moved the magnet or the coil. How can you use each of the following resources to move the magnet?
   a) wind
   b) water
   c) steam

Stretching Exercise

Find out about the 120-V (volt) AC used in home circuits. If household current alternates, at what rate does it surge back and forth? Write down any information about AC that you can find and bring it to class.

Activity 4 AC & DC Currents

GOALS

In this activity you will:

- Describe the induced voltage and current when a coil is rotated in a magnetic field.
- Compare AC and DC generators in terms of commutators and outputs.
- Sketch sinusoidal output wave forms.

What Do You Think?

In the last activity, you used human energy to produce motion to generate electricity.

- **What other kinds of energy can generate electricity?**

Write your answer to this question in your *Active Physics* log. Be prepared to discuss your ideas with your small group and other members of your class.

For You To Do

AC Generator

1. Your teacher will explain and demonstrate a hand-operated, alternating current (AC) generator. During the demonstration, make the observations necessary to gain the information needed to answer these questions:

   a) When the AC generator is used to light a bulb, describe the brightness of the bulb when the generator is cranked slowly, and then rapidly. Write your observations in your log.

   b) When the AC generator is connected to a galvanometer, describe the action of the galvanometer needle when the generator is cranked slowly, and then rapidly.

2. It is easier to understand the creation of a current if you think of a set of invisible threads to signify the magnetic field of the permanent magnets. The very thin threads fill the space and connect the north pole of one magnet with the south pole of the other magnet. If the wire of the generator is imagined to be a very thin, sharp knife, the question you must ask is whether the knife (the wire) can "cut" the threads (the magnetic field lines). If the wire moves in such a way that it can cut the field lines, then a current is generated. If the wire moves in such a way that it does not cut the field lines, then no current is generated.

   a) Look at the diagrams of the magnetic fields shown. In which case, I, II, or III will a current be generated?

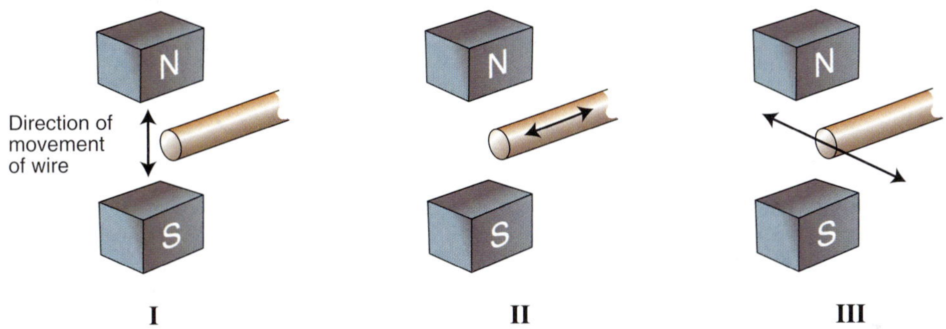

3. The following diagram shows the position of the rotating coil of an AC generator at instants separated by one-fourth of a rotation of the coil. Build a small model of the rectangular coil so that you can move the model to help you understand the drawings. The coil model can be constructed by carefully bending a coat hanger into the shape of the rectangular coil. Rest the coil between two pieces of paper—label the left paper N for the north pole of a magnet; label the right paper S for the south pole of a magnet.

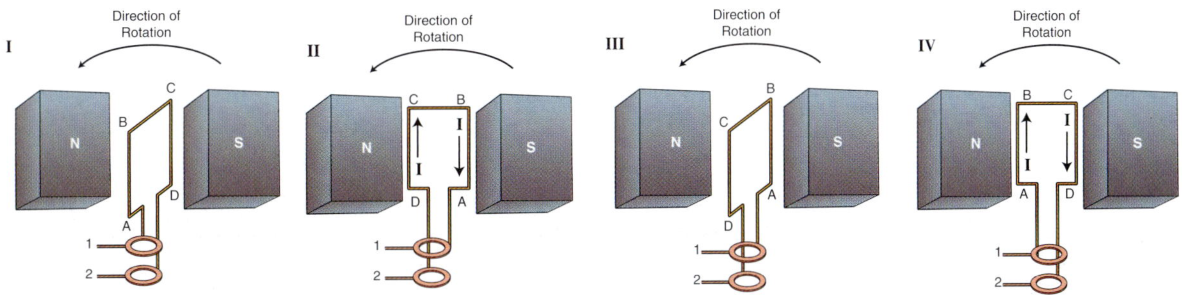

4. For the purpose of analyzing the rotating coil figure, the four sides of the rectangular coil of the AC generator will be referred to as sides AB, BC, CD, and DA. Side DA is "broken" to allow extension of the coil to the rings. The "brushes," labeled 1 and 2, make sliding contact with the rings to provide a path for the induced current to travel to an external circuit (not shown) connected to the brushes. The magnetic field has a left-to-right direction (from the north pole to the south pole) in the space between the magnets in the rotating coil figure. It is assumed that the coil has a constant speed of rotation.

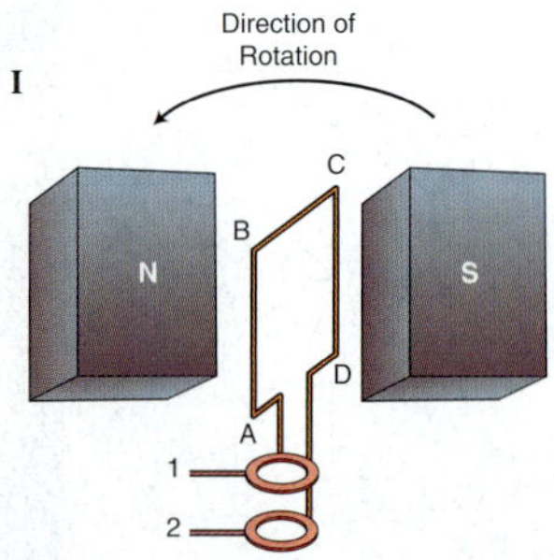

a) When the generator coil is in position I shown in the rotating coil, is a current being generated? A current is produced if the wire cuts the magnetic field lines. Record your answer and the reason for your answer in your log.

b) Use a graph similar to the one shown below. Plot a point at the origin of the graph, indicating the amount of induced current is zero at the instant corresponding to the beginning of one rotation of the coil.

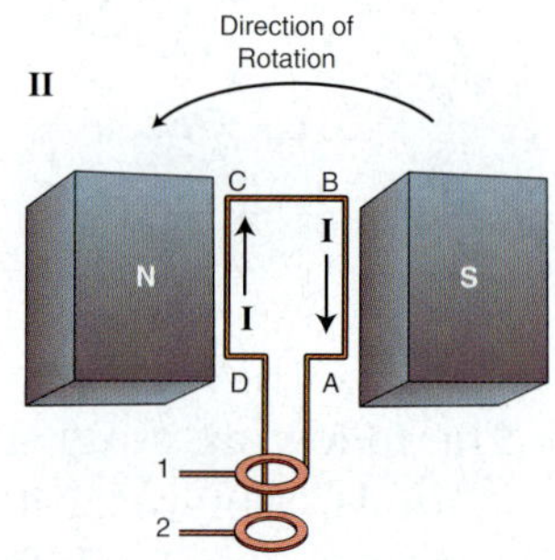

c) One-fourth turn later, at the instant when the rotating coil is in position II, is a current being generated? Record your answer and the reason for your answer in your log.

d) On your graph, plot a point directly above the $\frac{1}{4}$-turn mark at a height equal to the top of the vertical axis to represent maximum current flow in one direction.

The current induced by an AC generator during one rotation of the generator coil

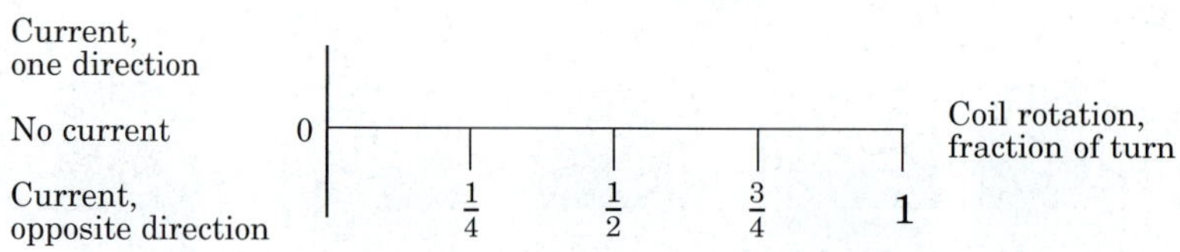

e) One-half turn into the rotation of the coil, at the instant shown in the rotating coil position III, the current again is zero because all sides of the coil are moving parallel to the magnetic field. Plot a point at the $\frac{1}{2}$ mark on the horizontal axis to show that no current is being induced at that instant.

III

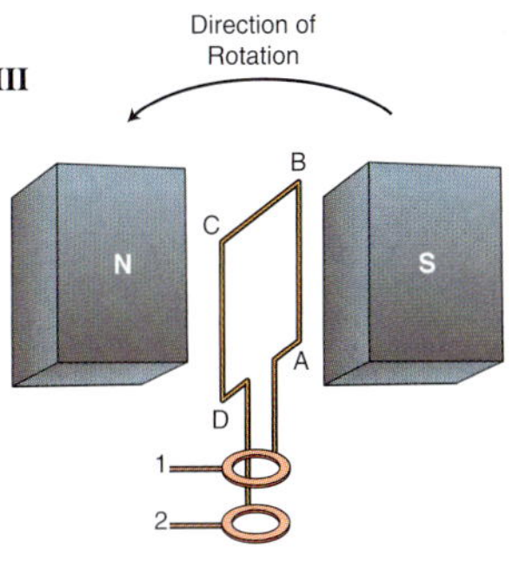

f) At the instant at which $\frac{3}{4}$ of the rotation of the coil has been completed, shown by the rotating coil in position IV, the induced current again is maximum because coil sides AB and CD again are moving across the magnetic field at maximum rate. However, this is not exactly the same situation as shown in the rotating coil position II; it is a different situation in one important way: the direction of the induced current has reversed. Follow the directions of the arrows which represent the direction of the current flow in the coil to notice that, at this instant, the current would flow to an external circuit out of brush 2 and would return through brush 1. On your graph, plot a point below the $\frac{3}{4}$-turn mark at a distance as far below the horizontal axis as the bottom end of the vertical axis. This point will represent maximum current in the opposite, or "alternate," direction of the current shown earlier at $\frac{3}{4}$-turn.

IV

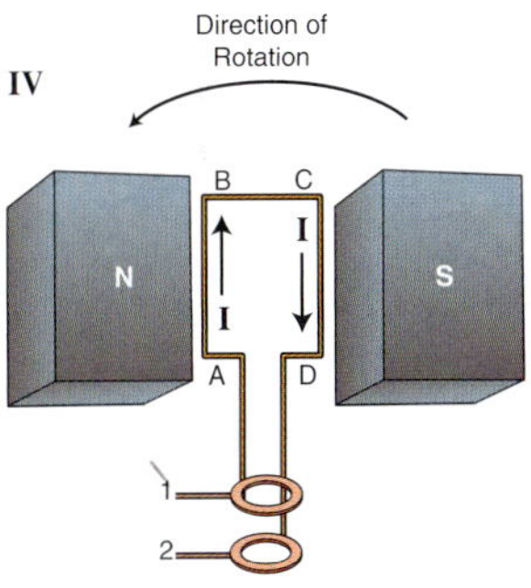

g) The rotating coil in position I is used again to show the instant at which one full rotation of the generator coil has been completed. Again, all sides of the coil are moving parallel to the magnetic field, and no current is being induced. Plot a point on the horizontal axis at the 1-turn mark to show that the current at this instant is zero.

5. You have plotted only 5 points to represent the current induced during one complete cycle of an AC generator.

a) Where would the points that would represent the amount of induced current at each instant during one complete rotation of the generator coil be plotted?

b) What is the overall shape of the graph? Should the graph be smooth, or have sharp edges? Sketch it to connect the points plotted on your graph.

c) What would the graph look like for additional rotations of the generator coil, if the same speed and resistance in the external circuit were maintained?

DC Generator

6. Your teacher will explain and demonstrate a hand-operated, direct current (DC) generator. During the demonstration, make the observations needed to answer these questions:

a) When the DC generator is used to light a bulb, describe the brightness of the bulb when the generator is cranked slowly, and rapidly. Write your observations in your log.

b) When the DC generator is connected to a galvanometer, describe the action of the galvanometer needle when the generator is cranked slowly, and rapidly.

7. The diagram shows important parts of a DC generator. As in **Step 3**, build a model of the generator to help you analyze how it works.

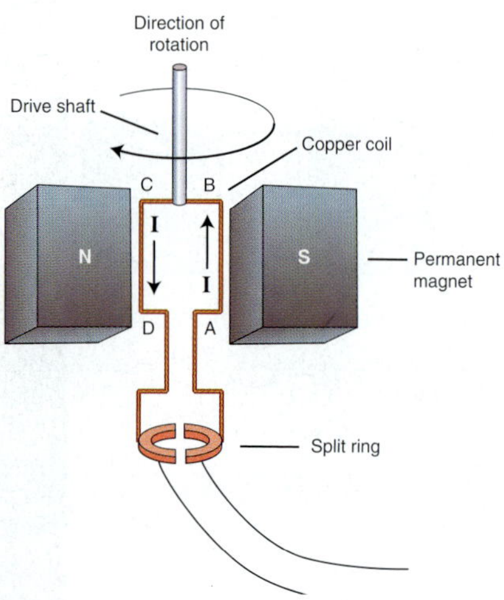

8. Use a graph similar to the one shown below. Complete the graph using the same pattern of analysis applied to the AC generator.

a) At the instant shown in the DC generator diagram, the induced current is maximum. The instant corresponds to the rotating coil II. Plot a point on the graph directly above the $\frac{1}{4}$-turn mark at a height equal to the top of the vertical axis to represent maximum current flow at that instant.

b) At the instant $\frac{1}{4}$-turn earlier than the instant shown in the DC generator figure, corresponding to the zero mark of rotation, the current would have been zero because all sides of the coil would have been moving parallel to the direction of the magnetic field. Therefore, plot a point at the origin of the graph.

c) Similarly, the induced current again would be zero at the instant $\frac{1}{4}$-turn later than the instant shown in the DC generator figure; therefore, plot a point on the horizontal axis at the $\frac{1}{2}$-turn mark.

The current induced by a DC generator during one rotation of the generator coil

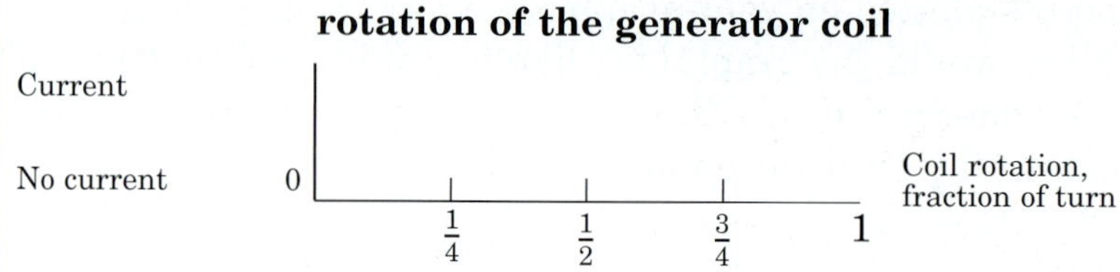

9. Notice the arrangement used to transfer current from the generator to the external circuit for the DC generator. It is different from the arrangement used for the AC generator. The DC generator has a "split-ring commutator" for transferring the current to the external circuit. Notice that if the coil shown in the DC generator figure were rotated $\frac{1}{4}$-turn in either direction, the "brush" ends that extend from the coil to make rubbing contact with each half of the split ring would reverse, or switch, the connection to the external circuit. Further, notice that the connection to the external circuit would be reversed at the same instant that the induced current in the coil reverses due to the change in direction in which the sides of the coil move through the magnetic field. The outcome is that while the current induced in the coil alternates, or changes direction each $\frac{1}{2}$-rotation, the current delivered to the external circuit always flows in the same direction. Current that always flows in one direction is called direct current, or DC.

   a) Plot a point on the graph at a point directly above the $\frac{3}{4}$-turn mark at the same height as the point plotted earlier for the $\frac{1}{4}$-turn mark.

   b) As done for the AC generator, find out how to connect the points plotted on this graph to represent the amount of current delivered always in the same direction to the external circuit during the entire cycle.

Reflecting on the Activity and the Challenge

It is time to begin preparing for the **Chapter Challenge**. Now that you know how a generator works, you should begin to think about toys that might generate electricity. You should also think about how you could assemble "junk" into a toy generator, or do some research on homemade generators and motors.

Physics To Go

1. What is the purpose of:

   a) An electric generator?
   b) An electric motor?

2. How does a direct current differ from an alternating current? Use graphs to illustrate your answer.

3. In an electric generator, a wire is placed in a magnetic field. Under what conditions is a current generated?

Stretching Exercises

1. What is the meaning of "hertz," abbreviated "Hz," often seen as a unit of measurement associated with electricity or stereo sound components such as amplifiers and speakers?

2. What does it mean to say that household electricity has a frequency of 60 Hz?

3. Have you ever heard 60 Hz AC being emitted from a fluorescent light or a transformer?

4. Look at a catalog or visit a store where sound equipment is sold, and check out the "frequency response" of speakers—what does it mean?

5. Heinrich Hertz was a 19th-century German physicist. Find out about the unit of measurement named after him, and write a brief report on what you find.

Activity 5 Building an Electric Motor

GOALS

In this activity you will:

- Construct, operate, and explain a DC motor.
- Appreciate accidental discovery in physics.
- Measure and express the efficiency of an energy transfer.

What Do You Think?

You plug a mixer into the wall and turn a switch and the mixer spins and spins—a motor is operating.

- **How do you think the electricity makes the motor turn?**

Write your answer to this question in your *Active Physics* log. Be prepared to discuss your ideas with your small group and other members of your class.

For You To Do

1. Study the diagram on the following page closely. Carefully assemble the materials, as shown in the diagram, to build a basic electric motor. Follow any additional directions provided by your teacher.

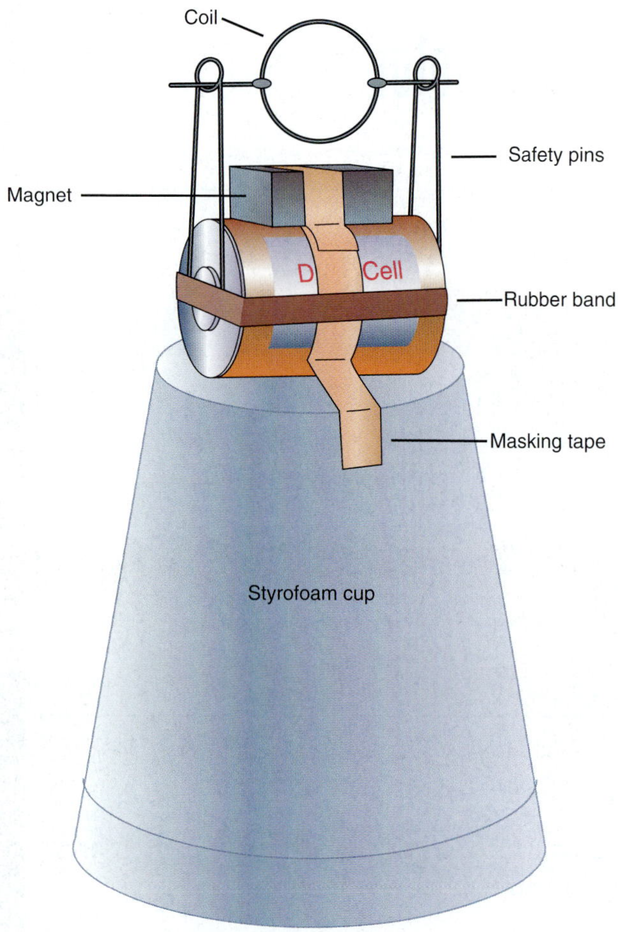

2. When your motor is operating successfully, find as many ways as you can to make the motor change its direction of rotation.
   a) Describe each way you tried and identify the ways that were successful.
3. Hold another magnet with your fingers and bring it near the coil from above, facing the original magnet, as the motor is operating.
   a) Describe what happens. Does the orientation of the second magnet make a difference?
4. Replace the single magnet with a pair of attracting magnets on top of the battery.
   a) What is the effect?
5. Think of other ways to change the speed of the motor. With the approval of your teacher, try out your methods.
   a) Describe ways to change the speed of the motor.
6. Use a hand generator as the energy source instead of the battery. You can disconnect the battery without removing it from the structure by placing an insulating material, such as a piece of cardboard, between the safety pin and the battery to open the circuit at either end of the battery. Then clip the wires from the generator to the safety pins to deliver current from the generator to the motor.
   a) Discuss what you find out.
7. Your motor turns! Chemical energy in the battery was converted to electrical energy in the circuit. The electrical energy was then converted to mechanical energy in the motor.
   a) List at least three appliances or devices where the motor spins.

8. The spin of the motor occurs because the current-carrying wire has a force applied to it. You know if something moves, a force must be applied. As you observed, when the battery connection was broken, the motor stopped turning. You know from a previous activity that a current-carrying wire creates a magnetic field. Pause for a bit to remind yourself of the behavior of magnets. Take a bar magnet and place its north pole near a compass. The compass is a tiny bar magnet that can easily turn.

   a) Draw a sketch to show the orientation of the compass.

9. Shift the compass to the south pole of the bar magnet.

   a) Draw a sketch to show the orientation of the compass.

10. The north pole of the bar magnet repelled the north pole of the compass. The south pole of the bar magnet attracted the north pole of the compass. This attraction and repulsion is the result of a force on the compass. You can now investigate the force between the poles of a magnet and the magnetic field of a current-carrying wire.

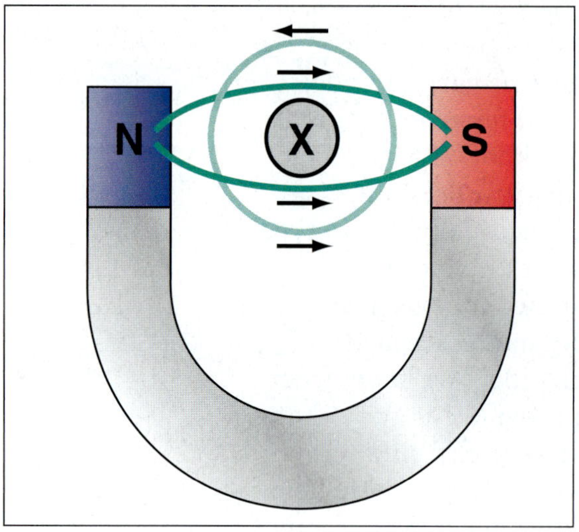

The magnetic field lines are drawn for a horseshoe magnet. The direction of the magnetic field lines is identical to a direction that a compass would point. The compass would point away from the north pole and toward the south pole. The magnetic field of the current-carrying wire is circular, as you investigated in an earlier activity. Compare the direction of this magnetic field to the direction of the magnetic field of the horseshoe magnet.

11. Think of the magnetic field lines above as small compasses.
    a) Write down whether the compasses above the wire attract one another or repel one another.
    b) Write down whether the compasses below the wire attract one another or repel one another.

12. This attraction/repulsion causes the wire to jump. There is a force on the wire. This force on the current-carrying wire is the basis for the electric motor that you built in this activity. The use of the loop of the wire allows the wire to rotate instead of jumping in the way a single wire would.

13. It is the moving electrons in the wire that create the current. In some TV sets, there is an electron beam that shoots the electrons from the back of the TV to the front. There are horseshoe magnets of a sort in the television. The moving electrons experience a force. The electrons' path is affected by the magnetic field. By varying the strength of the magnetic field, the electron beam can hit all parts of the screen and you receive a TV image.

FOR YOU TO READ

The history of science is filled with discoveries that have led to leaps of progress in knowledge and applications. This is certainly true of physics and, in particular, electricity and magnetism. These discoveries "favor" the prepared mind. Oersted's discovery in 1820 of the magnetic field surrounding a current-carrying wire already has been mentioned. Similarly, Michael Faraday

Michael Faraday

discovered electromagnetic induction in 1831. Faraday was seeking a way to induce electricity using currents and magnets; he noticed that a brief induced current happened in one circuit when a nearby circuit was switched on and off. (How would that cause induction? Can you explain it?) Both Oersted and Faraday are credited for taking advantage of the events that happened before their eyes, and pursuing them.

About one-half century after Faraday's discovery of electromagnetic induction, which immediately led to development of the generator, another event occurred. In 1873, a Belgian engineer, Zénobe Gramme, was setting up DC generators to be demonstrated at an exposition (a forerunner of a "world's fair") in Vienna, Austria. Steam engines were to be used to power the generators, and the electrical output of the generators would be demonstrated. While one DC generator was operating, Gramme connected it to another generator that was not operating. The shaft of the inactive generator began rotation—it was acting as an electric motor! Although Michael Faraday had shown as early as 1821 that rotary motion could be produced using currents and magnets, a "motor effect," nothing useful resulted from it. Gramme's discovery, however, immediately showed that electric motors could be useful. In fact, the electric motor was demonstrated at the very Vienna exposition where Gramme's discovery was made. A fake waterfall was set up to drive a DC generator using a paddle wheel arrangement, and the electrical output of the generator was fed to a "motor" (a generator running "backwards"). The motor was shown to be capable of doing useful work.

Reflecting on the Activity and the Challenge

Decision time about the **Chapter Challenge** is approaching for your group. In this activity you built a very basic, working electric motor. This is an important part of the **Chapter Challenge**. However, knowing how to build an electric motor is only part of the challenge. Your toy must be fascinating to children. You must also be able to explain how it works.

Physics To Go

1. Some electric motors use electromagnets instead of permanent magnets to create the magnetic field in which the coil rotates. In such motors, of course, part of the electrical energy fed to the motor is used to create and maintain the magnetic field. Similarly, electromagnets instead of permanent magnets are used in some generators; part of the electrical energy produced by the generator is used to energize the magnetic field in which the generator coil is caused to turn. What advantages and disadvantages would result from using electromagnets instead of permanent magnets in either a motor or generator?

2. Design three possible toys that use a motor or a generator or both. One of these may be what you will use for your project.

3. The motor/generator you submit for the **Chapter Challenge** must be built from inexpensive, common materials. Make a list of possible materials you could use to construct an electric motor.

4. In the grading criteria for the **Chapter Challenge**, marks are assigned for clearly explaining how and why your motor/generator works in terms of basic principles of physics. Explain how an electric motor or generator operates.

Activity 6 Building a Motor/Generator Toy

GOALS

In this activity you will:

- Design, construct and operate a motor/generator.

What Do You Think?

You may have heard the following expression used before: "The difference between men and boys is the cost of their toys."

- **What characteristics make an item a toy?**

Write your answer to this question in your *Active Physics* log. Be prepared to discuss your ideas with your small group and other members of your class.

For You To Do

1. Confer within your group and between your group and your teacher about whether you will pursue, as a basis for the motor/generator kit for the **Assessment**, the motor design presented in this activity, an alternate design, or both. Whatever

design(s) your group chooses to pursue, you are encouraged to be creative. Most designs can be improved in some way or another by substituting materials or making other changes. There is no single "best way" to go about designing the motor/generator and making it function within a toy or to produce electrical energy from another form of energy. The best way for your group is the way that the group can get the job done.

a) When you have decided on a design, submit your design to your teacher for approval.

2. In your group decide how you will make the motor/generator fascinating to children. You may wish to use some of the ideas you generated in answering the **What Do You Think?** question.

a) Record your ideas in your log.

b) Describe and make a sketch of your final design, and submit it to your teacher for approval.

3. Use the design for a DC motor as shown in the diagram as a basis to begin your construction. It can be adapted, as required, for the **Chapter Assessment**, to power a toy. Also as required, the motor could be adapted to be driven "backwards" by an external energy source to function as a DC generator.

[The motor design shown was adapted from the following public domain work: Educational Development Center, Inc., *Batteries and Bulbs II* (New York: McGraw-Hill, 1971), pp. 85-88.]

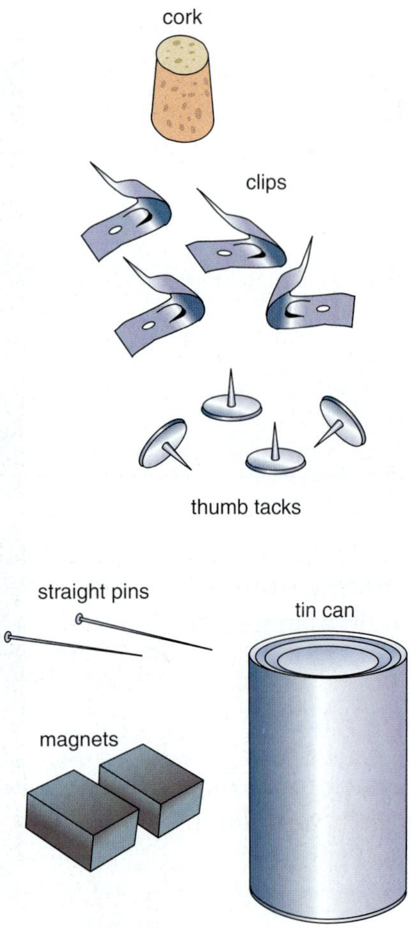

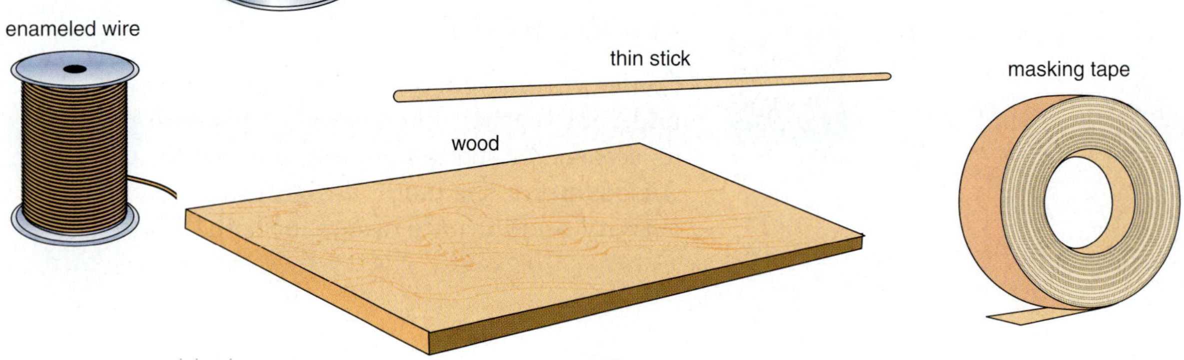

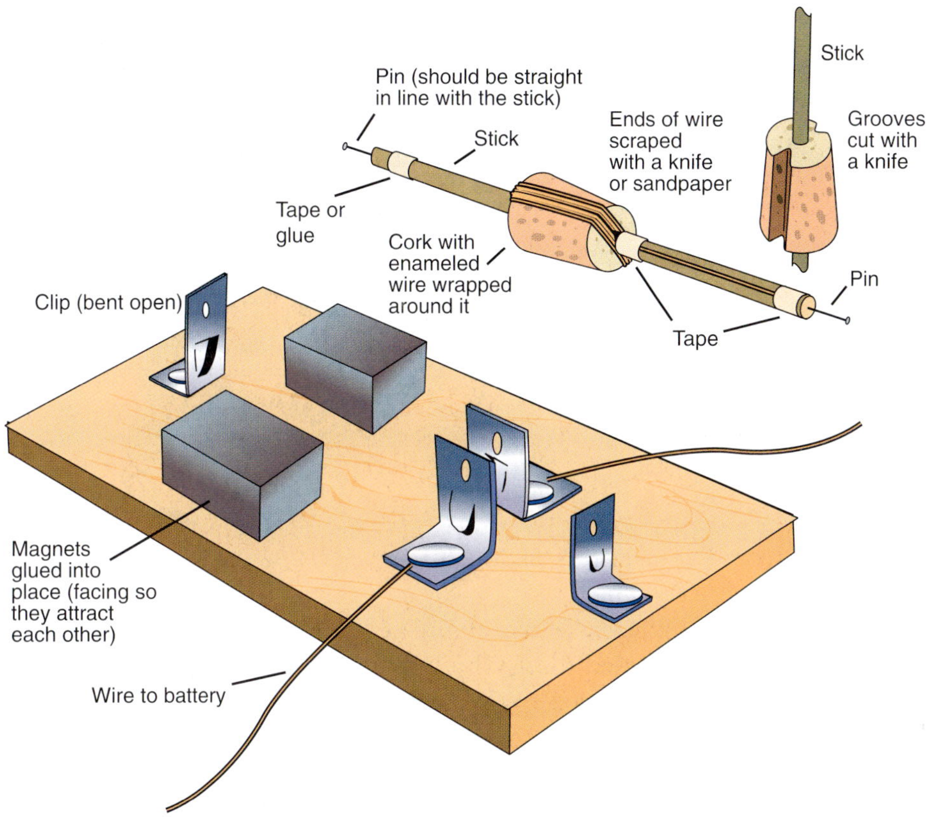

Reflecting on the Activity and the Challenge

You are now well on your way to completing the **Chapter Challenge**. You have decided on the design for your motor/generator and the toy it will power.

Physics To Go

Your assignment is to prepare to meet the criteria of the **Assessment** of the **Chapter Challenge**.

PHYSICS AT WORK

Uriah Gilmore

HEADED FOR THE STARS

Uriah Gilmore loved to take electric appliances apart when he was growing up. "I couldn't always get them back together," he admits, "but I was so curious I couldn't help myself. I just had to see how they worked." Fortunately, Uriah's parents supported his curiosity.

Uriah and his fellow teammates from Cleveland, Ohio's East Technical High School recently won first place at the National High School Robotics Tournament at Epcot Center in Orlando, Florida for building a robot. "We were counseled along the way by engineers from NASA," he enthusiastically explains. "We called our robot Froggy and painted it green," Uriah continues, "and we used noisemakers so it even sounded like a frog." During the final contest "Froggy" was put in a pit with two other robots and had to place balls of a certain color in a specified area. The robot who got the most balls in won the contest.

"In my sophomore year the school I was attending closed and I went to East Technical High School which was the best thing that happened to me." He entered the engineering program and became a member of the engineering team—a team that is more popular than any sports team in his school.

Uriah attended Morehouse College on a NASA scholarship. "But," he states, "it's not enough to be a good student. You also have to be involved with your school and your community." Uriah once led a march on the Cleveland, Ohio, City Hall to protest a law which threatened to fire certain teachers, including one who inspired Uriah and was responsible for the revitalization of East Technical High School.

"My ultimate goal is to travel in space and explore the galaxy," he states. A shorter term goal is to be as involved in college as he has been in high school.

Chapter 8 Assessment

Your task is to prepare a kit of materials and instructions that a toy company will manufacture. Children will use these kits to make a motor or generator, or a combination electric motor/generator. It will serve both as a toy and to illustrate how the electric motors in home appliances work or how electricity can be produced from an energy source such as wind, moving water, a falling weight, or some other external source.

Review and remind yourself of the grading criteria that you and your classmates agreed on at the beginning of the chapter. The following was a suggested set of criteria:

- **(30%) The motor/generator is made from inexpensive, common materials, and the working parts are exposed but with due consideration for safety.**
- **(40%) The instructions for the children clearly explain how to assemble and operate the motor/generator device, and explain how and why it works in terms of basic principles of physics.**
- **(30%) If used as a motor, the device will operate using a maximum of four 1.5-volt batteries (D cells), and will power a toy (such as a car, boat, crane, etc.) that will be fascinating to children.**

OR

- **(30%) If used as a generator, the device will demonstrate the production of electricity from an energy source such as wind, moving water, a falling weight, or some other external source and be fascinating to children.**

Physics You Learned

Motors

Generators

Galvanometers

Magnetic field from a current

Solenoids

Electromagnets

Induced currents

AC and DC generators

Chapter 9

ATOMS ON DISPLAY

Scenario

High school students in Oregon walk into a giant mouth and down a cavernous esophagus into a spherical and moist room filled with liquids and loud sounds. An elementary student in California melds a picture of his or her face with that of a friend to see how pretty or ugly a composite face would be. Middle school students in New York try to improve their "major-league" batting skills or their "professional" golf swing using computer video analysis. All of these students are visiting their science museums.

Science museums ensure that learning and fun are indistinguishable. They feature hands-on exhibits that stimulate, educate, and entertain. They provide exciting experiences that help visitors develop an understanding of science.

The visitors at a science museum can come and go as they please. Research has shown that there are only 30 seconds available at an exhibit to capture people's interest or they walk right by. Once they stop at the exhibit, you must get them involved. Often this involvement includes some kind of interaction with the exhibit. Some of the exhibits are targeted for a specific age group. Other exhibits are for a broad audience.

If you have never been to a science museum, it's worth a visit. If you are not able to get to a city that has a museum, you may want to visit one of the many virtual science museums on the Internet.

Challenge

Your *Active Physics* class has been asked to develop an exhibit that will provide visitors to a science museum with an understanding of the atom.

The exhibit must:

- **include distinct features of the structure of the atom**
- **communicate the size and scale of the parts of the atom**
- **provide information on how the atom is held together**
- **capture the visitor's attention within 30 seconds**
- **include written matter that will further illuminate the concepts**
- **have a model, a T-shirt, a poster, a booklet, or a toy that can be sold at the museum store**
- **have a building plan and cost estimate for the exhibit**
- **include safety features**
- **be interactive—visitors should not merely read**

Criteria

You will be presenting your museum-exhibit plan to your class. As a class, decide how your work will be evaluated. Some required criteria are listed in the **Challenge**. Are there other criteria that you think are worth including?

How should each of the criteria be weighted? Should the building plan and cost estimate be worth as many points as the item for the museum store? Should the "30-second" criterion be worth more than anything else? Since the museum exhibit is supposed to educate the visitor, how much should the content be worth? Work with your class to agree on how many points should be assigned for each criterion. The total points should add up to 100.

Once you decide on the point allocation, you will have to decide how you can judge the assigned point value. For instance, assume that you chose the 30-second criterion to be worth 15 points. How will your class decide on whether your exhibit gets the full 15 points or only 10 points? It is worth knowing how each criterion will be judged so that you can ensure success and a good grade.

You will probably want to assign some strict guidelines and also leave room for some extra points. Learning how to judge the quality of your own work is a skill that all businesses expect to see in their professionals.

Activity 1 Static Electricity and Coulomb's Law

GOALS

In this activity you will:

- Charge a ball with a positive charge.
- Charge a ball with a negative charge.
- Observe that like charges repel and unlike charges attract.
- Calculate the electrical force using Coulomb's Law.
- Apply the organizing principle of conservation of charge.
- Recognize the similarities and differences between Coulomb's Law and Newton's Law of Gravitation.
- Explain how Coulomb was able to measure Coulomb's constant using a torsion balance.

What Do You Think?

"Everyone" has heard of atoms, but no one has ever seen an atom. Look at the sketch below of an atom that you often see in advertisements and some science books.

- **How would you describe what is shown in the sketch of an atom?**
- **Are there any problems with this depiction of the atom? Explain your answer.**

Record your ideas about these questions in your *Active Physics* log. Be prepared to discuss your responses with your small group and your class.

For You To Do

1. Have you ever seen a tremendous lightning storm? Bolts of lightning ignite the sky as they streak toward the Earth. A tiny "lightning storm" also takes place when you get an electric shock. Think back to the last time you got a shock.

   a) Were you inside or outside?
   b) Was it winter or summer?
   c) What did you touch to get the shock?
   d) Include in your *Active Physics* log any additional details.

2. The study of lightning, shocks, and "static cling" can reveal important physics. Cut two strips of Scotch® Magic™ tape about 12 cm long. Bend one end of each strip under to form a tab. Place one strip sticky side down on a table and label the tab B, for "bottom." Place the other strip sticky side down on top of the first strip and label the tab T, for "top." With one hand, peel off the top strip using the tab. Then pick up the bottom strip with the other hand. Hold both strips apart, allowing them to hang down. Slowly bring the hanging strips toward each other, but do not let them touch.

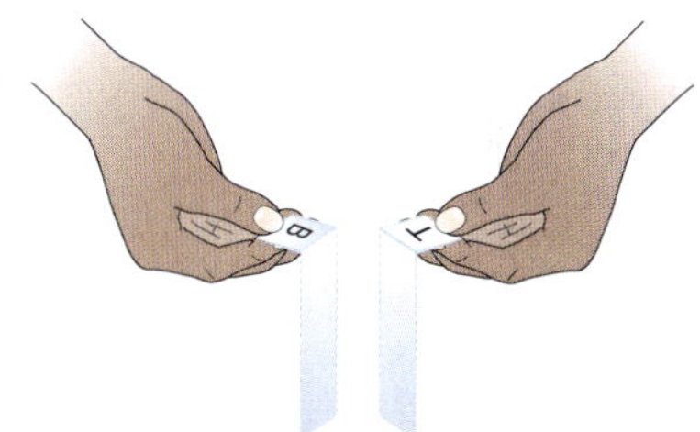

   a) Record your observations.
   b) If the strips accelerated (moved) toward or away from each other, Newton's Second Law tells you that there must be a force. (Newton's First Law tells you that an object at rest will tend to stay at rest. Newton's Second Law says that if an object accelerates, it's being acted on by a force.) Was there a force present?
   c) If a force was present, was it attractive or repulsive? Explain.

3. Make a second set of strips as in **Step 2**.

   a) Predict what you think will happen if the two top strips are picked up, one from each set, and brought toward each other. Record your prediction in your *Active Physics* log.

   Pick up the two top strips by the tabs, allowing both strips to hang down. Slowly bring them toward each other.

   b) Record your observations.
   c) Was the force attractive or repulsive? Explain.
   d) Predict what you think will happen if the two bottom strips of tape are picked up and brought toward each other. Record your prediction.

Pick up the two bottom strips by the tabs, allowing both strips to hang down. Slowly bring these toward each other.

e) Record your observations.

f) Was the force attractive or repulsive? Explain.

4. Charge up a rubber rod by vigorously rubbing it with a piece of wool. Use lots of friction to get the rubber rod charged up. Touch the rod to a small Styrofoam® ball coated with a conducting paint. Observe the interaction of the rod and the ball.

a) In your *Active Physics* log describe the interaction you observe.

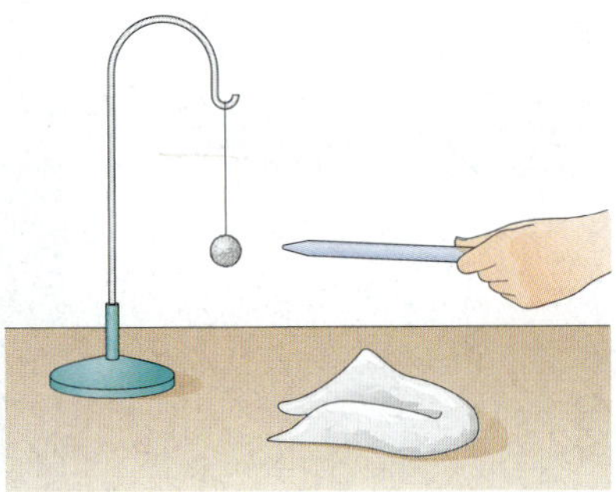

5. Charge up a glass rod by rubbing it with a piece of silk. Bring it toward the charged ball but try not to let them touch. Observe the interaction of the glass rod and the ball.

a) In your *Active Physics* log describe the interaction you observe.

6. Assume that the rubber rod gained electrons (negative charges) and became negatively charged when rubbed with the wool.

a) What do you think could have happened to some of the excess electrons on the rubber rod when it touched the Styrofoam ball? Use your observations to justify your answer.

b) If the rubber rod got the excess electrons from the wool, what would be the wool's charge?

c) Explain how you could check to see if your answer to **(b)** is correct.

7. Assume that the glass rod transferred some electrons to the silk.

a) What would be the charge of the glass rod? Use your observations to justify your answer.

b) If the silk gained electrons, what would be the silk's charge?

c) Explain how you could verify your answer to **(b)**.

8. Get rid of any excess electrons on the Styrofoam ball by touching it. (This is made more effective if you moisten your finger. The slight moisture helps remove the charge.)

Removing electrons to make an object neutral is called grounding. Adding electrons to make an object neutral is also called grounding. The "ground" is able to supply or remove electrons to make an object neutral. In this situation you are connected to the "ground."

9. Negatively charge the rubber rod. Bring it close to the neutral Styrofoam ball, but do not touch the rod to the ball.
   a) Describe your observations.
   b) Like charges repel, and unlike charges attract. How would you explain the interaction between the charged rubber rod and the uncharged ball? (Hint: The electrons in the ball are able to move about the ball so that some are able to move closer to the rubber rod or further from the rubber rod.) Use a diagram to help you explain your observations.

Physics Words

Coulomb's Law: the relationship among electrical force, charges, and the distance between the charges.

FOR YOU TO READ

Coulomb's Law

In physics, scientists often search for an equation that can provide a clearer, more precise description of what they observe. Charles Augustin de Coulomb experimentally determined the equation for electric forces in 1784.

$$F = \frac{kq_1q_2}{d^2}$$

where F is the force in newtons (N),

q_1 and q_2 are charges in coulombs (C),

d is the distance between the centers of the charges in meters (m), and

k is Coulomb's constant, always equal to 9×10^9 Nm^2/C^2.

Sample Problem

Two small charged spheres are placed 0.2 m apart. The first sphere has a charge of $+3.0 \times 10^{-6}$ C and the second sphere has a charge of -4.0×10^{-6} C. Calculate the force between them.

Strategy: You can use the equation for electric forces to calculate the force between the spheres.

Givens:

$q_1 = +3.0 \times 10^{-6}$ C

$q_2 = -4.0 \times 10^{-6}$ C

$d = 0.2$ m

Solution:

$$F = \frac{kq_1q_2}{d^2}$$

$$F = \frac{(9 \times 10^9 \text{ Nm}^2/\text{C}^2)(+3.0 \times 10^{-6}\text{ C})(-4.0 \times 10^{-6}\text{ C})}{0.2 \text{ m}^2}$$

$$F = -2.7 \text{ N}$$

The negative sign indicates that the force is attractive.

→

Static Charges and Forces

In this activity, you noticed that a force was present between the pieces of tape. It's an invisible force. You couldn't see the force, but you saw evidence that it was there because it moved the tape. It's tough to believe in invisible things, but since this force had (and consistently does have) visible and measurable effects, it would be tougher not to believe in it.

You also observed that the force could be attractive or repulsive. One explanation might be that there are two types of charges and that the interaction of the charges creates a force. If the two pieces of tape have the same charges, the force is repulsive. If the two pieces of tape have the opposite charges, the force is attractive. An alternative explanation is that there is one type of charge. If the two pieces of tape have an excess amount of charge, there is repulsion. If the two pieces of tape have a deficiency of charge (they've each lost some charge), there is repulsion. If one piece of tape has a deficiency and the other piece of tape has an excess of the charge, there is attraction.

You then found evidence for forces that you could attribute to excess charges. The rubber rod removed excess charges from the wool. These excess (negative) charges were shared with the Styrofoam ball. The rod and the ball now had the same charge and you observed a force of repulsion that pushed the ball and rod apart from each other. When the glass rod, which had a deficiency of (negative) charges, was placed near the (negatively) charged ball, there was a force of attraction.

In the 1700s Benjamin Franklin performed experiments similar to those you did in this activity. He defined which charge would be called positive and which would be called negative. A rubber rod rubbed in wool gains a negative charge. A glass rod rubbed in silk becomes positively charged. The glass rod did not gain positive charges but became positive by losing negative charges. Before Benjamin Franklin, investigators supposed that, since there was both attraction and repulsion, there were two kinds of electricity.

Conservation of Charge

A rubber rod gains electrons, while a piece of wool loses electrons. A glass rod loses electrons, while a piece of silk gains electrons. These situations are good examples of one of the major organizing principles of physics—the conservation of charge. In any system, the total amount of charge must remain constant. If negative charge is removed from the wool, that negative charge must go somewhere. For example, if 15 bits of negative charge are removed from the wool, then 15 bits of negative charge are transferred to the rubber rod. There are only a few quantities discovered that nature conserves—charge is one of them.

Perhaps without being aware of it, you used the concept of conservation of charge in your analysis of what you observed.

A few simple problems illustrate this conservation law. When two identical spheres touch, they will end up with identical charges. (If they did not end up with identical charges, then there would be a way to distinguish them. They would not be identical.) One sphere with a charge of –10, touches a neutral, identical sphere. After touching and separating, each sphere has a charge of – 5. The charge was transferred, but the total charge remained the same.

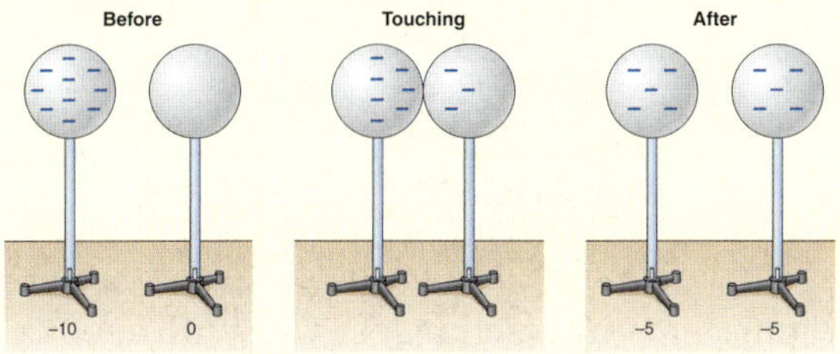

Keep in mind that neutral matter already contains enormous amounts of pluses and minuses, but in equal amounts. Thus the "excess" and "deficiency" are relative to a baseline of starting out neutral.

Coulomb's Law and Newton's Law of Gravitational Attraction

You found you could actually calculate the force of attraction or repulsion of charges by using **Coulomb's Law**:

$$F = \frac{kq_1q_2}{d^2}$$

Here's an incredible part of physics that you did not examine in this activity. Coulomb's Law for electrostatic attraction and repulsion is very similar to **Newton's Law of Gravitational Attraction**!

$$F = \frac{Gm_1m_2}{d^2}$$

where F is the force in newtons (N),

m_1 and m_2 are masses in kilograms (kg),

d is the distance between the centers of the masses in meters (m), and

G is the gravitational constant, always equal to 6.67×10^{-11} Nm^2/kg^2.

Look at the similarities:

- Both laws show forces that decrease with the square of the distance.
- Both laws show forces that depend on the product of the masses or charges.
- Both laws have constants.

Look at the differences:

- Electric forces are attractive and repulsive; gravitational forces are only attractive.
- Charges come in two varieties, + and –. Mass comes in one variety, +.
- The electric force constant is quite large, while the gravitational force constant is quite small.

If you look at the gravitational and electrical forces between two electrons, the gravitational force is much, much smaller. It is so small, you don't even take it into account when describing the behavior of the charges.

The experimental techniques to find the value of k and G are quite similar. In Coulomb's experiment, two spheres were attached to the ends of a rod and the rod was suspended by a wire. These spheres were charged, and similarly charged spheres were brought near the ends of the rods. The repulsive force caused the wire to twist. The twist was a measure of the force, and Coulomb was able to verify his law.

→

Cavendish's experiment was similar, but the attraction between the pairs of spheres was due to their gravitational attraction. This tiny force was measured by the twist in the wire.

This symmetry of what appears to be two unrelated forces provides a glimpse into a beauty of our world. Physicists remark on this beauty. It drives them to try to find out if there are other underlying understandings of the two forces because of that symmetry. This is what physicists are exploring when you hear about their work on unified theories.

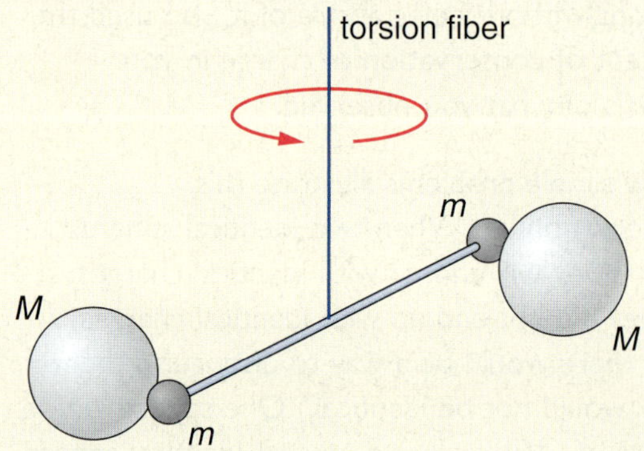

Reflecting on the Activity and the Challenge

You probably know that atoms are composed of electrons and other particles. These particles have charges. You will need to provide a description of the interaction between the charges when you provide a description of the atom. You may find a way to include in your exhibit the larger concepts of conservation of charge or the ability to actually calculate these forces of attraction and repulsion.

Physics To Go

1. Electrons are transferred from a rod to a piece of material.
   a) Which object will become negatively charged?
   b) Which object will become positively charged?

2. A rubber rod is negatively charged after being rubbed by wool. Explain how this happens.

3. Two identical spheres are mounted on insulated stands. The first sphere has a charge of –1. The second sphere has a charge of – 3. After the spheres touch, what will the charge on each be?

4. One of two identical metal spheres has a charge of +1 and

Physics Words

Newton's Law of Gravitational Attraction: the relationship among gravitational force, masses, and the distance between the masses.

the other sphere has a charge of – 5. Compare the total charge on the spheres before and after contact.

5. Charge A is $+2.0 \times 10^{-6}$ C and charge B is $+3 \times 10^{-6}$ C. The charges are 3 m apart. What is the force between them? Is it attractive or repulsive?

6. Charge A is -4.0×10^{-6} C and charge B is $+2 \times 10^{-6}$ C. The charges are 5 m apart. What is the force between them? Is it attractive or repulsive?

7. When the air is dry and you walk on a wool carpet with your shoes, you may experience a shock when you touch a doorknob. Explain what is happening in terms of electric charge. (Hint: Your shoes are similar to the rubber rod.)

8. Compare and contrast Coulomb's Law and Newton's Law of Gravitational Attraction. Provide at least one similarity and one difference.

9. Coulomb's Law states that the electric force between two charged objects decreases as the square of the distance increases. Suppose the original force between two objects is 60 N, and the distance between them is tripled; the new force would be 3^2, or 9 times weaker. The new force would be $60 \text{ N} \div 9 = 6.7$ N or 7 N.

   Find the new forces if the original distance was:

   a) doubled
   b) quadrupled
   c) halved
   d) quartered

10. Sketch a graph that shows how the electrostatic force defined by Coulomb's Law varies with the distance.

11. A single electron has a charge of 1.6×10^{-19} C. Show why it takes 6.25×10^{18} electrons to equal 1 C.

12. How could you depict the invisible electrostatic force in a museum exhibit?

Activity 2 Tiny and Indivisible

GOALS

In this activity you will:

- Determine the number of hidden pennies in a container.
- Explain why the masses of containers with pennies can only take on certain values.
- Describe the Millikan oil-drop experiment.
- Explain what we mean when we say that electric charge is quantized.

What Do You Think?

It's easy to share a box of popcorn with a friend at the movies. It's a bit tougher to share a slice of pizza equally, but it can be done with a knife.

- **Can you think of something that cannot be shared? Describe something that cannot be broken into two parts.**

Record your ideas about this question in your *Active Physics* log. Be prepared to discuss your responses with your small group and your class.

For You To Do

1. Your teacher will provide you with a set of film canisters or other containers that contain pennies. Do not open the containers. Your goal is to determine the mass of a single penny. You can find the mass of an empty container with a balance. You can also find the mass of each container, including the pennies, using a balance.

    a) Explain how you can find the mass of one penny using the measurements you are permitted to make. It's very possible that no container has only one penny. Write down your strategy.

2. Assume that each container has a mass of 5 g and each penny has a mass of 2 g.
   - a) Make a list of possible masses of containers that have 1 penny, 2 pennies, 5 pennies, etc. Do this for at least ten containers.
   - b) Given only the masses of the ten containers that you calculated in **(a)**, describe how you can find the mass of a single penny.

3. The container does not have a mass of 5 g, and each penny does not have a mass of 2 g. Using the balance and the containers of pennies, determine the mass of a single penny.
   - a) Record the mass of a single penny.
   - b) Compare your value of the mass of a penny with that of another group. How does your confidence in the value change as you compare your value with more and more groups? This is how the sharing and comparing of results of different teams of scientists make scientific progress.

4. Suppose you obtain a new set of containers with nickels in them. Assume that each container is 5 g and each nickel is 5 g.
   - a) What are some possible masses you would expect for four of the containers?
   - b) A lab group stated that they measured the mass of a container of nickels and it was 23 g. Your lab group thinks that they must have made an error. Explain to the first group in writing why you think that there is a problem.
   - c) When your lab group measured the mass of the container, you also found it to be 23 g. You now have a problem—a mystery—a puzzle. It is this kind of puzzle that challenges and intrigues physicists. How could this be? It would be great to open up the container, but this may not be possible. Can you solve the puzzle? Suggest at least three different solutions to this puzzle.

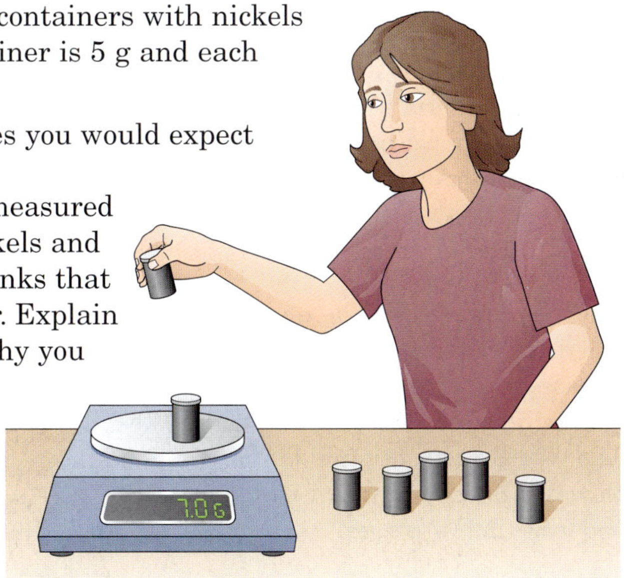

FOR YOU TO READ

In 1910, Robert A. Millikan completed an experiment very similar to the activity you just completed. He did not measure containers of pennies. He measured the forces on charged oil drops. This allowed him to calculate the charge on each drop. He made thousands of measurements. He always found that the oil drop had 1 charge, 2 charges, 5 charges, 23 charges, and other whole numbers of charges. He never found 3.5 charges or 4.7 charges or 11.2 charges. The Millikan oil-drop experiment shows that you never seem to have a piece of a charge. The experiment has been conducted many times. Nobody has found fractional charges. Millikan concluded from his oil-drop experiment that there is a basic unit of charge.

Robert A. Millikan taken in 1923 at age 55

J.J. Thomson had "discovered" the **electron**, a tiny negatively charged particle, about ten years before Millikan's experiment. He did this by analyzing electron beams in a tube very similar to the tube where electrons travel in your television. It is these electrons that hit the screen and make the TV images.

Your penny lab was easy compared to Millikan's oil-drop experiment. The oil drops are so small that in Millikan's experiment he had to view them through a microscope. To find their mass required some ingenuity as well. Millikan sprayed the oil droplets between a positively charged plate and a negatively charged plate. If the oil drop had a negative charge, it would be repelled from the negative plate and attracted to the positive plate. If the positive plate were on top, the electric force would be pulling the drop up, while gravity would be pulling the drop down. If the two forces were equal, the drop would come to rest and remain suspended (or travel at a slow constant speed). By comparing the electrical force and the gravitational force, the charge on the oil drop (i.e., the charge of the electrons) could be found. Millikan won the Nobel Prize for this experiment.

The diagram shows the oil drop, the electrical plates, and the battery voltage that provides the charge on the plates.

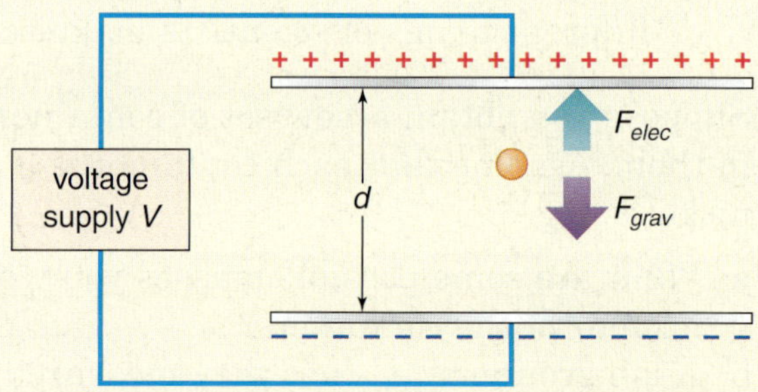

The weight of the oil drop (*mg*) can be found by observing the oil drop as it falls because of air resistance. You can calculate the electrical force from the voltage, charge, and distance between the plates. You can calculate the gravitational force from the mass and the acceleration due to gravity. Using these equations, you can find the charge on the oil drops. The voltage of the power supply and the distance between the plates can be measured:

$$F_{elec} = F_{grav}$$

$$\frac{qV}{d} = mg$$

$$q = \frac{mgd}{V}$$

From Millikan's and many additional experiments, the charge on an electron is determined to be 1.6×10^{-19} C (coulombs). You could expect to see twice this charge, three times this charge, nine times this charge, and any other whole multiple of this charge. If you never see a fractional part of this charge, you assume that charge is indivisible.

Current theories of physics state that a $\frac{1}{3}$ charge and a $\frac{2}{3}$ charge can exist. There is evidence for these fractional charges and the quarks (the tiniest components from which matter is made) associated with them. The Millikan oil-drop results led to the conclusion that quarks always join up to make a total charge of +1 or –1.

Reflecting on the Activity and the Challenge

Electrons are part of the atom. In your museum exhibit, you will have to include the charge on an electron. You may also find a way to include how the Millikan oil-drop experiment helped scientists find out about the charge and its indivisibility. You may choose to make a part of the exhibit deal with electrons and electric charge interesting by making it interactive.

Physics To Go

1. Two students are playing tug-of-war with a rope. How is this game similar to the two forces in the Millikan oil-drop experiment?
2. Describe how you could make Millikan's oil-drop experiment into an exciting interactive display.
3. A donut can be split into two pieces. Does the Millikan experiment prohibit the electric charge from being split into two pieces?
4. Assume that a container has a mass of 10 g and each penny has a mass of 3 g.
   a) Make a list of possible masses of five containers that have 1 penny, 2 pennies, 5 pennies, 10 pennies, and 12 pennies.
   b) List two masses you would not find for a container with pennies.
5. What is the net static electric charge on a metal sphere having an excess of +3 elementary charges (electrons)?
6. How many coulombs are equivalent to the charge of 100 electrons?

Physics Word

electron: a negatively charged particle with a charge of 1.6×10^{-19} coulombs and a mass of 9.1×10^{-31} kg.

7. Which electric charge is possible?
   a) 6.32×10^{-18} C b) 3.2×10^{-19} C
   c) 8.0×10^{-20} C d) 2.4×10^{-19} C
8. An oil drop has a charge of -4.8×10^{-19} C. How many excess electrons does the oil drop have?
9. J.J. Thomson, the discoverer of the electron, tried to describe the enormity of the discovery of this tiny particle: "Could anything at first sight seem more impractical than a body which is so small that its mass is an insignificant fraction of the mass of an atom of hydrogen, which itself is so small that a crowd of these atoms equal in number to the population of the whole world would be too small to have been detected by any means then known to science?" Create a quote of your own that captures the excitement of the discovery of the electron. Perhaps the quote could be displayed near the entrance to your proposed museum exhibit.
10. Quarks have charges $+\frac{1}{3}, -\frac{1}{3}, +\frac{2}{3}, -\frac{2}{3}$.
    a) Show how three quarks can combine to create a particle with a total charge of +1.
    b) Show how three quarks can combine to create a particle with a total charge of 0.
    c) Show how three quarks can combine to create a particle with a total charge of –1.

Stretching Exercise

The Millikan oil-drop experiment can be completed in a high school lab. It does require the use of a microscope and incredible patience to view the drops. Design a computer simulation or game that works like the Millikan oil-drop experiment. The simulation should have these features:

- The screen should look like the apparatus with a variable power supply and oil drops between the plates.
- The drop should be able to get a new charge.
- The drop should be able to move.
- The voltage should be able to be varied so that the net force on the drop is zero and the drop travels at constant velocity or is at rest.
- The velocity of the drop should be able to be measured to determine if it is traveling at a constant velocity.
- New drops should be able to be inserted between the plates.

Activity 3 How Big Is Small?

GOALS

In this activity you will:

- Measure the size of a penny using statistical techniques.
- Compare statistical measurements to direct measurements.
- Relate the Rutherford scattering experiment to the penny simulation.
- Describe the relative scale of the nucleus to the atom.

What Do You Think?

You may have heard that the salt you use on your food is made up of sodium chloride ions (NaCl).

- **If you could magnify a grain of salt so that the sodium ion was the size of a penny, where would you find the neighboring chloride ion? Where would you find the next sodium ion?**

Record your ideas about these questions in your *Active Physics* log. Be prepared to discuss your responses with your small group and your class.

For You To Do

1. Work with a partner. Use a ruler and a pencil to outline a square that is 10 cm × 10 cm on a card. Trace a penny as many times as you like within the square. Draw the circles so that they do not touch each other. Make the circles as close to the actual size of the penny as you can.

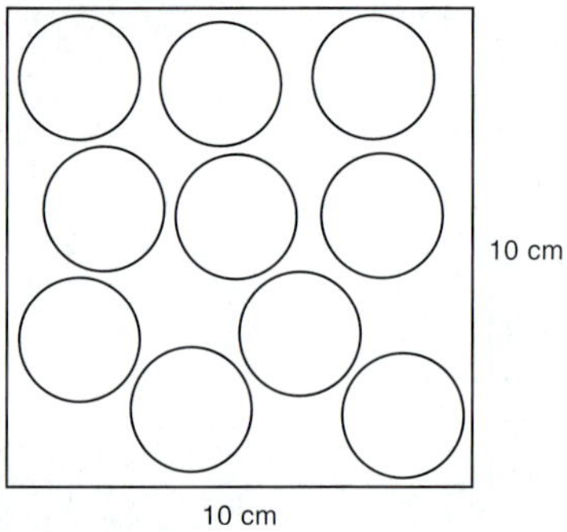

2. Drop a pencil onto the card so that the point hits within the square. Do not aim the pencil. It is actually better if you don't look. You want the drops to be as random as possible.
   a) Have your partner record whether you "hit" a circle, or "miss" a circle. If the pencil falls outside the square, ignore that drop.
   b) Make 50 "countable" drops. Then switch roles with your partner. Continue until 100 drops are recorded.

3. Find the area of all the pennies using the ratio:

$$\frac{\text{hits}}{\text{drops}} = \frac{\text{area of all pennies}}{\text{total area}}$$

   a) Show your calculations in your log.

4. Find the area of one penny by dividing the area of all the pennies by the number of penny outlines on your card.
   a) Record your calculation in your log.
   b) Compare your value with that of other lab groups. How close are the values?

5. You can also find the area of one penny using this equation: Area = πr^2, where $\pi = 3.14$ and r = the radius of the penny. To use this equation, you must find the radius of a penny.
   a) Measure the diameter of a one-penny circle on your card with a ruler. Record your measurement. (The radius is one-half the diameter.)
   b) Calculate the area of a penny and record your calculation in your log.
   c) If you wish to get a more accurate value of the diameter of a penny, you could line up 10 pennies before making the ruler measurement. Why would that give you a better result?

6. Compare the results you obtained using indirect measurement (dropping the pencil) and direct measurement (using a ruler).
   a) How close are the results you got using the indirect and direct methods?
   b) Compare your results with those of another lab group. How do your results compare?
   c) Which method is more accurate? Explain your answer.
   d) Why is it important not to aim the pencil in the indirect measurement?

7. If possible, average the results of the indirect measurements and direct measurements from the entire class.

   a) How close are the results the class got using the indirect and direct methods?

8. Extend the experiment to an unknown object. Suppose a student conducts a similar experiment, but replaces the penny with a *single* unknown object. If that student got 50 hits out of 100 drops, then you would conclude that the unknown object's area was approximately 50% or one-half of the total area of the 10 cm × 10 cm square. You don't know from the data the shape or the location of the object. All of the following are possible because one half of the total area of each square is shaded.

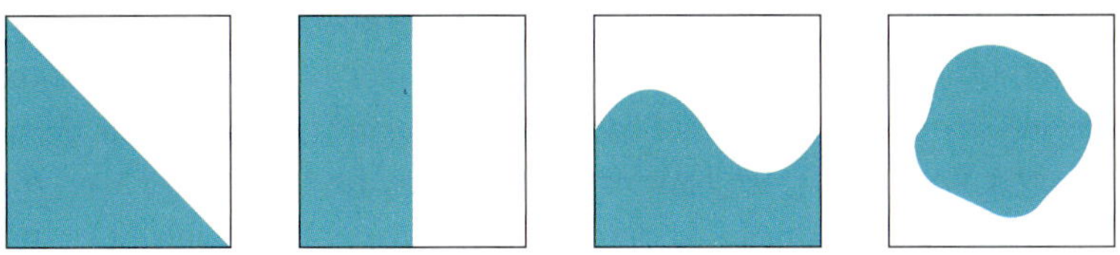

   a) What might the unknown object look like if it was reported to have 75 hits out of 100 drops? Draw the outer square and the unknown object's size within it.
   b) What might the unknown object look like if it was reported to have 25 hits out of 100 drops? Draw the square and the unknown object's size.
   c) What might the unknown object look like if it was reported to have one hit out of 100 drops? Draw the square and the unknown object's size.
   d) What might the unknown object look like if it was reported to have one hit out of 10,000 drops? Draw the square and the unknown object's size.

FOR YOU TO READ

Measuring the Size of the Nucleus

In this activity you used indirect measurements to find the area of a penny. Indirect measurement has been very useful in science. It continues to be useful. Finding the size of the penny without directly measuring it may seem like a good trick. However, you know that you can always verify the size by measuring a penny with a ruler. The **atom** cannot be measured with a ruler. Scientists must rely on indirect measurements.

A key scientific discovery—the discovery of the atomic **nucleus**—was made using a method similar to the one you used in this activity. Ernest Rutherford and his colleagues, Hans Geiger and Ernst Marsden, made the discovery. In the lab, the team bombarded a piece of thin gold foil with a beam of positively charged alpha particles. They originally thought that an atom had positive charges spread evenly through it, as shown in the diagram. If this idea were true, the alpha particles would go through the foil in nearly straight lines. The alpha-particle beam was their "dropping pencil." The foil was their card. Most of the particles did go straight through the gold foil. However, to the surprise of the team, Marsden saw that a very few of the alpha particles bounced back towards the source of the beam.

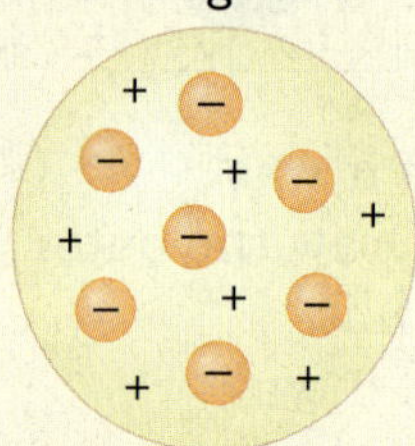

Rutherford thought deeply about these observations. He also thought about all of the ideas that scientists had about what was in atoms. He concluded that the positively charged alpha particles bounced back when they hit an area of concentrated positive charge. He also thought that the charge was in the center of the atom, the nucleus. Rutherford used his results to calculate the size of the nucleus of these atoms. He could have compared the number of "hits," particles that bounced back, with the total number of particles sent toward the foil. Then he could have used that ratio to determine the area of the foil where the atomic nuclei could be found. Rutherford's mathematics was a bit more complicated. In his situation the hit was not yes or no. It was an angle of deflection that could range from 0° to 180°. In future courses you may study those details. The result was that he calculated that the diameter of the atomic nucleus was 10^{-15} m. The diameter of the atom is 10^{-10} m. The nucleus is only $\frac{1}{100{,}000}$ the size of the atom.

RUTHERFORD'S EXPERIMENT

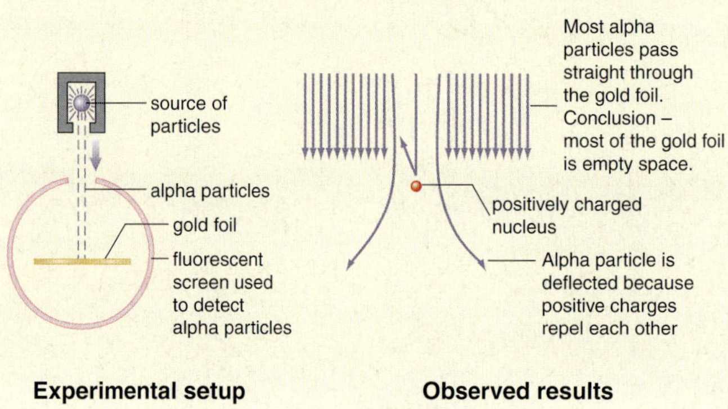

It would be great if it were possible to check Rutherford's indirect measurement. A direct measurement of the size of a nucleus is impossible. Only the indirect measurement is available. It is used as evidence of the existence of the nucleus and its size.

Rutherford's nucleus provides a view of matter as mostly empty space. Electrons surround a tiny, tiny nucleus. Why then do solids appear so solid? How can the empty space of your fist hurt so much when it hits the empty space of a table? Electrons whizzing around the nuclei of the atoms in your fist are repelled by the electrons whizzing around the nuclei of the atoms in the table. The closer you try to bring your fist to the table, the stronger the force of repulsion between the electrons. The force of repulsion can be greater than the force of attraction holding your fist together and bones can break.

How can empty space exert such forces? Imagine a thin propeller blade. If the blade is still, it is surrounded by mostly empty space. When the blade rotates, it seems to fill that empty space and create a solid wall. The tiny electrons, in rapid motion, create a similar effect. It is as if the electrons are everywhere at once and the empty space appears solid.

Your sense of touch is the repulsion of electrons. It follows Coulomb's Law: $F = \frac{kq_1q_2}{d^2}$. It's amazing that you are able to feel this force when it is as tiny as a kiss or as large as striking a table with your fist. The next time you kiss someone, remember that you are experiencing the repulsion of electrons.

Reflecting on the Activity and the Challenge

By shooting particles at thin foils and seeing how the particles scatter, you can investigate the structure of matter. Rutherford's scattering experiment revealed that a tiny nucleus contains all of the positive charge of the atom. This also implies that most of the atom is empty space. In your museum exhibit, you will certainly want to help visitors to get a sense of the structure of the atom with its tiny nucleus. Is there a way in which you can get visitors to your exhibit to explore the size of the nucleus? You may also want to help visitors to understand how "mostly empty space" can make hard, solid objects.

Probability was important in this indirect method of measuring the size of a penny or the size of the nucleus. If you had aimed the pencil, the experiment would not have given good results. The use of probability and indirect measurements may be something you wish to include in your museum exhibit.

The museum exhibit must capture a visitor's attention within 30 seconds. How can you "grab" the visitor?

Physics Words

atom: the smallest particle of an element that has all the element's properties; it consists of a nucleus surrounded by electrons.

nucleus: the positively charged dense center of an atom containing neutrons and protons.

Physics To Go

1. Determine the size of a quarter indirectly by repeating the penny experiment.
   a) Record your results.
   b) How close are the results you got using the direct and indirect method of measurement?
   c) Which method is more accurate? Explain your answer.

2. Repeat the quarter experiment, but this time aim at the card.
   a) Record your results.
   b) How does aiming change the results?

3. Which is greater, 10^{-15} m or 10^{-10} m? How many times greater?

4. Find the areas of circles with the following diameters:
   a) 4 cm
   b) 7 cm
   c) 10 cm

5. Why do you get better results when you drop the pencil 1000 times instead of 10 times?

6. You drop a pencil 100 times and get 23 "hits." There are seven coins on a card 10 cm $\times$ 10 cm. How large in area is each coin?

7. If the nucleus could be enlarged with a projector so that it was 1 cm in diameter, how far away would the next nucleus be? (Each nucleus is 10^{-15} m and each atom is 10^{-10} m.)

8. In the Rutherford scattering experiment, the alpha particles were deflected at different angles. It was not simply a hit or miss as it was in your penny simulation. How would you expect the deflection to be affected as the positive alpha particles came closer to a positive nucleus?

9. How might you show the proper scale of the nucleus and the atom in your museum exhibit?

Activity 4 Hydrogen Spectra and Bohr's Model

GOALS

In this activity you will:

- Observe the colors of light given off by tubes of hydrogen, helium and neon.
- Record the wavelengths of the light given off by the gases as measured by the grating spectrometer.
- Calculate the energy levels of the Bohr hydrogen atom.
- Describe how the electrons jump from one orbit to another and give off light of a specific wavelength.
- Match the theoretical, calculated values for the wavelengths of light with the observed values of the wavelength.
- Cite evidence for why we think that the Sun is composed of hydrogen and helium gases.
- Calculate the wavelengths of light emitted from transitions of electrons in the Bohr atom.

What Do You Think?

It is said that there are no two snowflakes alike.

- **How do you distinguish one person from another?**

Record your ideas about this question in your *Active Physics* log. Be prepared to discuss your responses with your small group and your class.

For You To Do

1. You are going to use your observation of colors to shed light on the structure of the atom. Your teacher will set up tubes of hydrogen, helium, and neon gases and connect each to a high-voltage power supply. Observe the light of each tube with the naked eye. Then observe the same light with a spectrometer.

a) Record your observations in your *Active Physics* log.

b) The spectrometer has a scale and values within it that are measurements of the wavelengths of light. Record the wavelengths that correspond to each color of light that you are observing. (The wavelengths are measured in nanometers (nm). The prefix *nano* means 10^{-9}.)

2. Each of the three spectrum tubes has a distinct pattern of colors that comprise it. These color bands correspond to specific wavelengths of light. Each gas has its own distinct set of wavelengths.

a) Write down three wavelengths of light from one gas tube. Pass this list onto someone else in your lab group. Have that person identify the name of the gas. Were you all successful?

Scientists try to determine characteristic properties of substances. A characteristic property is a unique attribute that can be used to identify that substance and distinguish it from other substances. Fingerprints or DNA patterns for humans are characteristic properties. No two people have been found who share the same fingerprint or the same DNA. The spectrum of light from a gas is a characteristic property of that gas.

Only your teacher will handle the power supply and the tubes of gas. The power supply uses high voltage, which can be dangerous. The tubes of gas are glass and can be broken, making sharp edges. Do not look into bright lights or the Sun with the spectrometer.

3. When viewing the Sun, a set of observed wavelengths had the following values: 434 nm, 471 nm, 486 nm, 588 nm, 656 nm, and 668 nm.

a) Which gas on Earth emits three of these wavelengths of light?

b) Which gas on Earth emits the other three wavelengths?

c) From these observations, what can you conclude about the gases that comprise the Sun?

4. In 1885, Johann Balmer, a Swiss high school physics teacher, created a simple equation to calculate the wavelengths of the hydrogen lines (the color bands you observed).

$$\lambda_n = 3646 \frac{n^2}{n^2 - 4} \times 10^{-10} \text{ m}$$

All wavelengths known to Balmer can be calculated by using this equation and substituting $n = 3, 4, 5, 6$.

Example:
Calculate the wavelength λ corresponding to $n = 3$.

$$\lambda_3 = 3646 \frac{3^2}{3^2 - 4} \times 10^{-10}\ \text{m}$$
$$= 6562 \times 10^{-10}\ \text{m}$$
$$= 656\ \text{nm}$$

a) Calculate the wavelengths corresponding to $n = 4$ and $n = 5$.

5. Niels Bohr explained the significance of Balmer's equation. In the Bohr model of an atom of hydrogen, the proton is the nucleus and the single electron orbits the proton. The electron is able to move in a circle about the proton because of the attractive Coulomb force.

Niels Bohr received the Nobel Prize in Physics in 1922.

a) Using Coulomb's equation:

$$F = \frac{kq_1q_2}{d^2}$$

calculate the force between the proton and electron (each has a charge of 1.6×10^{-19} C). The distance between them is 5×10^{-10} m.

6. Bohr was able to calculate the radii of the orbits and their corresponding energies. The energy levels derived from the Coulomb force equation can be found using this *monster* of an equation:

$$E = -\frac{2\pi^2k^2mq^4}{h^2}\frac{1}{n^2}$$
$$= -13.6\frac{1}{n^2}\ \text{eV}$$

(You don't have to memorize this equation. However, there are some features that you can understand.)

a) Write down which of the symbols in the equation you recognize and what they represent. You should be able to provide the meanings for π, q, m, and k. The n has the values of 1, 2, 3… and corresponds to the energy level. The h is Planck's constant and is a number always equal to 6.63×10^{-34} Js. This is a very, very tiny number. It has an exponent of – 34! It is one of the most important numbers in physics and is the key to all of quantum physics.

b) Since all of these numbers are constants, you can do the calculation and find that together they equal 13.6 eV. (The eV is the symbol for an electron volt. It is the energy an electron gains from a one-volt battery. It is much smaller than one joule and is very convenient to use when discussing these small energies. 1 eV = 1.6×10^{-19} J.) What are the values of the constants π, q, m, and k?

c) Calculate the energies of each Bohr orbit using the equation:

$$E = -13.6\frac{1}{n^2}\text{ eV}$$

for $n = 1, 2, 3, 4, 5, 6$.

7. In physics, when a particle is bound to another particle, it is defined to have "negative" energy. In hydrogen, an electron in orbit about the proton has an energy of – 13.6 eV. An electron in E_1 would have to be given 13.6 eV to free it.

a) Explain why an electron in E_2 would have to be given 3.4 eV to free it.

b) How much energy would have to be given to an electron in E_3, E_4, E_5 and E_6 to free it?

8. The energy required to free an electron from a nucleus is called its ionization energy. The ionization energy of an electron in the E_1 or ground state is 13.6 eV. The ionization energy of an electron in the E_2 state is 3.4 eV.

a) What are the ionization energies of the electron in the $n = 3, 4, 5, 6$ states?

9. Niels Bohr explained why light was given off by the hydrogen atom. The electron begins in one orbit and then jumps to a lower orbit. During that jump it loses energy. Energy is conserved. In an isolated system, if one object loses energy, another object must gain energy so that the total energy remains the same. The electron lost energy. The light was created with this energy.

Example:
Calculate the energy of the light when the electron jumps from E_3 to E_2.
The energy of E_3 = – 1.51 eV
The energy of E_2 = – 3.40 eV
An electron jumping from E_3 to E_2 would have a change of energy ΔE.

$$\begin{aligned}\Delta E &= E_{final} \text{ to } E_{initial}\\ &= E_2 \text{ to } E_3\\ &= E_2 - E_3\\ &= -3.40 \text{ eV} - (-1.51 \text{ eV})\\ &= -3.40 \text{ eV} + 1.51 \text{ eV}\\ &= -1.89 \text{ eV}\end{aligned}$$

The electron lost 1.89 eV of energy. Light was created with exactly this energy.

a) Calculate the change of energy ΔE when an electron jumps from E_4 to E_2, E_5 to E_2, E_6 to E_2.

10. Albert Einstein postulated the relationship between the energy of light and its frequency in 1905.

$$E = hf$$

where E is the energy of light and
f is the frequency of light.

Wavelengths of light (λ) can be found from the wave equation

$$c = f\lambda$$

where c is the speed of light (3.0×10^8 m/s).

It is possible to combine Bohr's calculation of the energy given to the light when the electron jumps from E_3 to E_2 and Einstein's calculation of the corresponding frequency (or wavelength). You must also convert from the energy unit of electron volts to joules so that the wavelength will be in meters.

$$\lambda = \frac{hc}{|\Delta E|}$$

$$= \frac{1.24 \times 10^6 \text{ (m)(eV)}}{|\Delta E|}$$

ΔE is the energy change in electron volts (eV).

Example:

When the electron jumps from E_3 to E_2, the change in energy is 1.89 eV. Calculate the corresponding wavelength of light.

$$\lambda = \frac{hc}{|\Delta E|}$$

$$= \frac{1.24 \times 10^6 (\text{m})(\text{eV})}{|\Delta E|}$$

$$= \frac{1.24 \times 10^6 (\text{m})(\text{eV})}{1.89 \text{ eV}}$$

$$= 654 \text{ nm}$$

This corresponds to the measured wavelength of one of the lines of hydrogen. It also gives the same value as Balmer's earlier equation. (See **Step 4**. Notice that there is a slight difference between the two numbers because some of the numbers in the calculation were rounded.)

a) Calculate the wavelengths of light when the electron jumps from E_4 to E_2, E_5 to E_2, E_6 to E_2.

b) How do these values compare with the ones you found in your observations of the hydrogen spectra?

FOR YOU TO READ

Discovery of Helium

When viewing the Sun, the following set of wavelengths can be observed: 434 nm, 471 nm, 486 nm, 588 nm, 656 nm, and 668 nm. From your observations in this activity, you concluded that these values match the wavelengths emitted by hydrogen and helium. This led you to the conclusion that the Sun is comprised of hydrogen and helium.

The history of these values, however, is a bit more interesting. The set of values corresponding to hydrogen were known from the lab. Nobody had ever observed the second set of wavelengths in a lab. In 1868, Pierre Jenson observed one line from the Sun during a total eclipse in India. J. Norman Lockyear interpreted this yellow line as being evidence of a new element. This set of wavelengths did not correspond to any gas on the Earth and so this unknown gas of the Sun was named "helium" after Helios, the Greek god of the Sun. Years later, in 1895, helium gas was discovered on Earth by William Ramsey of Scotland and independently by Per Cleve of Sweden.

Bohr's Model of an Atom

In this activity you observed the spectral lines of several gases. How are the spectral lines created? Are the lines telling you something about the structure of hydrogen, helium, neon, and the other elements? Are the lines a means by which nature is revealing its secrets?

Johann Balmer created a simple equation to calculate the wavelengths of the hydrogen lines.

$$\lambda_n = 3646 \frac{n^2}{n^2 - 4} \times 10^{-10} \text{ m}$$

Balmer's equation worked well. It illustrated the pattern in the wavelengths of light from hydrogen, but it opened up new mysteries. Where did this "magic" number 3646 come from? Why is n^2 used? These questions would be answered by Niels Bohr in the early 20th century.

Niels Bohr took the Rutherford model of the atom, which had a tiny nucleus in the center and orbiting electrons. From this, he created a new model for the emission of atomic spectra. Hydrogen has only one electron and one proton. In the Bohr model, the proton is the nucleus and the single electron orbits the proton in the same way that the planets orbit the Sun. The electron is able to move in a circle about the proton because of the attractive Coulomb force. This force keeps the electron moving in a circle of that radius about the proton. Using certain assumptions, Bohr was able to derive the following equation that gives the energy levels of the various orbits:

$$E = -\frac{2\pi^2 k^2 m q^4}{h^2} \frac{1}{n^2}$$

However, there were still two puzzles you need to understand:

- the – sign in the equation and
- the relationship between the energy levels and the wavelengths of light emitted in the spectra

You were told that when a particle is bound to another particle, in physics, it is defined to have "negative" energy. For instance, the planets orbiting the Sun are bound to the Sun and their total energy would be negative. All objects on the Earth are bound to the Earth and their total energy would also be negative. If you wanted to take an object and free it from the Earth, you would have to add energy to it. That is why incredibly large rockets are needed to lift spaceships off the Earth. When you throw a ball in the air, you also add energy, but the total energy is still negative and it comes back to Earth.

In hydrogen, an electron in orbit in the E_1 or ground state about the proton has an energy of – 13.6 eV. To free the electron, it must receive 13.6 eV of energy. This would bring its total energy to zero and it would be free. If it receives less than – 13.6 eV, it may move away from the proton nucleus but it will return to the nucleus. If it receives more than 13.6 eV, it is freed from the nucleus and has extra energy to move about.

Through calculations that you did in this activity you found that Balmer's equation worked as well as Bohr's equation. What then is gained from Bohr's work? Bohr provided a new theory, an explanation, and a mechanism by which the light is created. Balmer cleverly found a shortcut equation. Balmer invented a number to use to make the wavelengths turn out right. Bohr's theory explained that the number comes from combining Coulomb's Law with Einstein's relationship between the energy and frequency of light and the conservation of energy. Bohr's theory provides an insight into the structure of the atom. Electrons surround the nucleus. The electrons are in fixed orbits. →

When an electron drops from one orbit to another, light is given off.

Ionization energy is the energy required to free an electron. Energy can be provided to free an electron by having the electron absorb light. When the electron absorbs light, it jumps up to a higher energy state. If the electron absorbs enough light, it can be freed from the nucleus. This is the inverse of the emission process Bohr described to explain why light is emitted by an atom.

The only light that can be absorbed (or emitted) by the electron is the light that has just the right energy and just the right wavelength to get the electron to another fixed orbit. If the light has a little less energy or a bit more energy, it goes right past the electron and has no effect.

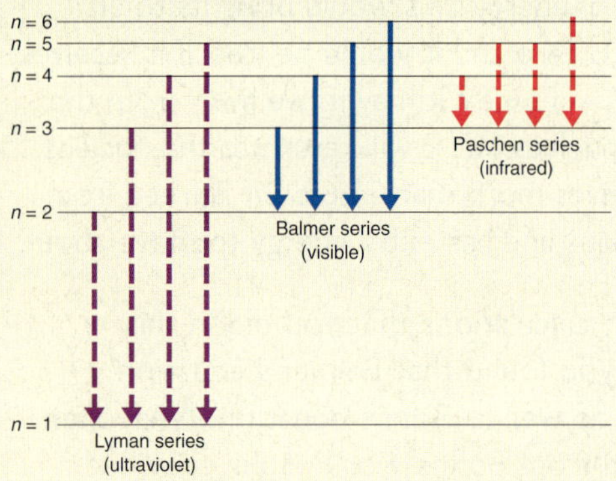

The energy transitions of the light emitted when the electron drops from E_3 to E_2, E_4 to E_2, and E_5 to E_2 can be calculated. You viewed these spectral lines using a spectrometer. You might expect light to be given off when the electron also jumps from E_2 to E_1, E_3 to E_1, and E_4 to E_1. These energy transitions can be calculated, but you were not able to see them. The light with those energies is not visible to the human eye. Visible light is only one part of the electromagnetic spectrum. If you use an ultraviolet light detector, you find that these additional wavelengths of light also exist. The visible light rays are called the Balmer series. The ultraviolet series is called the Lyman series.

Light should also be emitted when the electron drops from E_4 to E_3, E_5 to E_3, and E_6 to E_3. These energies can also be calculated. The light with those energies is also not visible to the human eye. If you use an infrared light detector, you find that these additional wavelengths of light also exist. They are called the Paschen series.

A good new theory should be able to explain what the old theory could explain. It should also be able to explain something that the old theory could not explain. Finally, it should be able to make a prediction of something that nobody had thought of previously. If that prediction turns out to be true, then we sense the power of the theory. The Paschen series had been observed before Bohr's theory. The Lyman series had never been observed. Bohr predicted the wavelengths of the Lyman series and then they were observed with the predicted wavelengths. Other spectra corresponding to transitions to the n = 4 state were also found later.

Reflecting on the Activity and the Challenge

The Bohr atom has electrons in fixed orbits surrounding the nucleus. Light is emitted when electrons jump from a higher energy orbit to a lower orbit. An electron that absorbs light can jump from a lower orbit to a higher energy orbit. The wavelengths of light can be calculated, observed, and measured. The values from Bohr's theory and your observations from hydrogen are almost exactly equal. In your museum exhibit, you may try to illustrate the Bohr model of the atom and the emission of light as electrons jump from one energy level to another. You may also wish to show how an atom becomes ionized when the electron absorbs enough energy to free it. Finally, you may choose to show that invisible light in the ultraviolet and infrared regions are also emitted. Electron jumps and emitted light can be an interactive museum display. Will your display create an immediate interest? Provide a means by which the museum visitor will want to stop and see what is going on with hydrogen.

Physics Word

ionization energy: the energy required to free an electron from an atom.

Physics To Go

1. Light of greater energy has a higher frequency. In the hydrogen spectrum, which visible line has the greatest energy? Which transition does it correspond to?
2. Compare the energy of the light emitted from the electron jump from $n = 3$ to $n = 2$ to the light emitted from $n = 5$ to $n = 2$.
3. Given that the speed of light equals 3.0×10^8 m/s and the wavelength of light is 389 nm (389×10^{-9} m), calculate the frequency of the light.
4. Calculate the energies of each Bohr orbit using the equation $E = -13.6 \text{ eV} \left(\frac{1}{n^2}\right)$ for $n = 1, 2, 3, 4, 5$.
5. Make a diagram showing the energies of each Bohr orbit as a vertical number line which goes from – 013.6 eV to 0 eV.
6. How could the electron transitions be creatively displayed in a museum exhibit? Describe a way in which the display could be interactive.
7. The hydrogen spectra is said to be a "fingerprint" for hydrogen. Explain why this is a useful metaphor (comparison).
8. Why can't light of 500 nm be given off from hydrogen?
9. Electrons can jump from the $n = 4$ state directly to the $n = 3$ or $n = 2$ or $n = 1$ states. Similarly, there are multiple jumps from $n = 3$. How many different wavelengths of light can be given off from electrons that begin in the $n = 4$ orbit?

Activity 5 Extending and Amending the Model

GOALS

In this activity you will:

- Observe the diffraction of light waves.
- Observe the interference of sound waves.
- Observe the interference of light waves.
- Create a model of wave interference.
- Measure the wavelength of light using a Young's double slit apparatus.
- Solve simple problems of the photoelectric effect.
- Describe the wave-particle duality of light.
- Describe the wave-particle duality of electrons.
- Represent an electron confined to a one-dimensional box and note the probability of where the electron will be located.

What Do You Think?

A boy asks one girl for a date, and she answers "yes." Another boy asks a second girl for a date, and she answers "no." The same "cause" did not produce the same "effect."

- **In what situations does the same "cause" always produce the same "effect"?**
- **In what other situations does the same "cause" not always produce the same "effect"?**

In your *Active Physics* log, write down one cause-effect relationship that always holds and one cause-effect relationship that only holds occasionally. Be prepared to discuss your responses with your small group and your class.

For You To Do

In order to better understand the structure of the atom, you need to better understand the behavior of electrons. To understand electrons, you will first look at the nature of light.

1. Shine a laser beam against a wall. Trace the beam onto a piece of paper and measure its thickness.
   a) Record your measurement.
   b) Place a single slit in front of the beam. Measure and record the thickness of the beam again.
   c) Place a thinner slit in front of the beam. Measure and record the thickness a final time.
   d) What happens to the width of the laser beam as it passes through a smaller and smaller opening?

Never look directly at a laser beam or shine a laser beam into someone's eyes. Always work above the plane of the beam and beware of reflections from shiny surfaces.

The process of light bending as it squeezes through a small opening is called diffraction. Light bending around an edge is also referred to as diffraction. Diffraction is one of the properties that water waves and sound waves exhibit. This leads us to the conclusion that light is a wave.

2. In the diagram shown, water waves have traveled through two openings. The solid lines represent the crests of the waves. The dotted lines represent the troughs of the waves.
   a) Sketch the diagram in your log.
   b) When waves meet, they interfere with one another. Constructive interference occurs when the crests of the first wave meet the crests of the second wave. Constructive interference is where you would see large disturbances in the water. On the sketch, mark with an X the regions of constructive interference where the crests intersect.
   c) Destructive interference occurs when the troughs of one wave meet the crests of the other wave. Destructive interference is where you would see undisturbed water. Mark with an O the regions of destructive interference where the crests of one wave interact with the troughs of another.
   d) There are lines of constructive interference which can be seen from the pattern of Xs. Add these lines. There are lines of destructive interference formed by the pattern of Os. Add these lines as well.

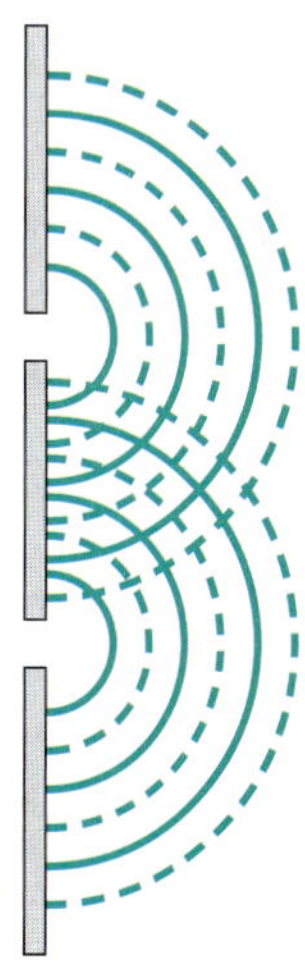

As you can see, there are positions of constructive interference (lots of movement of the water) and positions nearby of destructive interference (where the water is still).

3. Hit a tuning fork gently against a rubber stand. Place the tuning fork near your ear. Slowly rotate the tuning fork so that one prong gets closer to the ear while the other gets further away. Listen carefully to the volume of the sound.

Do not touch the vibrating tuning fork to your skin, especially near your ear.

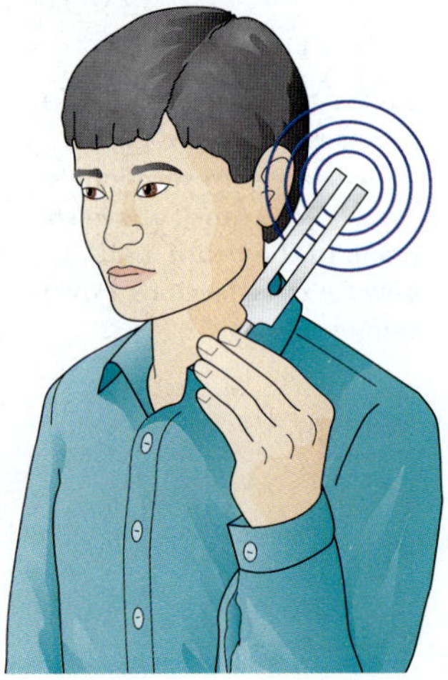

a) Record your observations in your log.

b) A tuning fork produces identical sounds from each prong. At one orientation from the prongs to your eardrum, you heard a loud sound. This was constructive interference. When the orientation from the prongs to your ear was different, the sound diminished. The interference of waves is another property of wave behavior. If you are unsure that you observed the interference of sound, listen to the tuning fork again.

4. Shine a laser beam through two slits or a diffraction grating made up of many slits. Direct the beam at a distant wall. Observe the pattern of the light on the wall.

a) Record your observations in your log.

b) From your observations, what can you conclude about the ability of light to interfere? Explain.

On the distant wall you will see places where the light from neighboring slits interfere constructively (there is maximum light) and places where the light interferes destructively (there is little or no light). Evidence of the interference of light would seem to prove that light behaves like a wave.

Never look directly at a laser beam or shine a laser beam into someone's eyes. Always work above the plane of the beam and beware of reflections from shiny surfaces.

c) Water and sound travel in waves. From your observations, what could you conclude about how light travels? Explain.

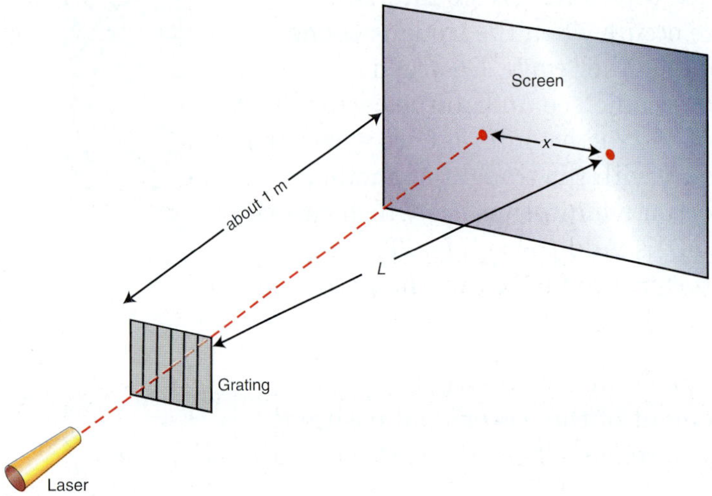

5. You can use the interference of light to measure the wavelength of light. Mount a diffraction grating in the path of a laser beam. Mount a screen several meters away from the grating as shown. Observe the pattern of spots on the screen.

a) Measure and record the separation between one spot and the next, x.

b) Measure and record the distance from the grating to the screen, L.

c) Measure and record the spacing between the lines of the grating, d. Alternatively, you can use the spacing given by the manufacturer.

6. From your measurements, find the wavelength of the light. You will use the following equation:

$$\lambda = \frac{dx}{L}$$

where λ is the wavelength of laser light,
L is the distance from the grating to the spot on the screen,
x is the separation between the spots,
d is the spacing between lines in the grating.

a) Show your calculations in your log.

The Photoelectric Effect

Now that you have found that light behaves like a wave and measured its wavelength, you can turn to the puzzle of the 20th century that Einstein solved. The puzzle had to do with the behavior of light when it freed an electron from a metal. In the photoelectric effect, beams of light shine on metals. Some wavelengths (and frequencies) of light will always free electrons, while other wavelengths (and frequencies) will never free electrons.

7. Why do some wavelengths (and frequencies) of light always free electrons, while other wavelengths (and frequencies) never free electrons? To understand the photoelectric effect, think of using a vending machine.
A vending machine sells potato chips. Suppose a bag of potato chips costs 10¢. The machine is not able to add coins.

- If you place 10¢ in the machine, a bag of potato chips comes out.
- If you place 5¢ in the machine, your 5¢ is returned.
- If you place 25¢ in the machine, a bag of potato chips comes out and you get 15¢ in change.
- If you place two nickels in the machine, the two nickels are returned (the machine cannot add coins).

a) What would happen if you placed a 50-cent piece in the machine?

b) What would happen if you placed 20 pennies in the machine?

c) What would happen if you placed a $1.00 coin in the machine?

d) An equation that could describe the behavior of the machine would be:
returned money = money inserted – cost of chips.
Does this equation work for the examples above?

In the photoelectric effect:

- Some frequencies of light will free electrons. The brighter the light, the more electrons freed. This is similar to having lots of dimes to place in the chip machine.
- Some frequencies of light will never free electrons, regardless of how much light there is. This is similar to having lots of pennies or nickels to place in the chip machine.
- Some frequencies of light will free electrons and give them lots of kinetic energy. This is similar to quarters being placed in the chip machine. Each quarter gets a bag of chips and some change.

An equation that describes the photoelectric effect is:
kinetic energy of freed electron =
energy of light – energy required to free an electron (work function).

$$KE_{electron} = E_{light} - w_o$$

The energy of the light E_{light} is equal to hf where h is Planck's constant (6.63×10^{-34} Js) and f is the frequency of light.

The photoelectric effect explanation makes sense only if light were behaving as a particle and there were a collision between the photon and the electron.

8. Light and electrons both exhibit wave and particle behavior. This wave-particle duality is the hallmark finding of the 20th century conception of matter. The electron's particle characteristics include its mass, its charge, and its ability to hit a small target.

If an electron had wave properties, then it would have a wavelength. Bohr's electrons are in fixed orbits because only specific circumference orbits would be equal in length to multiples of the wavelength of the electrons.

a) Draw a circle about a nucleus. Try to draw a wave so that it meets perfectly. See the diagram for one such example.

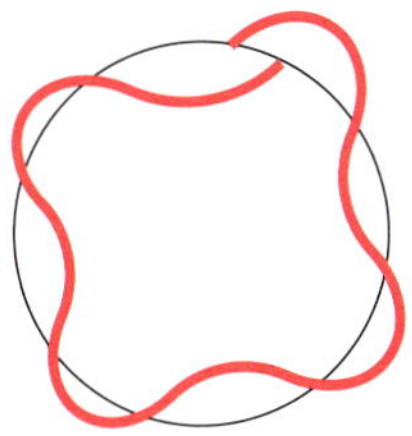

The wave does not fit.
This is not a permissible Bohr orbit.

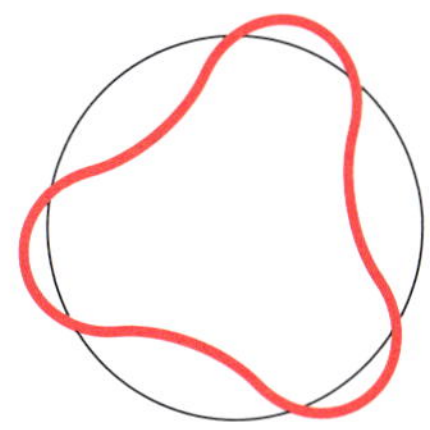

The wave fits.
This is a permissible Bohr orbit.

9. Imagine placing an electron in a one-dimensional box. A one-dimensional box is a line segment. The electron can move back and forth along the line segment. The electron that can exist in the box must have a specific wavelength and corresponding energy. This wavelength is dependent on the dimensions of the box. A few possibilities are shown.

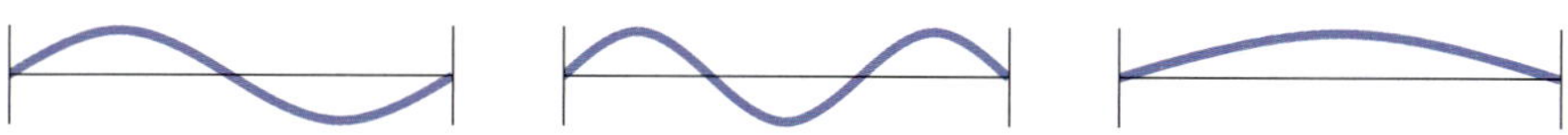

a) Draw two additional waves that could fit in the box.

10. The probability of finding the electron at any location is equal to the square of the amplitude of the wave that fits in the box.

- In the places where the square of the amplitude is large, there is a high probability of finding the electron.
- In the places where the square of the amplitude is small, there is a small probability of finding the electron.
- In the places where the square of the amplitude is zero, there is no probability of finding the electron.

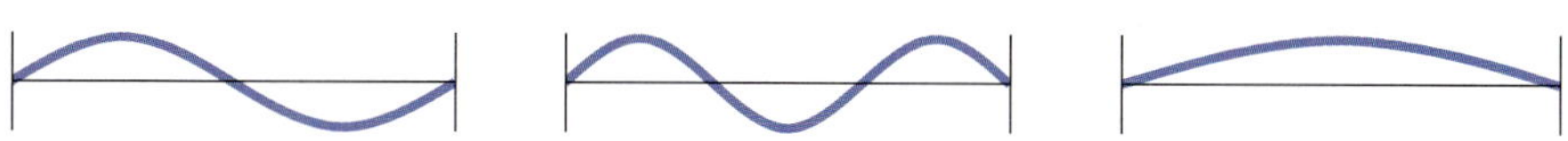

The amplitudes of three waves.

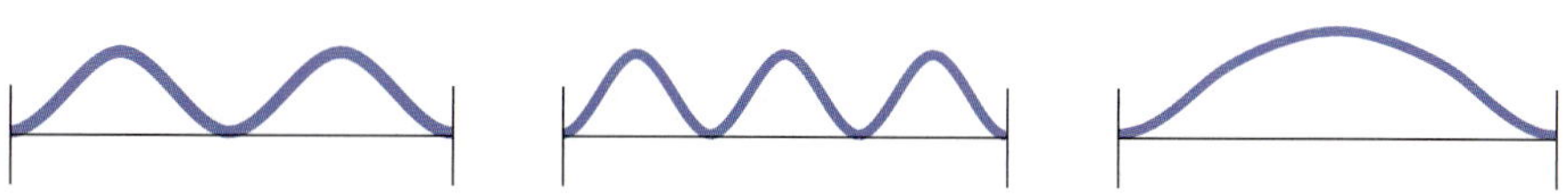

The square of the amplitudes of three waves.

a) Copy these diagrams in your log and describe where you would expect to find the electrons and where you would not expect to find the electrons.

Physics Words

diffraction: the ability of a wave to spread out as it emerges from an opening or moves beyond an obstruction.

constructive interference: the result of superimposing different waves so that two or more waves overlap to produce a wave with a greater amplitude.

destructive interference: the result of superimposing different waves so that two or more waves overlap to produce a wave with a decreased amplitude.

photoelectric effect: the emission of electrons from certain metals when light (electromagnetic radiation) of certain frequencies shines on the metals.

photon: a particle of electromagnetic radiation; a quantum of light energy.

model: a representation of a process, system, or object.

FOR YOU TO READ

Developing the Schrödinger Model

Light is everywhere. You can see because of light, but it is unusual to see light beams. When you do see light rays piercing through the clouds or a laser beam intensely moving forward, you recognize that light travels in straight lines. In this activity you found out that light may also travel a curved path as it squeezes through a small opening.

The process of light bending as it squeezes through a small opening is called **diffraction**. Light bending around an edge is also referred to as diffraction. Diffraction is one of the properties that water waves and sound waves exhibit. This leads toward the conclusion that light might also behave as a wave.

You then explored another property of water and sound waves. When water waves meet, they interfere with one another. **Constructive interference** occurs when the crests of a wave meet the crests of the second wave and there is lots of movement of the water. **Destructive interference** occurs when the troughs of one wave meet the crests of the other wave and the water remains still.

You used a tuning fork to investigate if sound waves also interfere. A tuning fork produces identical sounds from each prong. At one orientation from the prongs to your eardrum, you heard a loud sound. This was constructive interference. When the orientation from the prongs to your ear was different, the sound diminished. The sound from one source can meet with the sound from another source and produce silence! The interference of waves is another property of wave behavior.

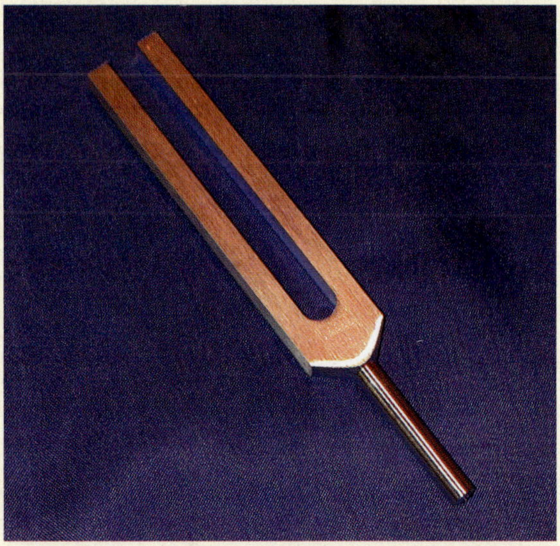

If light behaves like a wave, then it, too, must show interference. When you shone a laser beam through two slits onto a distant wall, you saw places where the light from neighboring slits interfered constructively (there was maximum light) and places where the light interfered destructively (there was little or no light). Evidence of the interference of light would seem to prove that light behaves like a wave. The colors you see when light reflects off a CD are interference effects. You were able to use the interference of light to measure its wavelength.

After you found that light behaves like a wave and measured its wavelength, you turned to the puzzle of the 20th century that Einstein solved. The puzzle had to do with the behavior of light when it freed an electron from a metal. In the **photoelectric effect**, beams of light shine on metals. Some wavelengths (and frequencies) of light will always free electrons, while other wavelengths (and frequencies) will never free electrons. Light of high frequency (ultraviolet light) frees many electrons and gives them kinetic energy. Light of low frequency (red light) is not able to free any electrons.

Einstein derived the following equation for the photoelectric effect in 1905:

$$KE_{electron} = E_{light} - w_o$$

The energy of the light E_{light} is equal to hf where h is Planck's constant (6.63×10^{-34} Js) and f is the frequency of light. It was this equation (not his Theory of Relativity) for which Einstein won the Nobel Prize in physics.

The photoelectric effect explanation makes sense only if light were behaving as a particle and there were a collision between the **photon** and the electron. The extraordinary conclusion is that light sometimes behaves as a particle (the photoelectric effect) and sometimes behaves like a wave (diffraction and interference effects).

Now consider the behavior of electrons. Electrons hit the front of your TV screen and make a temporary mark. Electrons have a mass (9.1×10^{-31} kg) and a charge (1.6×10^{-19} C). Electrons behave like particles. Electrons can also interfere. Clinton Davisson and Lester Germer demonstrated this by shooting electrons through a metal crystal foil. They observed some locations on a distant screen where many electrons landed and other locations where no electrons landed. This was explained as an interference effect of electrons. Electrons, like light, can sometimes behave like a particle and sometimes behave like a wave.

If electrons have wave characteristics, then you can expect some unusual behavior. If you place an electron in a one-dimensional box it can move back and forth along the line segment. When you try to locate the position of the electron, you never find it at the very edge of the box. This is not surprising since you would expect the electron to rebound off the walls at the end of the line segment. In the activity you found that there are times, surprisingly, where you also never find the electron in the middle of the box. This is quite peculiar.

You may be wondering how the electron gets from one side of the box to the other if it is never found in the middle of the box. The explanation is that the electron does not behave like any particle or billiard ball that you are familiar with. You need both the wave and particle languages to fully explain the electron. The electron is neither a wave nor a particle—it's an electron. This need for both models is called the wave-particle duality. A key to understanding wave-particle duality is to recognize that waves and particles are **models**. They are conceptual representations of real things. Wave and particle models are based on experience with water and billiard balls. The "real" nature of an electron is well beyond the range of immediate experience.

Electrons move about the nucleus and in particle language they move in a restricted orbit. However, the model for the structure of the atom also includes a wave model. The electron may move about the nucleus, but it is not restricted to fixed orbits. The electron has wave properties. The best one can do is to describe the probability of where the electron can be found and where it cannot be found. Bohr's orbits correspond to locations where the electron in three dimensions is most likely to be found. Each orbit corresponds to the most probable location for an electron of specific energy. The equation that describes the electron wave and its corresponding probabilities is the Schrödinger equation. It is out with the Bohr model and in with Schrödinger's model. In the Schrödinger model, light is given off or absorbed by the electron when the electron wave pattern changes.

Austrian physicist Erwin Schrödinger (1887–1961). He shared a 1933 Nobel Prize for new formulations of the atomic theory.

The quantum world is not the like the everyday world you experience. The classical view of the world makes sense, but does not provide a complete picture. The quantum world appears to contradict common sense, but it gives correct answers. All the numerical results from predictions based on the Schrödinger model are accurate. It is a dilemma. Do you go with the theory that gives all the right predictions but runs counter to common sense OR go with common sense that does not give accurate experimental results? What do you think?

Reflecting on the Activity and the Challenge

In this activity, you found out that light behaves like a particle in the photoelectric effect and like a wave in diffraction and interference effects. Similarly, electrons behave like particles when they hit a screen and like waves when they move about the nucleus. The atomic model has gotten more complex. You can no longer predict where the electron can be, but only where it is likely to be and where it will never be. Depicting the probability of electron location in an atom will be quite an undertaking for your museum exhibit. Light is given off when the wave pattern (and the probability locations) changes. Your museum exhibit may require you to help visitors explain what an interference effect is and how it is evidence of wave behavior. You may choose to focus on where the electron is found in a 1-dimensional box. Creativity and your imagination will be required to make this part of your exhibit interactive and scientifically correct.

Physics To Go

1. Describe two differences between particles and waves.

2. Someone decides that a laser beam is not thin enough. They decide to pass the beam through a very thin slit to slim it down. Will this work? Explain.

3. What was the principal understanding that emerged from Einstein's explanation of the photoelectric effect?

4. A baseball is bouncing back and forth between two walls. An electron is trapped in a similar "box." How does the ball's behavior differ from the electron's behavior?

5. Suppose the probability function (wave) for an electron in a box is represented below.

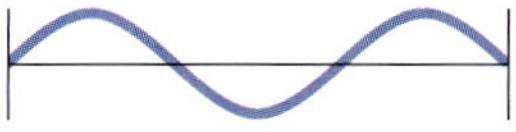

   a) In which locations would you likely find the electron?
   b) Where would you never expect to see the electron?
   c) If the electron were to lose energy, how would the probability function (wave) change?

6. Describe three things wrong with this familiar diagram of an atom.

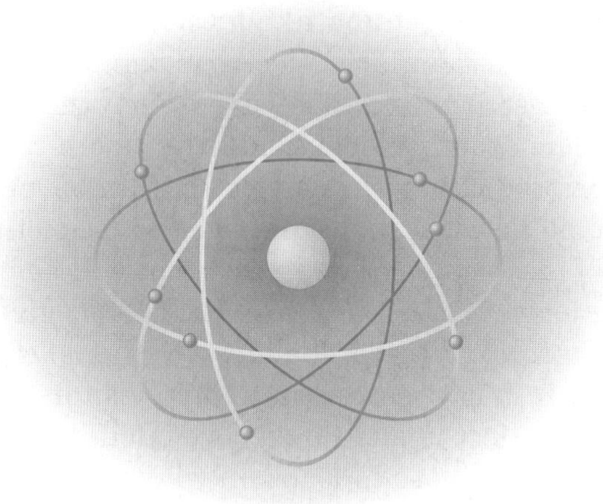

7. In the photoelectric effect, a 10 eV photon of light frees an electron from a metal with a work function of 4.2 eV. What is the energy of the emitted electron?

8. The equation for the photoelectric effect is:

$$KE_{electron} = E_{light} - w_o$$

Explain what each of the terms in the equation represent.

9. In designing your museum exhibit, what might be a creative way to show the unusual behavior of an electron in an atom?

10. For your museum exhibit or perhaps a product for the museum store, can you invent a photoelectric-effect bank? How would it work?

11. The great physicist Niels Bohr once suggested that if the behavior of the electron makes sense to you, then you *don't* understand it at all. Describe some strange behavior of the electron.

12. In the **For You To Read** section you were asked what you would do: go with a theory that gives all the right predictions but runs counter to your common sense OR go with your common sense that does not yield accurate experimental results. Explain your answer to this question and the reasoning you used. (Consider that common sense is based on experience.)

Activity 6 Inside the Nucleus

GOALS

In this activity you will:

- Determine the number of protons, electrons and neutrons in a neutral atom given the atomic number and atomic mass.
- Calculate the Coulomb force of repulsion between protons in a nucleus.
- Recognize that the nuclear force must be much greater than the Coulomb force.
- Draw Feynman diagrams for the interaction between charges.

What Do You Think?

The alchemist's dream has always been to turn worthless lead into valuable gold.

- **What is the difference between lead and gold?**
- **How can you distinguish one from the other?**

Record your ideas about these questions in your *Active Physics* log. Be prepared to discuss your responses with your small group and your class.

For You To Do

1. The nucleus has all the positive charge and almost all the mass of the atom. Hydrogen is the simplest atom. Hydrogen has one proton and one electron. The proton is the hydrogen nucleus. The mass of a proton is 1.7×10^{-27} kg and the mass of an electron is 9.1×10^{-31} kg.

   a) Calculate the ratio of the proton mass to the electron mass.

   b) Does the value you calculated convince you that most of the mass is in the nucleus? Explain your answer.

2. Carbon is more complicated than hydrogen. Carbon has a mass of 12, but has only 6 electrons surrounding the nucleus. One possibility is that there are 12 protons and 6 electrons within the nucleus and 6 electrons orbiting the nucleus.

   a) Show how this would account for a mass of 12.

   b) Show how this would account for the atom having a net charge of zero.

3. The neutron has a mass almost identical to the proton, but has no electrical charge. It was determined that the carbon atom has 6 electrons, 6 protons, and 6 neutrons.

   a) Show how this would account for a mass of 12.

   b) Show how this would account for the atom having a net charge of zero.

4. The neutral atom of chlorine has 17 electrons and 17 protons. It can be written in the notation $^{35}_{17}\mathrm{CL}$ where the lower number corresponds to the atomic number *and/or* the number of protons in the nucleus *and/or* the number of electrons surrounding the nucleus. The upper number corresponds to the atomic mass *and/or* the sum of the number of protons and neutrons.

   a) Calculate how many neutrons are found in the nucleus of $^{35}_{17}\mathrm{CL}$.

   b) Determine the number of protons, electrons, and neutrons in gold: $^{197}_{79}\mathrm{Au}$.

   c) Determine the number of protons, electrons, and neutrons in potassium: $^{39}_{19}\mathrm{K}$.

5. Protons repel other protons. The force can be calculated using Coulomb's Law:

$$F = \frac{kq_1q_2}{d^2}$$

a) Calculate the force between two protons in the nucleus. The charge on each proton is 1.6×10^{-19} C. The distance between them on the average is the size of the nucleus 1.0×10^{-15} m. Coulomb's constant k equals 9×10^9 Nm^2/C^2.

b) Calculate the acceleration of a proton from that force using Newton's Second Law $F = ma$. The mass of a proton is 1.7×10^{-27} kg.

6. This strong repulsive force should push the protons apart. There must be a force holding the nucleus together. This nuclear force is also called the "strong force." It has the following properties:

- strong at short distances (the size of the nucleus 10^{-15} m)
- weak at long distances (greater than 10^{-15} m)
- attraction of protons to protons, neutrons to neutrons, and protons to neutrons
- no effect on electrons

a) Why is it important that the force is strong at short distances and weak at long distances?

b) Draw a graph of the strong nuclear force versus the distance between protons. Note the size of the nucleus on the x-axis of your graph.

7. Neutrons and protons belong to a family of particles called "baryons." Baryons can "feel" the strong nuclear force. Electrons belong to a family called leptons. Leptons do not "feel" the strong nuclear force.

a) List the combination of particles between which the nuclear force can exist.

b) The Coulomb force is only between proton and proton in the nucleus. Why do you think that there is no Coulomb force between neutrons?

8. By what means do forces such as gravitational and Coulomb force act at a distance?

   a) You hold out your hand and drop a penny. How does the penny "know" which way is down? How is the gravitational force between the penny and the Earth communicated? Write down your thoughts on this profound question.

   b) You charge up a rod and attract a small Styrofoam ball. How does the ball know the directions to move? How is this electrostatic force different than the gravitational force? Write down your thoughts.

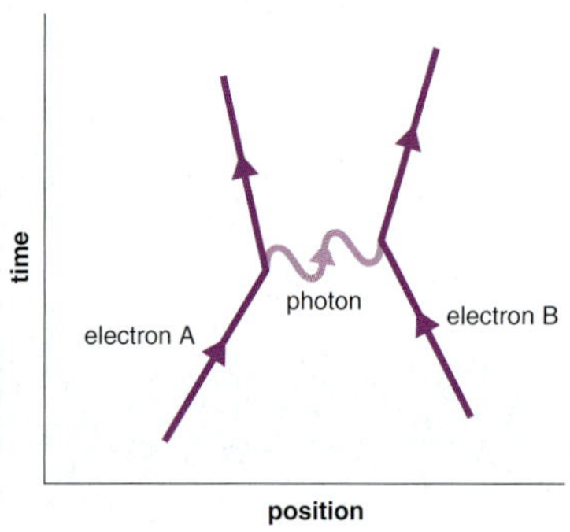

9. Forces act through an exchange of "virtual particles." This can be shown in a Feynman diagram of the repulsion of two electrons.

   Time is on the *y*-axis. Electron A is at first moving to the right as time progresses. Electron B at first is moving to the left. Electron A emits a virtual particle of light (a photon) and recoils to the left. Some time later, electron B absorbs the virtual photon of light and recoils to the right. The effect is that the electrons have repelled each other.

   a) Draw a Feynman diagram of an electron at rest.

   b) Draw a Feynman diagram of a proton and proton repelling.

   c) Draw a Feynman diagram of the attraction between a proton and an electron.

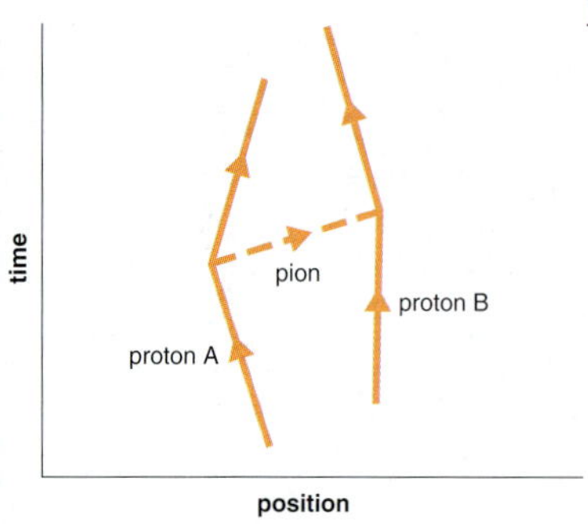

10. The nuclear force is due to the exchange of virtual particles called "mesons." The protons and neutrons exchange mesons; the protons and protons exchange mesons; the neutrons and neutrons exchange mesons.

    The diagram looks very similar to the exchange of virtual photons in the repulsion of two electrons. In this case there is the attraction of two protons. The pion is a type of meson. There are pions with a positive (+) charge, a negative (–) charge, and no (0) charge. It is the exchange that is most important in the diagram, not the direction of motion of the particles. The attractive strong nuclear force depicted here is mediated with a virtual pion and is shown with a straight line.

    a) Draw the Feynman diagram for the attractive strong nuclear force between two neutrons.

    b) Draw the Feynman diagram for the attractive strong nuclear force between a proton and a neutron.

11. In both of the previous Feynman diagrams, the mediating particle was uncharged. The photon is always neutral. The pion comes in three varieties – the neutral, the positive, and the negative pions. The attractive strong nuclear force between a proton and a neutron can also be mediated by a positive pion as shown in the diagram to the right.

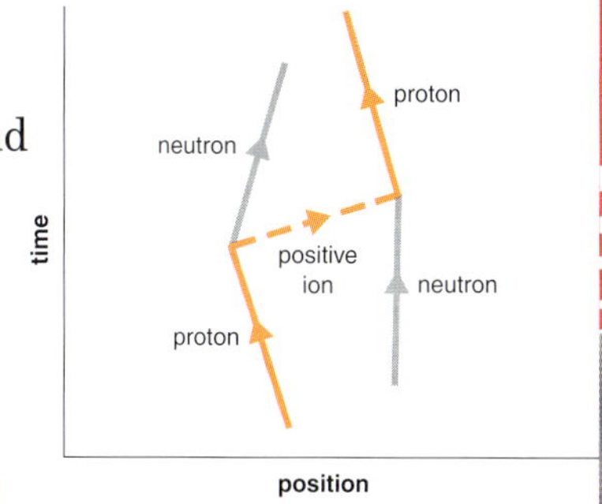

Notice that the proton emits a virtual positive pion. The pion carries the positive charge and the proton becomes a neutron. The neutron absorbs the virtual positive pion and becomes a proton. The strong nuclear force is still attractive. At every vertex of a Feynman diagram, the charge must be conserved.

a) Draw a Feynman diagram where a neutron emits a pion and becomes a proton. Note the charge of the pion on your diagram.

FOR YOU TO READ

Holding the Nucleus Together

Rutherford's scattering experiment provided evidence for a dense nucleus at the center of the gold atom. Early models of the nucleus had included the **proton**. However, when you examined a carbon atom you realized that the proton alone could not account for the mass of the nucleus. Rutherford addressed this problem when he suggested that another particle was present in the nucleus. He proposed a particle with about the same mass as the proton, but with no electric charge. He named this particle the **neutron**. In 1932 James Chadwick, a British physicist, actually discovered the neutron. This discovery added a great deal to the understanding of the nucleus of the atom. However, the model of the nucleus still posed a huge puzzle.

How can all of those protons and neutrons fit into a small space? Protons repel protons. You calculated this force of repulsion in the **For You To Do** section. Also, you calculated the acceleration due to this force. With these kinds of forces and corresponding accelerations, you would expect the nucleus to explode. There must be another force holding the nucleus together. This force must be very large. It must also be limited to very small distances. If the force were long-range, then all the nuclei would clump together. The force must be strong enough to hold all of the protons together but short-range so that one nucleus

Sir James Chadwick, a British physicist, was winner of the Nobel Prize in 1935. His discovery of the neutron led eventually to the development of the nuclear bomb.

does not affect the neighboring nucleus. There is such a force holding the nucleus together. It is called the **nuclear force** or the strong force. The strong nuclear force exists between neutrons and protons, protons and protons, and neutrons and neutrons. These particles belong to a family of particles called **baryons**. Baryons can "feel" the strong nuclear force. The Coulomb force is only between protons and protons in the nucleus. Neutrons have no charge and do not "feel" the Coulomb force. Electrons belong to a family called **leptons**. Leptons do not "feel" the strong nuclear force.

In physics, you have studied the effect of forces as explained by Newton's Second Law ($F = ma$). Net forces produce accelerations. You have calculated the strength of forces such as the gravitational force $F = \frac{Gm_1m_2}{d^2}$,

weight $F = mg$,

and the Coulomb's electrostatic force $F = \frac{kq_1q_2}{d^2}$.

In this activity, you were asked to consider by what means these forces act at a distance.

When you push on something, where does the force come from? Consider this question at the atomic level. There are electrons in your hand and there are electrons in the table. As the electrons get close, there is repulsion. This repulsion is the force you feel. This repulsion moves the table forward. The electrons in your hand are trying to get closer to the electrons in the table and the table moves because of the electrostatic repulsion as calculated by Coulomb's Law. The force can be huge because there are so many electrons.

How does one electron know the other electron is there? The 20th century breakthrough answer is that forces act through an exchange of "virtual particles." The electrons continuously exchange virtual photons of light so that the repulsion continues. The virtual photons of light are exchanged in a time too short to observe directly. You drew a Feynman diagram to illustrate this exchange.

You also drew Feynman diagrams to illustrate the nuclear force. In this case, the virtual particles that are exchanged are **mesons**. Specifically, they are pions that can have positive, negative, or no charge. The Feynman diagrams you drew depicted the forces and their exchange of particles. These diagrams can provide even more information. They can be used to calculate the strength of the forces. Richard Feynman showed that equations and calculations that at one time required days to complete could be completed in minutes using the diagrams he invented. Feynman was awarded the Nobel Prize for these achievements.

Reflecting on the Activity and the Challenge

The nucleus is a crowded place. It contains all of the protons and neutrons of the atom. These protons and neutrons are held together by the strong nuclear force. The force is short-range and very attractive. The strong nuclear force overwhelms the Coulomb repulsive force between protons. The force is mediated by the exchange of charged and uncharged pions. Your museum exhibit on the structure of the atom may focus on the nucleus and its teeming micro-world of protons, neutrons and virtual pions. Communicating the size of the nucleus and how it is held together will be quite a creative challenge.

Physics To Go

1. In the following notation for a carbon nucleus $^{13}_{6}C$, what do each of the numbers represent?
2. A proton is represented as $^{1}_{1}p$ and a neutron is represented as $^{1}_{0}n$.
   a) Why do they both have superscripts of 1?
   b) Why do they have different subscripts?
3. Isotopes are elements that have identical atomic numbers but different atomic masses. For the following sets of isotopes, list the number of protons, neutrons, and electrons:
   a) $^{12}_{6}C$, $^{13}_{6}C$, $^{14}_{6}C_6$
   b) $^{40}_{20}Ca$, $^{41}_{20}Ca$, $^{42}_{40}Ca$
   c) $^{232}_{92}U$, $^{235}_{92}U$
4. Calculate the electrostatic force between two protons that are separated by 6×10^{-14} m within the nucleus.
5. Calculate the electrostatic force between an electron and a proton that are separated by 8×10^{-9} m in an atom.
6. Sketch a graph showing how the nuclear force between two protons varies over distance.
7. Sketch a graph showing how the electrostatic force between two protons varies over distance.

Physics Words

proton: a subatomic particle that is part of the structure of the atomic nucleus; a proton is positively charged.

neutron: a subatomic particle that is part of the structure of the atomic nucleus; a neutron is electrically neutral.

nuclear (strong) force: a strong force that hold neutrons and protons together in the nucleus of an atom; the force operates only over very short distances.

baryon: a group of elementary particles that are affected by the nuclear force; neutrons and protons belong to this group.

lepton: a group of elementary particles that are not affected by the nuclear force; electrons belong to this group.

meson: a virtual particle that mediates the strong, nuclear force of an atom; the protons and neutrons exchange mesons; the protons and protons exchange mesons; the neutrons and neutrons exchange mesons.

8.Complete a chart indicating if the following pairs of particles interact by the electrostatic force and/or the nuclear force:

proton-proton; proton-electron; proton-neutron; neutron-electron; electron-electron; neutron-neutron.

9. Two protons are separated by 1×10^{-15} m in a nucleus.
   a) Calculate the gravitational force between them.
   b) Calculate the electrostatic force between them.
   c) Can the gravitational force be responsible for holding the nucleus together? Explain.

10. Draw a Feynman diagram of a single proton at rest.

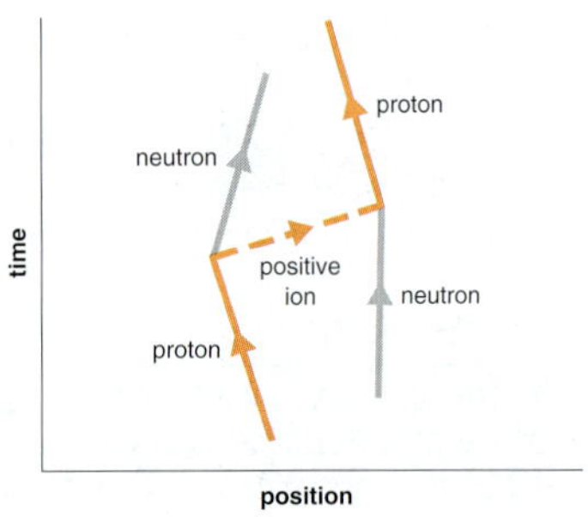

11. In the Feynman diagram on the left, show that the charge is conserved at each vertex (i.e., the charge entering the vertex is equal to the charge exiting the vertex).
12. Sketch a Feynman diagram for the interaction between two neutrons.
13. When your fingertips touch, your skin deforms. Explain why this happens and why the deformation increases when you press harder.
14. Could a museum exhibit help visitors understand how the nuclear force is able to hold the nucleus together but not pull neighboring nuclei together? Could it be interactive? Could it capture people's attention within 30 seconds? Describe such an exhibit.

Stretching Exercise

Prepare a research report and/or a simulation of the experiment that Chadwick performed to discover the neutron in 1932.

Activity 7 Radioactive Decay and the Nucleus

GOALS

In this activity you will:

- Create a graph representing radioactive decay.
- Use the concept of half-life to determine the age of an archaeological artifact.
- Write equations for alpha decay.
- Write equations for beta decay.
- Write equations for gamma decay.

What Do You Think?

Is nuclear radiation friend or foe? Radiation is a big plus for medical procedures and a big hazard from nuclear bombs.

- **What is radiation?**
- **Do we have any control over it?**

Record your ideas about these questions in your *Active Physics* log. Be prepared to discuss your responses with your small group and your class.

For You To Do

1. Mark one side of each of 100 sugar cubes with a dot. Place the sugar cubes in a container. Gently shake the sugar cubes so that they can land randomly on any side.

   Remove the cubes that have the dot face up and replace them with pennies.

Do not eat the sugar cubes. As with anything used in a lab activity, they are contaminated and not edible.

a) Record your result in a data table like the one shown.

Trial	Number of sugar cubes	Number of pennies	Total number of cubes and pennies
0	100	0	100
1			
2			
3			
Etc.			

2. Repeat with the remaining cubes. Continue with at least 10 additional trials.

a) Record your results each time.

b) Graph the information from the chart. Place the trials on the x-axis and the number of cubes on the y-axis. Draw the best-fit smooth curve through most of the cube data points.

3. Use your graph to answer the following questions:

a) How many tosses did it take to remove half the sugar cubes?

b) In this model, let each toss of the box represent a time interval of one hour. Change the "trials" label on your graph to "hours." How many hours did it take to remove half the sugar cubes? This unit of time is the half-life of your sample, the amount of time it takes for half of the sample to change.

4. Compare your results with that of other groups.

a) Are they similar?

b) Combine the data from all groups. Graph the class data.

c) How does the class data compare to your graph?

d) During each shake of the box, what was the probability that any one sugar cube would land with the dot facing up?

5. Consider what would happen if the sugar cube had a dot on two faces.

a) How would you expect the data to change?

b) During each shake of the box, what was the probability that any one sugar cube would land with a dot face up?

c) Sketch a graph that you might expect to get.

d) What may be the value of the half-life for this new set of sugar cubes with two dots?

6. Consider what would happen if the sugar cube had a dot on three faces.
   a) How would you expect the data to change?
   b) During each shake of the box, what was the probability that any one sugar cube would land with a dot face up?
   c) Sketch a graph that you might expect to get.
   d) What may be the value of the half-life for this new set of sugar cubes with three dots?

7. Consider what would happen if the sugar cube had a dot on five faces.
   a) How would you expect the data to change?
   b) During each shake of the box, what was the probability that any one sugar cube would land with a dot face up?
   c) Sketch a graph that you might expect to get.
   d) What may be the value of the half-life for this new set of sugar cubes with five dots?

8. The half-life of carbon-14 is 5730 years. That is, every 5730 years, half of the carbon-14 decays by emitting an electron and becomes nitrogen-14. By measuring the remaining carbon-14 in a long-dead sample, scientists can find out for how many half-lives the organism has been dead. If half the carbon-14 remains, then they know that 5730 years have passed. If $\frac{1}{16}\left(\frac{1}{2} \text{ of } \frac{1}{2} \text{ of } \frac{1}{2} \text{ of } \frac{1}{2}\right)$ of the carbon-14 remains, then scientists know that four half-lives, or 22,900 years, have passed.
   a) How many years have passed when the sample contains $\frac{1}{8}$ of the carbon-14?

9. Here is a graph showing the radioactive decay of a 500-g sample of C-14.

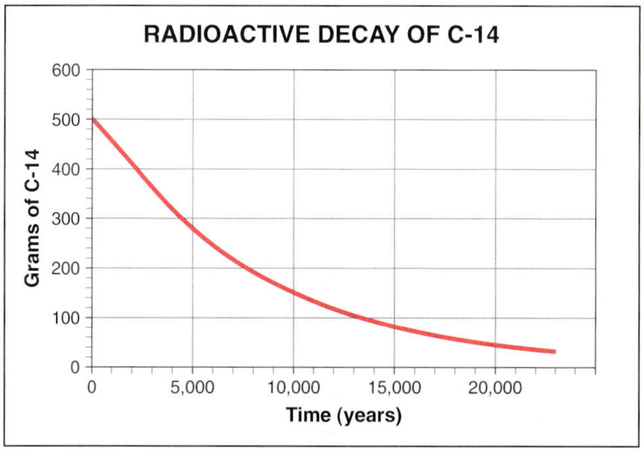

a) What is the half-life of the sample?
b) How many grams have undergone decay after 7000 years?
c) The original sample had 500 g. The sample now has 400 g of C-14. How much time has elapsed?

10. A 1600-g sample of iodine-131 is observed over the course of a few months. A graph of the remaining iodine-131 is shown below.

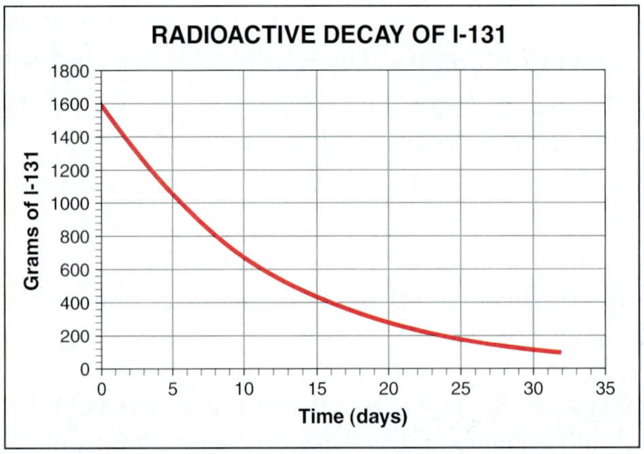

a) What is the half-life of the sample?
b) How many grams have undergone decay after 25 *days*?
c) The original sample had 1600 g. The sample now has 1400 g. How much time has elapsed?
d) The original sample had 1600 g. The sample now has 600 g. How much time has elapsed?

11. Carbon-14 (called the parent nucleus) decayed and became nitrogen-14 (called the daughter nucleus). In the last example, iodine-131 (parent nucleus) decayed and became xenon-131 (daughter nucleus). In the simulation using sugar cubes, the sugar cube (parent) "decayed" and became a penny (daughter). The total number of sugar cubes plus pennies always remained at 100. In radioactive decay, the total number of parent and daughter nuclei must also remain the same.

a) Construct a graph of the growth in the pennies over time from your earlier data.

12. Elements become other elements through radioactive decay. Three different decay processes can take place. The nucleus can emit an alpha particle. The alpha particle is a helium nucleus. It has two protons and two neutrons. It is written

as ${}^{4}_{2}He$. In all alpha decays, the number of protons must remain constant and the atomic mass (protons + neutrons) must remain constant. If an alpha particle is emitted from the parent nucleus, then the new daughter nucleus will have two fewer protons and two fewer neutrons. Writing the nuclear equation makes this much simpler.

$${}^{238}_{92}U \rightarrow {}^{4}_{2}He + {}^{234}_{90}X$$

where X is as yet unknown.

Notice that the numbers of protons (the numbers on the bottom) remain constant.

$$92 = 2 + 90$$

The total mass of protons and neutrons (the top number) also remains constant.

$$238 = 4 + 234$$

The new daughter element has an atomic number of 90. Referring to the periodic table, the element is thorium and its chemical symbol is Th (X = Th).

$${}^{238}_{92}U \rightarrow {}^{4}_{2}He + {}^{234}_{90}Th$$

Although the half-life of the element is not indicated in the decay scheme, the half-life of ${}^{238}_{92}U$ is 4.5 billion years.

a) Write the decay scheme for ${}^{235}_{92}U$ emitting an alpha particle (half-life = 700 million years).

b) Write the decay scheme for ${}^{232}_{90}Th$ emitting an alpha particle (half-life = 14 billion years).

c) ${}^{238}_{92}U$ and ${}^{232}_{92}U$ are isotopes of uranium. They both have 92 protons in the nucleus. They have different numbers of neutrons. It is the number of protons that determine the name of the element. The half-lives are quite different. Calculate the number of neutrons in these isotopes of uranium.

13. A second possible nuclear decay occurs with the emission of a beta particle from the nucleus. A beta particle is a high-speed electron or a high-speed positron (identical to the electron but with positive charge). A neutron within the nucleus becomes a proton and emits the beta particle. Potassium-40 undergoes beta emission. The decay process:

$${}^{40}_{19}K \rightarrow {}^{0}_{-1}e + {}^{40}_{20}X$$

The beta particle emission causes an increase in the number of protons (the neutron became a proton). There is no change in the atomic mass since the mass of a proton and neutron are almost identical.

The new daughter element has an atomic number of 20. Referring to a periodic table, the element is calcium and its chemical symbol is Ca.

$$^{40}_{19}K \rightarrow\ ^{0}_{-1}e +\ ^{40}_{20}Ca$$

a) Write the decay scheme for $^{14}_{6}C$ emitting a beta particle (half-life = 5730 years).

b) Write the decay scheme for $^{137}_{55}Cs$ emitting a beta particle (half-life = 30 years).

During some radioactive decays, additional energy is emitted in the form of gamma rays. Gamma rays are high-energy light. Their symbol in the radioactive decay is $^{0}_{0}\gamma$ since they have no mass.

Physics Words

probability: a measure of the likelihood of a given event occurring.

radioactive: a term applied to an atom that has an unstable nucleus and can spontaneously emit a particle and become the nucleus of another atom.

FOR YOU TO READ

Models

Lots of people go to Las Vegas or other gaming centers to gamble. They think that because dice are involved that it is all a matter of luck whether they win or lose. Have you ever looked at the size of the hotels in Las Vegas? From the size and numbers of these hotels, you have to assume that the "house" always wins and that there is less luck at the casinos than some people expect. **Probability** is not the same as luck. In this activity you studied probability as it relates to atomic structure

You modeled an event in the real world. Scientists make many kinds of models. Models are conceptual representations of real things. You are probably familiar with physical models of the planets, moons, and the Sun and their interactions in the solar system. This type of model can help you "see" something too big to see all at once in nature. The model of the atom that you will create for your museum exhibit is another example of a model. The model is based on many observations and experimental results. Many models in science are complex mathematical models.

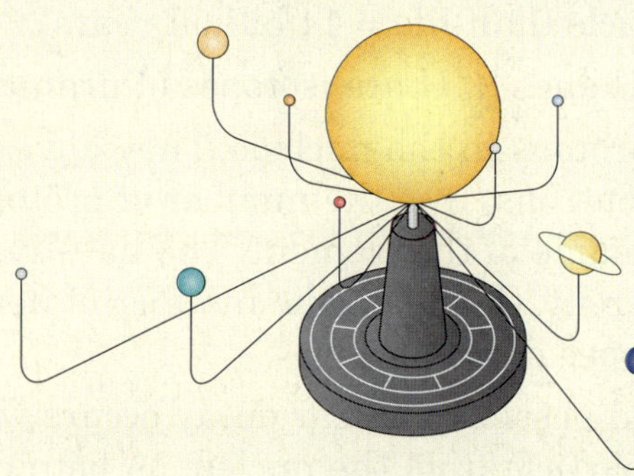

The **radioactive** model in this activity used sugar cubes to represent nuclei. When the sugar cubes were tossed and a "dot" emerged

face-up, the sugar cube "decayed" and became a penny. In radioactive samples like carbon-14 or iodine-131, the nuclei are not being tossed. As time goes on, some parent nuclei decay and become daughter nuclei. You had no idea which of the sugar cubes were going to land with the dot face-up. You also have no idea which of the carbon-14 nuclei will decay. You did get a good sense how many sugar cubes would land with the dot face-up. You have a very, very precise understanding of how many carbon-14 parent nuclei will decay in a given time.

Scientists use radioactive decay as precise clocks. Geologists want to know the age of rocks to provide an understanding of the Earth. Paleontologists want to know the age of fossils to better understand life on Earth. Archaeologists want to know the age of tools to better understand humans. Scientists in each of these fields make use of the known half-lives of radioactive elements for dating purposes.

One type of carbon atom, called carbon-14, is used in radioactive dating. As long as an organism lives, the ratio of carbon-14 to carbon-12 in the body stays the same. When organisms die, no new carbon-14 is added to their tissues. The decay of carbon-14 continues. If you know how much carbon-14 was in the organism when it lived, you can figure out how long ago it died by measuring the carbon-14 that remains.

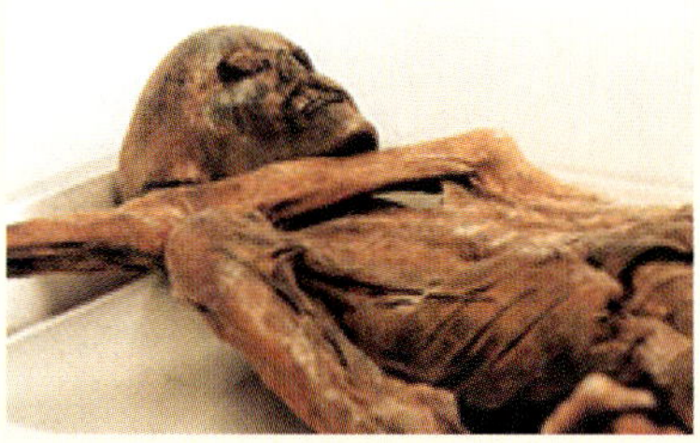

Entombed in ice for over 5000 years, the iceman was discovered in 1991 by hikers in the Tyrolean Alps in Austria. Carbon-14 dating, along with many other scientific tools, was used to investigate this archaeological find.

Other elements are important tools in radioactive dating. Potassium-40 has a half-life of 1.3 billion years. Uranium-238 has a half-life of 4.47 billion years. Thorium-232 has a half-life of 14 billion years. These radioactive elements, as well as others, are found in rocks. Often, all three of these elements are used to date a specific rock. By comparing the radioactive decays, the age of a rock can be determined with more precision than if only one element was used. Rocks that are several billion years old have been discovered, indicating that the Earth must be at least this old.

Reflecting on the Activity and the Challenge

Some nuclei are stable while some decay. The parent nuclei can decay through alpha, beta or gamma emission. The decay process is random in that you cannot tell which of the individual nuclei will decay. However, the number of nuclei that will decay is quite predictable. The decay rate is defined by the half-life of the element. The half-life is the time required for one-half of the parent nuclei to become daughter nuclei. You can show nuclei decaying as part of your museum exhibit. It may be a bit tricky to show the mechanism of decay and to adequately describe half-life, but these are both worthwhile topics to pursue.

Physics To Go

1. A box of 500 candies was dropped on a table. If the red side of the candy was facing up, the candy was eaten. The remaining candy was shaken and dropped on the table again. A data table was created.

Toss (every 10 minutes)	Candies with red face up and then eaten	Total candies eaten	Candies remaining
0		0	500
1	130	130	370
2	90	220	280
3	68	288	212
4	50	338	162
5	45	383	117
6	26	409	91
7	23	432	68
8	15	447	53

   a) Construct three graphs of the data.
   b) From the graph, determine the half-life of the candy in this tossing simulation.
   c) From the percentage that "decay" each toss, determine the shape of the candy.

2. There are 600 g of radioactive iodine-133. The half-life of iodine-133 is 21 h.
   a) How much of this radioactive substance would exist after 21 h?
   b) After 42 h?
   c) After 126 h?

3. The half-life for the radioactive gas radon-220 is 51.5 s. Suppose a sealed bottle contains 100 g of radon-220. What mass of radon-220 will still be in the bottle after:

   a) 51.5 s? b) 103 s? c) 257.5 s?

4. Radioactive decay can be measured. One type of measuring tool gives a reading called a counting rate. A sample of radioactive material initially produces 2000 counts per

second. Four hours later, the same sample produces 500 counts per second. What is the half-life of the sample?

5. Carbon-14 has a half-life of 5730 years. What fraction of the original amount of carbon-14 will remain after 23,000 years?

6. Suppose a sample of carbon is extracted from a partially burned log found in a prehistoric fire site. The sample is $\frac{1}{8}$ as radioactive as carbon extracted from a modern log. How old is the sample taken from the ancient log?

7. For what age materials is uranium dating more useful than carbon dating? Explain your reasoning.

8. Radioactive uranium-238 decays by alpha emission. The daughter product, thorium-234 then decays by beta emission. A chart of the decay series is shown. The first line shows the decay of uranium-238. The thorium-234 is then placed on the next line. It then produces Pa-234 by beta emission. The Pa-234 is then placed on the next line and its decay is written.

   Complete the chart. (The half-lives are given for interest and are not needed to complete the chart.) The periodic table will help you identify the element symbols from their atomic number. If the periodic table is not available, you can complete the chart with the atomic number and mass and leave a blank for the symbol.

Element	Half-life	Emission	Reaction
$^{238}_{92}U$	4.9×10^9 years	alpha	$^{238}_{92}U \rightarrow {}^4_2He + {}^{234}_{90}Th$
$^{234}_{90}Th$	24 days	beta, gamma	$^{234}_{90}Th \rightarrow {}^0_{-1}e + {}^0_0\gamma + {}^{234}_{91}Pa$
$^{234}_{91}Pa$	1.14 minutes	beta, gamma	$^{234}_{91}Pa \rightarrow {}^0_{-1}e + {}^0_0\gamma + {}^{234}_{92}U$
$^{234}_{92}U$	3×10^5 years	alpha	$^{234}_{92}U \rightarrow {}^4_2He + {}^{230}_{90}Th$
$^{230}_{90}Th$	83,000 years	alpha, gamma	
	1600 years	alpha, gamma	
	3.8 days	alpha	
	3.0 minutes	alpha	
	26.8 minutes	beta, gamma	
	19.7 minutes	beta, gamma	
	10^{-5} seconds	alpha, gamma	
	22 years	beta	
	5 days	beta, gamma	
	140 days	alpha	
	Stable		

Stretching Exercises

1. Iodine-133 is a common byproduct of nuclear fission in electric power plants. You are given a sample of material contaminated by iodine-133. Find out how long you have to wait before you can dump the iodine into the environment. How do you know it is "safe?" Report your findings to the class.
2. Find out about radon, a radioactive element found in your homes. Why is it considered a hazard? Report your findings to the class.
3. Medical doctors use radioactive materials in many ways. Research this and make a list of at least five uses for radiation in medicine. Create a poster of your findings.

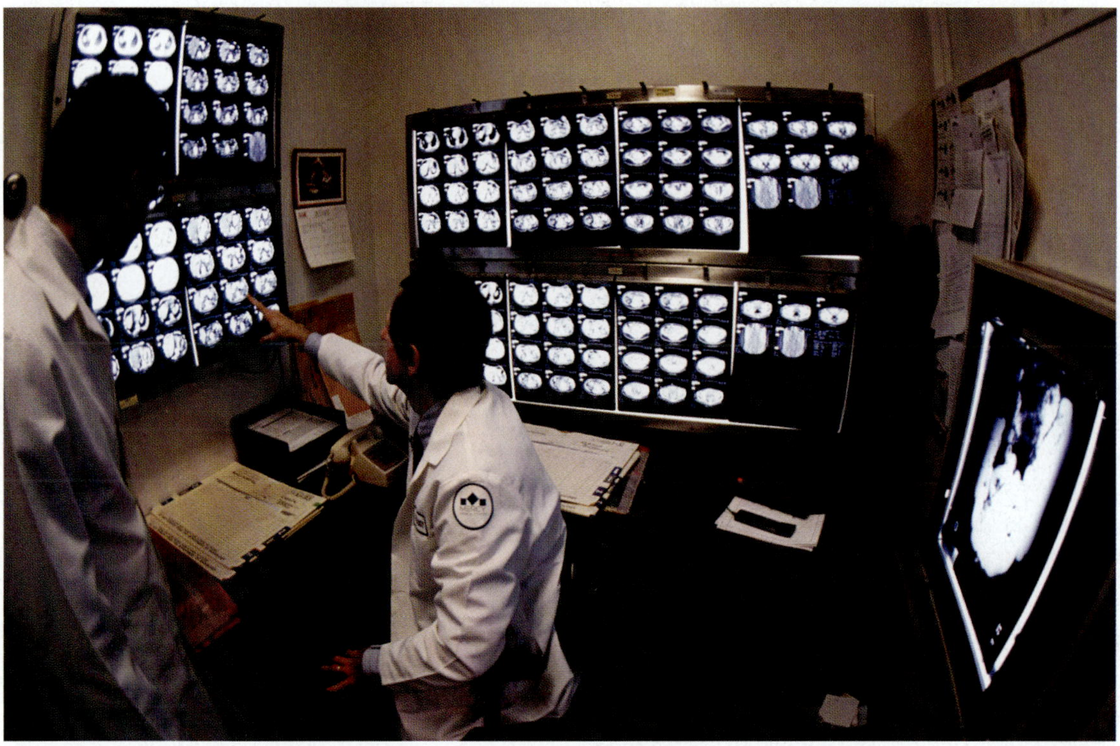

Activity 8 Holding the Nucleus Together

GOALS

In this activity you will:

- Explain the meaning of Einstein's equation: $E = mc^2$.
- Calculate the energy equivalent of different masses.
- Determine the binding energy of protons and neutrons in the nucleus.
- Compare the binding energy of protons in the nucleus with the binding energy of electrons in the atom.

What Do You Think?

Nuclear bombs, nuclear radiation, and nuclear medicine conjure up very different images than chemical explosions, chemical reactions, and chemical medicines.

- **How are nuclear reactions different from chemical reactions?**

Record your ideas about this question in your *Active Physics* log. Be prepared to discuss your responses with your small group and your class.

For You To Do

1. Many people can recite Einstein's famous equation:

   $E = mc^2$

   where E represents energy in joules (J), and
   m represents mass in kilograms (kg).

   Write down the approximate mass of the following objects:

   a) an eight-year-old child
   b) a bowling ball
   c) a box of spaghetti
   d) a compact car

2. The c in the equation $E = mc^2$ represents the speed of light, which is equal to 3.0×10^8 m/s. Use the equation:

$$d = vt$$

to calculate how long it takes light to travel from:

a) one side of a football field to the other (approximately 100 m)

b) from Canada to Texas (approximately 1000 km or 10^6 m)

c) from the Moon to Earth (approximately 380,000 km or 3.8×10^8 m)

d) from the Sun to Earth (approximately 150,000,000 km or 1.5×10^{11} m)

3. Consider the energy equivalent of a small mass. The mass of a pea may be 1 g or 1×10^{-3} kg.

a) Calculate the energy equivalent in joules of this mass using $E = mc^2$. (Assume 100% conversion.)

4. There are many forms of energy that can be calculated. Kinetic energy, associated with moving cars, people, planets, and subatomic particles can be calculated using the equation:

$$\text{KE} = \frac{1}{2}mv^2$$

where KE is kinetic energy in joules (J)
m is the mass in kilograms (kg), and
v is the velocity in meters per second (m/s)

a) If a sprinter with a mass of 50 kg is running at 8 m/s, calculate the kinetic energy in joules.

b) How fast would a sprinter have to be running to have the energy equivalent of one pea?

5. Gravitational potential energy is energy associated with position. It can be calculated using the equation:

$$\text{GPE} = mgh$$

where GPE is gravitational potential energy in joules (J)
m is the mass in kilograms (kg)
g is the acceleration due to gravity equal to about 9.8 m/s^2, and
h is the elevation in meters (m)

a) How high would a bowling ball (mass = 6 kg) have to be elevated to have the same gravitational potential energy as the energy equivalent of a pea?

6. In **Activity 3**, you studied electrons in the atom. The binding energy of an electron in hydrogen is the energy that holds the electron to the nucleus. In the $n = 1$ (ground) state of hydrogen, the binding energy was – 13.6 eV. The negative sign reminded you that the electron is bound to the nucleus. That electron requires +13.6 eV in order for it to be free of the nucleus.

   Neutrons and protons are also bound to the nucleus. A nitrogen-15 $\left({}^{15}_{7}\text{N}\right)$ nucleus has 7 protons and 8 neutrons. To free a nucleon (proton or neutron) requires an addition of energy equal to the binding energy of the nucleon. You can calculate that binding energy.

Example:

- *Calculate the mass of all the protons and neutrons.*
  The "atomic mass unit" is used for masses of atoms. It is defined in terms of the mass of carbon-12. Each atomic mass unit (u) is approximately equal to 1.7×10^{-27} kg. (You can see from this value why people prefer to use atomic mass units instead of kilograms.)

  Mass of 7 protons = 7 (1.007825 u) = 7.054775 u
  Mass of 8 neutrons = 8 (1.008665 u) = $\underline{8.069320 \text{ u}}$
  Total nucleon mass (protons + neutrons) = 15.124095 u

- *Compare this to the mass of the nitrogen-15 nucleus.*
  The mass of the nitrogen-15 nucleus (15.000108 u) can be found from a chart of the nuclides in a reference book. The nitrogen-15 mass will be less than the total mass of the protons and neutrons. Energy must be supplied to remove a neutron. That implies that energy was given off when the nitrogen nucleus was first formed.

  Total nucleon mass = 15.124095 u
  Mass of nucleus of ${}^{15}_{7}\text{N}$ = $\underline{15.000108 \text{ u}}$
  Mass difference = 0.123987 u

→

- *Calculate the total binding energy.*
  (1 u = 931.5 MeV) The explanation for 931.5 MeV is given in the **For You To Read** section.
  Total binding energy = 0.123987 × 931.5 MeV
  = 115.5 MeV

- *Calculate the average binding energy per nucleon.*
  Total number of nucleons = 15 (7 protons and 8 neutrons)
  Average binding energy $= \frac{115.5 \text{ MeV}}{15} = 7.7$ MeV

On average, it would require 7.7 MeV or 7.7 million eV to remove a proton or neutron from the nucleus of $^{15}_{7}N$. (The removal of the first nucleon actually requires more than this energy—this is an average.) Recall that removing an electron from hydrogen required only 13.6 eV.

a) Calculate the average binding energy per nucleon for $^{37}_{17}CL$.
The mass of chlorine-37 is equal to 36.965898 u. Use the masses of protons and neutrons given in the sample problem.

b) A neutron can be removed from $^{15}_{7}N$ to create $^{14}_{7}N$.
Compare the mass of $^{15}_{7}N$ to the masses of $^{14}_{7}N$ + 1 neutron to determine the actual energy required to free the neutron. The mass of nitrogen-14 equals 14.003074 u. The other values are given above.

c) Compare this binding energy to remove the neutron with the average binding energy calculated in the sample problem.

d) $^{15}_{7}N$ and $^{14}_{7}N$ are isotopes. What defines two elements as isotopes?

PHYSICS TALK

Converting Mass Units to Energy Units

The conversion of **atomic mass units** (u) to energy units in electron volts (eV) required the use of the equation $E = mc^2$ to find the corresponding energy from the mass defect.

This would require certain conversions:

- The mass difference between the nucleus and constituent parts is called the mass defect. It is given in atomic mass units (u).
- The mass units would have to be converted to kilograms ($1\ \text{u} = 1.7 \times 10^{-27}$ kg).
- The energy would then be calculated in joules ($E = mc^2$).
- The joules could then be converted to electron volts ($1\ \text{eV} = 1.6 \times 10^{-19}$ J).

All of these steps can be combined into one step:

$$1\ \text{u} = 931.5\ \text{MeV}.$$

Physics Words

atomic mass unit: a standard unit of atomic mass based on the mass of the carbon atom, which is assigned the value of 12.

FOR YOU TO READ

Energy and Matter

The speed of light is the "speed limit" of the universe. No material object can travel faster than this speed. In this activity you examined Einstein's equation:

$$E = mc^2$$

that equates energy and mass with the speed of light.

Einstein's equation can be interpreted in a number of different ways:

- It provides the energy equivalent of a piece of mass. (In the activity, you found the equivalent energy of a pea.)
- It tells you that energy and mass are equivalent entities but one is given in joules and the other is given in kilograms.
- It tells you that the conversion factor between energy and mass is the square of the speed of light (c^2). To change kilograms to joules, you have to multiply by c^2, equal to $9 \times 10^{16}\ \text{m}^2/\text{s}^2$.
- It explains how an electron and a positron (identical to the electron but with a positive charge) can annihilate each other and create light energy. This is called particle-antiparticle annihilation.

In **Activity 3**, you investigated the binding energy of electrons in an atom. In the $n = 1$ (ground) state of hydrogen, the binding energy was -13.6 eV. The binding energy of the

→

electron in the n = 2 state is – 3.4 eV. You would have to give that electron +3.4 eV to free it. Similarly, if the nucleus of hydrogen captures an electron to make a hydrogen atom and the electron drops into the n = 1 state, 13.6 eV is given off as a photon of light.

The conservation of energy requires that the energy before must equal the energy after the event. An electron absorbs a photon of light of +13.6 eV bringing its total energy to 0 and it is free. Or, a free electron of 0 energy becomes bound to the proton nucleus with an energy of –13.6 eV and gives off a light photon with energy equal to +13.6 eV.

In this activity, you calculated the binding energy on a **nucleon**. You found that to remove a proton or neutron from the nucleus of nitrogen-15 $\left({}^{15}_{7}\text{N}\right)$ requires 7.7 MeV or 7.7 million eV. Compare that to the binding energy of an electron in hydrogen, which is 13.6 eV, and you can begin to appreciate the difference between chemical and nuclear reactions. Chemical processes have to do with the exchange of electrons. Nuclear processes have to do with the exchange of nucleons. You can get a sense of the strength of these processes by comparing the binding energies. The nuclear binding energies are millions of times larger than the electron's binding energies. This is one reason why nuclear bombs are so much more devastating than dynamite.

Physics Words

nucleon: the building block of the nucleus of an atom; either a neutron or a proton.

Reflecting on the Activity and the Challenge

The nucleons (neutrons and protons) are held tightly in the nucleus by the strong nuclear force. You are now able to use the equation $E = mc^2$ to calculate the energy required to free a nucleon. The energy is of the order of a million electron volts while removing an electron requires only a few electron volts of energy. You might try to find a way to add numerical calculations to your atomic structure museum exhibit. You may want to compare the binding energy of electrons to the binding energy of nucleons. The nucleus is a wonderful place to introduce Einstein's famous $E = mc^2$ equation and its multiple interpretations.

Physics To Go

1. a) Calculate the energy equivalent of the following objects:

   i) an electron
   ii) a pea
   iii) a 50-kg student

b) Lifting a 4-kg shovel full of snow 1 m requires 40 J of energy. How many shovels full of snow could be lifted with the energies calculated on the preceding page?

2. a) A direct observation of the equivalence of mass and energy occurs when an electron and a positron (same mass as an electron, but opposite charge) annihilate each other and create light. Calculate the total energy of the light produced. (The mass of an electron is 9.1×10^{-31} kg.)

   b) Calculate the total energy of the light produced when a proton and an antiproton (same mass as a proton, but opposite charge) annihilate each other.

3. Lots of quantities can be measured using different units. Length can be measured in meters, centimeters, and miles. Volume can be measured in liters, milliliters, gallons, meters cubed, and cubic feet. In a similar way, mass and energy are related where mass is measured in kilograms and energy is measured in joules. How could this be depicted in a museum exhibit to make clear the notion that energy and mass are identical?

4. a) Calculate the total binding energy of phosphorus–31 with an atomic number of 15. The atomic mass is 30.973765 u.

   b) Calculate the binding energy per nucleon.

5. a) Describe binding energy in a way that a child visiting a science museum may understand. Is there some way that you can make the explanation visually appealing?

   b) Compare and contrast the binding energy of an electron to a nucleus with the binding energy of a proton with the nucleus.

6. In *Star Trek*, the matter-antimatter energy source provides all the energy the spaceship needs to operate. How might a matter-antimatter energy source work?

7. Two energy units are commonly used, the joule (J) and the electron-volt (eV). 1 eV is equivalent to 1.6×10^{-19} J. If both of these units are energy, why are two different units useful?

8. Sketch a graph showing the relationship between energy and mass. What would be the slope of the graph?

9. The mass difference between the nucleus and its constituent part (also called the mass defect) is greater for nucleus A than nucleus B. Compare their binding energies.

Activity 9 Breaking Up Is Hard to Do

GOALS

In this activity you will:

- Calculate the binding energy of various nuclei.
- Create a graph of average binding energy per nucleon versus the number of nucleons.
- Explain how a fusion reaction can release energy.
- Explain how a fission reaction can release energy.
- Describe a mousetrap model for a fission chain reaction.

What Do You Think?

Nuclear energy is used for nuclear power plants, for creating energy in the Sun, and for atomic weapons.

- **How is nuclear energy created?**
- **Is all nuclear energy produced the same way?**

Record your ideas about these questions in your *Active Physics* log. Be prepared to discuss your responses with your small group and your class.

For You To Do

In the previous activity you learned to calculate binding energy. The binding energy is the energy required to disassemble a nucleus into its separate protons and neutrons. The higher the binding energy, the more energy is required to free the nucleons. The higher the binding energy, the more stable the nucleus.

1. To see if there is a pattern to binding energies and the related stability of nuclei, you can calculate the binding energy per nucleon for all known elements and make a graph. In the spirit of learning the physics and making the most efficient use of your time, limit your calculations to some specific elements.

You know the mass of the proton and neutron. The only additional information that is needed to calculate the binding energy for a given nucleus is the mass of that nucleus. The masses of a number of elements are given below.

Element	Atomic number	Mass number	Atomic mass (u)	Element	Atomic number	Mass number	Atomic mass (u)
neutron	0	1	1.008665	Co	27	59	58.933190
proton	1	1	1.007825	Zn	30	66	65.926052
H	1	1	1.007825	Br	35	79	78.918330
He	2	4	4.002603	Zr	40	91	90.905642
Li	3	7	7.016004	Rh	45	103	102.905511
Be	4	9	9.012186	Sn	50	119	118.903314
B	5	11	11.009305	Cs	55	133	132.905355
C	6	12	12.000000	Nd	60	145	144.912538
N	7	14	14.003074	Yb	70	173	172.938060
O	8	16	15.994915	Hg	80	200	199.968327
F	9	19	18.998405	TI	81	205	204.974442
P	15	31	30.973765	Pb	82	207	206.975903
Ca	20	40	39.962589	Bi	83	209	208.981082
Mn	25	55	54.938051	Th	90	232	232.038124
Fe	26	57	56.935398	U	92	235	235.043915

Example:

Calculate the binding energy per nucleon for phosphorus-31 $\left({}^{31}_{15}\text{P}\right)$. (You will use the same method as you used in the previous activity.)

The atomic number of phosphorus is 15.

Number of protons = atomic number = 15
Number of neutrons = atomic mass – number of protons = 16

Mass of 15 protons = 15 (1.007825 u) = 15.117375 u
Mass of 16 neutrons = 16 (1.008665 u) = $\underline{16.138640\ \text{u}}$
Total nucleon mass = 31.256015 u
Mass of nucleus of ${}^{31}_{15}\text{P}$ = $\underline{30.973765\ \text{u}}$
Mass difference = 0.282250 u

Total binding energy = 0.282250 u × 931.5 MeV/u
= 262.9 MeV
Total number of nucleons = 31 (15 protons + 16 neutrons)
Average binding energy per nucleon = $\frac{262.9 \text{ MeV}}{31}$ = 8.48 MeV
On average, it would require 8.48 MeV or 8.48 million eV to remove a proton or neutron from the nucleus of phosphorus–31.

a) Calculate the binding energies for the elements assigned to you by your teacher.

b) Use combined class data to plot a graph of the binding energy per nucleon versus the number of nucleons for each element in the periodic table.

2. If you made a graph of all the elements, it would look like this:

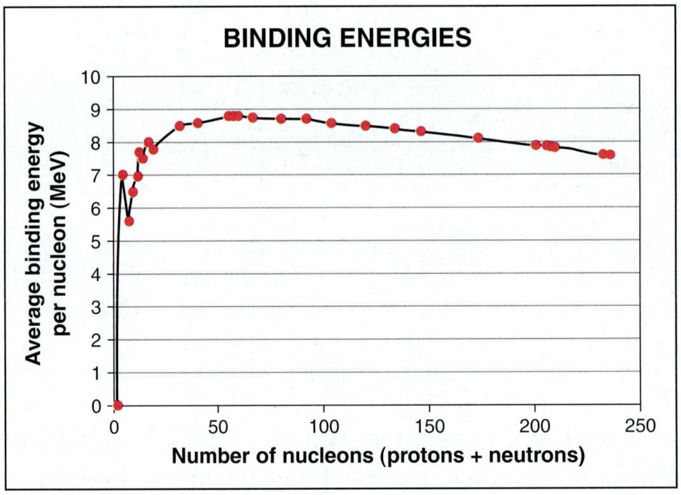

a) Sketch this graph in your log.

b) Describe three features of the graph that you took into account when you made your sketch.

3. Small nuclei can be put together to form larger nuclei.

a) Show evidence from your graph how the binding energy per nucleon increases if two small nuclei are put together to form one larger nucleus.

If the nucleon becomes more tightly bound to the nucleus, additional energy is given off. It takes additional energy to free this nucleon.

Larger binding energy means:

- more energy is given off
- the nucleon is more tightly bound
- more energy is needed to free the nucleon

b) Use the graph of nuclear binding energy versus atomic number to answer the following questions:
   i) Which element has the most tightly bound nucleons?
   ii) Which element is most stable?
   iii) Which element requires the most energy to free a nucleon?

c) Suppose two atoms of a lighter element (e.g., atomic number 5) are fused together to create a larger nucleus (e.g., atomic number 10). This process is called nuclear fusion. How does the binding energy of the nucleons in the larger nuclei compare with that of the smaller nuclei? Will energy be emitted or absorbed during this transition?

d) Not all elements can fuse and provide energy. In order to produce energy by nuclear fusion, the created nucleus must have a larger binding energy per nucleon. Draw a sketch of the binding energy graph and indicate the portion of the graph where smaller nuclei can fuse and produce a larger nucleus with an increase in binding energy per nucleon. Label this portion—fusion.

4. There is another means by which to create nuclei with larger binding energy. If a very heavy element like uranium were to break apart, the two daughter nuclei would have a greater binding energy per nucleon.

a) What is the average binding energy per nucleon for uranium-235?

b) From the graph, estimate the average binding energy per nucleon of barium-137.

c) From the graph, estimate the average binding energy per nucleon of krypton-84.

Since the daughter products of barium-137 and krypton-84 have more binding energy per nucleon than the parent uranium-235, then energy is given off. This energy release is due to nuclear fission. It happens when a large nucleus splits into smaller nuclei.

5. Uranium-235 will become unstable and undergo fission when it absorbs a neutron. The reaction can be written as:

$$^{235}_{92}\text{U} + ^{1}_{0}\text{n} \rightarrow ^{144}_{56}\text{Ba} + ^{89}_{36}\text{Kr} + 3\,^{1}_{0}\text{n}$$

a) Show that the number of protons on both sides of the reaction is equivalent.

b) Show that the total number of nucleons (the sum of protons and neutrons) is conserved.

6. Only the heaviest elements can undergo fission and provide energy. In order to produce energy by nuclear fission, the created nuclei must have a larger binding energy per nucleon. Refer to your sketch of the binding energy graph.

a) Indicate the portion of the graph where a larger nucleus can undergo fission and produce nuclei with an increase in binding energy per nucleon. Label this portion—fission.

7. Notice that the fission of uranium-235 with the absorption of a neutron yields three additional neutrons. Those three neutrons are responsible for our ability to create a chain reaction. The chain reaction permits two technologies – nuclear power and nuclear bombs.

In order to understand how the chain reaction takes place, imagine a set of 100,000 mousetraps. When a small marble is dropped onto any mousetrap, the mousetrap closes with a large snapping sound.

a) What will happen if 5 marbles are dropped?

b) Imagine that each mousetrap now has a marble balanced on it. When that mousetrap closes, its marble jumps in the air and can land on another mousetrap. What will happen if 5 marbles are now dropped?

c) Imagine now that each mousetrap has 2 marbles balanced on it. When that mousetrap closes, the 2 marbles jump in the air and can land on other mousetraps. What will happen if 5 marbles are now dropped?

d) The mousetraps get out of control very rapidly. The 5 marbles release 10 marbles that then release 20 marbles. Create a chart to show how many marbles are released in the first 10 cycles.

e) Assume that you have an unlimited supply of mousetraps. Continue your chart to show how many marbles are released in the first 20 cycles.

This enormous release of mousetraps is similar to the enormous release of energy in a nuclear chain reaction.

Physics Words

nuclear fusion: a nuclear reaction in which nuclei combine to form more massive nuclei with the release of a large amount of energy.

nuclear fission: a nuclear reaction in which a massive, unstable nucleus splits into two or more smaller nuclei with a release of a large amount of energy.

FOR YOU TO READ

Nuclear Fusion and Nuclear Fission

The structure of the atom has a nucleus made up of protons and neutrons. These nucleons are held tightly together with a specific binding energy per nucleon. To free a nucleon would require the addition of energy equal to this binding energy per nucleon. In this activity you calculated the average binding energy per nucleon for each nucleus. The resulting calculations were plotted on a graph of binding energy versus atomic number. The graph revealed that iron has the highest average binding energy per nucleon and is therefore the most stable nucleus.

Nuclei of smaller mass than iron have less binding energy. This makes **nuclear fusion** possible. In nuclear fusion two lighter nuclei fuse together to produce a larger nucleus. The larger nucleus has a greater average binding energy per nucleon. Energy is therefore released in the creation of the larger nucleus. This nuclear fusion is responsible for the Sun's energy. The Sun has provided the Earth's energy for over five billion years. It is expected to continue to produce energy through the fusion of hydrogen into helium for another five billion years.

Nuclei of larger mass than iron also have less binding energy. This makes **nuclear fission** possible. In nuclear fission a heavy nucleus can break apart into two smaller nuclei. The smaller nuclei have a greater average binding energy per nucleon and energy is therefore released during their creation.

You saw that the fission of uranium-235 with the absorption of a neutron yields three additional neutrons.

$$^{235}_{92}U + ^{1}_{0}n \rightarrow ^{144}_{56}Ba + ^{89}_{36}Kr + 3\,^{1}_{0}n$$

Those three neutrons have changed our politics, our culture, and our lives. Those three neutrons are responsible for our ability to create a chain reaction. The chain reaction permits two technologies—nuclear power and nuclear bombs.

You compared the release of mousetraps to the enormous release of energy in a nuclear chain reaction. The uranium-235 nucleus absorbs one neutron, but gives off three neutrons. Each of those three neutrons can be absorbed and more uranium-235 can undergo fission. With each fission reaction, more energy is released. In a matter of moments, a huge fission explosion can take place. The fission explosion can be a nuclear bomb. However, the fission chain reaction can also be controlled. By removing neutrons before the other uranium absorbs them, a controlled reaction takes place. In a nuclear power plant, the control rods remove these neutrons, so that the uncontrolled chain reaction does not take place.

Reflecting on the Activity and the Challenge

Both nuclear fusion and nuclear fission provide energy. Your museum exhibit on atomic structure can certainly include information about binding energy and nuclear fission and fusion. It will require some special insight to find ways to help museum visitors understand how breaking up a nucleus can provide energy while fusing together other nuclei can also produce energy. You may decide to draw people into your exhibit with something concerning fission or fusion, sunlight or bombs, endless energy sources, or political drama.

Physics To Go

1. a) Calculate and compare the binding energies per nucleon for these three elements:
   i) oxygen – 16
   ii) lithium – 7
   iii) calcium – 40

   b) Which nucleus in **Part (a)** is most stable?

2. a) Sketch the general shape of the binding energy curve where the number of nucleons is on the *x*-axis and the average binding energy per nucleon is on the *y*-axis.

   b) Label the peak of the graph as the most stable element. Identify that element.

   c) Label the part of the graph where fusion could occur.

   d) Label the part of the graph where fission could occur.

3. Should a single museum exhibit attempt to discuss both fusion and fission or should it focus on one? Discuss your reasons.

4. A simulation of a chain reaction can be constructed using dominoes.

   a) How would such a simulation be set up?

   b) Is there a way to set up such a simulation at a museum exhibit? You probably don't want everybody taking the time to set up the dominoes by hand. Is there a mechanical way of easing the setup time?

5. The Sun creates energy by the fusion of hydrogen into helium. Describe in detail (using equations and numbers) how this energy can be created. How long do you think that the Sun could continue to produce energy in this way?

6. In a nuclear reactor, control rods absorb excess neutrons. Why would the absorption of neutrons slow the reaction?

7. Nuclear medicine, nuclear energy, nuclear bombs... some people have very strong opinions about these applications of our knowledge of the nucleus. Could a museum exhibit poll

visitors to find their opinion? What would you like to know beyond their opinion? Describe an exhibit that could gather and display information.

Stretching Exercises

1. What research is presently being investigated to provide fusion as an energy source?
2. What are the major advantages and disadvantages of fission as an energy source? How does one decide whether the advantages outweigh the disadvantages?
3. The calculations in this activity were very repetitive. You could set up a spreadsheet to solve for all of the binding energies at one time. Constructing a spreadsheet allows many calculations of binding energy to be done effortlessly. You could then use the graphing function to display the relationship.

Element	Atomic number	Mass number	Number of protons	Number of neutrons	Atomic mass (u)	Total mass of protons	Total mass of neutrons	Total mass	Mass defect	Binding energy	Binding energy per nucleon
neutron	0	1			1.008665						
proton	1	1			1.007825						
H	1	1			1.007825						
He	2	4			4.002603						
Li	3	7			7.016004						

The first three lines of a sample spreadsheet are illustrated. The given information is atomic number, mass number, and atomic mass.

- You can set the cell for "Number of protons" to equal the "Atomic number" cell.
- You can set the cell for "Number of neutrons" to equal the "Mass number – atomic number."
- The "Total mass of protons" is equal to the "Mass of the proton" multiplied by the "Number of protons."

Determine equations for calculating the other cells in the spreadsheet. Fill down for all of the elements listed in the table of elements provided in the **For You To Do** section.

PHYSICS AT WORK

Dr. Allen Friedman

MAKES SCIENCE FUN

As director of the New York Science and Technology Center—an interactive museum in Queens, New York—it is Dr. Friedman's job to introduce complex scientific principles to people of all ages and levels of knowledge. "I have to communicate science to the general public and that is an extremely interesting intellectual puzzle."

Born in Brooklyn, Dr. Friedman earned his bachelor degree from the Georgia Institute of Technology and a Ph.D. in physics from Florida State University. He had every intention of becoming an experimental scientist. That all changed the day he walked into the Lawrence Hall of Science, an interactive museum at the University of California at Berkeley. As he walked from exhibit to exhibit he noticed something extraordinary: People were actually having fun.

Inspired by the museum's energy, he applied for a fellowship, received it, and has never looked back. "One of the very first exhibits we ever created was a three-dimensional illustration of the structure of an atom," he explains. "By projecting images on a moving screen we were able to simulate three dimensions. Quantum theory—which is how we understand atoms today—basically says that the traditional, rigid, mechanical model isn't nearly sophisticated enough to explain what we see in the physics of the atom. And that same principle can be transferred to the teaching of science."

"There are really three crucial aspects of every exhibit," he explains. "First, it has to attract an audience—and that means everyone from a small child with virtually no scientific knowledge to a high school science student and even practicing scientists. Then, you have to find ways to hold their attention long enough to communicate the information. And lastly, every good museum exhibit must be entertaining—the audience has to like it."

Chapter 9 Assessment

The time has come for you to complete your **Chapter Challenge**. It's time for your *Active Physics* group to put the final touches on your science museum exhibit on atoms. With what you learned in this chapter, you should be ready to meet the challenge.

You have probably already designed the item that will be for sale at the museum gift shop. However, before you finalize your exhibit, carefully review all the features that you were asked to include. You do not want to receive a poor grade just because you forgot to include something.

Your exhibit must:

- **include distinct features of the structure of the atom**
- **communicate the size and scale of the parts of the atom**
- **provide information on how the atom is held together**
- **capture the visitor's attention within 30 seconds**
- **include written matter that will further illuminate the concepts**
- **have a model, a T-shirt, a poster, a booklet, or a toy that can be sold at the museum store**
- **have a building plan and cost estimate for the exhibit**
- **include safety features**
- **be interactive—visitors should not merely read**

Before you began this chapter, you and your classmates decided on the criteria by which your **Chapter Challenge** will be evaluated. Revisit those criteria now.

- **What weight will be placed on the 30-second criterion?**
- **How many points will be allocated to the building plan and cost?**
- **How much should the physics content be worth?**

You may wish to revise your grading system in light of what you learned in this chapter. The better you understand how you will be graded, the better the likelihood of receiving a good grade.

Physics You Learned

Coulomb's Law

Conservation of charge

Millikan's oil-drop experiment

Size of nucleus

Neutron, proton, electron

Rutherford scattering experiment

Light spectra

Models of the atom

Conservation of energy

Diffraction of light

Interference of light

Photoelectric effect

Electron wavelength

Isotopes

Nuclear forces

Feynman diagrams

Radioactive decay

Binding energy

Nuclear fission

Nuclear fusion

Unit 2

Active Chemistry™

Chapter 10: Movie Special Effects
Chapter 11: The Periodic Table
Chapter 12: Cool Chemistry Show

Making Connections

You have just completed the unit related to physics. You are now going to start a unit of Active Chemistry. You investigate the world of chemistry every time you cook a meal. You mix chemicals together, heat them, and new chemicals appear. There are many different foods. How many different chemicals are there? In the Active Physics unit you learned about the properties of light waves and sound waves. Now you will explore properties of chemicals and try to make your complex world simpler to understand.

The laws of physics are also the laws of chemistry. In chemistry, you will continue to study forces, energy, and atomic structure. You will learn about these laws of nature from a new perspective.

What Do You Think?

- How do you think that colors are created on the atomic level?
- How is sugar different from salt?
- What makes some chemicals nutritious and others dangerous to eat?

Record your ideas about this in your log. Be prepared to discuss your response with your small group and the class.

For You To Try

- Look up a recipe for making a cake. When do you think that chemical reactions are taking place?
- Compare the labels on a box of cereal and a cleaning fluid. What similarities and differences are apparent?

Chapter 10

MOVIE SPECIAL EFFECTS

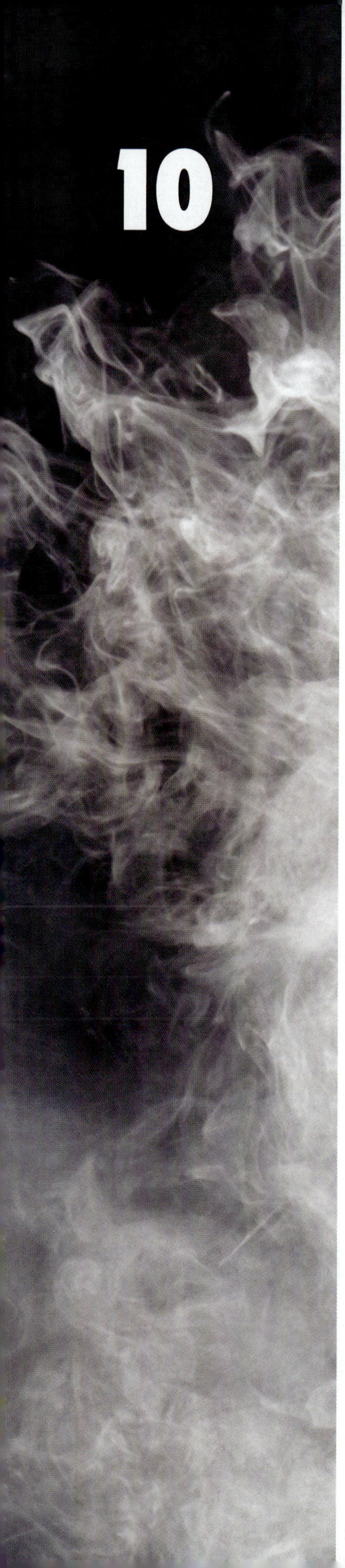

10 Movie Special Effects

Scenario

You have all been captivated and entertained by special effects in movies. The explosions, makeup, animation, and props in the "blockbuster" features take months and millions of dollars to put together. Many special effects are the result of the application of science and technology.

Your *Active Chemistry* class has been asked to participate in the production of a low-budget movie. However, to make the film exciting, the movie producer would still like to use some awesome special effects. You are being asked to write a simple scene in which you can incorporate some special effects. Of course, in addition to cost, safety is also a major concern.

Chapter Challenge

Your challenge is to create a story line and produce special effects based on the chemistry you have learned in your *Active Chemistry* class. You will need to demonstrate the special effects you created. Your special effects will be evaluated on their quality, interest, entertainment, and the knowledge of chemistry you exhibited in putting them together.

You will need to complete the following tasks:

- Write a script for a simple scene in a movie.
- Choose a special effect to include as part of your scene.
- Write a procedure on how your special effect is done.
- Demonstrate the special effect to the "producer."
- Write an explanation of how the special effect works, including the chemistry behind the demonstration.

Using more than one chemical principle will strengthen your presentation.

Criteria

How will your special effect be graded? What qualities should an effective special effect have? How will your demonstration and supporting documentation be graded? Discuss these issues in small groups and with your class. You may decide that some or all of the following qualities should be graded:

- Demonstration
- Safety
- Quality
- Interest and appeal to audience
- Supporting documentation
- Script — creativity
  - Procedure — clarity, safety, accuracy
    - Chemistry explanation — accuracy and quantity of chemical principles incorporated

Once you have determined the list of qualities for evaluating the documentation and demonstration, you and your class should also decide how many points should be given for each criterion. How many points should be assigned to the documentation and how many to the demonstration? Should more points go to the chemistry explanation than to the movie script? How many different chemical principles should be incorporated in your special effect? Determining grading criteria might be a time-consuming task, but knowing the point values in advance will help you focus your time and effort on the most important aspects of your special-effect documentation and demonstration.

Will each student produce his/her own special effect, will students be required to work in groups, or will both options be offered? Discuss the pros and cons of these possibilities. Keep in mind that if you are going to be working in groups, it is important to discuss before the work begins how each member of the group will be graded. Determine grading criteria that reward each individual in the group for his/her contribution and also reward the group for the final project. You should discuss different strategies and choose the one that is best suited to your situation. Make sure that you understand all the criteria as well as you can before you begin. Your teacher may provide you with a sample rubric to help you get started.

Activity 1 Elements and Compounds

GOALS

In this activity you will:

- Decompose water by electrolysis into the two elements from which it is composed.
- Test the two elements to determine their identities.
- Learn one way to determine the chemical formula of a material.
- Compare characteristic properties of a material to those of its constituent elements.
- Represent materials with chemical formulas using numbers and the symbols of elements.
- Practice safe laboratory techniques with flames and explosions.

What Do You Think?

Matter is the name for all the "stuff" in the universe. Anything that has mass and occupies space is called matter.

- How many kinds of matter are there in the universe: 1, 10, 100, 1000, 10,000, or more than 10,000?
- What makes up matter?

Record your ideas in your *Active Chemistry* log. Be prepared to discuss your responses with your small group and the class.

Investigate

1. Look around your classroom at the kinds of matter that make up the things you can see.
   a) Make a list of 10 kinds of materials that make up the objects you see. For example, you might list the wood that makes up your pencil, or the glass that makes the windows.
   b) To understand the nature of matter, it helps to know if it is simple or complex. Is it made from only one kind of material or is it a mixture of various materials? Classify each of the 10 materials you listed as pure or mixtures.

c) For each material you thought was a mixture, write your best guess about what materials make it up.

2. Sometimes the materials that make up a substance are not obvious. Early scientists thought that water was an element. In other words, they thought that there was only one kind of material in water. They had not discovered a way of breaking it down further. Water, however, can be broken down further.

   a) Carefully observe the characteristic properties of water. Record your observations in your log.

3. Put on your safety goggles and apron. Assemble the apparatus for decomposing water as shown in the diagram. Fill two test tubes with water. Submerge them in the water in the beaker and invert them. Make sure you do not allow any air to enter the test tubes. Make sure that the ends of the wires are stripped. Polish them with steel wool. Insert the ends of the wires into the test tubes. Add about 1 to 2 mL of sodium sulfate solution to the water in the beaker.

4. Plug in the battery. Let the reaction run until a test tube is full of gas. (Your teacher may decide to have you stop the reaction sooner.)

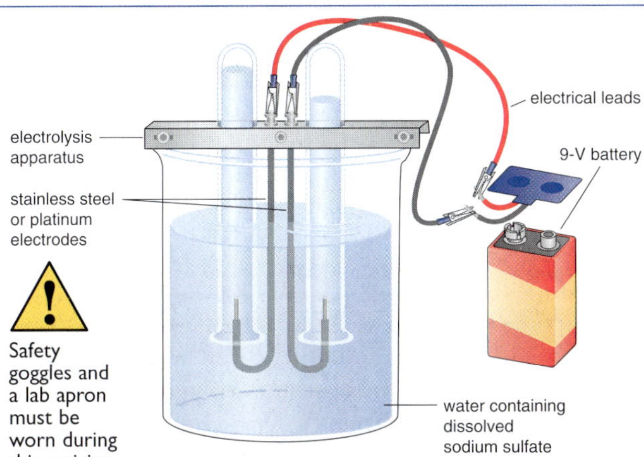

Safety goggles and a lab apron must be worn during this activity.

   a) Note what happens when the power is turned on. Record your observations in your *Active Chemistry* log.

   b) How do the relative amounts of gas formed in the test tubes compare?

5. Disconnect the battery. Place stoppers on the test tubes and remove them from the water.

   a) What gas do you think is contained in each test tube? (Hint: You've probably heard that water is H-2-O, written, H_2O.) Record your prediction, and give reasons for your prediction.

   b) Observe the physical properties of each gas. Record these properties in your log.

6. You are going to use a lighted wooden splint to identify the gas in each test tube. First, examine the test tube with the smaller volume of gas. Light a wooden splint. Blow out the flame, but leave the splint glowing. Hold the test tube with its mouth up. Remove the stopper. Quickly bring the glowing splint to the mouth of the test tube.

   a) Observe what happens to the splint, and record your observations.

   b) What gas do you think was produced in this test tube?

Be certain that the mouth of the test tube is pointed away from everyone.

Clean up spills immediately.

Safety goggles and a lab apron must be worn during this activity.

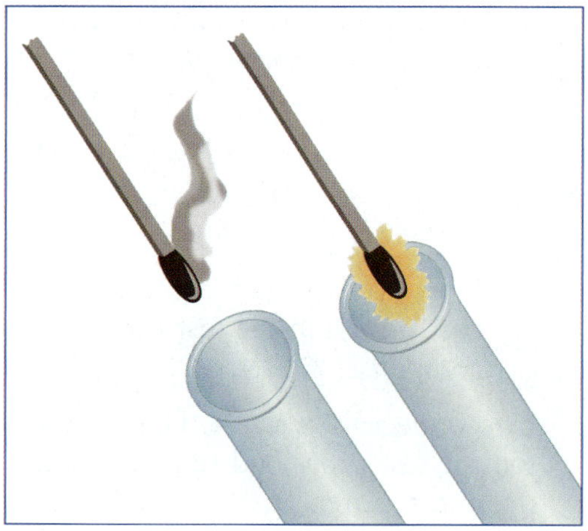

c) In your log record any additional properties of the gas that you discovered.

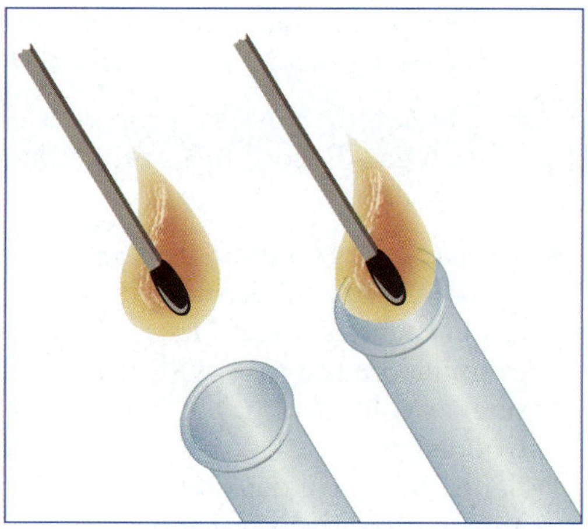

7. Next, examine the test tube that is full of gas (or contains the larger volume of gas). Light a wooden splint. Remove the stopper. Quickly bring the burning splint to the mouth of the test tube.

   a) Record your observations.

   b) What gas do you think was produced in this test tube?

   c) In your log record any additional properties of the gas that you discovered.

Both the teacher and the students must be shielded.

8. Return the splints and equipment as directed by your teacher. Clean up your work station. Wash your hands.

9. Your teacher will demonstrate a way you can use what you investigated to produce a special effect. Your teacher will set up a gas generator, similar to the one shown in the diagram. The test tube contains 10 mL of 6M HCl (hydrochloric acid — a compound of hydrogen and chlorine) and 3 g of mossy zinc.

   a) What gas do you think is being produced in the test tube? Give a reason for your answer.

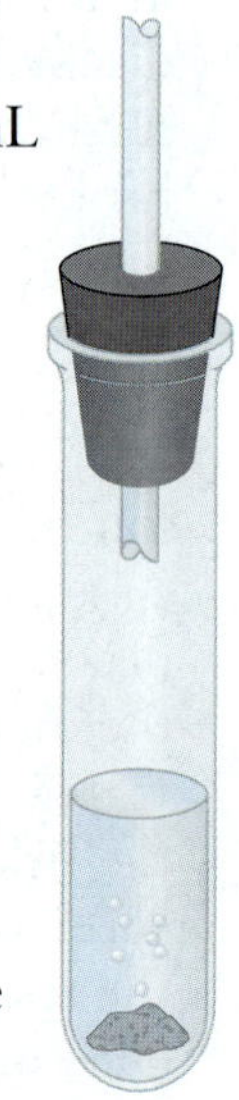

10. An egg has been emptied out. There is a small hole in the top and another in the bottom of the egg. One of the holes is taped. The egg will be placed over the gas generator, and gas will be collected in the egg for several minutes.

11. The egg will then be mounted in an egg holder behind a shield. If you have a video camera available, be prepared to start recording. Your teacher will remove the tape, and light the top of the egg with a burning splint. Begin recording or observing the reaction until it is over.

    a) Record your observations in your log.

    b) If the gas in the egg was hydrogen and air contains oxygen, what substance do you think may have been created in this process?

Chem Talk

THE STRUCTURE OF MATTER

In this investigation you used electricity to decompose water into two gases. You knew that the gases were different because they reacted differently to the burning and glowing splints. Since water is referred to as H_2O, a first guess would be that hydrogen (H) and oxygen (O) were created in the experiment. The test for hydrogen is a small explosion when exposed to a burning splint. The test for oxygen is igniting a glowing splint. If you look back on the results of the experiment, you find that the hydrogen gas filled one test tube while the oxygen filled half of the other test tube. There was twice as much hydrogen as oxygen. That's where the 2 comes from in the chemical formula H_2O.

Hydrogen and oxygen are elements. An **element** is any material that cannot be broken down into simpler materials by chemical means. You are probably familiar with many elements like hydrogen, oxygen, zinc, gold, or helium. Other elements like strontium and beryllium are more exotic and less likely to be familiar to you. Every kind of matter you observe in your everyday life is made up of the chemical elements. There are only about a hundred different kinds of chemical elements. This is an amazing discovery of chemistry — everything you observe in the world is made of different combinations of a hundred elements. Chemistry is the study of how these elements combine and the characteristics of these combinations.

Chem Words

element: any material that cannot be broken down into simpler materials by chemical means.

Elements are represented by symbols. The symbol is one, two, or three letters that represent the name. It's easier to write O than to write oxygen. It's easier to write H than to write hydrogen. The symbols come from many different sources. However, the same symbols are used for each element in all countries of the world.

Symbols for Some Elements

Name of Element	Symbol
aluminum	Al
bromine	Br
calcium	Ca
carbon	C
chlorine	Cl
copper	Cu
gold	Au
helium	He
hydrogen	H
iodine	I
iron	Fe
lead	Pb
magnesium	Mg
mercury	Hg
neon	Ne
nickel	Ni
nitrogen	N
oxygen	O
phosphorus	P
potassium	K
silicon	Si
sodium	Na
sulfur	S
tin	Sn
zinc	Zn

When elements combine they form new substances called **compounds**. These compounds have entirely new characteristics. It is like combining the letters of the alphabet to make words. Twenty-six letters can be combined to make thousands of different words.

Water is an example of a compound. A water molecule, H_2O, is composed of two atoms of hydrogen and one atom of oxygen. (For now, think of an atom as the smallest particle of an element and a molecule as the smallest unit of a compound.) In this activity you used electricity to decompose water into its elements, hydrogen and oxygen. This process is called **electrolysis**. You observed that oxygen gas made a glowing splint burst into flame, and that hydrogen gas was explosive. However, to extinguish a burning splint, you could use liquid water. The compound has very different characteristics from the elements from which it is made.

Compounds are represented by **chemical formulas**. A chemical formula shows the symbols of the elements that are combined to make the compound. If there is more than one atom of an element, a subscript is added after the symbol indicating how many atoms of that element there are. For example, as you discovered in this activity, the chemical formula for water is H_2O.

Chem Words

compound: a material that consists of two or more elements united together in definite proportion.

electrolysis: the conduction of electricity through a solution that contains ions or through a molten ionic compound that will induce chemical change.

chemical formula: the combination of the symbols of the elements in a definite numerical proportion used to represent molecules, compounds, radicals, ions, etc.

Examples of Some Chemical Formulas

Compound	Common Name	Chemical Formula
calcium carbonate	chalk	$CaCO_3$
carbon dioxide	dry ice	CO_2
hydrochloric acid	muriatic acid	HCl
hydrogen sulfide	rotten-egg gas	H_2S
sodium hydrogen carbonate (or sodium bicarbonate)	baking soda	$NaHCO_3$
sodium chloride	table salt	$NaCl$
sodium nitrate	fertilizer	$NaNO_3$
sulfuric acid	battery acid	H_2SO_4

From the table of chemical formulas, you can see that carbon dioxide is a compound of carbon and oxygen. There are two atoms of oxygen for every atom of carbon. Sodium hydrogen carbonate (sodium bicarbonate) is a compound of sodium, hydrogen, carbon, and oxygen. There are three atoms of oxygen for every atom of the other elements. Also, there are a total of three atoms in the carbon dioxide formula and a total of six atoms in sodium hydrogen carbonate.

To generate the gas to fill the empty eggshell in this activity (the teacher demonstration), zinc was placed in hydrochloric acid. Zinc is an element. Hydrochloric acid (HCl) is a compound of hydrogen and chlorine. The reaction of the zinc and hydrochloric acid created a gas. Given the explosion you observed, you can guess that the gas produced was hydrogen. The hydrogen gas came from the hydrogen in the hydrochloric acid.

There's much more to the structure of matter than you can discover in just one activity. However, this activity may have raised some new questions in your mind. For example:

- Can all compounds be decomposed into their elements?
- What techniques can be used to decompose compounds?
- What are elements made of?
- What are atoms?
- What are molecules?

These questions and many more will be explored in other *Active Chemistry* activities.

Checking Up

1. In your own words, explain the difference between an element and a compound.
2. Why are symbols useful in describing chemical elements?
3. What are the symbols for the following elements: carbon, copper, gold, and helium?
4. What information does a chemical formula of a compound provide?

Reflecting on the Activity and the Challenge

Part of the problem you are facing in creating a special effect is understanding what matter is made of and how it can change. In this activity you broke a chemical compound down into its component elements using electrolysis. In another part of the activity a compound was made from chemical elements through a fast and noisy reaction. There are only about one hundred elements, but there are many thousand compounds. You should begin thinking of ways in which some of the reactions you observe could be made to appear more dramatic on screen, without making them any larger in real life. You can now use the concepts of elements and compounds to provide the chemistry description of what is occurring.

Chemistry to Go

1. The table on the right contains several common compounds that are probably familiar to you.

   For each compound:

   a) List the names of the elements present.

   b) State the number of atoms of each element present.

   c) Give the total number of atoms present in each compound.

Common Name	Formula
sugar	$C_{12}H_{22}O_{11}$
marble	$CaCO_3$
natural gas	CH_4
rubbing alcohol	C_3H_8O
glass	SiO_2

2. Write a chemical formula for nitrous oxide (laughing gas) that is made up of two atoms of nitrogen and one atom of oxygen.

3. Choose one compound from the table in **Question 1**.

   a) Describe the properties of each element in the compound.

   b) Explain how the property of the compound is different from the property of each element.

Preparing for the Chapter Challenge

In a short paragraph, summarize the difference between an element and a compound and describe how the properties of a compound can be very different from the properties of the elements that make it up. Explain why knowing these differences is important when designing special effects for a movie set.

Inquiring Further

How is electrolysis used in industry?

Use the reference materials available to you to explore how electrolysis is used in industry to produce hydrogen gas and other elements from compounds.

Activity 2 States of Matter: Solid, Liquid, and Gas

GOALS

In this activity you will:

- Create an animation to illustrate the behavior of particles in different phases of matter, and as the material changes phase.
- Observe changes of state of water and describe the process graphically.
- Observe a change of state of carbon dioxide and describe the energy transformations involved.
- Describe the energy transformations and the roles of kinetic and potential energy as heat energy is transferred to or away from a material.
- Describe the behavior of gas particles, based on your observations of how the temperature, pressure, and volume of the gas are affected as heat energy is transferred to or away from the gas.
- Characterize materials by their unique phase-change temperatures.
- Practice safe laboratory techniques in working with temperature extremes.

What Do You Think?

You know that materials can exist as solids, liquids, or gases. Each state of matter has its own characteristics.

- Draw three circles. In the first circle draw what you think particles of material look like in the solid state. In the next circle draw the particles of the same material as a liquid. In the final circle illustrate the same material as a gas.

Record your ideas in your *Active Chemistry* log. Be prepared to discuss your responses with your small group and the class.

Investigate

Part A: The Heating Curve of Water

1. Put on your safety goggles and apron. Half-fill a 250-mL beaker with crushed ice.
2. Set up your equipment as shown in the diagram. The hot-plate dial should be in the off position. When you clamp the thermometer into place, be sure that the bulb is in the center of the ice and is not touching the bottom of the beaker.

Clean up spills immediately. Report any broken or cracked glassware to your teacher.

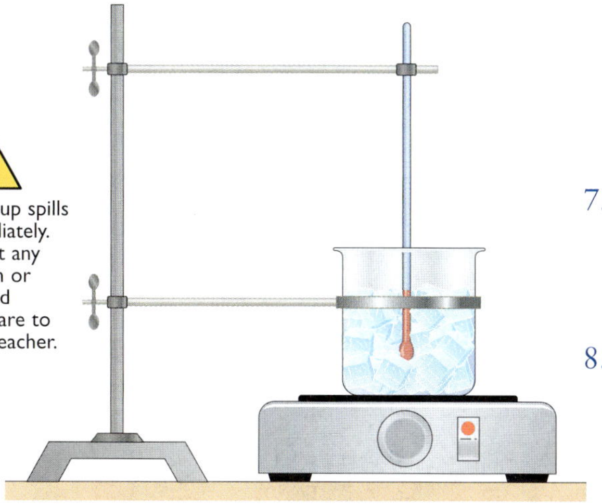

3. Observe the thermometer closely until it appears to have reached its lowest reading.

   a) Prepare a data table similar to the one shown below. Record the minimum temperature at time 0 min.

4. Turn on the hot plate. Use a medium-low setting, or one suggested by your teacher.

5. Gently stir the ice and water mixture with a stirring rod.

   a) Record the temperature every minute.

6. Continue to record the temperature until the water has been at a full boil for 5 min.

   a) Name all the changes of state that you observed. A change of state refers to a change from a solid to a liquid, a liquid to a gas, a solid to a gas, and vice versa.

7. Turn off the hot plate. When the water has cooled discard it. Return all equipment as directed and clean up your station.

8. Use your data to answer the following:

   a) What was the temperature at which all the ice had melted?

   b) What was the boiling point of the water?

   c) Plot a graph of the data with time along the x-axis and temperature on the y-axis. You may wish to use a microcomputer or a calculator to plot your graph.

   d) Describe your graph. Consider: What is happening at the various points along the graph? Heat energy is being continually transferred to the system by the hot plate. At which point is the heat energy causing the temperature to increase? What is the heat energy doing if it is not acting to raise the temperature of the water?

Hot plates may remain hot for some time after they are turned off.

Never use a thermometer as a stirring rod.

Time (min)	Temperature (°C)	Observations
0		
1		

Part B: Making Particles Shake

1. Your teacher will assemble or have you assemble a small booklet with many blank pages.
2. Use only the right-hand pages of the booklet. At the very bottom right of the last page of the booklet draw a dot. To make the dot appear to move from the top left to the bottom right at a constant speed, draw another dot on the previous page slightly to the left and up. Continue this until you reach the first page of the booklet. Now as you flip through the booklet from the front to the back, the dot will appear to move. The smaller the movements from one page to the next, the smoother the animation effect will appear.
3. Matter can be in a solid state, a liquid state, or a gaseous state. Each state of matter has its own typical motion of particles. The physical properties of each state are a result of how the particles move relative to one another. Use animation to model the movement of particles in each state of matter. Consider using different colors to keep track of the particles.
   a) In a solid the particles stay in the same position but vibrate. Use your flipbook to model the movement of particles in a solid.
   b) In a liquid the particles are about the same distance apart as in the solid, but they can move more freely. Use animation to model a liquid.
4. Gas particles are very, very far apart and they move very quickly.
   a) What problems would you have to make a flipbook for gases?

Part C: Volume Changes

1. Draw up a water plug about 1 cm long in a 30-cm long glass tube. Set the glass tubing into a test tube, as shown in the diagram below.

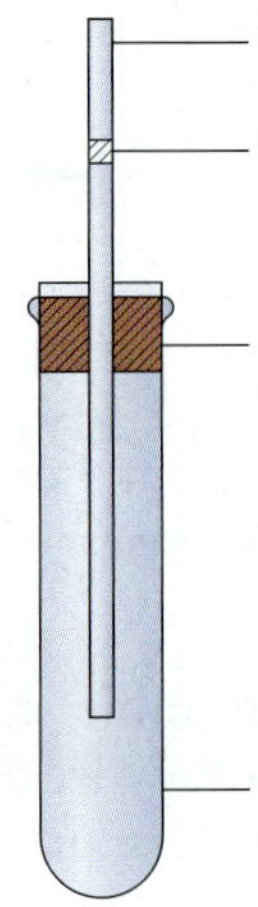

   a) Observe and describe the movement, if any, of the water plug.
2. Place the test tube into a beaker of warm water.
   a) Record your observations in your *Active Chemistry* log.
   b) How does the warm water affect the volume of the air in the test tube? What evidence did you observe that suggested a volume change?
   c) The warm water was a source of heat energy. What effect did this heat energy have on the volume of air?

The water should not be too hot to touch. Glassware is slippery when wet. Handle with care.

3. Place the test tube into a beaker of ice water.

   a) Record your observations in your *Active Chemistry* log.

4. In physics you've learned that for the drop of water to stay in place, the forces must be equal and opposite. The force of gravity pulling down on the water plug must be equal to the upward force of the gas.

   When the air in the test tube was heated, the upward force on the water plug must have increased, because the water rose in the tube. Since no additional air molecules were added, the molecules must have moved faster as a result of the additional heat. As the temperature of the air increased, the molecules of air increased their speed and therefore applied a greater force to the drop of water. Once the drop of water rose high enough, the larger volume and fewer air molecules hitting the drop per second compensated for the increased speed of the molecules. The forces of gravity and pressure were equal once again.

   a) Pressure is force per area. What happened to the pressure on each wall of the test tube as you heated up the air (gas)?

   b) Draw a box with a moveable piston, as shown in the diagram at right.

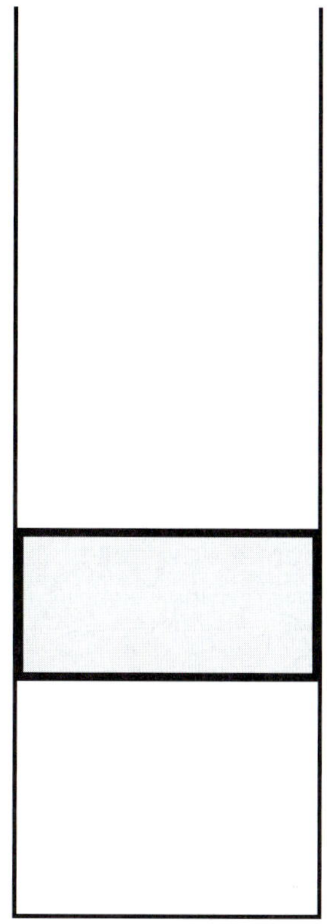

   c) Use animation to show what would happen to the piston if the temperature of the gas inside the cylinder and below the piston were increased.

Part D: "Special" Ice

1. Your teacher will place a small piece of dry ice (solid carbon dioxide) in an empty beaker.

   a) Record your observations in your *Active Chemistry* log.

   b) Is heat energy being transferred to or away from the dry ice by the surrounding air?

   c) What change of state is taking place?

Chem Talk

CHANGES OF STATE

All matter is made up of tiny particles. Different materials are made of different kinds of particles. These particles are always moving, and there are spaces between them. The more energy the particles have, the faster they move. There are also attractive forces among the particles. The closer the particles are together, the greater are the attractive forces.

Temperature

You get an intuitive sense of **temperature** by how hot or cold something feels to your skin. Your body is at 37°C (98.6°F). When something with a higher temperature comes in contact with your skin, you know that it is "hot." When something with a lower temperature comes in contact with your skin, you know that it is "cold." As you observed in the activity, when the temperature of air increased, the drop of water lifted. This drop of water could be a crude thermometer. As the drop rises, you know that the temperature of the air is higher. Liquids like alcohol and mercury expand when they get hot and are used for the thermometers with which you are most familiar.

Chem Words

temperature: the measure of the average kinetic energy of all the particles of the material.

kinetic energy: a form of energy related to the motion of a particle ($KE = \frac{1}{2}mv^2$).

The movement of the water drop gives you an insight into another interpretation of temperature. The air molecules inside the tube were moving faster as the temperature of the air increased. The temperature of the air is a measure of the speed of the molecules. In physics, you learned that kinetic energy is related to speed. **Kinetic energy** is equal to one-half the mass times the square of the speed of the particles ($KE = \frac{1}{2}mv^2$). Observing the behavior of many gases, scientists have concluded that temperature is a measure of the average kinetic energy of the molecules.

Melting and Boiling Points

In this activity you started with a beaker of crushed ice that was at a temperature less than 0°C. As you heated the ice it did not initially melt, but the temperature of the ice began to rise. As the temperature of any solid increases, the average kinetic energy (energy of motion) of the particles of the material increases.

This motion is mainly a vibration-type motion where the molecules vibrate around a fixed location. As heat energy continued to be transferred, the temperature of the ice increased until it reached 0°C. This is called the **normal melting point** of water. It is the temperature at which water changes from a solid to a liquid state at 1 atm (atmospheric pressure at sea level). It is also the **normal freezing point** of water, when water changes from a liquid to a solid at 1 atm. Each material has its own characteristic normal melting/freezing point.

The temperature then remained at 0°C as the solid water changed to a liquid. Since the temperature remained constant, the average kinetic energy did not change. All the heat energy that was transferred caused a phase change during which the molecules of water were rearranging or decomposing. There was a change in the **potential energy**. If there is a change in kinetic energy you will see a change in temperature and if there is a change in potential energy, the temperature will remain constant while heat energy is transferred to a material.

When all of the water had melted at 0°C, the temperature of the liquid water increased until it reached 100°C. This is called the **normal boiling point** of water at 1 atm.

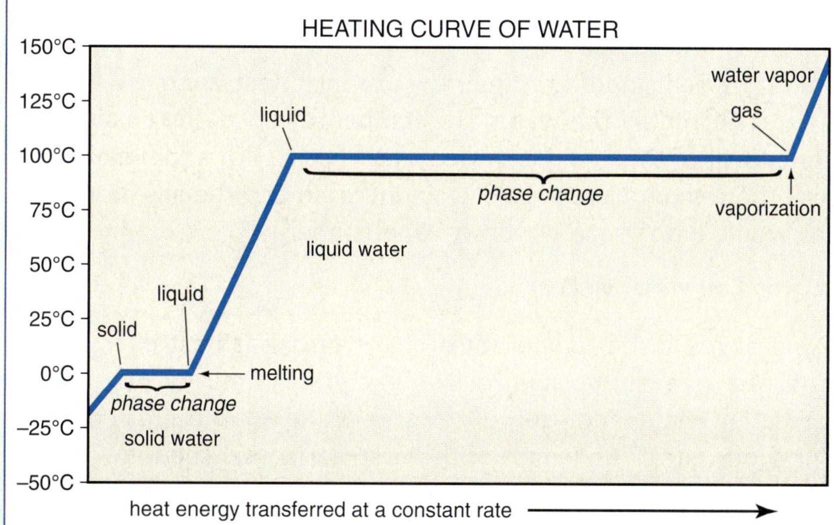

Chem Words

normal melting point: the characteristic temperature, at 1 atm, at which a material changes from a solid state to its liquid state.

normal freezing point: the characteristic temperature, at 1 atm, at which a material changes from a liquid state to its solid state.

potential energy: stored energy of the material as a result of its position in an electric, magnetic, or gravitational field.

normal boiling point: the temperature at which the vapor pressure of the pure liquid equals 1 atm.

If the atmospheric pressure is less than 1 atm, then the water will boil at less than 100°C and is just called the boiling point of the liquid. For example, on Mt. Rainier in Washington State, at an altitude of about 4393 m (14,411 ft), the atmospheric pressure is much less than 1 atm, and you would find that water boils at a lower temperature.

When the water arrived at the boiling point, you again noted that the temperature remained the same, even though heat energy was still being transferred to the water. The temperature would remain the same until all of the liquid is **vaporized.** Then, with additional heat energy, the temperature would again increase and the gas molecules of water would have greater average kinetic energy.

Chem Words

vaporization: the change of state from a liquid to a gas.

sublimation: the change of state of a solid material to a gas without going through the liquid state.

Heating Curve of Water

These changes in the temperature of a material, as heat energy is transferred to it, can be summarized in a graph, similar to the one you constructed. The heating curve of water is shown in the diagram on the previous page. The length of the first horizontal section corresponds to the amount of heat energy required to make the material change from solid to liquid.

Dry ice (solid carbon dioxide) that your teacher used in the demonstration, does not have a normal melting point; instead, it has a normal sublimation point (-78.5°C at 1 atm). **Sublimation** is the process where the solid goes directly to the gaseous state. The changes of state are summarized in the diagram below.

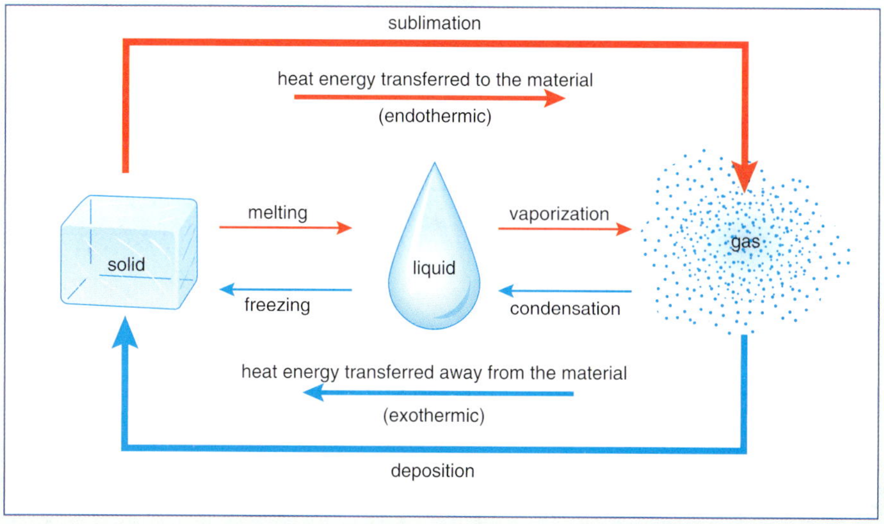

Checking Up

1. What does temperature measure?
2. Describe what is happening to particles of a material when heat energy is transferred to the material and the temperature increases.
3. What happens to the temperature of a material when it is undergoing a change of state?
4. What is the difference between the normal boiling point of water, and the temperature at which water might boil?

Reflecting on the Activity and the Challenge

In this exercise you focused on very simple chemistry, the motions of particles in solids, liquids, and gases. You can use the techniques learned in this activity to animate more complicated chemical systems. Consider how you could illustrate a phase change, like boiling or freezing. With research you might be able to use animation to explain the chemistry you use in staging your special effect.

Chemistry to Go

1. Copy and complete the following table summarizing the changes of state in your *Active Chemistry* log.

Change of State	From	To	Heat (added or removed?)
boiling	liquid	gas	added
condensation	gas	liquid	removed
evaporation			
freezing			
melting			
deposition	gas		
sublimation	solid		
vaporization			

2. Copy and complete the following table in your *Active Chemistry* log.

Definite or Indefinite?	Solid	Liquid	Gas
shape			
volume			

3. The heating curve for water was given in the **ChemTalk** reading section. Create the cooling curve for water and describe each part of the curve.
4. Create a heating curve for water when you have twice the amount of water you used in the investigation. How does this heating curve compare with the original curve?
5. The melting and boiling points of three materials are given on the next page.
   a) Draw a heating curve for each material.
   b) From graphs, explain why, at room temperature (22°C), copper is solid, mercury is liquid, and oxygen is a gas.
6. When water is in a pressure cooker the pressure is greater than 1 atm. What will the boiling point of water be compared to the normal boiling point of water? What information did you use to support your answer?

Material	Melting Point (°C)	Boiling Point (°C)
copper	1083	2336
mercury	- 39	357
oxygen	- 218	- 183

7. A certain material has a normal freezing point of 10°C and a normal boiling point of 70°C. What state is the material in at room temperature (about 22°C)? Explain your answer.
8. The normal boiling point of water is 100°C and the normal boiling point of ethanol is 78.5°C. These two liquids are soluble in each other. What technique would you use to separate these two liquids from the solution?
9. A certain material is white in the solid phase and clear (transparent) in the liquid phase. It has a normal melting point of 146°C. Which of the following materials could this be: water, sugar, or carbon dioxide? Explain your choice. (Hint: You will have to base your argument on concepts you have learned and observations made in this activity, as well as common sense.)

Preparing for the Chapter Challenge

1. Make an animation of 10 water particles changing from the solid state at 0°C to the liquid state at 0°C.
2. Animate the change of state of 10 dry-ice particles placed at room temperature.

Inquiring Further

1. Video animation

Making the leap from flipbook animation to video animation is not difficult. You will need access to a video camera that can record one frame at a time. Set the video camera so that it can record an image on a flat table or desk. To animate the process you can either record a series of drawings on paper, putting one after another on the table and taking a frame or two of each image.

Go to the Internet and see if you can find animations on the Internet. You might find a computer helpful in making the drawings or you could investigate using animation software to make more elaborate animations.

2. Design a thermometer

Use what you learned in this activity to design a thermometer that you could use to measure temperature. If your teacher approves your design, try out your thermometer.

Activity 3 Solutions, Suspensions, and Colloids

GOALS

In this activity you will:

- Explore different ways that materials can be mixed together to make new materials.
- Test some materials to determine what kinds of mixtures they are.
- Determine why certain kinds of mixtures are manufactured for commercial use in particular situations.

What Do You Think?

One way to get different types of materials is to just mix them together. Lots of different things can happen when materials are mixed. Each kind of mixture has its own characteristics.

- Is it easier to separate milk from coffee or milk from a bowl of cereal?

Record your ideas about this question in your *Active Chemistry* log. Be prepared to discuss your responses with your small group and the class.

Safety goggles and a lab apron are required for this activity. Wipe up spills immediately. Report any broken or cracked glassware to your teacher.

Investigate

1. Half-fill six large test tubes with water. Number the test tubes.

   a) In your *Active Chemistry* log, prepare a table for your observations. You may wish to use a table similar to the one on the following page.

No.	Materials mixed with water	Observations before mixing	Observations after mixing	Kind	Filter
#1	water only				
#2	0.5 g sugar				
#3	drops of milk				
#4					

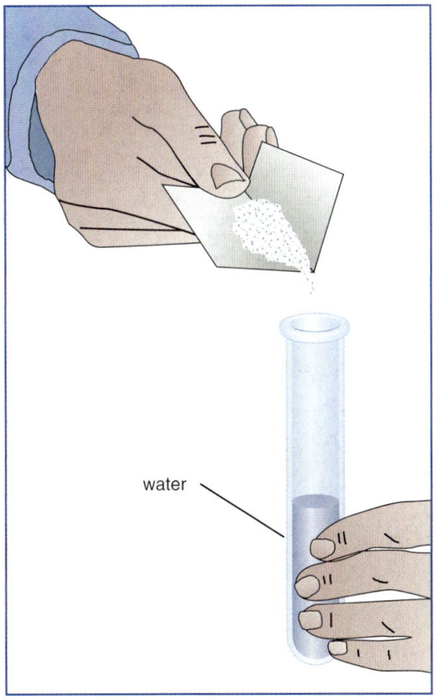

2. Add the following materials to the test tubes:

   #1—nothing
   #2—0.5 g sugar
   #3—a few drops of milk
   #4—0.5 g $CuSO_4$ (copper sulfate)
   #5—2 mL olive oil
   #6—0.5 g soil

   a) Describe each material before mixing.

3. Stopper each test tube. Place your finger over the stopper and shake each for several minutes to make a mixture.

   Observe each mixture.

4. Consider the following characteristics of the mixtures:

- What is the appearance of each mixture after the vigorous mixing? Which ones have visible particles suspended in them? Which ones look totally uniform (homogeneous) throughout?
- Which mixtures separate (are heterogeneous) after sitting a few moments after vigorous mixing? Which remain mixed?
- Shine a laser pointer through each mixture. In which mixtures is the laser beam clearly visible? In which mixtures does it pass through with little effect?

Never look directly at a laser beam.

- For each mixture place a small beaker below a funnel to catch the filtrate, as shown in the diagram on the next page. Pour the contents of each test tube into a funnel with filter paper. Which mixtures pass through? Which leave part behind on the filter paper?

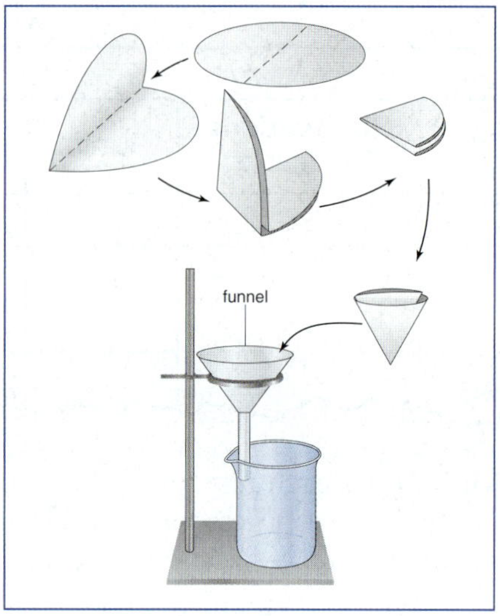

a) Record all your observations in the table in your log.

b) Using your data and the descriptions of the kinds of mixtures in the **ChemTalk** reading section, classify each of the mixtures as a solution, colloid, or suspension.

5. Discard materials and return all equipment as directed by your teacher. Clean up your station and wash your hands.

ChemTalk

CLASSIFYING MIXTURES

In this activity you mixed together water and several different materials to produce different kinds of mixtures. In some cases the materials you used were **pure substances.** A pure substance contains only one kind of particle throughout. For example, sugar is a pure substance. A mixture contains at least two pure substances. You may think of water as a pure substance, however most water found in nature has different materials mixed with it and is in fact a mixture.

Mostly all materials that you find in nature, as well as most human-made materials are mixtures of one or more pure substances. You made one kind of mixture, called a **solution**, when you added sugar to water. In a solution, the particles that dissolve are so tiny they can't be seen with the naked eye. The mixture is said to be homogeneous. The dissolved particles (called the **solute**) remain mixed forever with the solvent. The water, since it is doing the dissolving in this case, is the **solvent**. If a solution is filtered, everything passes through. Light passing through a solution has no special effect.

When you added milk to water, the water appeared cloudy. However, the tiny drops of milk remained suspended in the water and did not

Chem Words

pure substance: a substance that contains only one kind of particle.

solution: a homogeneous mixture of two or more substances.

solute: the substance that dissolves in a solvent to form a solution.

solvent: the substance in which a solute dissolves to form a solution.

settle out over time. You could see the laser beam as it passed through the mixture, and when you filtered the mixture, it all passed through the filter paper. This kind of mixture is a **colloid**. In colloids, the dispersed particles are larger than those in solution and may be visible on close inspection with a microscope. The particles will also stay suspended indefinitely. All parts of the colloid will pass through a filter. When light passes through a colloid it is scattered and you can see where the light beam passes through. This is known as the **Tyndall Effect.**

When you added soil to water, you created a suspension. **Suspensions** have the largest of all the dispersed particles. The particles are visible to the eye and will settle out in time. The suspended particles can be separated by filtration. The mixture is said to be heberageous. A light beam shining through a suspension may be scattered, but the suspension is definitely not transparent.

Chem Words

colloid: a mixture containing particles larger than the solute but small enough to remain suspended in the continuous phase of another component. This is also called a colloidal dispersion.

Tyndall Effect: the scattering of a light beam as it passes through a colloid.

suspension: heterogeneous mixture that contains fine solid or liquid particles in a fluid that will settle out spontaneously. By shaking the container they will again be dispersed throughout the fluid.

Checking Up

1. In your own words describe how you would distinguish among a solution, a colloid, and a suspension.
2. What is the Tyndall Effect?

Reflecting on the Activity and the Challenge

In this activity you made mixtures with solid or liquid solutes and water as the solvent. The same basic principles apply regardless of the states of matter involved. For example, it is possible to have solid solutions. Metal alloys such as brass or bronze are such solutions. Fog, smoke, and clouds are mixtures that show the Tyndall Effect. A common stage effect is to produce smoke or fog to give an eerie setting. Many movies have used the Tyndall Effect to show the path of laser beams or flashlights. One easy way to produce this effect is to use a spray bottle of water to mist the air in a darkened room (problem with allergies/asthma). When a flashlight or laser pointer is directed through the mist, it shows up nicely, thanks to the Tyndall Effect.

Chemistry to Go

1. Classify each of the following as a suspension, colloid, or solution. Explain your reasoning. (Hint: In some cases more than one answer may be possible.)
   a) A mixture is poured through a filter, and the entire mixture passes through.
   b) A mixture is left to stand for a while and small particles settle out.
   c) When viewed under a microscope, small particles are visible in the mixture.
   d) A beam of light passed through the mixture is scattered.
   e) The mixture is blue and transparent.
2. Suggest a method by which you could separate the various materials in each of the following mixtures:
   a) solutions b) colloids
   c) suspensions
3. On some days it is possible to see rays of light from the Sun distinctly coming through breaks in clouds. What in the atmosphere might account for this difference?
4. The term homogeneous means uniform or the same throughout. Heterogeneous means that different parts of the mixture are clearly visible. Classify each of the mixtures in this activity as homogeneous or heterogeneous.
5. Look in your kitchen at home and choose five products. Make your best guess as to the type of mixture they represent. Elaborate on the evidence that you used to classify the products into the respective category. Explain what would happen to each product that you have chosen if it were in a different kind of mixture. For example, milk would settle if it was a suspension instead of a colloid.

Preparing for the Chapter Challenge

Consider how you could use the properties of a solution, colloid, or suspension to produce a special effect for a movie scene. In a few sentences establish the setting for the scene, and describe the mood you want to create.

In a paragraph or two describe the chemistry required to understand the differences between the different types of mixtures and how they produce the special effect.

Inquiring Further

Cooking and mixtures

Cooking is a practical application of mixtures.

One such application is the recipe for making mayonnaise. Mayonnaise is classified as a colloid. A liquid—liquid colloid is also called an emulsion. If you have access to the following materials, try making mayonnaise at home.

A Recipe for Mayonnaise

Combine in a blender:

2 tablespoons beaten egg

1 large egg yolk

$\frac{1}{4}$ teaspoon dry mustard

1 teaspoon lemon juice

Do this only with the approval of a responsible adult.

Blend the mixture for 20 seconds. While the blender is still running, add 3/4 cup of vegetable oil slowly in the thinnest stream you can manage. Serve immediately or keep for 1 to 2 days in the refrigerator.

Investigate the properties of the ingredients of mayonnaise that allow them to form this colloidal emulsion. What other substances are classified as emulsions? How could emulsions be used in a movie special effect?

Metals in cooking utensils are also examples of mixtures. Investigate the kinds of materials used in making metal cooking utensils. Why are some mixtures more suitable for cooking than others?

Activity 4 Properties of Matter

GOALS

In this activity you will:

- Make modeling dough from common kitchen materials.
- Adjust the properties of the modeling dough by adding another material to it.
- Compare the properties of an emulsion to those of a composite material.
- Recommend whether an emulsion or a composite would be best for certain applications.
- Consider the advantages and disadvantages of using composites for industrial applications.

What Do You Think?

Frequently in movies, models of sets or characters are built and then enlarged on the screen. A giant marshmallow or dough man that appeared to be as tall as a 10-story building was created for a popular movie that used many special effects.

- Estimate how many marshmallows it would take to build an actual-size giant marshmallow man.
- What do you think would be the cost of the marshmallows?

Record your ideas about these questions in your *Active Chemistry* log. Be prepared to discuss your responses with your small group and the class.

Investigate

1. Use the ingredients and directions on the following page to prepare a batch of modeling dough.

Safety goggles and a lab apron must be worn during this activity.

Ingredients:

1 cup flour
$\frac{1}{2}$ cup salt
3 tablespoons oil
2 tablespoons cream of tartar
1 cup water

Do not overheat the oil.

Directions:

- Warm the oil in a saucepan or large beaker on a hot plate set at medium-low.
- Add the other ingredients and cook 3 to 5 minutes while stirring constantly.
- Using tongs, drop the mixture onto waxed paper or aluminum foil.
- Cool until the modeling dough is easy to handle.
- Knead the dough until the texture is consistent.
- Add a drop of food coloring, if desired.

2. Divide the dough into two samples. Set aside one sample.

   Add a new material (tissue paper or paper towel strips, sand, gravel, pieces of twine, marbles) to the other sample, as directed by your teacher. Knead the mixture well until the new material is well incorporated into the dough.

3. You will now observe and compare the properties of the original sample and the new sample that you made. In your group, discuss how you will test for each of the following properties:

   texture,
   elasticity,
   uniformity,
   bounce,
   strength,
   malleability.

   If you are not sure what characteristics you should be testing for, each property is explained in the **ChemTalk** reading section.

   a) Record in your log the procedure you will use to test for each property.

   b) Make a table in your *Active Chemistry* log to record your observations.

4. When your teacher has approved your procedure, test each property.

   a) Record your observations in your log.

   b) In a few sentences, describe the differences between the two kinds of mixtures.

   c) Speculate on why your new mixture behaved differently from the original modeling dough.

5. Compare the properties of your new mixture with the mixtures of other groups.

   a) What new or useful properties did each mixture have?

6. Dispose of materials and return all equipment as directed by your teacher. Clean up your station and wash your hands.

Chem Talk

PHYSICAL PROPERTIES

In this activity you examined a variety of physical properties of a mixture. In addition to the state and color of the mixture, you also observed the **texture** of the material. Texture is defined as the feel or appearance of a surface or substance. For example, if you were looking at a photograph, you may be interested in the texture of the surface of the paper used, or if you were looking at cloth, you might be interested in how the material is woven together. **Uniformity** describes how consistent a material is throughout. Does everything in the material seem to be evenly distributed? **Strength** determines how durable the material is. How well the material withstands the application of a force establishes how strong it is. **Elasticity** determines how well the material will resist deformation and return to its normal size or shape after a force has been applied to it. **Bounce** refers to the material's ability to return to its original position when it is dropped from a given height. Another way to think about bounce is to consider how much it behaves like a ball. **Malleability** determines how easy it is to roll or hammer out the material without breaking it apart. Can the material be reshaped without breaking it apart? Lead is an example of a metal that can easily be shaped into other forms without breaking. Brittle objects shatter easily.

Chem Words

texture: the characteristics of the surface of a material, like how smooth, rough, or coarse it is.

uniformity: the property of how consistent a material is throughout.

strength: the property of how well a material withstands the application of a force.

elasticity: the property of a material to resist deformation and return to its normal size or shape after a force has been applied to it.

bounce: the ability of an object to rebound to its original position when dropped from a given height.

malleability: the property of a material to be able to be hammered into various shapes without breaking.

COMPOSITE MATERIALS

The properties of a mixture can be changed by adding different materials. The modeling dough you made in this activity was an **emulsion**. When you added one of the other materials, you made a **composite**. Composites are heterogeneous mixtures that use the characteristics of the components to make useful substances. Another example of a composite is paper maché.

The composite industry has been growing at a very rapid pace. Think about it. Composites are used as the skin on jets, rotor blades of a helicopter, bulletproof clothing for law enforcement officers, and special armor for tanks. Buildings, cars, boats, trains, and planes all take advantage of different types of composites. If you were trying to define what a composite is, you would probably say that it is a solid that consists of two or more materials. It may be the process by which fibers are embedded within another material. Usually, this will lead to a stronger material and provide the best qualities of both materials. Composites not only help make stronger and lighter materials but they also help in extending the life of materials. Composites are used for materials that cover electrical cables and protect them from chemicals.

A major drawback to composites is the initial cost of research and the use of special raw materials in fabricating the composite. However, overall composites are materials of the future and their manufacture could be a great career choice.

Chem Words

emulsion: a colloid or colloidal dispersion of one liquid suspended in another.

composite: a solid heterogeneous mixture of two or more substances that make use of the properties of each component.

Checking Up

1. Two pieces of cloth have the same color and texture. Name another property of cloth that you might use to distinguish between the two materials.
2. Provide a situation in which each of the following properties of a material might be important:
   a) elasticity,
   b) uniformity,
   c) strength,
   d) bounce.
3. What is a composite material?
4. Give an advantage and a disadvantage of using a composite material.

Reflecting on the Activity and the Challenge

One part of making a special effect for a movie is finding materials that could be used for making models. In this activity you learned to make a material that has many characteristics that make it ideal for making models. Consider how each of the properties you investigated is important to a special effect. A malleable material is easy to shape. To construct a large building, you will want to use a composite that is strong. You could also construct a small model of a tall building, and then make a model of a character that can bounce 10 stories high.

Chemistry to Go

1. A composite when tested bounces half as high as the original material. How can you create another composite that bounces only one-quarter as high?
2. Pure, 24-K gold is very malleable. By mixing gold with other metals it becomes less malleable. In what situations would you want to use gold that is not as malleable?
3. Plastics are made of one type of material. Why are they not classified as a composite?
4. When road contractors are laying cement, they first place a series of steel rods in a specific pattern. Then they pour the cement on the rods and allow it to cure. The composite is called reinforced concrete. What is the purpose of the steel rods?

Preparing for the Chapter Challenge

1. Each kind of modeling dough that you made in this activity could be used in making sets for a movie special effect. In your log list the kinds of structures each might be useful in constructing.

2. Give two examples of special effects in movies where each of the following properties is important: strength, elasticity, bounce, uniformity, malleability.

Inquiring Further

1. Other modeling materials

Many other types of modeling materials, such as plaster of Paris or Fimo® dough, are available in crafts and art stores. Research these substances and consider their usefulness as modeling materials for your movie special effect.

2. Different composites

Investigate different composites. Learn about the special properties of these materials. List the items of products that are constructed from composites.

3. Investigating other techniques with modeling dough

Color and appearance of models are crucial to making them look authentic in a movie scene. With the supervision of an adult, try some alternative means of coloring your modeling dough. Try painting the surface of the dough, or using other substances to add color and texture. How do the various techniques affect the characteristics of the dough?

Activity 5 Mass and Volume

GOALS

In this activity you will:

- Determine the densities of various liquid and solid materials.
- Make measurements in the laboratory to the precision of the instruments used.
- Learn the difference between accuracy and precision in experimental measurements.
- Retain significant figures in calculations involving experimental measurements.
- Use density measurements to determine the identity of a material.
- Locate sources of the variation in the class's experimental results.

What Do You Think?

A piece of steel sinks in water, but a steel boat floats. A tiny rock sinks in water, but a large log floats.

- Since a kilogram of feathers and a kilogram of lead have the same mass, how do they appear different and why?

Record your ideas about this question in your *Active Chemistry* log. Be prepared to discuss your responses with your small group and the class.

Investigate

Part A: Mass and Volume of Liquids

1. In your *Active Chemistry* log create a table to record your data for this part of the activity. You may wish to use a table similar to the one on the opposite page.

a.

Volume and Mass of Water

Mass of graduated cylinder (g)	Volume of water (mL)	Mass of graduated cylinder and water (g)	Mass of water (g)	$\frac{\text{Mass}}{\text{Volume}}$ (g/mL)

2. Measure the mass of an empty, dry, graduated cylinder.

   a) Record the mass of the cylinder in your *Active Chemistry* log.

3. Add 10 mL of water to the graduated cylinder. Remember when reading the volume, take the reading at the lowest part of the meniscus, as shown in the diagram.

   a) Record the volume of water in your table. Remember to consider the precision of your measurement when recording your data.

4. Measure the mass of the graduated cylinder and 10 mL of water.

   a) Record the measurement in your log.

   b) Calculate the mass of the water and record this in your table.

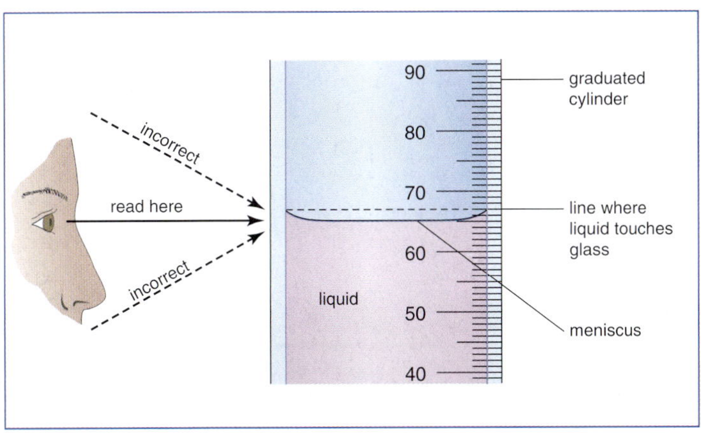

5. Add another 10 mL to the graduated cylinder and measure the mass. Calculate the mass of 20 mL of water.

   Repeat this step for 30 mL, 40 mL, 50 mL, and so on up to 100 mL.

   a) Record all your measurements and calculations in the table in your log.

6. Use the data you obtained.

   a) Plot a graph of the mass versus the volume of water. Plot volume on the *x*-axis (horizontal axis) and mass on the *y*-axis (vertical axis).

   b) As the volume of the water increases, what happens to the mass?

   Since the graph you created is a straight line (or close to a straight line), you should draw the best fit line through the data points. Do not connect the points with small segments but draw one line that comes closest to all of the individual points.

   c) From your graph predict the mass of 55 mL of water. What would be the volume of 75 g of water? Predicting values from within a graph is called interpolation.

   d) An important attribute of a straight line graph is its slope. How steep is the graph? Calculate the slope of the graph you plotted. Remember to calculate slope you divide the "rise" by the "run." What does the "rise" of the graph represent? What does the "run" represent?

   e) Divide the mass of each sample of water by the volume. What do you notice about the relationship between the mass and the volume?

   How does the slope of the graph compare to the values you calculated in this step?

7. Your teacher will provide you with a sample of a liquid.

   Use the procedure you used to find the mass and corresponding volumes of water to determine the slope of this liquid's mass/volume graph.

   a) Record all your data and calculations in your *Active Chemistry* log.

8. Dispose of your liquid sample as directed by your teacher. Clean up your work station.

Part B: Mass and Volume of Solids

1. Your teacher will provide you with three samples of two different solid materials.

2. As a group, decide on a procedure to calculate the mass/volume ratio and slopes of the graph of each material.

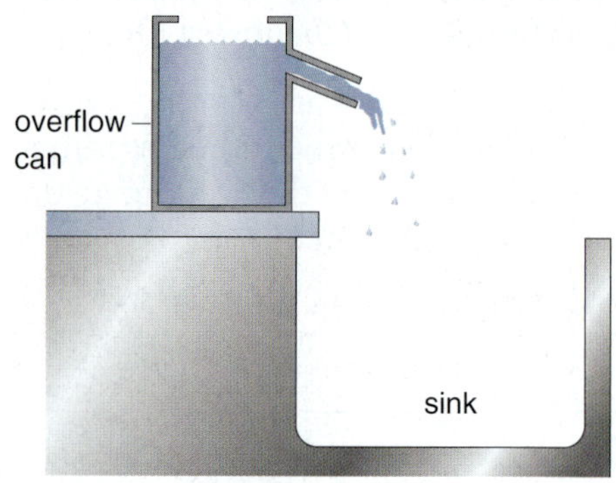

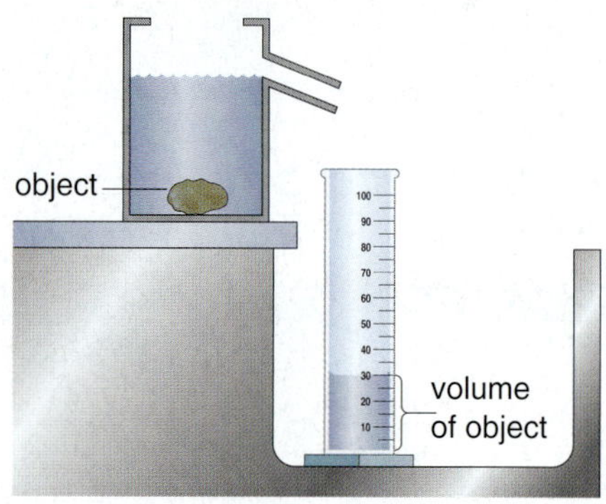

You can consider using either method shown in the diagrams on the preceding page and at right for measuring the volume of each solid. Volume of solids is usually expressed in cubic centimeters. One milliliter is equivalent to one cubic centimeter ($1 \text{ mL} = 1 \text{ cm}^3$).

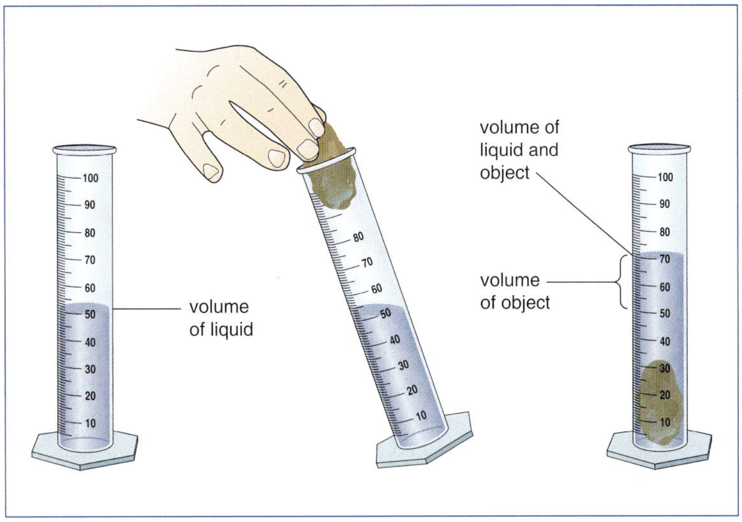

a) Record your procedure in your *Active Chemistry* log. Be sure to include what measurements you need to make, what equipment you will need, what safety precautions you must use, and what calculations you have to do.

3. When your teacher has approved your procedure, carry out your activity.

   a) Carefully record all your data.

4. Use the data you collected.

   a) Plot a mass versus volume graph for each solid. Plot both solids on the same graph.

   b) How do the slopes for the two solids compare? Which solid is more dense?

   c) Use the table in the **ChemTalk** reading section to identify the two samples of materials.

5. The mass of a unit volume of a material is called its density. You found the density of water by calculating the slope of the mass versus volume graph. You can also calculate density by dividing the mass of a sample of a material by the volume.

$$\text{Density } (D) = \frac{\text{Mass } (m)}{\text{Volume } (V)}$$

   a) Find the densities of water, the other liquid, and the two solid materials.

   b) Compare your answers with another lab group.

Part C: Density and Special Effects

1. Your teacher will display a set of four colored liquids that float on one another. The densities of each of the liquids were measured. The top layer has a density of 0.8 g/mL. The next layer has a density of 0.9 g/mL. The following layer has a density of 1.1 g/mL. The bottom layer has a density of 1.3 g/mL.

   a) What do you notice about the densities of the liquids and their position in the display?

2. Your teacher will drop an ink-pen barrel into the liquids.

   a) What would you predict will happen to the ink-pen barrel? Write your prediction in your *Active Chemistry* log.

   b) Observe the movement of the ink-pen barrel as your teacher places it in the liquid. Record your observation in your log.

3. You will now make an ink-pen barrel float in liquid.

   Place the ink-pen barrel in a beaker of ethanol.

   Next place the ink-pen barrel in a beaker of distilled water.

   Return the ink-pen barrel to the beaker of ethanol. Slowly add distilled water to the ethanol until the ink-pen barrel floats.

   a) Before you begin, predict what you think will happen in each part of this step. Give a reason for your prediction.

Chem Talk

DENSITY

Density as a Property of Matter

If you were to compare a 1 cm^3 cube of iron to a 1 cm^3 cube of wood, you would probably say that the iron is "heavier." However, if you compared a tree trunk to iron shavings, the tree trunk is obviously heavier. As you discovered in this activity, a "fair" comparison of the "heaviness" of two materials is a comparison of their densities. **Density** is the mass per unit volume of a material. In this activity you measured the density of water and other liquids. You found that each sample of the same liquid had the same density and each different liquid had its own characteristic density. You also found that each solid material you investigated had its own characteristic density. Density can be expressed in grams per milliliter (g/mL) or grams per cubic centimeter (g/cm^3). The table on the next page shows the densities of some common liquids and solids.

Chem Words

density: the mass per unit volume of a material.

You used the slope of the mass versus volume graph of a material to calculate density. You also calculated density using the equation:

$$\text{Density } (D) = \frac{\text{Mass } (m)}{\text{Volume } (V)}$$

Density and Flotation

In this activity you further observed that materials with a greater density than a given liquid will sink, and materials with less density than a given liquid will float. In the column of colored liquids, the liquid with the highest density was on the bottom, and the liquid with the lowest density was on the top. The ink-pen barrel sank in ethanol and floated in water. When you added ethanol to the water you created just the right density to have the ink-pen barrel float within the liquid. This position of floating is where the density of the ink-pen barrel is equal to the density of the ethanol/water. The ink-pen barrel "found" the place where the density of the liquid was identical to the density of the ink-pen barrel.

Approximate Densities of Some Common Liquids and Solids

Material	Density (g/cm^3)
wood (balsa)	0.12
wood (birch)	0.66
gasoline	0.69
isopropanol	0.79
vegetable oil	0.92
distilled water	1.00
glycerol	1.26
magnesium	1.70
aluminum	2.70
iron	7.90
copper	8.90
nickel	8.90
silver	10.50
mercury	13.50
gold	19.30

The most famous story about density is when Archimedes jumped out of the bath, ran through the town naked, and shouted "Eureka!" As the story supposedly goes, Archimedes was asked by the king to determine if his crown was solid gold. Archimedes knew the density of gold. He also knew that he could correctly determine if the crown were gold if he knew the density of the crown. The mass of the king's crown was easy to measure. The volume posed a real problem because it had such an unusual shape, and of course the king did not want his crown altered. When Archimedes submerged himself in the bathtub, he realized that the displacement of water would provide him with the volume. *Eureka* is Greek for "I found it."

MAKING MEASUREMENTS AND USING THE MEASUREMENTS TO MAKE CALCULATIONS

Uncertainty of Measurements

Every measurement that you make involves some uncertainty. When you measured the volume of water using a graduated cylinder, you used the division of units marked on the side of the cylinder to make your measurements. Suppose the smallest precision division marked on the graduated cylinder was a milliliter. This means that you can estimate the measure to the nearest tenth of a milliliter, because you can see if the level of the water is at, above, or below the mark. When you record your measurement of volume, you can record it as 10.0 mL, because you can see whether the level of the water is at, above, or below the 10 mL mark.

Remember to always look at the instrument that you are using and determine the smallest precision mark it has. When you make your measurement using the instrument you can only estimate to the next place. If you are using an electric balance to measure mass, it will do the estimating work for you. Most school balances will measure to the tenth, or the hundredth of a gram.

Calculations

When you perform calculations using the measurements that you made in an investigation, you need to express the result of your calculations in a way that makes sense of the certainty of the measurements you made. For example, when calculating the density of a 10.1 mL sample of liquid with a mass of 9.8 g, you may obtain a value of 0.9702770… g/mL using a calculator. This value does not seem reasonable when considering limitations of your measurements.

There are rules that you can use when making your calculations:

Adding and Subtracting

When adding or subtracting numbers, arrange the numbers in columnar form lining up the decimal points. Retain no column that is to the right of a column containing a doubtful digit.

Multiplying and Dividing

In multiplication and division, the result should have no more significant digits than the factor having the fewest number of significant digits. To determine how many digits are significant, count all the digits excluding zeroes at the beginning or end (e.g., 0.00326 and 71800 each have 3 significant digits). (Exception: a zero at the end of a number is significant if the number contains a decimal point — for example, 0.00326 has 3 significant figures, 0.003260 has 4, and 3260 has 3.)

Checking Up

1. Explain the meaning of density.
2. Explain the difference between feathers and lead, using the concept of density.
3. Why is balsa instead of birch wood used in the construction of model airplanes?
4. In an investigation the volume of a material is measured as 80.0 cm^3 and its mass is measured as 253 g. Which calculation of density correctly uses the precision rules: 3.1625 g/cm^3, 3.163 g/cm^3, 3.16 g/cm^3, 3.2 g/cm^3, or 3 g/cm^3?

Reflecting on the Activity and the Challenge

In this activity, you discovered that the ratio of mass to volume (m/V) is a special number associated with each material. The ratio m/V is called the density and is a characteristic property of matter. You can identify whether a piece of metal is gold or gold-plated by measuring the density. You can distinguish one material from another by comparing densities. Objects of greater density than a liquid will sink, while objects of lesser density will float. Objects of the same density will appear suspended. You can make use of these conclusions in your challenge. For instance, you may want to have a movie special effect where a material appears suspended in space. The concept of density will help you create this special effect. You can compare the density of different materials and decide which materials will float and which materials will sink, and also how to make them appear to be suspended somewhere between the top and bottom of a liquid.

Chemistry to Go

1. Look at the table in the **ChemTalk** reading section. Use density to identify the liquid and solid samples you investigated in this activity.
2. Calculate the density of a solid from the following data:

Volume of water	48.4 mL (or cm^3)
Volume of water and solid	62.7 mL (or cm^3)
Mass of solid	123.4 g

3. Determine the density of a liquid from the following data:

Mass of the graduated cylinder	33.79 g
Mass of the cylinder and liquid	40.14 g
Volume of liquid	13.3 mL

4. Methanol has a density of 0.79 g/mL. How much would be the mass of 589 mL of methanol?
5. Copper has a density 8.90 g/cm^3. What would be the volume of a 746 g sample of copper?
6. In a well-known movie, there is a famous scene in which the hero tries to outwit the designers of a trap by replacing a gold statue with a bag of sand of about the same volume.
   a) Given the density of gold is 19.3 g/mL and sand is 3.1 g/mL, does this seem like a scientifically reasonable plan?
   b) In the movie, the hero grabs the gold statue with one hand and appears to handle it quite easily. Given that the volume of the statue appears to be about 1 liter, what would be the mass of the statue?
   c) A mass of 454 g has a gravitational weight of about 4.45 N (newtons) which is about 1 lb. How many pounds would the statue weigh?
   d) One gallon of milk has a mass of 3.7 kg and a weight equivalent of approximately 8 pounds. How many gallons of milk would be equivalent to the gold statue?
7. In each of the following pairs, which has the greater mass?
   a) 1 kg lead or 1 kg feathers?
   b) 1 L gold or 1 L water?
   c) 1 L copper or 1 L silver?

8. Which of the following has the greater volume:
   a) 1 kg lead or 1 kg feathers?
   b) 1 kg gold or 1 kg water?
   c) 1 kg copper or 1 kg silver?
9. Review the measurements you made for mass and volume. How certain were your measurements? If you were to make the measurements again, could you be more certain? Explain your answer.
10. In calculating density you divided the mass of the material by the volume. Review the calculations you made. Adjust the accuracy of your answers using the rule for division given in the **ChemTalk** reading section.

Preparing for the Chapter Challenge

Design a special effect in which an object is suspended in a liquid. Consider the density of the material you will suspend, and the density of the liquid you will use. Show the calculations that you used to make your choice of materials.

Inquiring Further

1. Is it real gold?

The new United States dollar coin has a golden color. Could it be made of real gold? Devise a method to determine if the new golden coin is any of the metals in the table in the **ChemTalk** reading section.

2. Density of gas

Devise an investigation that you could do to determine the density of air.

Activity 6 Metals and Nonmetals

GOALS

In this activity you will:

- Observe some chemical and physical properties of various materials.
- Classify the materials as metals or nonmetals.
- Identify the metals that make up common alloys and learn about some special properties and uses of the alloys.
- Make generalizations about the properties that differentiate metals from nonmetals.
- Explore how heat treatments can alter the properties of metals.

What Do You Think?

If you look around your house you will see hundreds of objects made from dozens of kinds of materials. Have you ever wondered why the manufacturer chose the materials they did for each item?

- Why are frying pans made of metal and baking dishes often made of glass or ceramic?
- Could a baking dish be made of metal? Could a frying pan be made of glass or ceramic?

Record your ideas about these questions in your *Active Chemistry* log. Be prepared to discuss your responses with your small group and the class.

Investigate

1. Your teacher will provide you with samples of a number of materials: iron, copper, zinc, magnesium, tin, aluminum, brass, solder, chalk, graphite or charcoal, wood, glass, plastic, and concrete.

Choose two of the materials that are obviously different. In your group, brainstorm at least five characteristics or properties of each material.

a) In your *Active Chemistry* log, use the characteristics to describe each material. Could someone else reading your log be able to identify the material?

2. Chemists use specific characteristics or properties to describe and distinguish among materials. You investigated some of these properties in **Activity 4** and **Activity 5**. Additional properties used by chemists are described below.

    a) In your log, prepare a table for recording your observations.

3. **Luster:** Is the material shiny or dull in appearance? Does it look more like a mirror or more like mud? If a material has lots of luster it reflects light and you may be able to see images reflected from the surface. Polished metal has high luster. Dull surfaces don't reflect as much light. They appear flat and no images can be reflected.

    a) In the table in your log, record whether the material has a high or low luster.

4. **Electrical conductivity:** Test each substance with a conductivity tester or multi-meter. To test each substance, touch the two leads to each end of the sample. Do not allow the leads to touch each other, or it will give a false reading. In the example shown in the diagram, the bulb will glow if the material is conductive. Your teacher will demonstrate how the specific conductivity tester used in your lab works.

    a) In the table in your log, record whether the material is conductive. You can use words to reflect this like nonconductive, slightly conductive, or very conductive. If you are using a meter, you can record the reading.

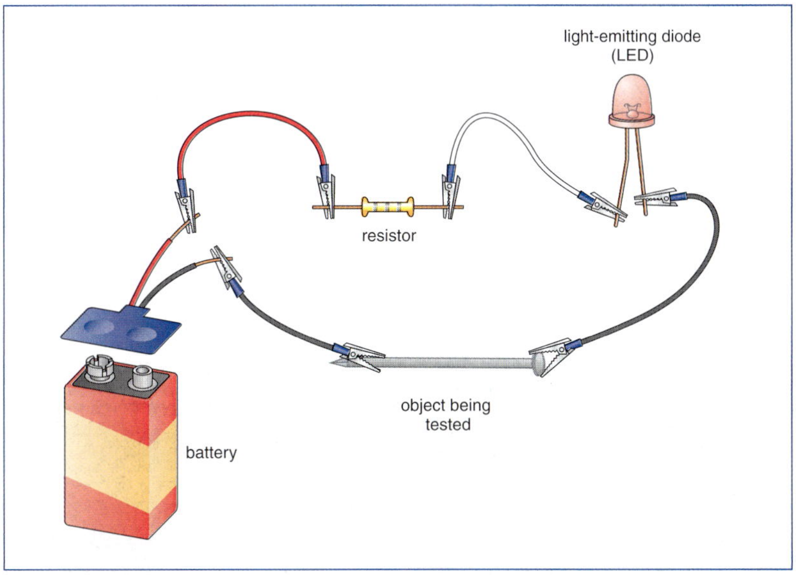

Safety goggles and a lab apron must be worn.

5. **Malleability:** Wrap the material being tested in heavy plastic or a cloth to prevent pieces from flying off the sample. Place the material on a hard, flat surface. Using a hammer, try to pound the material flat. If the sample can be pounded into a flatter shape it is called malleable. If it breaks or doesn't change it is called nonmalleable.

   a) In the table in your log, record whether the material is malleable or nonmalleable.

6. **Reactivity:** Try scraping or sanding a small part of each sample. Is the surface underneath the same in appearance or different? If the surface is different, that means the sample has reacted with the air.

   a) In the table in your log, record whether the material is highly reactive, slightly reactive or nonreactive.

7. **Ductility:** Ductility refers to how easily the substance can be pulled out into a wire or how bendable it is. Try bending each piece to determine how ductile it is.

   a) In the table in your log, record whether or not the material is ductile.

8. **Color**

   a) Record the color of each sample material in the table in your log.

9. Once you have completed the table, compare your list of characteristics of each substance with those recorded by the other students in your class. Have each member of your group pair off with a student from another group. If there is a difference in the results go back to the material and review your observations until everyone agrees on the most accurate list of properties for each material.

   a) Be sure to record any changes you make in your log.

10. Classify the substances into two groups. Use any property you have observed to divide the samples into groups that have the most in common. For example, you could divide the materials into those that do have a luster and those that do not.

    a) Record your classification of the materials in your log.

11. Metals have a shiny and lustrous surface. They conduct electricity and heat. They are malleable and ductile and they are often relatively reactive. Nonmetals have characteristics that are generally opposite to metals in every way. Instead of being lustrous their surfaces are dull in appearance. They are nonconductive, brittle, and nonductile. Now separate your samples into metals and nonmetals. If you have any samples that do not fit clearly as either a metal or a nonmetal, set them aside.

    a) Make a list of the samples in each category in your *Active Chemistry* log.

Chem Talk

METALS, NONMETALS, AND METALLOIDS

In this activity you investigated specific properties of materials. You then used your observations to classify a material as a **metal** or a **nonmetal**. Metals have **luster**. They exhibit **conductivity**. They conduct electricity and heat. They are malleable and **ductile** and they are often relatively **reactive**. Many metals form a compound on their surface that results from reactions with air. When you scrape or sand a piece of metal you are removing that coating of metal compound. Sometimes that natural coating can prevent further reacting and will preserve the metal underneath.

Looking at the drawing at the right, you can see that in solid copper metal the centers of the copper atoms are in fixed locations but they are surrounded by a sea of electrons. If an electric circuit is set up, the electrons are free to move. This is the basis of the metallic property of electrical conductivity.

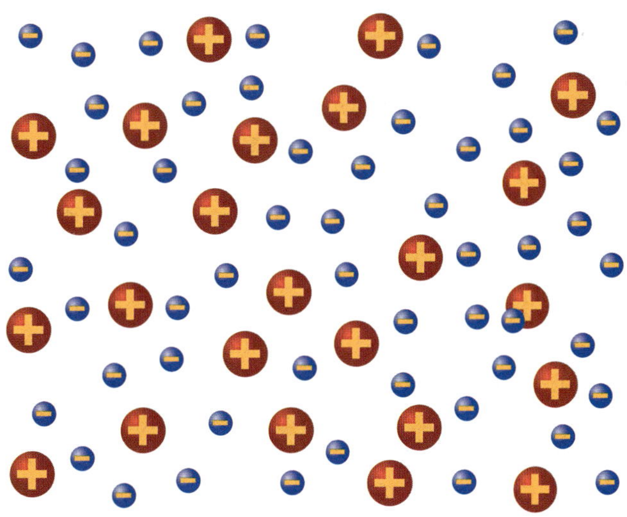

Electron-Sea Model for Copper Metal

Chem Words

metal: classes of materials that exhibit the properties of conductivity, malleability, reactivity, and ductility. Metal elements readily lose electrons to form positive ions.

nonmetal: elements that do not exhibit the properties of conductivity, malleability, reactivity, and ductility. These elements tend to form negative ions. The oxides of the elements are acidic.

luster: the reflection of light from the surface of a material described by its quality and intensity.

conductivity: the property of transmitting heat and electricity.

ductility: a property that describes how easy it is to pull a substance into a new permanent shape, such as, pulling into wires.

reactivity: a property that describes how readily a material will react with other materials.

On the other hand, silicon dioxide is an amorphous solid; you know it as glass. In glass, electrons are fixed into position and are held tightly by each atom due to covalent bonding (sharing of electrons) between silicon and oxygen atoms. Since the electrons are not mobile, the glass does not conduct an electric current like copper metal does. Glass is a nonconductor of electricity.

Preserving metal and preventing its reaction with some of the components in the air is a major task. When metals react with oxygen in the air it is called **oxidation**. This type of reaction is what happens when things rust. Preventing rust is important. While a metal like steel is very strong and makes excellent building material, once it rusts it loses all strength and flakes away. Millions of structures, tools, and vehicles are made primarily of metal. Preventing rust (also called corrosion) is essential if they are to remain in good operating condition. In order to prevent oxidation, metal surfaces can be painted, coated, or combined with another metal to make them less reactive.

Nonmetals have characteristics that are generally opposite to those of metals in every way. Instead of being lustrous, their surfaces are dull in appearance. They are nonconductive, brittle, and nonductile. Over the past 150 years, chemists have developed a chart for classifying and organizing the chemical elements. Elements are classified as metals and as nonmetals. Some other elements are called metalloids. They share some characteristics of metals and some of nonmetals.

Brass and solder are not elements but they are still classified as metals. They are commonly called **alloys**. Alloys are materials that contain more than one metal element and still maintain the characteristic properties of metals. Many metals are not practically useful because they may be too soft and are hard to work with. Gold is a good example of a metal that is too soft for jewelers to work with so they make an alloy of gold, silver, and copper. The alloy is harder and will hold its shape. Iron combined with chromium, nickel, and carbon makes the alloy called steel. This gives it the strength that it needs in construction. The brass that you investigated contains 67% copper and 33% zinc. Solder contains 67% lead and 33% tin. Alloys are classified as metal solutions and if they are uniformly mixed then they are homogeneous mixtures called solution alloys.

Chem Words

oxidation: the process of a substance losing one or more electrons.

alloy: a substance that has metal characteristics and consists of two or more different elements.

Checking Up

1. List five properties of metals and five properties of nonmetals.
2. Why is it important to prevent the oxidation of metals used in construction?
3. Explain the meaning of an alloy.
4. Why are alloys used?

Reflecting on the Activity and the Challenge

As you are creating your movie special effect, you may have to build a stage set, model, or prop. It is important to consider the nature of the materials you choose before you start construction. You need to match the characteristics of the material you choose with the object you are trying to build. You need to decide if the building material should be heavy or light, flexible or rigid. Each of the characteristics of the substance is important. The characteristics required will probably vary with each construction project. Whichever materials you do choose, you will strengthen your report by discussing the specific properties of the materials and classifying them as metals, nonmetals, or metalloids.

Chemistry to Go

1. a) List the names of three metals you are familiar with in your daily life.

   b) For each metal you listed in (a), describe two different uses for each.

2. a) List the names of three nonmetals you are familiar with in your daily life.

   b) For each nonmetal you listed in (a), describe two different uses for each.

3. Examine the following objects or machines that you use every day:

   a) backpack,

   b) bicycle,

   c) car.

   Make a list of the metals and nonmetals that make up the major components of the object. List the function and characteristics of each component.

4. a) List two properties of a material that you can observe using your senses.

   b) List two properties of a material that require tests to observe.

5. Classify each of the following as a metal or nonmetal:

   a) aluminum (Al),

   b) iron (Fe),

   c) oxygen (O),

   d) carbon (C),

   e) mercury (Hg).

6. The way materials are used can change with time. Milk was originally delivered in glass bottles. Now cartons made from wax-coated paper and plastic jugs are used for milk. Snow skis used to be made of wood. Now they are made from fiberglass or graphite. What factors go into decisions about changing what materials should be used when building a product?

Inquiring Further

The effect of heat treatment on the property of a metal

It is possible to change the characteristics of a material by treating them in various ways. Determine the effect of heat on a metal. Obtain a few paper clips or bobby pins from your teacher. As a control, determine how many times it takes to bend the clip or pin back and forth in order to break it. The stress at the point of bending causes the paper clip or bobby pin to break.

Now try some various types of heat treatment to see what effect they have on the metal. Try heating the piece in the flame of a burner until it is red. To do this, hold the piece with tongs or forceps. Allow the piece to cool on its own until it is safe to handle. Then try to break it again, being careful to bend it back and forth exactly as you did with the control. Record your results and try another heating scheme. Consider treatments such as heating to redness and then cooling by plunging into water, or heating and cooling several times, or heating, but not to redness, only heating to a moderate amount.

Record your observations and summarize what seems to be the relationship of heat treatment to the characteristics of the metal.

What characteristic of metals do you think is demonstrated by bending the metals back and forth until breaking?

Activity 7 Polymers

GOALS

In this activity you will:

- Make a polymer-based material that has properties different from other states of matter that you have studied.
- Observe the material's properties and compare them to those of solids and liquids.
- Describe the process of cross-linking in polymeric materials.
- Discuss and invent new commercial uses for water-soluble polymers.
- Compare the viscosities of two non-Newtonian fluids.

What Do You Think?

In ancient times, humans made most of the objects they needed for daily life from natural substances. Relatively little processing was done between harvesting and use. For much of history, people have used metals, cotton, wood, and other natural materials for building, clothing, and tools. In the 1900s things changed dramatically as scientists were able to create new materials.

- What materials are polyester and rayon made from?
- How are they manufactured?

Record your ideas about these questions in your *Active Chemistry* log. Be prepared to discuss your responses with your small group and the class.

Investigate

1. Measure 50 mL of polyvinyl alcohol (PVA) solution into a beaker. Observe closely the properties of the PVA solution. Refer to the different type of properties you investigated in previous activities to make your observations.

a) Record your observations in your *Active Chemistry* log.

2. Measure 10 mL of sodium tetraborate (borax) solution into another beaker. Observe the properties of borax.

   a) Record your observations in your *Active Chemistry* log.

3. Add one drop of food coloring to the sodium tetraborate solution.

4. Add the sodium tetraborate solution to the PVA solution while stirring with a wooden stick. Keep stirring until the mixture thickens.

   When the mixture has thickened, remove the stick. Place a few paper towels on your desktop and use your hands to remove the thickened mixture.

   Mold and stretch the new material while you observe its characteristics.

   a) Record your observations in your log.

5. Test your new "slime" and see if it behaves more like a solid or more like a liquid. Try holding the slime in your fingers and dangling it downward. Wait for a few minutes and record what happens.

   Place your slime back in the beaker and see what happens as it sits for a few minutes.

   Try pulling the slime out slowly and see what happens. Now try the same thing but pull quickly.

   Roll the slime into a ball and try bouncing it gently on the tabletop.

   a) Record all your observations in your log.

6. It was not difficult to label the PVA solution as a solid or a liquid. Similarly it was easy to label borax. Liquids have very different characteristics from solids. Liquids flow while solids have a rigid shape. Liquids assume the shape of their container and solids have their own shape. Slime blurred the line between the definitions. Liquids spread out when force is applied, but solids break. Liquids splatter when dropped, while solids may bounce.

   a) In what ways does your slime behave like a liquid?

   b) In what ways does it behave like a solid?

7. The PVA molecules are relatively long, slender molecules. The sodium tetraborate molecules are shorter and can form bonds on both ends.

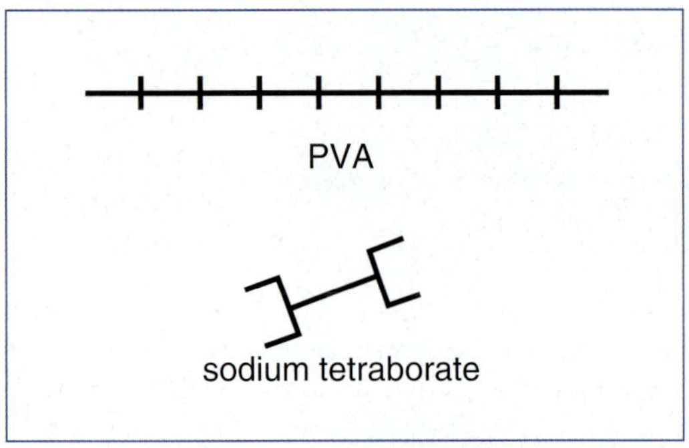

   a) Draw a sketch to represent how the substances act in this reaction. Use the model in the diagram to draw PVA molecules alone, and then draw sodium tetraborate molecules alone. Finally, draw the reaction of sodium tetraborate cross-linking the PVA molecules. Remember your drawing is just a representation of the molecules.

Chem Talk

POLYMERS

In this activity you made "slime." This substance is a **polymer** that has unique characteristics. It is classified as a non-Newtonian liquid. Liquids resist flow. This phenomenon is known as **viscosity**. Newton devised a simple model for fluid flow. You will learn more about this model in later chemistry and physics courses. Liquids like water and gasoline behave according to Newton's model. They are called Newtonian fluids. Ketchup, blood, yogurt, gravy, pie fillings, mud, and slime do not follow the model. They are classified as non-Newtonian liquids.

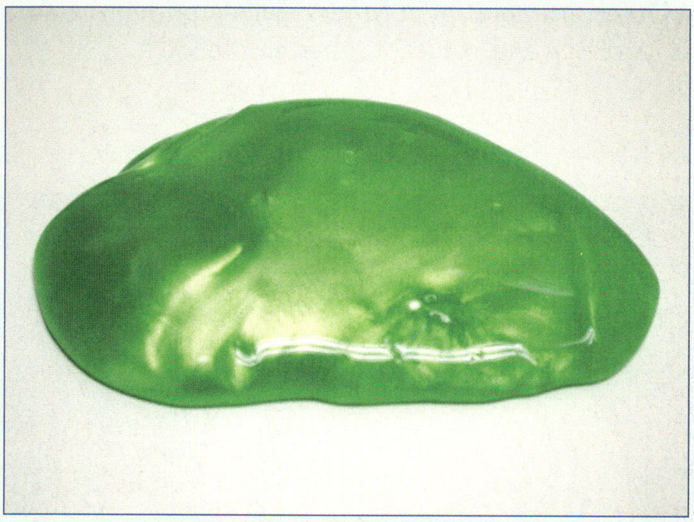

Slime has some characteristics of liquids such as being fluid and taking the shape of its container. On the other hand it bounces and breaks when pulled quickly, as solids do. The unique characteristics of slime are due to the two substances that make it up. The borax solution forms cross-links among the polyvinyl alcohol molecules. These cross-links make the resulting polymer slower to flow and change shape.

The formation of very large molecules from many smaller molecular units is called **polymerization**. Polymers are materials that are made up of many similar small molecules, called monomers, linked together in long chains. These materials have "always" existed in nature, but they have only been produced by industrial processes in the 20th century.

Chem Words

polymer: a substance that is a macromolecule consisting of many similar small molecules (monomers) linked together in long chains.

viscosity: a property related to the resistance of a fluid to flow.

polymerization: a chemical reaction that converts small molecules (monomers) into large molecules (polymers).

Proteins are natural polymers. They are the basic structural unit of plants and animals. There are more than ten thousand different proteins found in nature, yet they are all made up of combinations of about 20 different amino acids. Starch and cellulose are examples of carbohydrates that are polymers. The human digestive system is able to break apart the bonds that form starch molecules, releasing glucose, which the body uses as a source of energy. Humans cannot digest cellulose.

Polyethylene (polyethene) is a polymer made up of many ethane molecules. It is one of many chemicals manufactured from crude oil. It is used to make plastic milk bottles as well as a variety of other objects. Other examples of polymers include polyvinyl chloride, Plexiglas®, polystyrene, natural rubber, and Teflon®.

Checking Up

1. Describe polymers and polymerization in your own words.
2. Name two natural-occurring polymers.
3. Give examples of five polymers that are manufactured using technological processes.

Reflecting on the Activity and the Challenge

What monster movie is complete without slime? Now you know how to make slime. Supplies for making slime are relatively expensive. Keep costs in mind as you plan for your special effect.

Chemistry to Go

1. List three ways you could modify the properties of your slime. If the slime you made in this activity is not quite correct for the special effect you have in mind, how might you modify it for use in your movie?
2. List five other uses that might be possible for the slime substance, in addition to its use as a movie special effect.
3. Thin sheets of solid PVA slime have been used in several new commercial products. Since the sheets are soluble in water, they are ideal recyclable containers. Seed manufacturers have made long tapes of PVA sheets with seeds imbedded inside. When the tapes are planted in the ground the PVA dissolves and the seeds are free to grow. Another use is collecting dirty laundry in hospitals. The solid laundry is placed into PVA bags. When it is time to wash the clothing the entire bag is thrown into the washing machine. It is not necessary for workers to handle the dirty clothes again. What two other uses can you think of for PVA sheet material?

Inquiring Further

1. Self-siphoning slime

Spread your slime on a piece of plastic wrap or waxed paper and let it dry completely. This may take a day or two. Compare the characteristics of the dry slime to the wet slime.

PVA slime has been reported to be "self-siphoning." To siphon most substances, a tube is placed in a liquid and the siphon is started by sucking the liquid out of the container, over the edge and down the tube. The mass of the water flowing out of the tube creates a pressure difference and draws the remainder of the liquid with it. Slime is supposed to have the same effect, but without the tube! Try demonstrating this self-siphoning effect with your slime.

2. Viscosity of non-Newtonian liquids

Ketchup is also classified as a non-Newtonian liquid. Determine which non-Newtonian liquid is the most viscous (resistance to flow).

- Set up a retort stand with a ring to hold a long-stem funnel.
- On the funnel place two marks with a pen at about 2 cm apart.
- Pour one of the liquids into the funnel and allow it to flow through the stem into a beaker.
- Let the liquid level come down to your top mark on the side of the funnel and stop the flow with your finger.
- Using a stopwatch, time how long it takes for the liquid to flow before it reaches the second line after you remove your finger.

Activity 8 Identifying Matter

GOALS

In this activity you will:

- Produce colored flames.
- Identify the metal ions present in materials by the colors of light a material gives off when held in a flame.
- Describe how atoms create the colored light.
- Investigate ways of producing new colors not among those produced by the materials you test.
- Practice safe laboratory techniques in working with laboratory burners.

What Do You Think?

You may have read how royalty, suspicious of enemies, had their food tested for poison by having someone else taste it first. That's not a job anyone would enjoy!

- How else could you test for a poisonous substance?

Record your ideas about this question in your *Active Chemistry* log. Be prepared to discuss your response with your small group and the class.

Investigate

1. Your teacher will supply you with seven numbered wooden splints that have been soaked in various solutions. Some are harmful, and some are not.

   a) How could you distinguish one splint from another?

2. Light your Bunsen burner. Immediately make observations about the color of the burner flame.

Safety goggles and a lab apron are required for this activity.

a) Record your observations in your *Active Chemistry* log.

3. Take one of the wooden splints and hold it in the hot part of the flame using forceps, tongs, or a fireproof glove. Note the color of the flame when the solution-soaked wooden splint is heated.

   As soon as the flame from the splint is no longer strongly colored, extinguish the splint by placing it in a beaker of water.

   a) Organize a table to record your data.

   b) Record your observations in your table.

4. Repeat **Step 3** for the remaining six splints.

   a) Record all your observations.

5. Refer to your data table. The salt solutions in which the splints were soaked are metal compounds of the elements lithium, barium, sodium, strontium, potassium, and calcium. One splint was soaked in water.

   a) What do you notice that was different about each splint?

   b) Your teacher will identify the solution each splint was soaked in. Record this information in your data table.

6. Obtain another splint soaked in an unknown solution.

   Hold the wooden splint soaked in the unknown solution in the burner flame. Observe what happens.

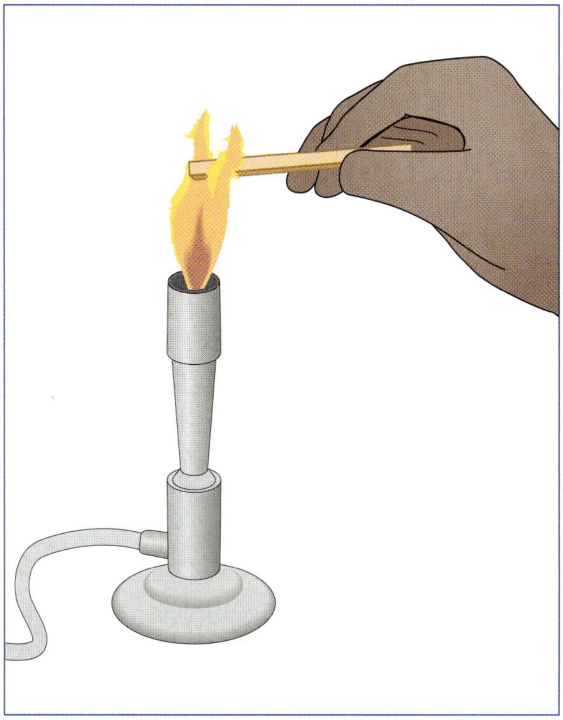

   a) Record your observations in your log.

   b) Use your observations to decide what metal ions are on the splint. Confirm your results with your teacher.

   c) Did you correctly identify the metal ion?

   d) In your log describe how you decided what the identity of your compound was, and how sure you are that you were correct.

7. Clean up your work station and return all your equipment as directed by your teacher. Wash your hands thoroughly.

Chem Talk

IDENTIFYING ELEMENTS USING ELECTRONS

The process you observed in this activity is not a chemical reaction. When the metal **ions** are placed in the hot flame their **electrons** absorb the energy and move to higher energy levels around the nucleus. When these electrons fall back to their original level they give off the light you see in the **flame test.** Since each atom has a unique arrangement of electrons, each gives a unique color. This experimental technique is called a flame test, and is the basis for identifying the composition of the salt solutions. This same effect is responsible for the glowing light of neon signs and the burst of color in fireworks. In this experiment only the metal ions are changed; the negatively charged ion (chloride) is kept constant.

The flames produced in this activity have interesting colors. It is not obvious where they come from. The metal compounds are made from atoms that have **nuclei** surrounded by electrons. When electrons fall from a higher energy level to a lower energy level, they give off light. In this case the electrons were able to get to a higher energy level

Chem Words

ion: an electrically charged atom or group of atoms that has acquired a net charge, either negative or positive.

electron: a negatively charged particle with a charge of 1.6×10^{-19} coulombs and a mass of 9.1×10^{-31} kg.

flame test: an experimental technique or process in identifying a metal from its characteristic flame color.

nucleus: the very dense core of the atom that contains the neutrons and protons.

by absorbing heat energy from the Bunsen burner flame. Each compound has a different configuration of electrons and electron energy levels. The change in energy levels represents the color of the light given off. If you refer to the higher energy level as E and to the lower energy level as E_l, then you can state:

The energy of the emitted light is equal to $E_h - E_l$. The amount of energy remains the same as it is conserved. The energy the electron gains from the heat energy of the Bunsen burner flame raises the electron to a higher energy state. When the electron drops down to the lower energy state, it gives off that energy as light energy.

Checking Up

1. Explain what is meant by a flame test.
2. Explain how energy is conserved during the flame test.

Reflecting on the Activity and the Challenge

Most of the common fuels are primarily hydrocarbons, which do not contain metal ions. Therefore, you are most familiar with flames that are yellow-orange or occasionally a light blue. The bright colors of the excited metals are a surprise and tend to look "out-of-this-world."

Colors could play an interesting part in a plot line of a movie script. Perhaps an unusually colored flame could be taken as a mystical sign. A sudden change in the color of campfires could be a sign that danger is near. Consider how you might use flame tests to create unusually colored flames for your movie special effect.

Chemistry to Go

1. Compare the colors you observed in your flame tests to the colors you have seen in fireworks displays. Identify what metals are used in producing the different colors in fireworks.
2. Develop a series of sketches showing how an electron of an atom can give off light. Your first sketch should have the electron in the ground state. The next sketch should show the electron excited when you apply energy to the atom and the third sketch should show how light is emitted when the electron falls to a lower energy state.

Preparing for the Chapter Challenge

Name three new colors of light that could be produced by combining some of the metal salts that you tested in the lab, and identify which metal salts you would combine to produce these colors. If time and your teacher permit it, test your predictions.

Inquiring Further

Fireworks

Fireworks have been around for centuries. They were used long before anyone knew why they produced the stunning effects they do. Investigate the manufacturing of fireworks and find the specific compounds used to produce the many colors presented during fireworks displays.

Are all colors equally represented in fireworks displays, or are some colors easier to obtain than others? Research what colors are available and which, if any, do not seem possible given the range of substances readily available.

Activity 9 Organic Substances

GOALS

In this activity you will:

- Combust a material present in fruit rinds.
- Represent the combustion of hydrocarbons as chemical equations.
- Learn the formulas and proper names for simple hydrocarbons.
- Make two-dimensional drawings showing the chemical bonding structure in simple hydrocarbons.
- Classify materials as organic or inorganic.
- Explore various definitions of the term "organic."

What Do You Think?

At a grocery store or market, "regular" apples might cost about 79 cents per pound, while "organic" apples may cost as much as or even more than \$1.93 per pound.

- Why do people buy organic apples if they are so much more expensive?
- Does organic mean one thing to shoppers and another to chemists? Explain your answer.

Record your ideas about these questions in your *Active Chemistry* log. Be prepared to discuss your responses with your small group and the class.

Investigate

1. Obtain a piece of orange peel from your teacher. Make observations of its texture, color, smell, and any other characteristics you notice.
   a) Record your observations in your log.

2. Take one piece of the orange peel and hold it in your hand between your thumb and index finger. Fold the orange peel in half.

   Using forceps or tongs, hold it with the smooth fold pointing away from you.

   Hold the surface of the orange peel near the flame of a candle. Pinch or squeeze hard.

   Repeat using fresh pieces of orange peel.

   a) Record your observations in your *Active Chemistry* log.

3. Repeat **Step 2** with other types or fruit and vegetable skins. Use pieces that are the same approximate size and shape as the orange peel.

   a) Record your observations in your log.

4. Organic chemistry is the study of molecular compounds of carbon. The gas and liquid that you ignited in the previous steps are carbon compounds. They belong to a class of organic compounds called hydrocarbons.

That means it is a compound composed only of carbon and hydrogen.

   a) Draw what the molecules of some simple hydrocarbons might look like. Each carbon molecule can form four chemical bonds. Each hydrogen can form one bond. Carbon can link to another carbon atom or to hydrogen. Draw structures for hydrocarbons containing one through 10 carbons, adding hydrogen as necessary to fill the bonds. The first two hydrocarbons are done as examples.

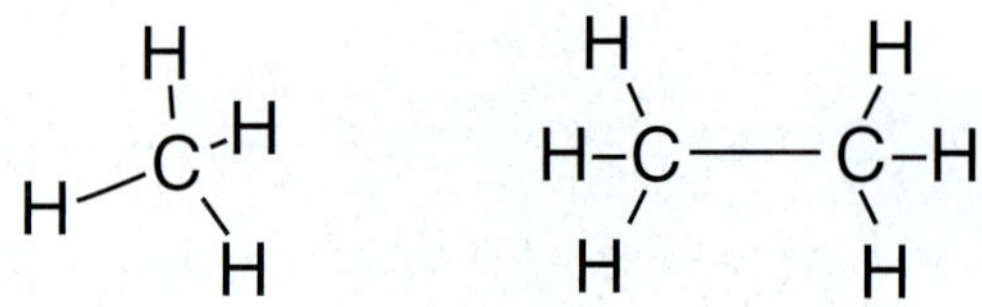

   b) In the case of ethylene, a chemical that stimulates the ripening of fruit, the carbons share two bonds between them. Write some formulas for carbon compounds containing two, three and four carbons that contain at least one carbon—carbon double bond. Remember that each carbon can still only have a total of four bonds. If it shares two bonds with another carbon atom that leaves only two others to share.

Chem Talk

THE CHEMISTRY OF MOLECULAR COMPOUNDS OF CARBON

Chemistry attempts to explain macroscopic events through an understanding of microscopic phenomenon. Chemists rationalize how the world they see works, by describing how things they cannot see function. In this activity you found evidence of a gas. This gas is called ethylene and is responsible for the ripening of fruits.

As plants age, they release ethylene gas (C_2H_4) which acts as a plant hormone to begin the ripening process. Green fruit can be made to ripen faster by bathing it in ethylene gas. Absorbing the gas from fruit storage areas can stop the ripening process. This activity demonstrated the presence of the ethylene gas, which is flammable, in the skin of the fruit.

Ethylene is one of many materials classified as organic. To a chemist the term organic means "based on carbon molecules." Everything that is not organic is termed inorganic. In the very early 19th century, compounds were classified into two categories: **organic**, which were those compounds that came from living (or once living) organisms; and **inorganic**, which were obtained from mineral sources. A theory, called "vitalism," proposed that a "vital force"

Chem Words

organic compound: a molecular compound of carbon.

inorganic compound: a compound not based on molecular compounds of carbon.

was required to produce organic compounds. This theory was disproved when an organic compound was synthesized in the laboratory. Today, organic chemistry refers to the study of carbon compounds. This is the largest area of study in chemistry. In the natural world alone, plants and animals synthesize millions of carbon compounds. Many manufactured chemicals are facsimiles of these natural products. Chemists have also invented processes that synthesize similar, new chemicals for use in society and technology. Examples of synthesized chemicals include gasoline, artificial sweeteners and flavors, and a variety of medicines.

The popular use of the word, "organic" implies totally natural or not human-made. Chemists do not use the word in this way, and would consider a molecule organic if it came either from a test tube or a tomato plant. They would also argue that all apples are organic, no matter how they are cultivated.

Ethylene belongs to a class of organic compounds called **hydrocarbons**. That means it is a compound composed only of carbon and hydrogen. Most fuels, such as gasoline, kerosene, candle wax, propane, and natural gas are hydrocarbons too. The burning of a hydrocarbon in the presence of oxygen is an example of **combustion**. In combustion, a substance is combined with oxygen and accompanied by the generation of heat and sometimes light.

Chem Words

hydrocarbon: a molecular compound containing only hydrogen and carbon.

combustion: the rapid reaction of a material with oxygen accompanied by rapid evolution of flame and heat.

Checking Up

1. What does a chemist mean by an organic compound?
2. How did the meaning of "organic" change during the 19th century?
3. What is a hydrocarbon?

Reflecting on the Activity and the Challenge

When you held the orange peel near the flame and squeezed it, you forced a natural material made by the orange to be released. The material that comes from the skin of the orange peel is a hydrocarbon fuel. It is similar to the fuel used in barbecues and portable lighters. This material burns very quickly and is quite like the fuels used in movies to make huge explosions and fires. Of course, explosions and fires can be quite dangerous. Often, movie directors will film an explosion on a small scale, by using models, and then make it appear huge on the screen. Think of some ways you could use this very small explosion to represent a larger effect on the screen.

Chemistry to Go

1. Write two or three examples of the use of the word organic to mean natural.
2. Classify each of the following materials as inorganic or organic:

salt (NaCl)	sugar ($C_{12}H_{22}O_{11}$)	vinegar (CH_3COOH)
gasoline (C_8H_{18})	sand (SiO_2)	oxygen (O_2)

3. Draw an example of a simple hydrocarbon that contains five carbon atoms, all with single bonds.
4. a) Draw an example of a hydrocarbon compound containing seven carbons that contain at least two carbon—carbon double bonds.

   b) Is there more than one answer possible in **Part (a)**? Explain your answer.
5. Acetylene (C_2H_2) is also a hydrocarbon. The carbons in this case share three bonds between them. Again, remember that a carbon atom can only have a total of four bonds. Draw the structure for acetylene.
6. In addition to chains, carbon atoms can bond together to form rings and other shapes. Draw a hydrocarbon that is made up of six carbon atoms, all with single bonds, that forms a chain.
7. Some products or processes claim to be "chemical free" or say "no chemicals added." Based on your understanding of chemistry, what are the arguments for and against such statements?

Inquiring Further

Organic produce and federal regulations

The federal government has proposed standards for food products bearing the label "organic." The National Organic Standards Board, of the United States Department of Agriculture, has issued the following definition of an organic food:

"Organic is a labeling term that denotes products produced under the authority of the Organic Foods Production Act. The principal guidelines for organic production are to use materials and practices that enhance the ecological balance of natural systems and that integrate the parts of the farming system into an ecological whole."

- How does this definition of the term "organic" compare with the chemical definition?
- Investigate the other proposed USDA standards and evaluate them relative to the chemical definition of organic.

Movie Special Effects Assessment

Your Chapter Challenge is to create a story line and produce special effects for a low-budget movie. In this chapter, you have investigated chemistry phenomena which can be adapted to create those special effects.

The following tasks are required:

- Write a script for a simple scene in a movie.
- Choose a special effect to include as part of your scene.
- Write a procedure on how your special effect is done.
- Demonstrate the special effect to the "producer."
- Write an explanation of how the special effect works, including the chemistry behind the demonstration. Using more than one chemical principle will strengthen your presentation.

To begin, you should probably review all of the activities that you have completed. You may wish to construct a chart with three columns. The first column would include a brief description of the activity. The second column would include what kinds of special effects might be done with this activity. The third column would include chemistry principles involved in the activity. As you review the chapter in this manner, you should pay particular attention to the section **Reflecting on the Activity and the Challenge.** This section will help you to see the connection between these three elements — the activity, the special effects show and the chemistry concept. You may also use the end-of chapter summary chart **Chemistry You Learned** as a review tool.

There are two different ways in which to continue — you can create a story that will require some special effects or you can decide which special effects you will use and then make up the story. Special effects do not have to include explosions and fire. Some of the best special effects are more subtle (e.g., where something floats in space or when a young person looks old).

If you choose to write the story line first, you may start the creative process by brainstorming ideas in your group. In a brainstorming process, every team member contributes ideas as the thought arises. After brainstorming for 10 minutes, your team can then choose what may be the most valuable ideas and then put together the story. If your team chooses to decide on the special effects first, you can refer to your completed list of the activities. Which activities do you wish to repeat as your special effects? The **Chapter Challenge** states that more than one chemical principle will strengthen your proposal. Will it be possible to use a few activities together in your special effects presentation?

Compromise will be an important ingredient for your success. You may not be able to include the best special effect because of limitations of time, money or knowledge. You may also have to limit your script ideas for some of the same reasons. A tight and well prepared script and special effects show will take into account these limitations and still be entertaining, interesting and appealing.

Once you have agreed on the script and the special effects, each team member can be assigned one of the remaining responsibilities. Who will write the detailed instructions on how to complete the procedure? Who will demonstrate the special effect? Who will write the explanation of the chemistry behind the demonstration? Writing the procedure will require a closer review of the specific activity that you are using. Writing an explanation of the chemistry will also require you to carefully review the activity with particular emphasis on the **ChemTalk** section.

If there are 4 members of the team, each team member may be responsible for one special effect and its procedure, demonstration and chemistry. Alternatively, one team member may be writing all of the procedures while another team member writes all of the chemistry explanations. However you divide the work, it will be important for each team member to share the contribution with the entire group before the due date. Be sure to leave some time to provide feedback and improve upon each part before the final project is due.

You should practice your demonstration before the presentation. Does every team member know their part? Is every team member aware of the time limitations? Are safety issues noted and taken into account? Review all of the criteria listed at the beginning of the chapter and the grading criteria agreed to by your teacher and class. Create a checklist and perform a final review of your checklist. Grade yourself before your presentation to ensure that your team is heading for a great grade.

Chemistry You Learned

- Elements and symbols
- Chemical formulas
- Compounds
- Electrolysis and synthesis of water
- Chemical properties of elements and compounds
- Physical properties of elements and compounds
- Phase changes and properties of matter
- Sublimation
- Kinetic energy and potential energy
- Freezing points and boiling points
- Temperature
- Solutions
- Suspensions
- Colloids
- Composites
- Tyndall Effect
- Density of liquids, solids, and gases
- Instrumental measurements (precision and accuracy)
- Properties of metals and nonmetals
- Polymers
- Hydrocarbons
- Organic compounds vs. inorganic compounds

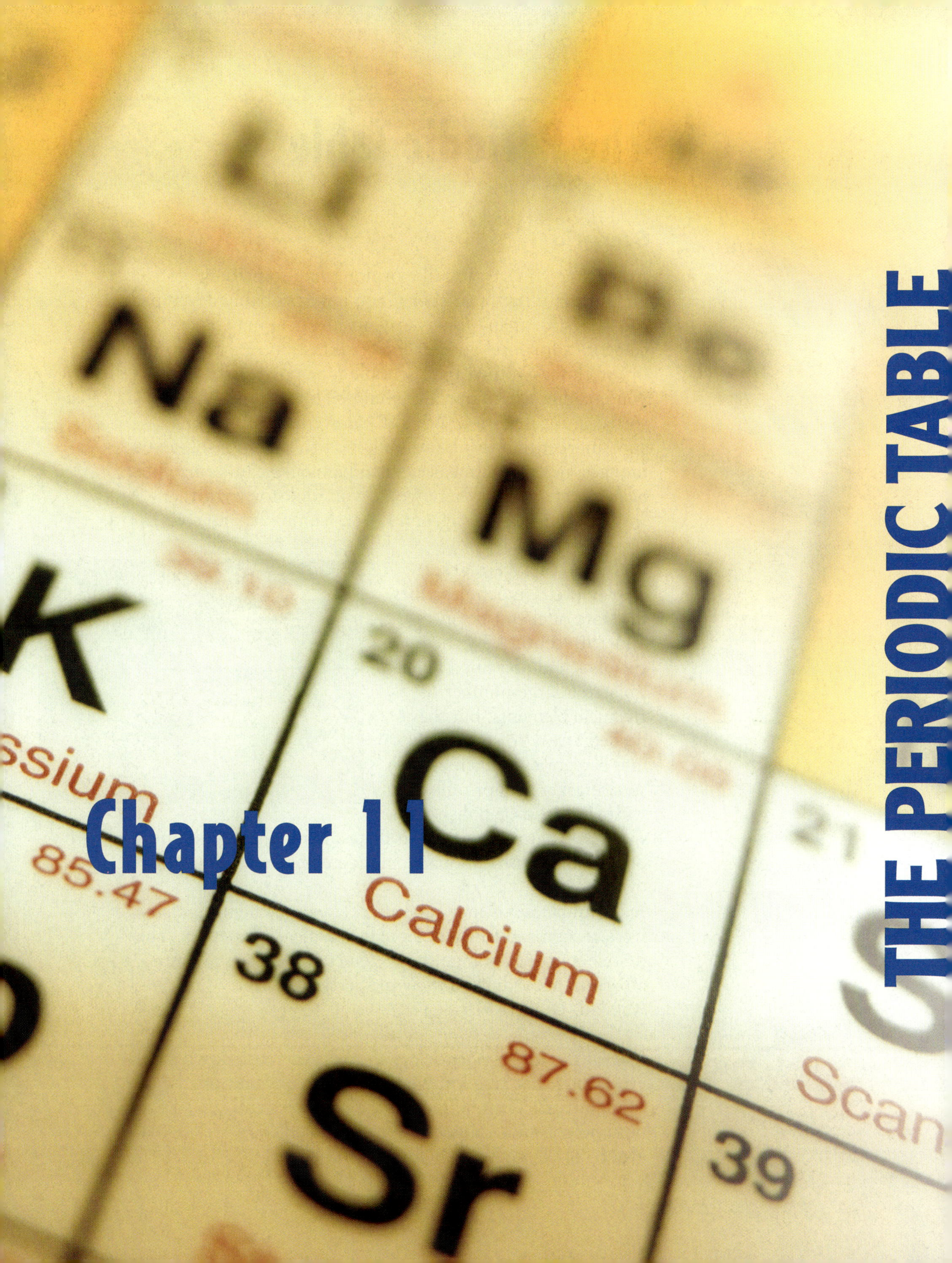

Chapter 11

THE PERIODIC TABLE

11 The Periodic Table

Scenario

Every time you say you like or don't like something, you are putting it into a category. You have probably developed categories for many things in your life. You may have categories for food you eat for breakfast, as opposed to dinner, or for clothes you wear to school, as opposed to at home. Can you imagine what your life would be like if nothing were sorted into categories? What if you went shopping in a supermarket that displayed milk next to shoe polish, next to oranges, next to oatmeal, next to hams, next to orange juice, next to detergent? Where would you look for yogurt, shoelaces, corn flakes, ground beef, lemonade, and soap?

That kind of supermarket display pretty much describes the state of chemistry in the mid-19th century. By then chemists had identified and isolated a large number of chemical elements, but they needed a way to sort them into categories—much as a supermarket groups milk with yogurt, shoe polish with shoelaces, oatmeal with corn flakes, ham with ground beef, orange juice with lemonade, and detergent with soap.

Like similar items in a supermarket, some chemical elements were recognized to share similar chemical properties. The first chemist to arrange these elements successfully into a pattern according to their properties was the Russian, Dimitri Mendeleev.

One of the things Mendeleev did was to write down everything that was known about each element on a small card. Then he moved the cards around until he got an arrangement that showed the groups of elements with similar properties.

Dimitri Mendeleev

In Mendeleev's time the periodic table was developed as a way to arrange elements according to their chemical behavior. Surprisingly, it then revealed information about the structure of the atoms of those elements as well.

By writing the properties of the elements onto separate cards and arranging them, Mendeleev created a puzzle, and he solved that puzzle when he arranged the first version of what is now known as the Periodic Table of the Elements. The table was independently created at the same time by the German, Julius Lothar Meyer.

Chapter Challenge

Your challenge in this chapter is to develop a game related to Mendeleev's Periodic Table of the Elements.

How the game is played, whether on a table, with cards, on a computer, or with equipment that only you might choose, is up to you. You might even choose to emphasize some aspects of the periodic table over others, or to focus on some types of information presented by the table rather than others. However, you need to keep in mind the criteria you and your teacher establish.

Criteria

How will your game be graded? What qualities should a good game have? Discuss these issues in small groups and with your class. You may decide that some or all of the following qualities should be graded:

- how well the game shows your understanding of the periodic table
- how well the game enables players to learn about the periodic table
- how interesting the game is to play
- how long the game takes to play
- whether the game is sequential or can be continued

Once you have determined the list of qualities for evaluating the game, you and your class should also decide how many points should be given for each criterion. Make sure that you understand all the criteria as well as you can before you begin. Your teacher may provide you with a sample rubric to help you get started.

Activity 1 Organizing a Store

GOALS

In this activity you will:

- Plan the arrangement of the items for sale in a store.
- Analyze trends in the arrangement of the store.
- Relate the arrangement of items in the store to the arrangement of elements in the periodic table.

What Do You Think?

Some supermarkets now sell books, flowers, and prescription drugs in addition to eggs, meat, and cereal.

- How many different items do you think that a supermarket has in its inventory?

Record your ideas about this question in your *Active Chemistry* log. Be prepared to discuss your responses with your small group and the class.

Investigate

1. Suppose that you decided to go into the business of opening and running a supermarket grocery store. In your group, brainstorm a list of between 50 and 100 items you would sell at your supermarket.

   A member of your group should volunteer to record the items suggested by all members of the group. Everyone, including the person serving as recorder, should participate in suggesting items to be sold.

   a) Make a map showing the locations of all of the items in your store. Give some thought to what will be at the front of each aisle, and what will be at the back, and how the store will be arranged from left to right.

b) Keep in mind which items you want shoppers to see as they enter the store and which should be near as they approach the cash register. Would either of these factors alter your arrangement?

c) Consider the arrangement of items going from left to right across your store. Why did you choose to arrange the items that way?

Reflecting on the Activity and the Challenge

Organizing 50 to 100 items in your store is not unlike the problem faced by Mendeleev when he organized about the same number of chemical elements into the periodic table. In the following activities you will learn about the properties of chemical elements that led Mendeleev to arrange the elements the way he did and the information about them provided by the periodic table. This activity, in which you were asked to organize a group of items familiar in your everyday experience, was designed to acquaint you with some of the problems Mendeleev faced in the hope that you can better appreciate what he did. You may wish to build this experience into the game you design.

Chemistry to Go

1. What is the pattern or arrangement in your store's aisles?
2. Choose one aisle in your store. Describe the arrangement of items going from the front of the store to the back of the store. What is the trend (or general drift) in that aisle?
3. A new item is brought into the store — chocolate covered peanuts. Where would you place this item? Provide an explanation for your decision.
4. Your store decides to sell napkins, plates, and other decorations for Thanksgiving. How will you adapt your store arrangement to accommodate these items?
5. You would like people to purchase a certain item because it gives you a big profit. Where would you place it in your store and why?
6. One of the characteristics of Mendeleev's original periodic table was a series of blank spots. Since As and Se didn't have anything in common with Al and Si, but do with P and S, Mendeleev decided there must be a couple of other elements yet to be discovered. He left spaces for them and put As under P and Se under S where they belong. What would such a "blank" correspond to in your store?

Activity 2 Elements and Their Properties

GOALS

In this activity you will:

- Apply ancient definitions of elements to materials you believe are elements.
- Test some properties of several common chemical elements.
- Classify elements as metals, nonmetals or neither.
- Learn to differentiate between chemical and physical properties of materials.
- Organize a table of the elements you tested based on their properties.
- Practice safe handling of corrosive chemicals in the laboratory.

What Do You Think?

Throughout the ages of history, philosophers and scientists have talked about "the elements." Reference to elements is most frequent today in the field of chemistry.

- What is a chemical element?

Record your ideas about this question in your *Active Chemistry* log. Be prepared to discuss your responses with your small group and the class.

Investigate

1. Work individually first and then in your group.

   a) Make a list of four or more substances you use or encounter in your everyday experience that meet your definition of element.

2. The ancient Greeks believed that the four elements were: earth, air, fire, and water.

   The alchemists of the early Renaissance identified three elements: mercury, sulfur, and salt.

   a) Does each of the above "elements" satisfy your definition of an element? Why or why not?

3. Your teacher will provide a series of jars containing several common chemical elements: aluminum, copper, iodine, iron, magnesium, silicon, sulfur, and zinc.

   You will investigate the properties of these elements. By observing common properties, you may gain an insight into how an organizational chart can be created for all of the known elements. Observe the sample of the chemical element in each jar (without removing any).

Element	Initial observations	Conducts electricity	Reacts with HCl	Metal or nonmetal
aluminum				
copper				
iodine				
iron				
magnesium				
silicon				
sulfur				
zinc				

   a) Record your observations in a table. You may wish to use a table similar to the above to record your observations in this activity.

4. One of the properties of the chemical elements on Mendeleev's cards was the ability of the element to conduct electric current.

   Insert the terminals of the electrical conductivity apparatus into the jar containing each element. If the light on the apparatus goes on, that means that a complete circuit is created, and an electric current is passing through both the light bulb and the sample of the element in the jar.

   It is important to make sure that the part of the apparatus immersed into the elements stays dry and is not contaminated by any of the other elements it has been immersed in. Also, use steel wool to polish the metal strips before you test them.

Safety goggles and a lab apron must be worn during this activity.

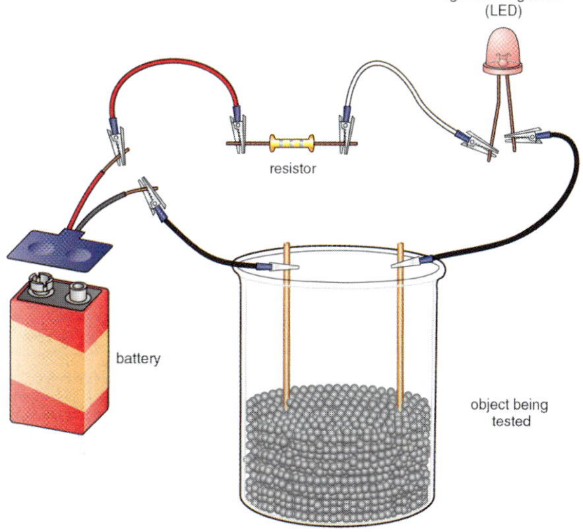

a) Test the samples of each element with the electrical conductivity apparatus and record whether they conduct electric current (yes) or not (no).

5. Another of the properties of each chemical element known to Mendeleev was how it reacts with acid.

You must still be wearing your safety goggles. Pour 5 mL of 1 M hydrochloric acid into each of eight small test tubes. Use a chemical scoop or tongs to remove a small portion of each element from the jar and add it to the hydrochloric acid. It is important to add the hydrochloric acid to the test tube first so that you will not be surprised by a reaction by pouring acid over a reactive chemical element. Place a piece of white paper in the background behind the test tube and observe the reaction between the element and hydrochloric acid by looking through the side of the test tube.

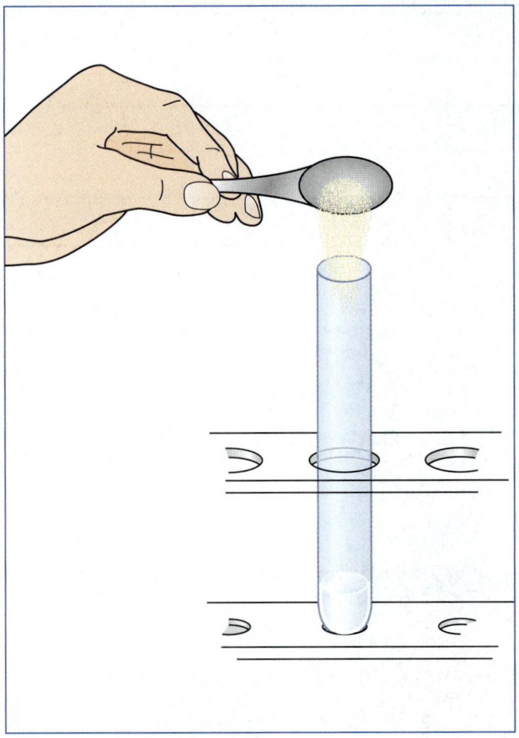

a) Test small samples of each element for their reaction with hydrochloric acid and record whether they react with the acid (yes) or not (no).

b) For those that do react, try to determine whether all exhibit the same type of reaction (do they all do the same thing?) and compare the relative vigor of the reactions. If the reaction is vigorous, include a + sign next to your "yes." If the reaction is weak, place a – sign next to your "yes."

6. Dispose of the contents of the test tubes and clean the test tubes as directed by your teacher. Wash your hands.

7. A metal is generally a solid that is shiny, malleable, and a good conductor of heat and electricity. A nonmetal is generally dull, brittle, and is a poor conductor of heat and electricity.

Classify each of the elements you observed as either a metal or nonmetal.

a) Record your observations in the table in your *Active Chemistry* log.

8. Create index cards for each of the elements and find a way to sort them

based on their properties. You may try arranging them and/or color coding them. Your method of sorting will be successful if you can quickly find an element and know from its position whether it:

- conducts electricity
- reacts with HCl
- is metallic or nonmetallic

a) Record your method of sorting the cards in your *Active Chemistry* log.

Chem Talk

PHYSICAL AND CHEMICAL PROPERTIES

Classifying Elements Using Properties

You began this activity by trying to define the meaning of a chemical element. The ancient Greek philosopher Aristotle defined an element as "a body into which other bodies may be analyzed . . . and not itself divisible into bodies different in form." The first modern definition of chemical element, which is not much different, is from Robert Boyle: "Bodies, which not being made of any other bodies, or of one another, are the ingredients of which all those . . . mixed bodies are . . . compounded." We now state that an element is any material that cannot be broken down by chemical means into simpler materials.

Before the mid-19th century, scientists were preoccupied with discovering elements and observing and recording their properties. Then they tried to organize the elements they had discovered in a useful way. At first, they listed the elements alphabetically. However, every time a new element was discovered, the whole list had to be changed. They tried other methods. Could elements be organized by properties like state, color, or taste? None of these methods appeared practical or safe! However, chemists worldwide were sure that elements existed in families that had similar physical and chemical properties. To the Russian scientist, Dimitri Mendeleev (1843 - 1907), the development of a tool to organize the elements began the same way that so much of science inquiry begins, with a simple question. The question Mendeleev wanted answered was: "What is the relationship of the elements to one another and to the chemical families to which

Checking Up

1. Define a chemical element.
2. What question did Mendeleev use to guide his science inquiry?
3. In your own words, describe the difference between a physical and a chemical property.

they belong?" At that time there were 63 known elements. To help him with his organization, he developed a card game, much the same as you did in this activity. He wrote the properties of each known element on a different card and then spent many hours arranging and rearranging the cards. He was looking for patterns or trends in the data in front of him. Mendeleev, however, had more information on his cards than you presently have. In the following activities, you will look at additional properties of elements that will help you organize your game.

Physical and Chemical Properties

In this activity you observed several properties of the elements you were provided. You probably initially observed the color and the state of the element. You then investigated whether or not the chemical element conducted electricity. You could have also observed the luster, measured the density or the strength, or determined the malleability of each element. In each case, you would not have changed the element itself. In this investigation the element still looked the same in the jar after you removed the electrical conductivity apparatus as it did when you initially inserted it. If measuring a property of a substance does not change the chemical identity of one substance, it is called a **physical property**.

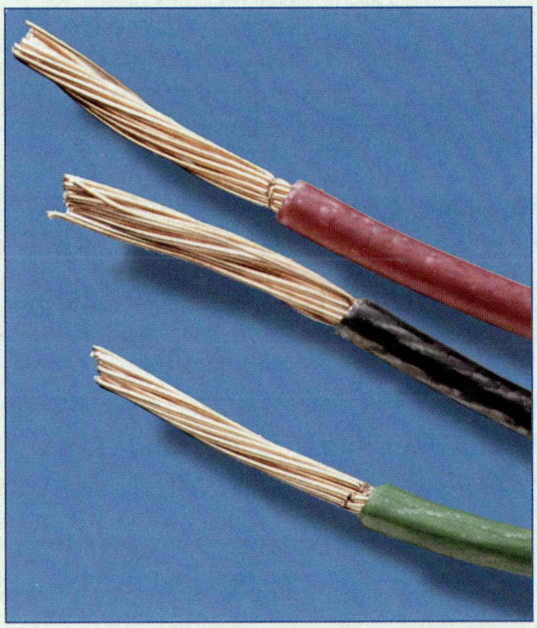

On the other hand, when you observed whether the chemical element reacted with hydrochloric acid, the element clearly changed. A **chemical property** is the kind of reaction that a substance undergoes. Measuring chemical properties changes the chemical composition of a substance.

Chem Words

physical property: a property that can be measured without causing a change in the substance's chemical composition.

chemical property: a characteristic that a substance undergoes in a chemical reaction that produces new substance(s).

Reflecting on the Activity and the Challenge

In this activity you learned not only the definition of a chemical element but also some of the properties of chemical elements. Measuring these properties not only enabled Mendeleev to place the elements in his periodic table but also allowed other chemists to identify the elements. You have tried to sort the cards of elements in the same way that Mendeleev did. Perhaps your periodic table game can have sorting cards as one part of the strategy.

Chemistry to Go

1. Make a list of three more physical properties of a chemical element that you could observe.
2. Make a list of three more chemical properties of a chemical element that you could measure.
3. Why did you want the metals to be clean or polished before you tested them for electrical conductivity?
4. What criteria did you use to differentiate metals from nonmetals in this investigation?
   a) Is this a valid statement of a trend you saw: as the color of the element becomes darker, the element is less metallic? Support your assessment of this statement with evidence that you observed in your investigation.
   b) Is this a valid statement of a trend you saw: the elements react with hydrochloric acid more as you move down a list of the elements listed in alphabetical order? Support your assessment of this statement with evidence that you observed in your investigation.

Preparing for the Chapter Challenge

Prepare a set of index cards for each of the elements with which you are familiar. Record as many properties of each element you know on the card. Use your observations in the following activities and any research you complete on your own to add information to each card.

Activity 3 Atoms and Their Masses

GOALS

In this activity you will:

- Explore the idea of atoms by trying to isolate a single atom.
- Measure how many times greater mass a copper atom has than an aluminum atom.
- Practice careful laboratory technique with measuring masses and filtration.
- Locate sources of the variation in the class's experimental results.
- Compare Dalton's experimental results to the masses of atoms known today.
- See that atoms react in definite proportions of mass when forming a compound.
- Relate the mole concept to real quantities.
- Use scientific notation in calculations.

What Do You Think?

Atoms are the smallest, indivisible part of an element.

- When did you first hear of atoms? What did they mean to you then, and what do they mean to you now?
- Use the following headings: Things I Know about Atoms, Things I Think I Know about Atoms, Things I Would Like to Know about Atoms. Include at least one entry under each heading.

Record your ideas in your *Active Chemistry* log. Be prepared to discuss your responses with your small group and the class.

Investigate

1. One way to think about an atom is to imagine trying to isolate a single atom from a large number of atoms.

   Take a piece of aluminum foil and cut it in half. Take one of the resulting pieces and cut it in half again. Repeat this process with each successive half until you cannot make another cut.

   a) In your *Active Chemistry* log, record how many cuts you were able to make. How does the size of the smallest piece of aluminum compare to the size of the original piece? Does the smallest piece of aluminum have the same properties as the original piece of aluminum foil?

2. An atom is the smallest part of an element. Since you can still cut the aluminum in pieces, you have not reached the size of a single atom.

   Now imagine that there may be a way to cut the smallest piece of aluminum you have into even smaller and smaller pieces.

   a) How small can the smallest piece be and still retain the properties of aluminum? Could you cut the piece in half and half again 10 more times? 100 more times? 1000 more times?

   Using your imagination of cutting and cutting, you will eventually get to one atom of aluminum.

3. Chemists combine elements to form new substances. By measuring the amounts of elements used and substances formed, they are able to draw conclusions about the properties of the elements involved. You will study the reaction between aluminum metal and a solution of copper (II) chloride.

   Read through the procedure below.

   a) Make a table in your *Active Chemistry* log for the data you will be collecting. You will need room for measurements (mass) and observations. You can use a table similar to the one provided below.

4. Check your balance to make sure that it reads zero with nothing on it. Then measure the mass of a 50-mL beaker.

   a) Record the mass in your *Active Chemistry* log.

Safety goggles and a lab apron must be worn during this activity.

Finding the mass of aluminum	
1. Mass of empty 50-mL beaker	g
2. Mass of beaker and aluminum foil	g
3. Calculate the mass of aluminum from 1 and 2 above.	g
Finding the mass of copper (II) chloride	
4. Mass of weighing paper	g
5. Mass of paper and copper (II) chloride	g
6. Calculate the mass of copper (II) chloride from 4 and 5 above.	g
Finding the mass of the product	
7. Mass of dry filter paper	g
8. Mass of filter paper with product, after drying	g
9. Calculate the mass of the product material from 7 and 8 above.	g

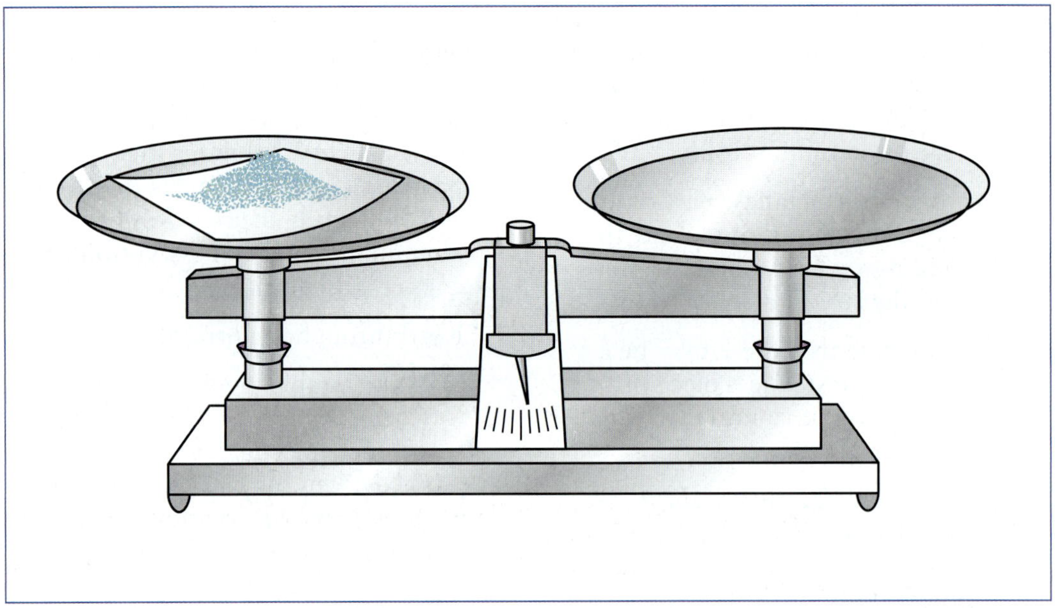

5. Measure out approximately 0.20 g of aluminum foil into the empty beaker. Try to get your mass measurement close to the assigned value.

   If you have a centigram balance, you'll need to adjust the balance to read 0.20 more grams than the beaker alone. Then add pieces of aluminum foil until it rebalances.

   If you have an electronic balance, simply add pieces of aluminum foil until the display indicates 0.20 g more than the empty beaker.

   a) Record the value that you obtain, even though you might not hit the target value.

Wipe up any spills immediately.

6. Measure the mass of a piece of weighing paper.

   Place approximately 2.00 g of copper (II) chloride on the weighing paper. (The chemical elements you would expect to find in copper (II) chloride are copper and chlorine.) Again, remember that your target value is 2.00 g and that you may be slightly over or under this value.

   a) Record the masses in your *Active Chemistry* log.

7. Add the copper (II) chloride to the beaker.

   Next add water to the beaker until it is approximately half full.

   a) Record your observations in your *Active Chemistry* log. Consider including the following: how the beaker feels when you touch it; what you hear when you listen closely to the beaker; what you see happening in the beaker.

   b) What color forms on the aluminum foil? What do you think is responsible for this color? Where is the color coming from?

8. You will now need to find the mass of the substance formed in the chemical reaction. You can filter out this substance and then find its mass.

   First measure the mass of a piece of dry, clean filter paper.

   a) Record the mass in your *Active Chemistry* log.

9. Set up a filtration system, as shown in the diagram below.

10. Wait until you no longer see or hear any reaction between the aluminum foil and copper (II) chloride and the liquid begins to clear up. Pour the contents of the 50-mL beaker into the funnel. You should rinse the beaker a couple of times with some water to be sure that all of the contents of the beaker are transferred.

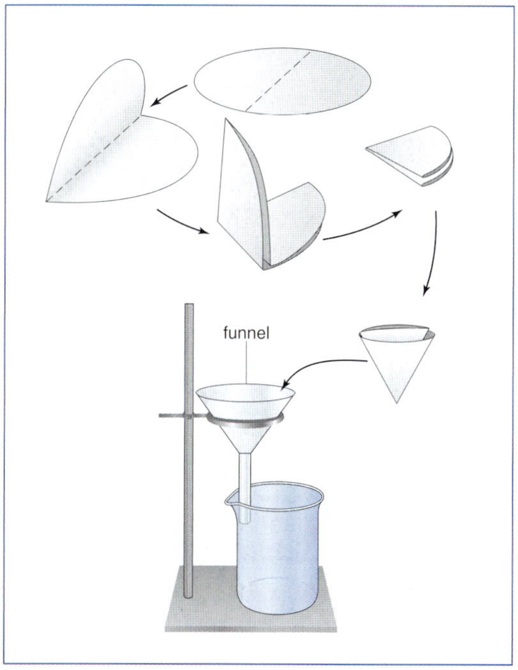

Remove the filter from the funnel and place it on a piece of folded paper towel and allow it to dry overnight. Label the paper towel so that you can identify your filter paper.

Clean and put away your equipment and dispose of your chemicals as directed by your teacher. Wash your hands.

11. When the filter paper is dry, measure the mass of the filter paper and its contents. Dispose of the filter paper and its contents as directed by your teacher. Wash your hands.

    a) Record the mass in your *Active Chemistry* log. Determine and record the mass of the contents of the filter paper.

    b) What element is inside the filter paper?

12. The reaction you witnessed is called a single-displacement reaction. In this reaction, a single element (aluminum) replaces another element (copper) in its combined form (copper (II) chloride). As a result of this reaction, the copper leaves its combined form to become an uncombined, or free element. The aluminum leaves its uncombined form to join with the chlorine to form a new compound. It's time to look at your data and see if you can make some sense of the numbers.

    a) How many grams of aluminum did you start with? How many grams of copper did you end up with (contents of the dry filter paper)?

    b) How many times as great is the mass of the copper as compared to the mass of aluminum you originally used? (What is the ratio of mass of copper to mass of aluminum?)

c) If one atom of copper is released for each atom of aluminum that becomes combined (with chlorine), the masses of copper and aluminum determined in (a) should contain the same number of atoms. What does this tell you about the relative masses of copper and aluminum atoms? The number of atoms in 0.20 g of aluminum EQUALS the number of atoms in 0.71 g of copper. How many times more massive is a copper atom than an aluminum atom? (You may wish to compare objects that you are familiar with: a dozen bowling balls of mass 60.0 kg can be compared to a dozen eggs of mass 2.4 kg. Since there are a dozen bowling balls and a dozen eggs, you can find that the bowling balls are 25 times as heavy as the eggs (60.0 kg/2.4 kg.)

13. Recall that every group in your class reacted the same mass of aluminum with the same mass of copper (II) chloride. Discuss the similarities and differences in the data and calculations among the groups in the class.

a) Record your thoughts on how and why the results are similar and/or different.

Chem Talk

ATOMIC MASS

Atoms

In **Activity 2**, you defined the term element and explored the properties of some common elements. In this activity, you focused on atoms. An **atom** is the smallest representative part of an element. The ancient Greek philosopher Aristotle did not believe in the existence of atoms. In his thinking, if atoms did exist, there would have to be empty space between them.

Aristotle

Chem Words

atom: the smallest representative part of an element.

Aristotle did not believe it was possible to have empty space. Not everyone agreed with Aristotle. Another ancient Greek named Democritus believed that matter was made up of tiny particles that could not be broken down further. He called the particles atoms, from the Greek word *atomos*, meaning indivisible.

If you could have continued cutting the aluminum foil until it could no longer be cut, by any means, you would have reached one atom of aluminum. A mind-expanding fact is that if you started with 27 g of aluminum, you would find that there are 6.02×10^{23} atoms of aluminum. Nobody has ever counted this nor could they. Scientists have determined this number by other means and are very confident that it is correct.

Masses of Elements and Compounds in a Reaction

By the turn of the 19th century, chemists were combining elements to form new substances. The new substance was called a **compound**, because the atoms of the elements were believed to combine to form what they called a compound atom. The chemists were also particularly interested in measuring the amounts of elements used and substances formed. Their first attempts in determining masses were wrong, possibly due to the equipment that they had available at that time.

John Dalton

John Dalton, an early 19th century chemist who did much to advance the belief in the existence of atoms, expected that atoms combined in the simplest possible relationship. He reported that seven pounds of oxygen reacted with one

Chem Words

compound: a material that consists of two or more elements united together in definite proportion.

pound of hydrogen to form water. Accurate modern experiments give eight pounds to one. We will use modern values rather than historical ones to avoid confusion. Dalton reported that five pounds of nitrogen reacted with one pound of hydrogen to form ammonia. He also reported that seven pounds of oxygen reacted with five pounds of nitrogen to form a compound he called nitrous gas.

In 1809, Joseph Gay-Lussac reported that the hydrogen reacting with oxygen to form water occupied twice as much volume as the oxygen. He also noted that the hydrogen reacting with nitrogen to form ammonia occupied three times as much volume as the nitrogen. Furthermore, he found that equal volumes of nitrogen and oxygen reacted to form nitrous gas (now known as nitric oxide or nitrogen monoxide, NO).

Gay-Lussac's data was inconsistent with Dalton's assumption that water, ammonia, and nitrous gas are formed from one atom of each of the combining elements. This inconsistency was subsequently resolved by Amadeo Avogadro, who furthered the understanding of the correct chemical formulas and atomic masses.

Relative Mass of Atoms

Eventually, chemists determined a scale of relative masses of atoms through the systematic study of chemical reactions. By measuring the masses of two elements reacting with each other and knowing the formula for the compound that was formed, the relative mass of the two elements was determined. In this way, chemists were able to determine, for example, that one element has twice the mass of a second element. Relative mass does not tell you the exact mass measured in kilograms. It does provide a relative scale. Comparison of many reactions resulted in a scale of relative masses. Atoms of carbon were found to have a mass 12 times greater than the mass of hydrogen atoms, whereas oxygen atoms were found to have a mass 16 times greater than the mass of hydrogen. The units for this scale are called **atomic mass units**, defined in such a way that the mass of one type of carbon (carbon-12) is exactly 12 atomic mass units. The average mass of an atom of a given element in atomic mass units is known as the **atomic mass**. Atoms of hydrogen have an atomic mass of one unit. In addition to the physical and chemical properties of

Chem Words

atomic mass unit (amu): a unit of mass defined as one-twelfth of the mass of a carbon-12 atom.

atomic mass: atomic mass is determined by the mass of the protons and neutrons of the atom.

law of definite proportions: the composition of a pure substance is always the same or the elements of the compound always combine in the same proportion by mass.

Element	Relative atomic mass
Aluminum	26.98
Copper	63.55
Iodine	126.90
Iron	55.85
Magnesium	24.31
Silicon	28.09
Sulfur	32.06
Zinc	65.38

elements, the relative mass (incorrectly called weight) of each element was known and used by Mendeleev as he organized his table. The atomic mass is still one of the most prominent pieces of information provided for each element on the periodic table. The table gives the relative atomic masses of the eight elements that you observed in **Activity 2**.

The Law of Definite Proportions

Chemists at the beginning of the 19th century noted that eight pounds of oxygen always reacted with one pound of hydrogen to form nine pounds of water. This observation is an example of **The Law of Definite Proportions**. This law, first articulated by Joseph Proust in 1799, states that whenever two elements combine to form a compound, they do so in a definite proportion by mass. Proust based this statement on his observations that 100 pounds of copper, dissolved in nitric acid and precipitated by carbonates of soda (sodium) or potash (potassium), invariably gave 180 pounds of green carbonate. The Law of Definite Proportions is not direct proof of the existence of atoms. However, if you believe in the existence of atoms it does make it easier to explain why the Law of Definite Proportions should hold. The existence of atoms can also help explain why a given mass of aluminum reacting with sufficient copper (II) chloride in solution should always produce the same mass of copper.

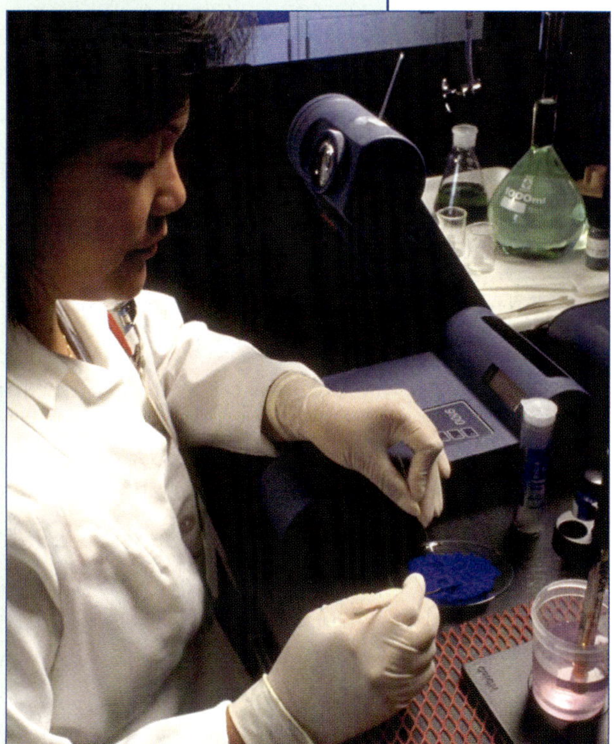

Checking Up

1. What is the difference between an element and a compound?
2. What is an atom?
3. How is an atomic mass unit defined?
4. How can the existence of atoms help to explain the Law of Definite Proportions?

Reflecting on the Activity and the Challenge

In this activity you learned how chemists measured elements in chemical reactions to determine the relative masses of atoms and how these masses were assembled into a scale of atomic masses. The atomic mass is one of the most important pieces of information listed for each element on the periodic table. How will you incorporate it into your game about the periodic table? Will you test players' ability to simply identify the atomic mass from a periodic table, or will you require that players understand how the relative scale was determined?

Chemistry to Go

1. John Dalton believed that water was formed from the simplest combination of hydrogen and oxygen atoms—one of each. Observations today show that eight pounds of oxygen react with one pound of hydrogen to form water. We now know that Dalton's values were wrong but his attempt was based on the data that he had available at the time.

   a) Based on these two statements, what conclusion could Dalton draw about the relative masses of oxygen and hydrogen atoms? How many times more massive is an oxygen atom than a hydrogen atom?

   b) The atomic mass of oxygen is 16 and the atomic mass of hydrogen is 1. How do the current atomic masses of oxygen and hydrogen compare to Dalton's?

   c) You know that water molecules are not made from one atom of hydrogen and one atom of oxygen. A water molecule is made up of 2 atoms of hydrogen and one atom of oxygen. If the pound of hydrogen reacting with eight pounds of oxygen is due to twice as many hydrogen atoms as there are oxygen atoms, how many times more massive is an oxygen atom than a hydrogen atom?

   d) Are the values of these revised masses closer to the current atomic masses of oxygen and hydrogen atoms?

2. In Dalton's time, ammonia was observed to be formed when nitrogen reacted with hydrogen. Today's values show that fourteen pounds of nitrogen react with three pounds of hydrogen.

   a) If ammonia were formed from Dalton's simplest formula of one atom of each element, what would he have concluded about the relative masses of nitrogen and hydrogen atoms?

   b) Ammonia molecules are made not from one atom each of hydrogen and nitrogen but from three atoms of hydrogen and one atom of nitrogen. If the three pounds of hydrogen reacting with fourteen pounds of nitrogen is due to three times as many hydrogen atoms as there are nitrogen atoms, how many times more massive is a nitrogen atom than a hydrogen atom?

3. A student uses magnesium instead of aluminum in this activity and obtains the following data:

Mass of beaker: 30.20 g

Mass of beaker + magnesium strip: 30.40 g

Mass of beaker + magnesium + copper (II) chloride: 32.40 g

Mass of beaker + magnesium + copper (II) chloride + water: 57.40 g

Mass of dry filter paper: 0.67 g

Mass of new beaker: 30.50 g

Mass of beaker + wet filter and residue + solution: 58.37 g

Mass of dry filter + residue: 1.19 g

a) How many grams of magnesium did the student use in this experiment?

b) How many grams of copper did the student measure in the dry filter paper?

c) How many times as great is the mass of the copper as the mass of magnesium the student originally used? (What is the ratio of mass of copper to mass of aluminum?)

d) If one atom of copper was released for each atom of magnesium that becomes combined (with chlorine), the masses of copper and magnesium determined in (a) and (b) should contain the same number of atoms. What does this tell you about the relative masses of copper and magnesium atoms? How many times more massive is a copper atom than a magnesium atom?

4. Look at the table in the **ChemTalk** reading section that gives the atomic masses of the eight elements that you observed in **Activity 2**. Add this data to your element cards. Can you now improve upon the way you sorted the cards in the previous activity taking this new information about relative masses into account?

Preparing for the Chapter Challenge

In a paragraph, explain how the relative scale of atomic masses is determined.

Inquiring Further

Avogadro's number and a mole

Chemists are interested in keeping track of quantities of particles. However, the particles are very small so chemists use a particular quantity that is convenient for counting particles. The quantity is called a mole. The quantity of particles in a mole is 602,000,000,000,000,000,000,000. The mole can be represented more easily in scientific notation as 6.02×10^{23}. This is a very large number because many, many small particles (atoms or molecules) make up a mole. The number 6.02×10^{23} is sometimes called Avogadro's number.

Research to find the significance of Avogadro's number and a mole. Record your findings in your *Active Chemistry* log.

Then, to appreciate how huge a mole is, answer the following question:

Imagine you have 7 billion people (7×10^9 people, which is approximately the human population of the world) and they are given the task of dropping $1-bills once every second into a large hole. How long will it take 7×10^9 people to drop 1 mole of dollar bills into the hole? How old will you be when they complete this task?

Activity 4 Are Atoms Indivisible?

GOALS

In this activity you will:

- Observe the behavior of a cathode ray in the presence of a magnet.
- Discuss Thomson's conclusions from 1897 about cathode rays.
- Simulate an experiment from 1911 by Rutherford in which he learned more about the structure of atoms.
- Organize your understanding of some of the different particles that comprise matter.

What Do You Think?

Ever since Democritus from ancient Greece hypothesized the existence of atoms, a major question was how atoms of different elements were different.

- If you could observe a single atom of gold and a single atom of lead, how do you think they would be different? How could they have something in common?

Record your ideas about these questions in your *Active Chemistry* log. Be prepared to discuss your responses with your small group and the class.

Investigate

1. Your teacher will demonstrate the behavior of what were called cathode rays a hundred years ago. They were called cathode rays because they were emitted from the negative terminal, or cathode of what was known as a cathode-ray tube, a forerunner of the television or the computer monitor tube.

a) What happens to the path of the cathode rays when a horseshoe magnet is placed near the tube? Record your observation in your *Active Chemistry* log.

b) Record what you think will happen to the path of the cathode rays when the orientation of the horseshoe magnet is reversed.

c) Observe the path of the cathode rays as your teacher reverses the magnet. Record what does happen.

Do not to try to see the effect of a magnet on an actual television or computer monitor; you may damage the monitor.

2. Magnets exert a force on moving electrically charged particles. The effect of the magnet on the cathode rays therefore shows that these rays are moving electrically charged particles. Cathode rays, which have a negative electric charge, are made up of electrons. In 1897, Joseph John (J. J.) Thomson showed that identical rays (electrons) were emitted from the cathode of a cathode-ray tube, regardless of the metal of which the cathode was made.

Discovery of electrons emerging from the atoms of the cathode gave scientists new information about the atom. The atom is not indivisible. It has internal parts, one of which is the electron.

a) In a sentence or two, describe the relationship between cathode rays, the electron, and the structure of atoms.

3. In order to investigate the other components of an atom, you will take part in the following simulation, similar to the game *Battleship*. You will work with a partner for this activity.

You and your classmate should each construct a grid of squares, 8 by 10. Without letting your classmate see your grid, color in a section of ten squares. The squares must touch each other. To make the simulation relatively simple, begin with a fairly compact design. This shape (colored region) represents your target.

You and your partner will try to guess the shape of each other's targets by sending "missiles" onto any of the 80 squares in this array. For the purpose of this description, designate one person to be Player X and the other person to be Player Y. To begin, Player X will tell Player Y the destination (number and letter) of

Cathode-ray tube with electrons being deflected by a magnetic field

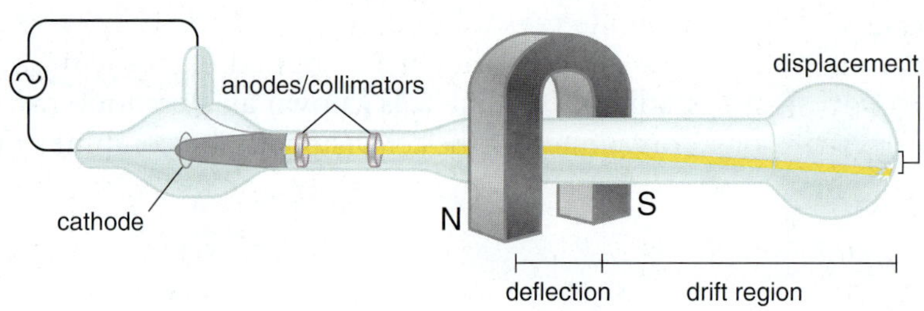

	A	B	C	D	E	F	G	H	I	J
1										
2										
3										
4										
5										
6										
7										
8										

the missile being sent. Player Y will respond, indicating that the missile "hit" or "missed" the target shape. Player X will make note of the response. Then Player Y sends the next missile, noting the response. Continue this process until one player identifies the other player's target.

a) Record the number of turns taken to complete the game.

b) Repeat the game with a target of only 2 adjacent squares. Record the number of turns taken.

4. Now do a thought experiment. The same-size game grid is divided into smaller squares—100 squares across and 100 squares down. There are now 10,000 squares in the same size board as before. A target of only one square is chosen.

   a) Record an estimate of how many turns will be required to identify the target square amongst the 10,000 squares in the game grid.

5. Now modify the thought experiment. The same-size grid is now 1000 rows across and 1000 squares down. That is a total of 1,000,000 squares.

   a) Record an estimate of how many turns will be required to identify the target square among the 1,000,000 squares in the game grid.

6. An experiment similar to your game of "*Battleship*" was carried out in 1911 by Lord Ernest Rutherford. Rutherford sought to learn something about the structure of the atom by bombarding gold atoms with energetic particles given off by certain atoms.

   In Rutherford's game of "*Battleship*," it seemed that he was required to send an incredible number of missiles to get a "hit." He concluded that the grid of the atom must be composed of tiny, tiny cells and only one cell contains all of the positive charge of the atom.

   a) Explain why you think he concluded this.

Chem Talk

THE CHANGING MODEL OF AN ATOM

J.J. Thomson's Model of an Atom

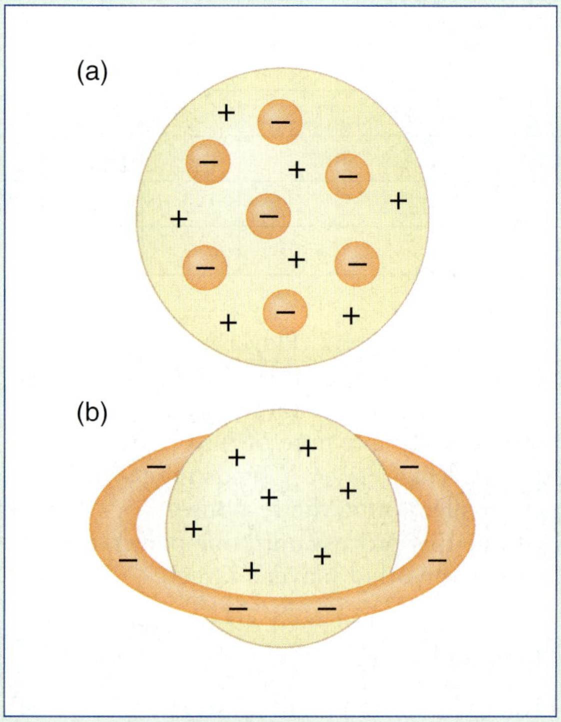

As you noted in this activity, in the late 1800s J.J. Thomson, an English physicist, found evidence for the existence of negatively charged particles that could be removed from atoms. He called these subatomic particles with a negative charge **electrons.** Using this new information, Thomson then proposed a model of an atom that was a positive sphere, with electrons evenly distributed and embedded in it, as shown in the diagram. Using the same evidence, H. Nagaoka, a Japanese scientist, modeled the atom as a large positively charged sphere surrounded by a ring of negative electrons.

Chem Words

electron: a negatively charged particle with a charge of 1.6×10^{-19} coulombs and a mass of 9.1×10^{-31} kg.

Rutherford's Discovery of the Nucleus

For several years there was no evidence to contradict either Thomson's or Nagaoka's atomic models. However, in the early 1900s, Ernest Rutherford, a New Zealand-born scientist, designed experiments to test the current model of an atom. In Rutherford's experiment, alpha particles were sent as "missiles" toward a thin sheet of gold. Gold was used because it is malleable and could be hammered into a thin, thin sheet. Most of the alpha particles went through the sheet and were not deflected. It is as if they missed the target. This was expected since it was assumed that the atom's charge and mass

Ernest Rutherford

was spread evenly throughout the gold. Occasionally one of the alpha particles that "hit" the gold sheet bounced back. This was the big surprise. The conclusion: there must be tiny places containing lots of charge and mass. Since the bouncing back was so unusual, it was assumed that the places where all the charge and mass were concentrated were only 1/100,000 of the area of the gold. Rutherford concluded that almost all the mass and all of the positive charge of the atom is concentrated in an extremely small part at the center, which he called the **nucleus**. He also coined the term **proton** to name the smallest unit of positive charge in the nucleus.

The story of Rutherford's discovery of the atomic nucleus is best told by Rutherford himself. Examining the deflection of high-speed alpha particles as they passed through sheets of gold foil, Rutherford and his student Hans Geiger noticed that some particles were scattered through larger angles than predicted by the existing theory of atomic structure. Fascinated, Rutherford asked Geiger's research student Ernest Marsden to search for more large-angle alpha scattering. Rutherford did not think that any of the alpha particles in his experiment would actually bounce backward. "Then I remember two or three days later Geiger coming to me in great excitement and saying, 'We have been able to get some of the alpha particles coming backwards ... ' It was quite the most incredible event that has ever happened to me in my life. It was almost as incredible as if you fired a 15-inch shell at a piece of tissue paper and it came back and hit you."

Chem Words

nucleus: the very dense core of the atom that contains the neutrons and protons.

proton: a positively charged subatomic particle contained in the nucleus of an atom. The mass of a proton is 1.673×10^{-24}g and it has a charge of +1.

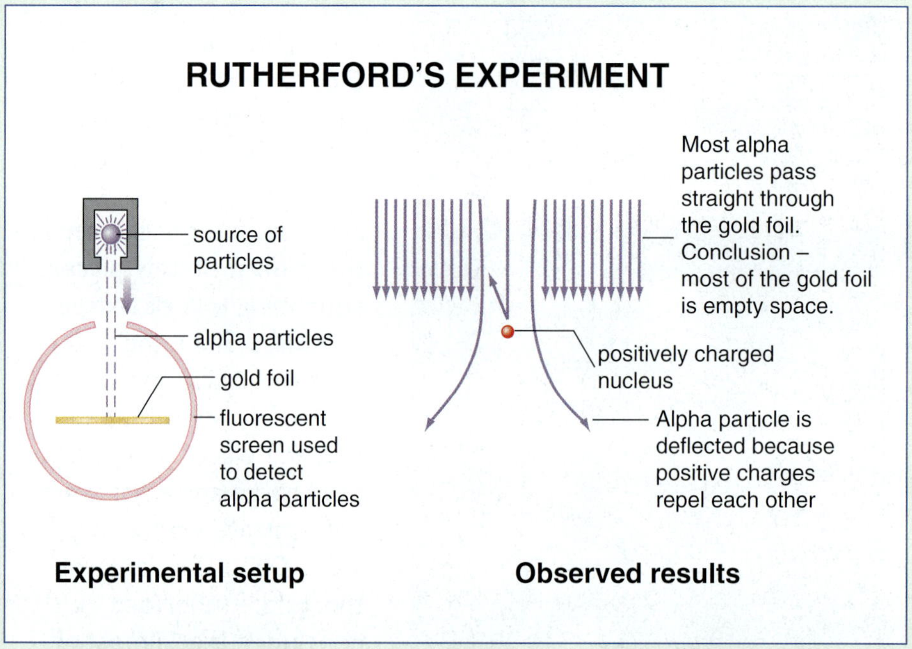

Checking Up

1. What is an electron?
2. Thomson's model of an atom is sometimes referred to as the "plum-pudding" model. (A plum pudding is a heavy pudding with raisins mixed into it.) Explain why this is an appropriate comparison.
3. Why was Rutherford surprised that some alpha particles bounced back from the gold foil?
4. What is the nucleus of an atom?

A Physics Connection

What was responsible for the wide-angle scattering of the alpha particles and their bouncing back? The force between the positive nucleus and the positive alpha particle is the coulomb force. Positive charges repel one another according to the coulomb force law.

$$F = \frac{kq_1q_2}{d^2}$$

where k is Coulomb's constant
($k = 9.0 \times 10^9$ N m^2/C^2),
q is the charge in coulombs, and
d is the distance between the charges.

The closer the alpha particle gets to the nucleus, the larger the force and the larger the deflection of the alpha particle.

Reflecting on the Activity and the Challenge

In this activity you learned of evidence that atoms are made of a positively charged nucleus and negatively charged electrons. The nucleus contains most of the atom's mass and its positive charge is balanced by the combined negative charge of the electrons, resulting in an atom that is electrically neutral. The number of protons of the neutral atom plays a very important role in the periodic table. Called the atomic number, it supports the order in which Mendeleev arranged the elements in his periodic table, long before anything was known about the structure of the atom or atomic number. How will your game reflect your new knowledge about atomic structure and its relationship to the periodic table?

Chemistry to Go

1. Since the electron has a negative electric charge and the nucleus has a positive electric charge, where would you expect to find electrons in atoms?
2. Are atoms indivisible? Support your answer using information from this activity.
3. Construct a chart or diagram to summarize what you have learned in this activity about the particles that make up an atom. Include electric charge and location of the particles.
4. Lead has an atomic number of 82; iron has an atomic number of 26; and copper has an atomic number of 29. How do the charges of the nuclei of these three elements compare?
5. The element chlorine has an atomic number of 17. How many electrons does chlorine have? Support your answer with a logical explanation of how you could arrive at this answer.
6. Sketch the outline of three grids. Pretend that each grid has 100,000 squares.
   a) If the target was 50,000 squares, draw the target.
   b) If the target was 25,000 squares, draw the target.
   c) If the target was only 1 square, draw the target.

   Which grid most closely relates to Rutherford's experiment? Explain your answer.

Inquiring Further

1. An atomic timeline

Construct a timeline that reflects how scientists' views of the atom have changed through the ages. Identify significant scientists, their beliefs, and experimental findings as mentioned in this chapter. You may also wish to consult other resources. Add information to your timeline as you continue to work through this chapter.

2. John Dalton's Atomic Theory

John Dalton, an English scientist, developed his atomic theory in the early 1800s. This theory was based on the Greek concept of atoms and the studies of Joseph Proust's Law of Definite Proportions or Law of Constant Composition. Dalton's atomic theory contained a series of postulates based on the data of his time and his observations:

- Matter consists of small particles called atoms.
- Atoms of one particular element are identical and the properties are identical.
- Atoms are indestructible. In chemical reactions, the atoms rearrange or combine, but they are not destroyed.
- Atoms of different elements have a different set of properties.
- When atoms of different elements combine to form compounds, they combine in a fixed numerical ratio.

From his postulates, the Law of Conservation of Mass would be supported. Since his postulates state that atoms cannot be destroyed but they can be moved around and combine with other atoms to form compounds, then the mass of the compound must be the sum of the atoms of the compound. This law still exists with a slight modification for nuclear reactions. So, you can conclude that if water has a certain mass today, it will have the same mass any other day (unless evaporation occurs).

Investigate whether all of Dalton's postulates are presently accepted or describe how some have been modified based on our current understanding.

Joseph Proust

Activity 5 The Chemical Behavior of Atoms

GOALS

In this activity you will:

- View the spectrum of hydrogen.
- Interpret changes in electron energies in the hydrogen atom to develop an explanation for where the colored light in the hydrogen spectrum comes from.
- Use Bohr's model of the atom to predict parts of the hydrogen atom spectrum.
- Calculate and compare the wavelengths, energies, and frequencies of light of different colors.
- Identify regions in the electromagnetic spectrum.

What Do You Think?

A neon sign uses electricity and a gas-filled tube to produce a colored light. A fluorescent bulb uses electricity, a gas-filled tube and a phosphor coating to produce white light.

- How is the color produced in a neon sign?

Record your ideas about this question in your *Active Chemistry* log. Be prepared to discuss your response with your small group and the class.

Investigate

1. In order to directly observe the behavior of atoms, you can observe the spectrum of light given off when atoms are excited by a high-voltage, electric-power supply. You probably have already seen this effect in the familiar red color of neon signs.

   For this demonstration, your teacher will set up a tube of hydrogen gas connected to a high-voltage power supply. This light can be viewed through a spectroscope or a diffraction grating lens, as shown in the diagram on the next page. When the slit at the end of the spectroscope is aimed toward the light, the colors of the spectrum appear separately off to the sides of the slit.

a) What colors do you see in the spectrum of light given off by hydrogen gas?

b) Make a colored diagram in your *Active Chemistry* log of what you see inside the spectroscope. Make sure to draw the colors with the order and spacing between them that you observe.

2. When you observed the spectrum of light given off by hydrogen gas, you probably saw three or four distinct lines, each having a different color. The color of light is determined by its frequency; the greater the energy in the light, the greater the frequency.

The diagram on the right shows the visible lines of the line spectrum of hydrogen.

The wavelengths are measured in nanometers (nm). The prefix *nano* means 10^{-9}. Thus, 656.5 nm is 656.5×10^{-9} m. The frequency of each line can be determined using the equation:

$$f = \frac{c}{\lambda}$$

where,

f is the frequency (waves/s or cycles/s or hertz),

c is the speed of light and is a constant (2.998×10^8m/s), and

λ is the wavelength measured in nanometers.

Example:

Given red light with a wavelength $\lambda = 656.5 \times 10^{-9}$ m and the speed of light $c = 2.998 \times 10^8$ m/s, calculate the corresponding frequency and energy of the light.

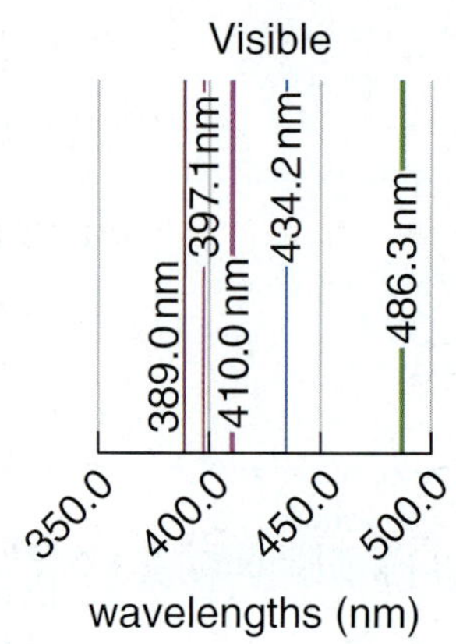

First calculate the frequency.

$$f = \frac{c}{\lambda}$$

$$= \frac{2.998 \times 10^{8} \text{ m/s}}{656.5 \times 10^{-9} \text{ m}}$$

$$= 4.567 \times 10^{14} \text{ cycles/s (or Hz)}$$

$$= 4.567 \times 10^{14} \text{ Hz (Hertz)}$$

To calculate the corresponding energy of a photon of light, you can use an equation called Planck's equation that relates the frequency to the energy. A photon is a tiny, indivisible packet of light. Planck's constant is h and has the value of 6.63×10^{-34} J•s.

$$E = hf$$

$$= (6.63 \times 10^{-34} \text{ J}\cdot\text{s})(4.567 \times 10^{14} \text{ cycles/s})$$

$$= 3.03 \times 10^{-19} \text{ J}$$

a) Calculate the frequencies and energies for the other visible wavelengths of light in the hydrogen spectra. Complete a table like the one below in your *Active Chemistry* log.

3. In 1913, Niels Bohr, a Danish physicist, tried to explain the line spectrum of hydrogen by hypothesizing that the electron in the hydrogen atom is allowed to have only certain amounts of energy. Bohr believed that an electron in a higher energy level would give off light when it fell to a lower energy level, and the amount of energy in the light would be the difference in these energy levels. Bohr pictured the atom to be similar to the solar system, with electrons orbiting around the nucleus in the way that planets orbit around the Sun. If you picture the electron to be in orbit about the nucleus of the hydrogen atom, as Bohr did, allowing the electron to have only certain amounts of energy would mean that the electron could be allowed in orbits of only certain distance from the nucleus. As the electron "jumps" from one energy level to another, it behaves something like a ball falling down a flight of stairs. It's allowed to rest only on one of the steps, nowhere in between.

The light was the result of electron energy jumps. These jumps should be able to produce the corresponding energies of the light. Bohr calculated the lowest energy level to be -2.18×10^{-18} J. Each of the other

Wavelength λ (nm)	**Frequency f (cycles/s or Hz)**	**Energy E**
410.3		
434.2		
486.3		
656.5	4.567×10^{14}	3.03×10^{-19} J

Niels Bohr

energy levels can be computed from this first one with the equation:

$$E_n = \frac{1}{n^2} E_1$$

The second energy level:

$$E_2 = \frac{1}{2^2} E_1$$
$$= \frac{1}{4} (-2.18 \times 10^{-18}\ \text{J})$$
$$= -5.45 \times 10^{-19}\ \text{J}$$

The third energy level:

$$E_3 = \frac{1}{3^2} E_1$$
$$= \frac{1}{9} (-2.18 \times 10^{-18}\ \text{J})$$
$$= -2.42 \times 10^{-19}\ \text{J}$$

a) Calculate the energy levels for the fourth, fifth, and sixth levels and record all these values in your log.

4. In Bohr's model, the electrons jump from E_3 to E_2 and give off light with that energy.

$$E_3 - E_2 = (-2.42 \times 10^{-19}\ \text{J}) - (-5.45 \times 10^{-19}\ \text{J})$$
$$= 3.03 \times 10^{-19}\ \text{J}$$

This is equivalent to the energy of the red light. The model is successful, so far.

a) Calculate the energy of light when the electron of hydrogen jumps from:

E_4 to E_2

E_5 to E_2

E_6 to E_2

b) Do these energies correspond to the energies of the light that is emitted from hydrogen that you calculated earlier?

c) If the model is a truly good model, it should be able to predict something unknown. Bohr's model does just this. Calculate the energy of light when the electron of hydrogen jumps from E_2 to E_1.

The light corresponding to this transition was not observed because it is not in the visible range of light. It is in the ultraviolet spectrum. When an ultraviolet detector is used, the wavelength of the ultraviolet light is just as predicted by Bohr's model. The light emitted from transitions of E_3 to E_1 and E_4 to E_1 are also in the ultraviolet.

When the electron jumps from a higher energy level to a lower energy level, light is emitted. The light emitted from a neon sign is based on this phenomenon.

The opposite can also occur. If the electron absorbs light of just the right energy, the electron can jump up to a higher energy level. For instance, an electron in the E_2 energy level can absorb light of 3.02×10^{-19} J and jump up to the E_3 energy level. If enough energy is applied to the hydrogen atom, the electron can be totally removed. This is commonly called the ionization energy; for one electron of hydrogen, the ionization energy is 2.18×10^{-19} J. If the electron in the E_1 level absorbs light of energy 2.18×10^{-19} J, then its new energy will be 0 J. This means that it is no longer bound to the hydrogen nucleus. The atom is ionized (the electron has been removed).

Chem Talk

BOHR'S MODEL OF AN ATOM

The Electromagnetic Spectrum

Visible light is only one part of the **electromagnetic spectrum**. You've probably heard of some of the other parts including ultraviolet, infrared, x-rays, gamma rays, microwaves, and radio waves. As you demonstrated in calculations using Bohr's model, the light from some of the transitions is in the ultraviolet. Infrared light is also emitted as the electron jumps from E_4 to E_3 and E_5 to E_3 and other higher energy levels.

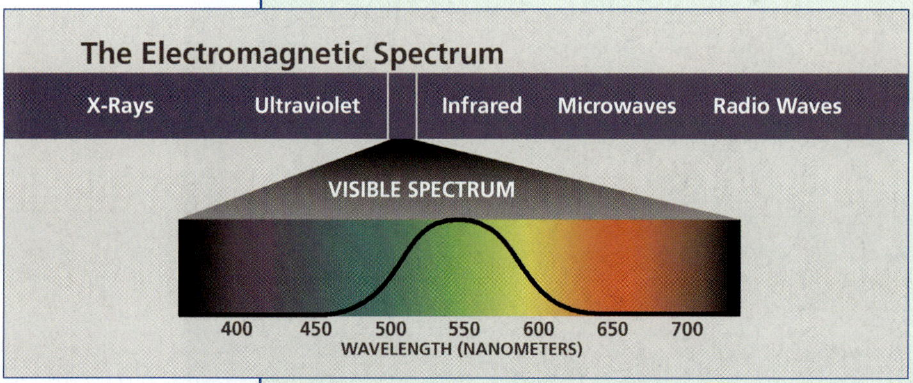

Many people do not think of radio waves as being light waves, but they are part of the electromagnetic spectrum. The problem is that you can only see a very small part of the electromagnetic spectrum. Often, you hear radio announcers say that they are broadcasting at a certain **frequency**. Your FM radio dial may have MHz (megahertz) printed on the side. This tells you that the numbers correspond to frequencies in units of MHz or 10^6 Hz. Frequency tells you the number of cycles or waves that are being produced per second. The unit for frequency is a hertz (Hz). 1 Hz = 1 cycle/s = 1 s^{-1}. Normally, frequency is read as per second and the cycles are dropped from the terminology.

Wavelength (λ), where λ is the Greek letter *lambda*, is the distance from crest to crest of a wave. All light waves travel at the same speed. The speed of electromagnetic radiation is constant and it is called the speed of light (c), $c = 2.998 \times 10^8$ m/s or 3.00×10^8 m/s. From this information you can calculate the frequency of light of a given wavelength. The equation that is used for this is:

$$f = \frac{c}{\lambda}$$

Chem Words

electromagnetic spectrum: the complete spectrum of electromagnetic radiation, such as radio waves, microwaves, infrared, visible, ultraviolet, x-rays, and gamma rays.

frequency: the number of waves per second or cycles per second or hertz (Hz).

wavelength: the distance measured from crest to crest of one complete wave or cycle.

As an example, if the wavelength is 434.2 nm, then the frequency is:

$$f = \frac{2.998 \times 10^8 \text{m/s}}{434.2 \times 10^{-9} \text{ m}}$$

$$= 6.905 \times 10^{14} \text{ cycles/s or } 6.905 \times 10^{14} \text{ Hz.}$$

As you go across the electromagnetic spectrum you should note that the wavelength continues to get smaller as the frequency increases. Also, you should understand that the energy of the spectrum increases as you go from radio waves to x-rays or gamma rays. Max Planck, a German physicist, found that the energy of a wavelength could be calculated. The equation that he developed was based on measuring the change in energy from one level to another level like you did in the **Investigate** section. The equation he developed is

$E = hf$,

where h is Planck's constant and is 6.63×10^{-34} J • s and f is the frequency.

The corresponding energy of the red light above would be:

$$E = hf$$

$$= (6.63 \times 10^{-34} \text{ J} \cdot \text{s})(4.567 \times 10^{14} \text{ Hz})$$

$$= 3.03 \times 10^{-19} \text{ J.}$$

So the next time that you are standing around a campfire, you can inform your fellow campers that red light has less energy than blue light and you can also tell them how to calculate these values.

Bohr's Atomic Theory

Niels Bohr, a brilliant Danish physicist, was aware that his theory of electron jumps had incredible success but also raised some problems.

Bohr proposed a "planetary" model of the atom. He theorized that electrons travel in nearly circular paths, called **orbits**, around the nucleus. Each electron orbit has a definite amount of energy, and the farther away the electron is from the nucleus, the greater is its energy. Bohr suggested the revolutionary idea that electrons "jump" between energy levels (orbits) in a quantum fashion. That is, they can never exist in an in-between state. Thus, when an atom absorbs or gives off energy (as in light or heat), the electron jumps to higher or lower orbits. Electrons are the most stable when they are at lower energy levels closer to the nucleus. Bohr's theory could only account for the spectrum of hydrogen and not for the spectra of any other element. Bohr's theory could not explain why only certain orbits were allowed, nor how the electron could jump from one orbit to another. Other scientists improved on Bohr's model as they discovered more about the atom and quantum mechanics.

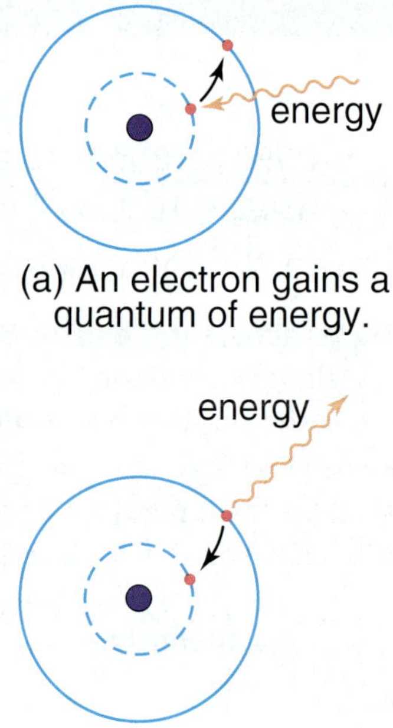

(a) An electron gains a quantum of energy.

(b) An electron loses a quantum of energy.

Chem Words

orbit: the path of the electron in its motion around the nucleus of Bohr's hydrogen atom.

Checking Up

1. How are visible light, ultraviolet light, infrared light, x-rays, gamma rays, microwaves, and radio waves related?
2. Explain the meaning of wavelength.
3. Why is "planetary" model an appropriate name for Bohr's model of the atom?
4. How do the energy levels of different electron orbits compare?
5. Why do elements produce certain color light when heated?

Reflecting on the Activity and the Challenge

In this activity, you've recognized that it is the electrons in an atom that are responsible for the colors of light emitted. Bohr explained the spectrum of light given off by excited hydrogen atoms by hypothesizing that each hydrogen atom's electron was allowed in only certain energy levels. Although Mendeleev knew nothing about electrons or energy levels when he first developed the periodic table (though he did live long enough to know of the electron's discovery by Thomson), the periodic table today is seen as a reflection of the number of electrons in an atom of each element and the energy levels those electrons occupy. Many versions of the periodic table today list the energy levels occupied by each electron in an atom of a given element. You will learn more about this in **Activity 6.** The periodic table is a way of organizing information and knowledge about the atom. Electron structure is a crucial part of that knowledge. Continue to think about how this information could be included in the game you are designing.

Chemistry to Go

1. In this activity you were told that light with greater energy has greater frequency.
   a) Which color in the visible spectrum of hydrogen has the greatest energy? The least energy?
   b) Which color in the visible spectrum of hydrogen has the highest frequency? The lowest frequency?
2. If an electron were to fall down to the E_1 level from the E_3 level, how would its energy compare to one that fell to the E_2 level? Explain.
3. What is the difference measured in energy when an electron falls from E_3 to E_1? How many times greater is this value as compared to the difference of the electron falling from E_3 to E_2?
4. A wavelength of light is 389.0 nm and its frequency is 7.707×10^{14} Hz. Show how this frequency value was calculated.
5. Show that the wavelength of 389.0 nm has an energy of 5.11×10^{-19} J.
6. a) Microwave radiation is absorbed by the water in food. As the heat is absorbed by the water it causes the food to get hot. If the λ of a microwave is 10 cm, calculate the frequency of the microwave and the energy of each photon.

b) The red light you observed in the hydrogen spectrum produced a $\lambda = 656.5$ nm. The energy of the red light was 3.02×10^{-19} J. How many times greater is this value when compared to the energy value that you found for the microwave energy?

Inquiring Further

Balmer, Lyman, and Paschen series

In this activity, you learned that the visible light emitted by the hydrogen lamp resulted when an electron moved from a higher energy level to the second lowest energy level in Bohr's scheme. This series of visible spectral lines is said to comprise the **Balmer series.**

What do you think happens when an electron in a hydrogen atom moves from a higher energy level to the lowest level? This series is known as the **Lyman series** (ultraviolet light).

What if a transition from a higher energy level leads to an electron in the third lowest energy level in Bohr's scheme? How do these energy levels compare with visible light? This series is known as the **Paschen series** (infrared light).

Activity 6 Atoms with More than One Electron

GOALS

In this activity you will:

- View the spectra of various materials.
- Graphically analyze patterns in the amounts of energy required to remove electrons from different kinds of atoms.
- Compare trends in stability of atoms in the periodic table.
- Compare the structure of the periodic table with the patterns of levels and sublevels to which electrons can be assigned.
- Develop a shorthand notation to describe the configuration of electrons in an atom.

What Do You Think?

In **Activity 5** you learned that Niels Bohr was able to explain the spectrum of light emitted by hydrogen using a model that assigned the electron to specific energy levels. Bohr was awarded a Nobel Prize in 1922 for expanding the understanding of atomic structure. He worked with hydrogen, the simplest atom, which contains only one electron. However, the atoms of other elements contain more than one electron.

- How do you think an increase in the number of electrons would impact the spectrum of an atom?
- What modifications in Bohr's model would need to take place to accommodate the extra electrons?

Record your ideas about these questions in your *Active Chemistry* log. Be prepared to discuss your responses with your small group and the class.

Investigate

1. In **Activity 5** you observed the spectrum of hydrogen gas as its electron moved from a higher energy level to a lower energy level. You also explored a model that used Bohr's theory to explain this spectrum. Now it's time to look at the spectra of some other elements.

   a) Your teacher will connect a tube containing an element other than hydrogen to a high voltage supply. Record the name of the element in your *Active Chemistry* log. Look at the spectrum of light of this element through the spectroscope.

   b) What colors do you see? Make a diagram in your log of the spectrum (pattern of colors) you see inside the spectroscope.

   c) Record how this spectrum is similar to and different from the hydrogen spectrum you observed in **Activity 5**.

   d) Repeat **Steps** (**a**), (**b**), and (**c**) for as many samples as your teacher demonstrates.

2. Although the spectra of such elements as helium and neon are very beautiful, they cannot be explained by Bohr's simple theory for the single electron in the hydrogen atom. The basic idea is still true—light is emitted when electrons jump from a higher energy level to a lower energy level. The energy levels, however, are more complex if there are additional electrons. A more elaborate labeling of electron energy levels is necessary. In this activity you will explore the pattern of electron energy levels in

Atomic Number	Element (Symbol)	1st Ionization Energy J ($\times 10^{-19}$)	2nd Ionization Energy J ($\times 10^{-19}$)
1	H	21.8	
2	He	39.4	87.2
3	Li	8.6	121.2
4	Be	14.9	29.2
5	B	13.3	40.3
6	C	18.0	39.1
7	N	23.3	47.4
8	O	21.8	56.3
9	F	27.9	56.0
10	Ne	34.6	65.6
11	Na	8.2	75.8
12	Mg	12.3	24.1
13	Al	9.6	30.2
14	Si	13.1	26.2
15	P	16.8	31.7
16	S	16.6	37.4
17	Cl	20.8	38.2
18	Ar	25.2	44.3
19	K	7.0	50.7
20	Ca	9.8	19.0
21	Sc	10.5	20.5
22	Ti	10.9	21.8
23	V	10.8	23.5
24	Cr	10.8	26.4
25	Mn	11.9	25.1
26	Fe	12.7	25.9
27	Co	12.6	27.3
28	Ni	12.2	29.1
29	Cu	12.4	32.5
30	Zn	15.1	28.8
31	Ga	9.6	32.9
32	Ge	12.7	25.5
33	As	15.7	29.9
34	Se	15.6	34.0
35	Br	18.9	34.9
36	Kr	22.4	39.0

atoms containing more than one electron.

When multiple electrons are present, some are easier (i.e., require less energy) to remove from the atoms than others. The chart on the left provides information about the amount of energy required to remove the electrons in the two highest energy levels. These are the electrons that are easiest to remove. These energies are called the 1st and 2nd ionization energies, and are given in units of joules. Notice that all values are multiplied by 10^{-19}.

a) Make a graph that shows how the ionization energies vary with atomic number. Since the atomic numbers range from 1 to 36, label the x-axis with atomic numbers from 1 to 36. Since the ionization energies range from 7 to 122, label the y-axis with ionization energies from 0 to 130. Plot the first ionization energy data from the chart in one color, connecting the data points as you go along.

b) Plot the values for the second ionization energies in a different color.

c) Include a title and legend on your graph.

3. Look at the graph of the first ionization energies and answer the following questions:

a) What kinds of patterns do you see? How could you quickly relate the shape of the graph to someone who had not seen it?

b) Where are the ionization energies the largest? The smallest?

c) What happens to the ionization energies as the atomic number increases?

d) Group the elements by their ionization energies into four consecutive "periods." List the range of atomic numbers in each group.

e) Is there any interruption in the general trend of ionization energies as the atomic number increases for a "period"? If so, describe it.

4. Look at the second colored graph line you drew.

a) Describe how the two graphs are alike and/or different. Do you see similarities between the two graphs?

5. If a large amount of energy is needed to remove an electron from an atom, the arrangement of electrons in that atom is considered to be especially stable. Thus, a high first ionization energy means that a lot of energy must be supplied to remove an electron from an atom and that the electron arrangement in that atom is especially stable. Any element that has a larger first ionization energy than its neighboring elements has an electron arrangement in its atoms that is more stable than its neighboring elements.

a) Which element in the first period (atomic numbers 1 and 2) has the most stable arrangements of electrons in its atoms? (Remember, you are looking for elements that have larger ionization energies than their neighbors. In reality you are looking for peaks in your graph, not just those elements with higher values.)

b) Which elements in the second period (atomic numbers 3 through 10) of the periodic table have the most stable arrangements of electrons in their atoms?

c) Which elements in the third period (atomic numbers 11 through 18) of the periodic table have the most stable arrangements of electrons in their atoms?

d) Which elements in the fourth period (atomic numbers 19 through 36) of the periodic table have the most stable arrangements of electrons in their atoms?

6. As mentioned earlier, the Bohr model was not able to account for the spectrum of an element containing more than one electron. A more elaborate model was needed. In this new model, the energy levels are broken down into sublevels. When these sublevels are filled, the atom exhibits a higher degree of stability. In this model, the sublevels are designated by the four letters *s*, *p*, *d*, and *f* .

The periodic table shows the atomic number, the chemical symbol, and how many electrons in an atom of each element are in each sublevel. The total number of electrons is equal to the atomic number of the element. This is because the atoms are neutral and therefore have a number of electrons equivalent to the number of protons. This arrangement of the electrons in each sublevel will be referred to as the electron assignment or electron configuration of the element. Use this periodic table to answer the following questions:

a) In what sublevel (include number and letter) are the electrons in hydrogen (1 electron) and helium (2 electrons) found?

Periodic Table of the Elements

KEY: 1 – Atomic Number; 1.0 – Electronegativity; 1 – Oxidation Number; H – Symbol; 1.00794 – Average Atomic Mass; $1s^1$ – Electron Configuration; Hydrogen – Name

Gases at room temperature
Liquids at room temperature
Solids at room temperature
Metals
Metalloids
Nonmetals

PERIOD \ GROUP	1	2	3	4	5	6	7	8	9	10	11	12
	IA/1A Alkali Metals	IIA/2A Alkaline Earth Metals	IIIB/3B	IVB/4B	VB/5B	VIB/6B	VIIB/7B	VIIIB/8B	VIIIB/8B	VIIIB/8B	IB/1B	IIB/2B
			Transition Metals									
1	1 2.1 1 H 1.00794 $1s^1$ Hydrogen											
2	3 1.0 1 Li 6.941 $1s^2 2s^1$ Lithium	4 1.5 2 Be 9.012182 $1s^2 2s^1$ Beryllium										
3	11 0.9 1 Na 22.98977 $[Ne]3s^1$ Sodium	12 1.2 2 Mg 24.3050 $[Ne]3s^2$ Magnesium										
4	19 0.8 1 K 39.0983 $[Ar]4s^1$ Potassium	20 1.0 2 Ca 40.078 $[Ar]4s^2$ Calcium	21 1.3 3 Sc 44.95591 $[Ar]4s^2 3d^1$ Scandium	22 1.5 3,4 Ti 47.867 $[Ar]4s^2 3d^2$ Titanium	23 1.6 2,3,4,5 V 50.9415 $[Ar]4s^2 3d^3$ Vanadium	24 1.6 2,3,6 Cr 51.9961 $[Ar]4s^2 3d^4$ Chromium	25 1.5 2,3,4,6,7 Mn 54.93805 $[Ar]4s^2 3d^5$ Manganese	26 1.8 2,3 Fe 55.847 $[Ar]4s^2 3d^6$ Iron	27 1.9 2,3 Co 58.93320 $[Ar]4s^2 3d^7$ Cobalt	28 1.9 2,3 Ni 58.6934 $[Ar]4s^2 3d^8$ Nickel	29 1.9 1,2 Cu 63.546 $[Ar]4s^2 3d^9$ Copper	30 1.6 2 Zn 65.39 $[Ar]4s^2 3d^{10}$ Zinc

As you move to the second period (second row on the periodic table) each new element has one more proton in its nucleus and one more electron. The electrons must find a place to reside — an energy level and a sublevel within that energy level.

As you move along in the periodic table to increasing atomic numbers, you see that the additional electrons fill the sublevel. A completed sublevel is one that is holding the maximum number of electrons allowed to it before electrons must be placed in the next higher sublevel.

b) In what region of the periodic table are electrons added in an *s* sublevel? What is the greatest number of electrons found in any *s* sublevel?

c) In what region of the periodic table are electrons added in a *p* sublevel? What is the greatest number of electrons found in any *p* sublevel?

d) In what region of the periodic table are electrons added in a *d* sublevel? What is the greatest number of electrons found in any *d* sublevel?

e) In what region of the periodic table are electrons added to an *f* sublevel? What is the greatest number of electrons found in any *f* sublevel?

f) Select a column in the periodic table. (A column of elements on the periodic table is called a family or group.) Look at the electron configuration for each element within the column. Take special note of the last entry, the sublevel to which the last electron in an atom of each element in that column is added. What do all of these sublevels have in common? How many electrons are in these particular sublevels?

g) Mendeleev assigned elements to the same column of the periodic table because the elements had similar properties, both physical and chemical. How, then, does the number and location of the electrons in the outermost sublevel relate to chemical properties? We can now acknowledge that electrons (as opposed to the nucleus) are the key to the chemical properties of elements.

13	14	15	16	17	18
					VIIIA/8A or 0 Noble Gases
IIIA/3A	IVA/4A	VA/5A	VIA/6A Chalcogens	VIIA/7A Halides	2 He 4.002602 $1s^2$ Helium
5 2.0 3 B 10.811 $1s^2 2s^2 2p^1$ Boron	6 2.5 -4,2,4 C 12.011 $1s^2 2s^2 2p^2$ Carbon	7 3.0 -3,2,3,4,5 N 14.00674 $1s^2 2s^2 2p^3$ Nitrogen	8 3.5 -2 O 15.9994 $1s^2 2s^2 2p^4$ Oxygen	9 4.0 -1 F 18.998403 $1s^2 2s^2 2p^5$ Fluorine	10 Ne 20.1797 $1s^2 2s^2 2p^6$ Neon
13 1.5 3 Al 26.981539 $[Ne]3s^2 3p^1$ Aluminum	14 1.8 2,4 Si 28.0855 $[Ne]3s^2 3p^2$ Silicon	15 2.1 -3,3,4,5 P 30.973762 $[Ne]3s^2 3p^3$ Phosphorus	16 2.5 -2,2,4,6 S 32.066 $[Ne]3s^2 3p^4$ Sulfur	17 3.0 -1,1,3,5,7 Cl 35.4527 $[Ne]3s^2 3p^5$ Chlorine	18 Ar 39.948 $[Ne]3s^2 3p^6$ Argon
31 1.6 3 Ga 69.723 $[Ar]4s^2 3d^{10} 4p^1$ Gallium	32 1.8 4 Ge 72.61 $[Ar]4s^2 3d^{10} 4p^2$ Germanium	33 2.0 -3,3,5 As 74.92159 $[Ar]4s^2 3d^{10} 4p^3$ Arsenic	34 2.4 -2,4,6 Se 78.96 $[Ar]4s^2 3d^{10} 4p^4$ Selenium	35 2.8 -1,1,5,7 Br 79.904 $[Ar]4s^2 3d^{10} 4p^5$ Bromine	36 Kr 83.80 $[Ar]4s^2 3d^{10} 4p^6$ Krypton

Chem Talk

A PERIODIC TABLE REVEALED

Ions and Ionization Energy

In the table in the **Investigate** section the amount of energy required to remove an electron from an atom was called **ionization energy**. Atoms are neutral. That is, the number of electrons is equivalent to the number of protons. However, atoms can gain or lose electrons. Atoms that have lost or gained electrons are called **ions** and thus the energy used to remove the electrons is known as the ionization energy. The energy required to remove a single electron from the highest occupied energy level is called the first ionization energy, and the energy needed to remove a second electron from the same atom, after the first one has already been removed, is called the second ionization energy.

Chem Words

ionization energy: the energy required to remove an electron from a gaseous atom at ground state.

ion: an electrically charged atom or group of atoms that has acquired a net charge, either negative or positive.

electron configuration: the distribution of electrons in an atom's energy levels.

Electron Configuration and Energy Levels

As you discovered, the Bohr model was not able to account for the spectrum of an element containing more than one electron. In the new model you investigated, the energy levels are broken down into sublevels. This arrangement of the electrons in each sublevel is called the electron assignment or **electron configuration** of the element. When these sublevels are filled, the atom exhibits a higher degree of stability. The sublevels are designated by the four letters *s*, *p*, *d*, and *f*. The letters come from the words, *sharp*, *principal*, *diffuse*, and *fundamental*. The early scientists used these words to describe some of the observed features of the line spectra. They are governed by the following rules:

(i) The first energy level (corresponding to E_1 in **Activity 5**) has only one type of orbital, labeled 1*s*, where 1 identifies the energy level and *s* identifies the orbital.

(ii) The second energy level (corresponding to E_2 in **Activity 5**) has two types of orbitals (an *s* orbital and three *p* orbitals) and are labeled as the 2*s* and 2*p* orbitals.

(iii) The third energy level (corresponding to E_3 in **Activity 5**) has three types of orbitals, (an *s* orbital, three *p* orbitals, and five *d* orbitals) and are labeled as the 3*s*, 3*p*, and 3*d* orbitals.

(iv) The number of orbitals corresponds to the energy level you are considering. For example: E_4 has four types of orbitals (*s*, *p*, *d*, and *f*); E_5 has five types of orbitals (*s*, *p*, *d*, *f*, and *g*).

(v) The maximum number of electrons that can be contained in an orbital is two. Three *p* orbitals could contain a maximum of six electrons. The number of the type of electrons is indicated by superscript following the orbital designation. For example, $2p^5$ means five electrons in the $2p$ orbitals.

Stability is an important feature for all matter. Remember the excited electron of the hydrogen atom? If the electron were in energy level 3, it would drop down to energy level 2 and give off a specific wavelength of light. Alternatively, the electron in energy level 3, could drop down to energy level 1 and give off a different, specific wavelength of light. The word "excited" is used to describe an electron that has been promoted to a higher energy level, before it falls back down to its original state. The electron in the excited state was unstable and lost energy to go to a more stable form. Particles arranged in an unstable way will move to a more stable arrangement.

The Periodic Table

In previous activities you tried to organize elements by their properties and then by their atomic number. When elements are arranged according to their atomic numbers a pattern emerges in which similar properties occur regularly. This is the periodic law. The horizontal rows of elements in the periodic table are called **periods**. The set of elements in the same vertical column in the periodic table is called a **chemical group**. As you discovered, elements in a group share similar physical and chemical properties. They also form similar kinds of compounds when they combine with other elements. This behavior is due to the fact that elements in one chemical group have the same number of electrons in their outer energy levels and tend to form ions by gaining or losing the same number of electrons.

Chem Words

period: a horizontal row of elements in the periodic table.

chemical group: a family of elements in the periodic table that have similar electron configurations.

Checking Up

1. What is an ion?
2. What is ionization energy?
3. What are the horizontal rows of the periodic table called?
4. Explain the term chemical group.
5. Name three elements in a chemical group.

Reflecting on the Activity and the Challenge

In this activity you learned that electrons in atoms are assigned not only to energy levels but also to sublevels, labeled *s*, *p*, *d*, and *f*. You have also learned that the electron configuration of atoms of all elements in the same column of the periodic table end with the same sublevel and number of electrons in that sublevel. Mendeleev organized elements into columns based on similar chemical properties. Thus, electron energy sublevels are clearly associated with chemical properties of elements and their position on the periodic table. How will you incorporate the information about electron configuration in your game to meet the **Chapter Challenge**?

Chemistry to Go

1. Write the complete sequence of electron energy levels, from 1*s* to 4*f*.
2. Consider the element boron (B) as an example.
   a) What is boron's atomic number?
   b) How many electrons does boron have?
   c) What is the complete electron sequence for boron? (Be sure to include the number and letter of the appropriate sublevels, as well as the number of electrons in each sublevel.)
3. Answer the following questions for the element zinc:
   a) What is zinc's atomic number?
   b) How many electrons does zinc have?
   c) What is the complete electron sequence for zinc? (Be sure to include the number and letter of the appropriate sublevels, as well as the number of electrons in each sublevel.)
   d) What is the last sublevel (number and letter, please) to which electrons are added? How many electrons are in this sublevel?
   e) Where would you expect to find zinc on the periodic table? Support your prediction with your answers from (d).
   f) What other elements might you expect to have chemical properties similar to zinc? Explain your choices.
4. Answer the following questions for the element calcium:
   a) What is calcium's atomic number?
   b) How many electrons does calcium have?

c) What is the complete electron sequence for calcium? (Be sure to include the number and letter of the appropriate sublevels, as well as the number of electrons in each sublevel.)

d) What is the last sublevel (number and letter, please) to which electrons are added? How many electrons are in this sublevel?

e) Where would you expect to find calcium on the periodic table? Support your prediction with your answers from (d).

f) What other elements would you expect to have chemical properties similar to calcium? Explain your choices.

5. A chemist has synthesized a heavy element in the laboratory and found that it had an electron configuration:

$1s^2 2s^2 2p^6 3s^2 3p^6 4s^2 3d^{10} 4p^6 5s^2 4d^{10} 5p^6 6s^2 4f^{14} 5d^{10} 6p^6 7s^2 5f^{14} 6d^8$.

a) What is the number of electrons in this element?

b) What is the atomic number?

c) What might you predict about this element?

6. If the electron configuration is given you should be able to determine what element it is. Identify the following element from its electron configuration:

$1s^2 2s^2 2p^6 3s^2 3p^6 4s^2 3d^6$.

Preparing for the Chapter Challenge

Write a sentence or two to explain in words the pattern you noticed between any group and the electron configurations of the elements belonging to that group.

Inquiring Further

Determining electron configuration

In this activity, you were able to look at the electron configuration for a given element provided in the periodic table. Research other ways that the electron configuration can be determined.

Activity 7 How Electrons Determine Chemical Behavior

GOALS

In this activity you will:

- Investigate more patterns in the electron arrangements of atoms.
- Relate the positions of elements on the periodic table, their electron arrangements, and their distances from the nearest noble gas, to chemical properties of the elements.
- Relate electron arrangements to ionization energies.
- Assign valence numbers to elements and organize the periodic table according to valence numbers.

What Do You Think?

You have learned that electron configurations determine an atom's chemical behavior. You have also learned how these electrons are labeled according to a series of energy sublevels.

- How does the arrangement of electrons in an atom determine its chemical behavior?

Record your ideas about this question in your *Active Chemistry* log. Be prepared to discuss your responses with your small group and the class.

Investigate

1. In **Activity 6** you learned that elements with relatively high ionization energies have stable arrangements of electrons. One particular group of elements, located in the column at the extreme right of the table, exhibit high ionization energies and therefore have stable arrangements of electrons. They are called the noble gases.

Look at the periodic table on the inside back cover for the assignment of electrons to energy sublevels for atoms of each of these elements. Focus on the sublevel at the end of the electron arrangement where the last electron is assigned.

a) Make and complete a chart like the one below in your *Active Chemistry* log. An example has been provided for you.

	Column A	Column B	Column C	Column D
Element	Energy level (number) to which the last electron is assigned	Sublevel (letter) to which the last electron is assigned	Number of electrons in the sublevel to which the last electron is assigned	Total number of all electrons of the energy level in Column A
Helium				
Neon				
Argon	3	*p*	6	8
Krypton				
Xenon				
Radon				

b) Look at the numbers in Column A in your chart. How are these numbers related to the respective rows of the periodic table in which each of the elements is located?

c) What pattern do you notice in Columns B and C?

d) What pattern do you notice in Column D?

2. The chemical behavior of an element can be understood by looking at the electron assignment of an atom of the element as compared to the electron assignments of neighboring noble gas atoms. The chemical inactivity of noble gas atoms reflects the stable arrangement of their electrons, one which other atoms cannot easily disturb. In the following questions you will compare the electron assignments in atoms with those of noble gases.

a) Make and complete a chart like the following one in your *Active Chemistry* log. In this chart you will compare the electron assignments of lithium, beryllium, and boron to the electron assignment of helium ($1s^2$), the closest noble gas. An example has been provided for you.

Element being compared	Number of electrons more than those found in closest noble gas (He)	Energy level (number) to which the last electron is assigned	Energy sublevel (letter) to which the last electron is assigned	Location of element (row) in the periodic table	Location of element (column) in the periodic table
Lithium	*1*	2	s	*Row 2*	*Column 1*
Beryllium					
Boron					

b) Make and complete another chart like the one above in your *Active Chemistry* log. This time you will compare the electron assignments of sodium, magnesium, and aluminum to the electron assignment of neon ($1s^2 2s^2 2p^6$), the closest noble gas.

Element being compared	Number of electrons more than those found in closest noble gas (Ne)	Energy level (number) to which the last electron is assigned	Energy sublevel (letter) to which the last electron is assigned	Location of element (row) in the periodic table	Location of element (column) in the periodic table
Sodium					
Magnesium					
Aluminum					

c) Describe any patterns you notice in the charts in 2(a) and/or 2(b) above.

d) Make and complete a chart like the one at the top of the next page in your *Active Chemistry* log. In this chart you will compare the electron assignments of nitrogen, oxygen, and fluorine to the electron assignment of neon ($1s^2 2s^2 2p^6$), the closest noble gas. An example has been provided for you.

e) Make and complete another chart like the one at the top of the next page in your *Active Chemistry* log. This time you will compare the electron assignments of phosphorus, sulfur, and chlorine to the electron assignment of argon ($1s^2 2s^2 2p^6 3s^2 3p^6$), the closest noble gas.

Element being compared	Number of electrons more than those found in closest noble gas (Ne)	Energy level (number) to which the last electron is assigned	Energy sublevel (letter) to which the last electron is assigned	Location of element (row) in the periodic table	Location of element (column) in the periodic table
Nitrogen	*3*	*2*	*p*	*Row 2*	*Column 5*
Oxygen					
Fluorine					

Element being compared	Number of electrons more than those found in closest noble gas (Ar)	Energy level (number) to which the last electron is assigned	Energy sublevel (letter) to which the last electron is assigned	Location of element (row) in the periodic table	Location of element (column) in the periodic table
Phosphorus					
Sulfur					
Chlorine					

f) Describe any patterns you notice in the charts 2 (d) and/or 2(e) above.

3. In your *Active Chemistry* log, draw a simplified periodic table that contains only the first three rows of the periodic table. Your table should have 3 rows and 8 columns, and should contain the elements with atomic numbers 1 through 18.

a) Write the symbol for each element in each appropriate box.

b) In the columns headed by lithium, beryllium, and boron, indicate how many more electrons are found in atoms of those elements than in an atom of the nearest noble gas. Place a plus sign in front of these numbers to indicate that these elements contain more electrons than their nearest noble gas element.

c) In the columns headed by nitrogen, oxygen, and fluorine, indicate how many fewer electrons are found in atoms of those elements than in an atom of the nearest noble gas. Place a minus sign in front of these numbers to indicate that these elements contain fewer electrons than their nearest noble gas element.

d) The carbon column was not listed in the table. How many fewer electrons are found in atoms of these elements than in an atom of the nearest noble gas? How many more electrons are found in atoms of these elements than in an atom of the last noble gas? What can you conclude about the position of the carbon column in respect to the other columns that you have examined?

4. The questions in **Step 2** asked you to compare the electron assignments in atoms of the second and third rows with the electron assignments of the noble gas that was closest in atomic number to those elements. You noted that electrons are added to the *s* and *p* sublevels of the energy level whose number is the same as the row of the periodic table the elements are found in. For instance, lithium has one more electron than its closest noble gas element. That additional electron is added to the *s* sublevel of the second energy level, corresponding to the second row, where lithium is located.

   The number of an electron energy level is significant, because the higher the number, the greater the average distance between the nucleus and the electron. The electrons in the energy levels with the highest number are, on average, the farthest from the nucleus. Because differences in electrons located in the outermost level distinguish an atom from its nearest noble gas, these are the electrons responsible for the atom's chemical behavior. These electrons are often called valence electrons. The general definition of valence electrons are those electrons found in the outermost energy level *s* and *p* orbitals. The maximum number of valence electrons that you can have is 8 for the representative elements (2 in the *s* and 6 in *p* sublevel). For example, sodium contains one valence electron since its outermost level is 3 and sublevel is *s*. It does not have any electrons in the *p* sublevel. Bromine would have seven valence electrons, since its outermost energy level is 4 and it contains 2 electrons in *s* sublevel and 5 electrons in *p* sublevel. Use the assignment of electrons to energy sublevels in the periodic table on the inside back cover and your answers to the questions in **Steps 1** and **2** to answer the following questions:

   a) How many valence electrons are in an atom of helium? Neon? Argon? Krypton? Xenon? Radon?

   b) How many valence electrons are in an atom of lithium? Sodium? Potassium? Rubidium? Cesium? This family of elements is known as the alkali metals.

   c) How many valence electrons are in an atom of beryllium? Magnesium? Calcium? Strontium? Barium? This family of elements is known as the alkaline earth metals.

   d) How many valence electrons are in an atom of boron? Aluminum?

   e) How many valence electrons are in an atom of carbon? Silicon?

   f) How many valence electrons are in an atom of nitrogen? Phosphorus?

   g) How many valence electrons are in an atom of oxygen? Sulfur?

   h) How many valence electrons are in an atom of fluorine? Chlorine? Bromine? Iodine? This family of elements is known as the halogens.

Chem Talk

THE NOBLE GASES

The Discovery of Argon

Imagine that you prepared two samples of nitrogen, each one liter in volume, and found one to have a mass of 1.250 g (grams) and the other a mass of 1.257 g. You might be tempted to attribute the difference to experimental error. Lord Rayleigh didn't! In 1892 he decomposed ammonia to generate one liter of nitrogen with a mass of 1.250 g. In another preparation method, he isolated one liter of nitrogen with a mass of 1.257 g by removing what he thought were all the other gases from a sample of air.

What accounted for the difference in masses of the two liter samples? Could there be yet another gas in the air that Rayleigh didn't know about? William Ramsay, a colleague of Rayleigh, looked to the experiments conducted by Henry Cavendish a hundred years earlier. Henry Cavendish (the discoverer of hydrogen) had been puzzled by a small bubble of gas remaining after he had chemically absorbed all of a sample of nitrogen he had similarly extracted from the atmosphere.

As Cavendish had done, Ramsay extracted a sample of nitrogen from the atmosphere and then chemically absorbed all of the nitrogen in that sample. He looked at the spectrum of the remaining bubble of gas, just as you have looked at the spectra of various gases in **Activity 4** and **5**. The new spectral lines of color he saw showed him that the bubble of gas was a new element, which today we know as argon. It had previously escaped notice because of its rarity and lack of chemical activity.

A New Family

When Mendeleev formulated the first periodic table argon had not been discovered, and therefore it had not been placed in the periodic table. There was no obvious place for argon. A new column was created. A prediction was therefore made that there would be other elements with similar properties to argon. The other elements of this family (He, Ne, Kr) were subsequently discovered by the end of the 19th century.

Checking Up

1. Why did one liter of nitrogen prepared by Lord Rayleigh appear to have a greater mass than the other liter prepared by a different method?
2. What are two reasons that the noble gases had escaped notice?

Chem Words

noble gas (also rare or inert gas): a family of elements (Group 18 or VIIIA) of the periodic table.

Unlike atoms of the other chemical elements, atoms of elements in this column are so stable that they either do not react at all, or they react only in unusual circumstances, with other elements. For this reason, this family has been known as **rare** gases (because they are rare in abundance), **inert** gases (because they are not chemically reactive), or **noble** gases.

Reflecting on the Activity and the Challenge

In this activity you have learned that the electrons in the energy sublevels with the highest number are, on the average, the farthest from the nucleus of an atom. These electrons, also known as valence electrons, determine the atom's chemical behavior. This chemical behavior is best understood in relationship to the arrangement of electrons in energy sublevels in atoms of noble gases, which, by virtue of their chemical inactivity, have a stability which is unmatched by other chemical elements. The key is an atom's excess or deficiency of electrons compared to an atom of the nearest noble gas on the periodic table. This excess or deficiency is readily indicated by the position of an element on the periodic table. How will you include this in your game about the periodic table?

Chemistry to Go

1. From the periodic table on the back inside cover, identify the excess or deficiency of electrons in an atom of a given element relative to an atom of the closest noble gas. Be sure to indicate both the number of electrons and a sign (plus or minus) to indicate whether the electrons are in excess or deficiency.

   a) calcium
   b) arsenic
   c) potassium
   d) iodine

2. Listed below are groups of three elements. For each group determine which two elements have more in common in terms of electron arrangement and therefore exhibit more similar chemical behavior. Give a reason for your selection.

   a) carbon, nitrogen, silicon
   b) fluorine, chlorine, neon
   c) sulfur, bromine, oxygen
   d) sodium, magnesium, sulfur
   e) helium, neon, hydrogen

3. Listed below are pairs of elements. For each pair determine which of the two has the most stable arrangement of electrons. You may refer to the table of first ionization energies in **Activity 6** and the periodic table on the back inside cover. Provide a statement explaining your choice in terms of ionization energy and electron arrangement.

   a) helium and lithium
   b) lithium and beryllium
   c) magnesium and chlorine
   d) magnesium and argon
   e) neon and krypton

4. a) Write the electron configuration for magnesium.

   b) Determine how many valence electrons magnesium contains.

   c) Write the electron configuration for barium.

   d) Determine how many valence electrons barium contains.

   e) How do the number of valence electrons of magnesium compare to the number of valence electrons of barium?

   f) What general statement can you make about the number of valence electrons of each element of the alkaline earth metals?

5. a) Write the electron configuration for cobalt.

   b) How many valence electrons does cobalt contain?

Inquiring Further

1. Valence electron of transition elements

How many valence electrons are there in an atom of iron (Fe, atomic number 26 and called a transition element)? When you look at the periodic table on the back inside cover, you see that iron has all the electrons in an atom of calcium plus six additional electrons in the $3d$ sublevel. Relative to the 2 electrons in the $4s$ sublevel in calcium, do these $3d$ electrons qualify as valence electrons? Explain your thinking. What can you say about the number of valence electrons in the other transition elements, from scandium to copper?

2. Ionization energies of beryllium (Be) atoms

The first three ionization energies of beryllium atoms are as follows:

1st = 1.49×10^{-18} J

2nd = 2.92×10^{-18} J

3rd = 2.47×10^{-17} J

Explain the magnitudes of the energies in terms of electron configurations and from this information determine how many valence electrons are contained in beryllium.

Activity 8 How Atoms Interact with Each Other

GOALS

In this activity you will:

- Relate patterns in ionization energies of elements to patterns in electron arrangements.
- Use your knowledge of electron arrangements and valence electrons to predict formulas for compounds formed by two elements.
- Contrast ionic bonding and covalent bonding.
- Draw electron-dot diagrams for simple molecules with covalent bonding.

What Do You Think?

You have learned that the chemical behavior of an atom is determined by the arrangement of the atom's electrons, specifically the valence electrons. The salt that you put on your food is chemically referred to as NaCl—sodium chloride.

- How might the valence electrons of sodium (Na) and chlorine (Cl) interact to create this bond?

Record your ideas about this question in your *Active Chemistry* log. Be prepared to discuss your responses with your small group and the class.

Investigate

1. In **Activity 3** you read that John Dalton assumed that chemical compounds formed from two elements combined in the simplest possible combination—one atom of each element. In **Activity 6** you began to see that an atom's chemical behavior reflects its excess or deficiency of electrons relative to an atom of the closest noble gas on the periodic table. Use the list of ionization energies in **Activity 6** to answer the following questions:

a) Which atoms have the smallest values for first ionization energies? (Remember, the first ionization energy is the amount of energy required to remove the first electron.)

Where are these atoms located on the periodic table?

b) What do you observe about the amount of energy required to remove the second electron from atoms of the elements identified in (**a**) above?

c) Use your understanding of the arrangement of electrons in this group of elements to suggest a reason for the pattern you noted in (**b**).

d) Which atoms have the smallest values for second ionization energies? Where are these atoms located on the periodic table?

e) Use your understanding of the arrangement of electrons in this group of elements to suggest a reason for the pattern you noted in (**d**).

2. Once you recognize the role of an atom's electron arrangement—especially the valence electrons—in an atom's chemical activity, you can often predict formulas for compounds formed by two chemical elements. (Recall that valence electrons are the electrons located in the highest energy level, the levels designated by the sublevel having the highest numbers.)

Sodium (Na) has one valence electron in the 3*s* sublevel. By losing that electron, the sodium atom becomes a sodium ion and it has the same stable electron arrangement as neon. What is the electric charge on the resulting Na ion?

Consider a chlorine atom.

a) How many valence electrons does a chlorine atom have?

b) How many electrons does a chlorine atom need to gain to have the same number of electrons as an argon atom?

c) When a chlorine atom gains an electron, a chloride ion is formed. Since the original chlorine atom was electrically neutral and it gained a negative electron to form the ion, what is the electric charge (sign and value) on the resulting ion?

d) Each chlorine atom is capable of accepting one electron. Describe how you think the compound sodium chloride (NaCl) is formed?

3. Consider the reaction between aluminum and zinc chloride (similar to the reaction in **Activity 3**). The zinc atoms in zinc chloride have two valence electrons located in the 4*s* sublevel. You can note the two valence electrons in the electron arrangement marked on the periodic table.

In order to acquire the electron configuration of argon atoms a rather stable arrangement, the zinc atoms give up their two valence electrons to form zinc ions. Since the original zinc atom was electrically neutral and it lost two negative electrons to form the ion, the resulting ion has a positive charge with a magnitude two times the charge on the electron. It has a plus two charge.

a) Each chlorine atom is capable of accepting one electron. How many chlorine atoms are needed to accept the 2 electrons that zinc atoms have to give?

b) When writing the formula for a compound, the number of atoms necessary to balance the loss and gain of electrons can be designated through the use of a subscript, such as the 2 in H_2O. How would you write the formula for the compound zinc chloride?

4. In a reaction between aluminum and zinc chloride, aluminum replaces the zinc in the zinc chloride, forming aluminum chloride and zinc.

a) Consider an atom of aluminum. How many valence electrons does an aluminum atom have?

b) How many electrons does an aluminum atom need to give up to reach the same chemical stability as a neon atom?

c) What are aluminum atoms called after they give up their valence electrons? What is their electric charge (sign and value)?

d) How many chlorine atoms are needed to accept the electrons given up by an aluminum atom?

e) How would you write the formula for the compound aluminum chloride?

Chem Talk

FORMING COMPOUNDS

The Octet Rule

In this activity, you explained the formation of the compounds that you investigated by how the electrons are transferred or shared between atoms. Some scientists explain these observations using the octet rule. The octet rule works well with the representative elements and is stated as follows: atoms tend to gain or lose electrons during chemical reactions so that the atoms have an outer shell configuration of 8 electrons. The exceptions to this are the transition elements. They can form compounds that do not have 8 electrons in their outer shell. For example, when chlorine, one of the **halogens** (Group 17), gains one electron it now has 8 electrons in its outermost *s* and *p* sublevels (octet of electrons). Also, you should note that the name of all of the compounds that you have studied always started with the name of the metal and then followed with the nonmetal part. The second thing that you should note is that all of these compounds are binary (meaning two parts). **Binary compounds** always end with the suffix *ide* (except for a few compounds with common names like water and ammonia).

Chem Words

halogens: group VIIA (17) on the periodic table consisting of fluorine, chlorine, bromine, iodine, and astatine.

binary compound: a compound, formed from the combining of two different elements.

ionic bond: the attraction between oppositely charged ions.

covalent bond: when two atoms combine and share their paired electrons with each other.

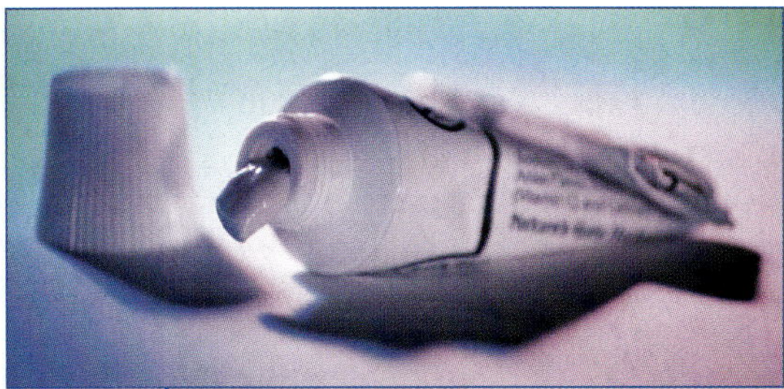

Covalent and Ionic Bonds

You may have noticed that the column headed by carbon in the periodic table has not received a lot of attention so far. In **Activity 7** you learned that atoms of these elements contain 4 valence electrons. Atoms with a small number of valence electrons give up electrons, and atoms with a large number of valence electrons gain additional electrons to have the same electron arrangement as an atom of the nearest noble gas. Except for helium, with 2 valence electrons, the noble gases each have 8 valence electrons. What do atoms of carbon do: give up their 4 valence electrons, or gain 4 more?

In actuality, atoms of carbon do neither. Instead of giving or taking electrons, carbon atoms share electrons with atoms of other elements. Instead of giving or taking electrons to form what are called **ionic bonds**, carbon atoms share electrons with atoms of other elements to form what are called **covalent bonds**. In fact, all nonmetallic elements whose atoms have four or more valence electrons can form covalent bonds by sharing electrons. This sharing results in a situation in which each atom is associated with 8 valence electrons, as is characteristic of an atom of a noble gas.

Covalent bonding can be illustrated using electron-dot diagrams, in which each valence electron is indicated by a dot (or other appropriate symbol) around the chemical symbol for the element in question. Consider the following covalent compound illustrations:

hydrogen chloride contains one hydrogen atom and one atom of chlorine:

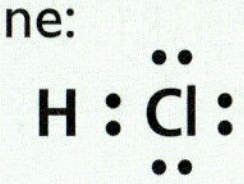

water contains 2 hydrogen atoms and one oxygen atom:

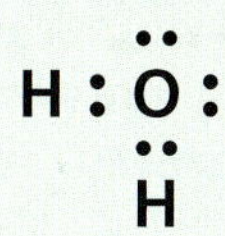

and carbon dioxide contains one atom of carbon and 2 oxygen atoms:

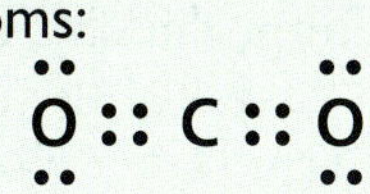

You can count 8 electrons around the chlorine in hydrogen chloride (note that hydrogen can only have 2 electrons). In water the oxygen has 8 electrons around it and hydrogen again has only 2 electrons. Carbon dioxide shows that oxygen contains 8 electrons and carbon has 8 electrons as well. These shared electrons in these examples all produce stable covalent compounds.

Checking Up

1. When naming a binary compound, which element is named first, the metal or the nonmetal? Give an example to explain your answer.
2. Explain the difference between an ionic and a covalent bond.
3. Draw electron-dot diagrams showing covalent bonding in the following compounds:
   a) water (two atoms of hydrogen, one atom of oxygen). Note: Since the noble gas nearest hydrogen is helium, with only two valence electrons, hydrogen atoms need be associated with only two valence electrons.
   b) methane (four atoms of hydrogen, one atom of carbon).
   c) ammonia (three atoms of hydrogen, one atom of nitrogen).
   d) carbon tetrachloride (four atoms of chlorine, one atom of carbon).

Reflecting on the Activity and the Challenge

In this activity you learned that atoms of two chemical elements will interact with each other in order to achieve a stable electron arrangement like that of nearby noble gases. The way in which atoms interact is based on their excess or deficiency of electrons relative to atoms of the closest noble gas on the periodic table. An atom's excess or deficiency of electrons relative to the closest noble gas is readily indicated by the position of an element on the periodic table. In this way the periodic table can be used to predict the chemical formulas when two elements interact to form a compound. This information can be deduced from the periodic table. Perhaps you can invent a way to make this more explicit as you create your periodic table game. How might you incorporate this information into your game to meet the **Chapter Challenge?**

Chemistry to Go

1. From the periodic table on the back inside cover, identify the excess or deficiency of electrons in an atom of each of the following pairs of elements relative to an atom of the closest noble gas, and then predict the formula of the compound formed when these elements interact:

   a) sodium and oxygen (to form sodium oxide),

   b) magnesium and chlorine (to form magnesium chloride),

   c) aluminum and oxygen (to form aluminum oxide).

2. a) In your *Active Chemistry* log, draw a simplified periodic table that contains only the first three rows of the periodic table. Your table should have 3 rows and 8 columns, and should contain the elements with atomic numbers 1 through 18.

   b) In the box for each element, write the symbol for the element.

   c) In the columns headed by lithium, beryllium, and boron, indicate the charge (sign and value) of the ion formed when atoms of these elements give up electrons to attain a more stable electron arrangement.

   d) In the columns headed by nitrogen, oxygen, and fluorine, indicate how many electrons are gained by atoms of those elements to attain a more stable electron arrangement.

3. You know the formula for sodium chloride (NaCl) and from this knowledge you can also write the formula for potassium chloride, cesium bromide, lithium iodide and sodium fluoride.

a) What tool makes it possible for you to do this, even though you may have never investigated these compounds?

b) Using the periodic table, explain what information you used to help explain how you arrived at the formulas.

4. The formula for calcium chloride is $CaCl_2$ and from this knowledge you should be able to write the formulas of the alkaline earth metals (Group 2) and halogens combining.

a) Write the formula for magnesium bromide, strontium iodide, beryllium fluoride, barium chloride and calcium iodide.

b) What information from the periodic table did you use to support your writing the formulas of these compounds?

5. A compound that was used previously is aluminum chloride ($AlCl_3$). Use your understanding of why elements are grouped together and are called a family, to write the formula for each of the following compounds:

a) boron fluoride,
b) aluminum bromide,
c) gallium iodide,
d) indium (III) chloride,
e) thallium (III) bromide.

6. You have explored how the alkali, alkaline earth metals, and IIIA group (boron, aluminum, gallium, indium, and thallium) combine with the halogens, but how do these metals combine with the group VIA elements (oxygen, sulfur, selenium, tellurium, and polonium)? Look at a few compounds such as sodium oxide that has a formula of Na_2O. Alkali metals like sodium "want" to lose one electron. Oxygen "wants" to gain 2 electrons. Therefore, when the alkali metals combine with oxygen or other elements of group VIA, they should have a similar formula.

a) Write the formula for each of the following compounds:

i) potassium sulfide
ii) rubidium selenide
iii) lithium telluride
iv) sodium sulfide
v) cesium oxide

You can do the same thing with the alkaline earth metals combining with the group VIA elements. Calcium oxide has the formula CaO. The alkaline earth metals want to lose 2 electrons and the group VIA elements want to gain 2 electrons. They will combine in a 1:1 ratio.

b) Write the formula for each of the following compounds:

i) magnesium sulfide
ii) strontium selenide
iii) barium oxide
iv) beryllium telluride
v) calcium sulfide

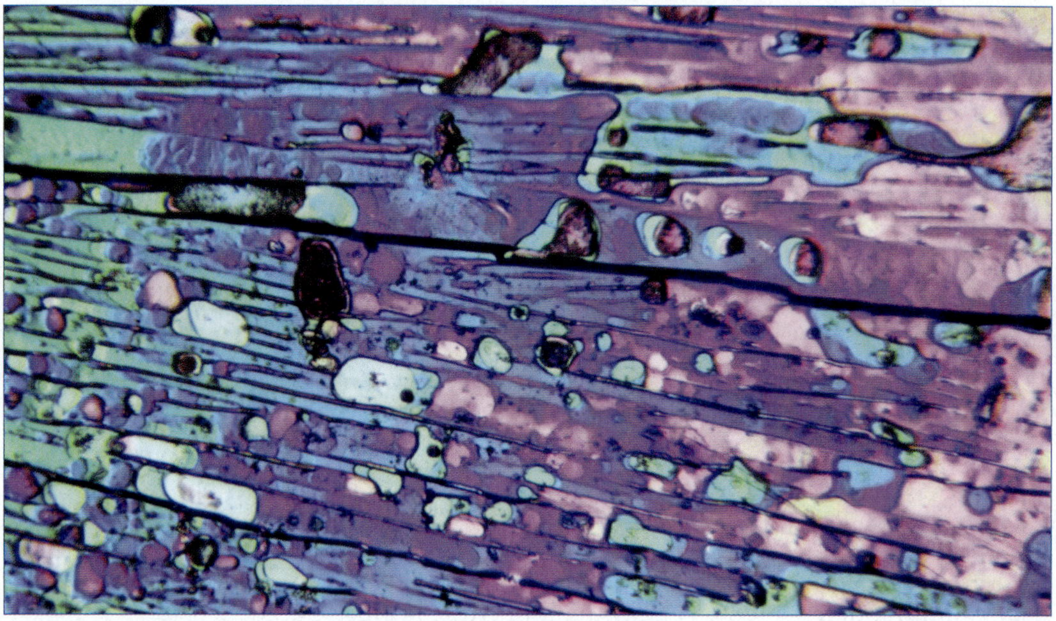

Magnification of sulfur crystal

Finally, look at the group IIIA elements combining with the group VIA elements. Aluminum oxide has the formula Al_2O_3. The group IIIA elements want to lose 3 electrons and the group VIA elements want to gain 2 electrons. In order to make a whole number exchange 6 electrons must be transferred. (Remember to make sure that your formula has the correct number of electrons being transferred.)

c) Write the formula for the following compounds:

i) boron sulfide
ii) aluminum selenide
iii) gallium telluride
iv) indium (III) oxide
v) thallium (III) sulfide

Inquiring Further

Creating compounds with "inert" gases

For a long period of time the noble gases were called the inert elements, because it was assumed that they were non-reactive and did not want to gain or lose electrons. Research to find out if compounds can be formed with the noble gases. Record your finding in your *Active Chemistry* log. If time permits, share your findings with the class.

Activity 9 What Determines and Limits an Atom's Mass?

GOALS

In this activity you will:

- Investigate the composition of the atom's nucleus.
- Explain why the atomic masses of some elements are not whole numbers.
- Use symbols to represent different isotopes of an element.
- Determine the composition of the nucleus of an atom from its isotope symbol.
- Calculate the average atomic mass of an element from the percent abundance of its isotopes.

What Do You Think?

In **Activity 4** you learned that the structure of an atom includes a nucleus surrounded by electrons. Most of the mass of an atom is concentrated in the small nucleus that has a positive electric charge equal in magnitude to the negative charge of all the electrons surrounding the nucleus.

- What do you think makes up the nucleus of the atom?

Record your ideas about this question in your *Active Chemistry* log. Be prepared to discuss your responses with your small group and the class.

Investigate

Part A: What's in the Nucleus?

1. Atomic mass is the average mass of atoms of each element. Atomic number indicates the number of electrons in the atom and the number of protons located in the nucleus needed to produce an electrically neutral atom. Refer to the periodic table to answer the following questions:

a) How many protons are there in a hydrogen atom?

b) To the nearest whole number, what is the atomic mass of a hydrogen atom?

c) How many protons are there in a helium atom?

d) Since the mass of an electron is negligible, compared to the nucleus, what would you expect the atomic mass of a helium atom to be? Explain your answer.

e) To the nearest whole number, what is the atomic mass of a helium atom?

2. In **Step 1**, you found that the helium atom has a mass that is four times the mass of a hydrogen atom, while the electric charge on the helium nucleus is only twice that of the hydrogen atom. This suggests the presence of another particle in the nucleus, with about the same mass as the proton but no electric charge. This particle is called a neutron.

Sample:

Boron has atomic number 5. This informs you that there are 5 electrons and that the nucleus contains 5 protons. The average atomic mass of boron is 10.811 atomic mass units. Most boron has 11 atomic mass units and some has 10 atom mass units. Since the mass is the sum of the protons and the neutrons (electrons have very, very little mass) then you can conclude that most boron nuclides have 5 protons and 6 neutrons in the nucleus.

Refer to your table of atomic numbers and atomic masses to answer the following questions:

a) How many protons would you expect to find in the nucleus of a helium atom? (Recall that the number of protons needs to balance the number of electrons.)

b) How many neutrons would you expect to find? (The atomic mass is a combination of the mass of the protons and the mass of the neutrons.)

c) How many protons and neutrons would you expect to find in the nucleus of an atom of each of the following elements:

- lithium
- beryllium
- boron
- carbon
- nitrogen
- oxygen
- fluorine
- neon

3. Refer again to the periodic table.

a) What are the atomic masses of magnesium and chlorine? What are the atomic masses of sodium and fluorine? Which set is closer to whole numbers?

b) We expect protons and neutrons to exist in whole numbers. You cannot have part of a proton in the nucleus. What would you expect the atomic masses of most magnesium, chlorine, sodium, and fluorine atoms to be? Explain your answer.

4. The fact that some atomic masses are not close to whole number multiples of the atomic mass of hydrogen is now explained by the fact that the number of neutrons is not the same in all atoms of a given element. Only the number of protons, the atomic number, is the same in all atoms of a

given element. Atoms of the same element with different number of neutrons in the nucleus are known as isotopes (meaning "same number of protons"). Isotopes are identified by their mass number, the sum of the number of neutrons plus protons.

Sample:

Lithium has an atomic number of 3 and an average atomic mass of 6.941. All lithium atoms have 3 protons in the nucleus. A neutral atom of lithium always has 3 electrons to balance the charge of the three protons. The average atomic mass of a lithium atom is 6.941 atomic mass units, indicating that some lithium atoms have 3 neutrons, to make a total atomic mass of 6 and other lithium atoms have 4 neutrons, to make a total atomic mass of 7. These 2 isotopes are designated lithium-6 and lithium-7. Since there are so many more lithium-7 atoms, the average of all of the atoms is very close to 7.

Refer to your list of atomic masses to answer the following questions:

a) What isotopes (as indicated by their mass numbers) do you expect to account for the known atomic masses of the following elements?

- carbon (carbon-12 atoms with 6 neutrons and carbon-13 atoms with 7 neutrons; more carbon-12 atoms)
- hydrogen
- boron
- magnesium
- beryllium
- sodium

b) In the notation below, the mass number is written at the upper left of the chemical symbol of the element. The atomic number is written at the lower left of the chemical symbol of the element. How many neutrons and protons are present in the following isotopes?

i) ${}^{3}_{2}\text{He}$ and ${}^{4}_{2}\text{He}$

ii) ${}^{6}_{3}\text{Li}$ and ${}^{7}_{3}\text{Li}$

iii) ${}^{12}_{6}\text{C}$ and ${}^{13}_{6}\text{C}$

iv) ${}^{14}_{7}\text{N}$ and ${}^{15}_{7}\text{N}$

Part B: Forces within the Atom

1. There are two very different forces acting on the electrons, protons, and neutrons in the atom. In order to better understand the atom, you must first understand these forces.

   Cut two strips of transparent tape about 12 cm long. Bend one end of each strip under to form a tab. Place one strip sticky-side down on a table and label the tab "B," for "bottom." Place the other strip sticky-side down

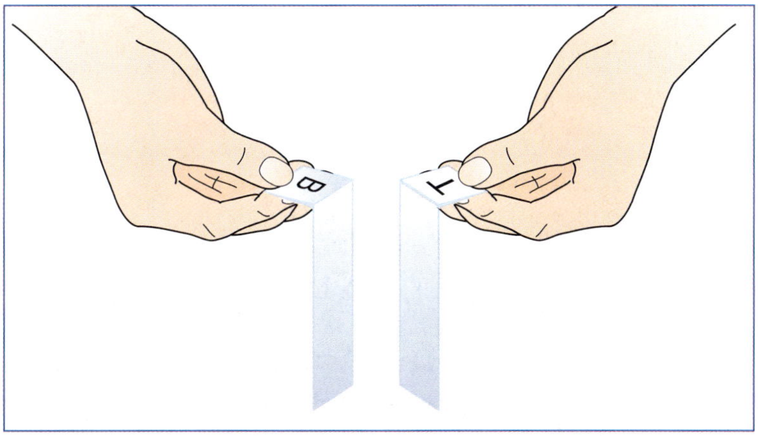

on top of the first strip and label the tab "T," for "top."

Peel off the top strip, using the tab, with one hand and then pick up the bottom strip with the other hand. Hold both strips apart, allowing them to hang down.

Slowly bring the hanging strips toward each other, but do not let them touch.

a) Record your observations.

b) If the strips accelerated toward or away from each other, Newton's Second Law tells you that there must be a force. Is the force between the two strips of tape attractive or repulsive?

2. Make a second set of strips as in **Step 1**.

a) Predict what you think will happen if the two top strips are picked up, one from each set and brought toward each other. Record your prediction in your *Active Chemistry* log.

Pick up the two top strips by the tabs, allowing both strips to hang down. Slowly bring them toward each other.

b) Record your observations.

c) Was the force attractive or repulsive? Explain.

d) Predict what you think will happen if the two bottom strips of tape are picked up and brought toward each other. Record your prediction.

Pick up the two bottom strips by the tabs, allowing both strips to hang down. Slowly bring them toward each other.

e) Record your observations.

f) Was the force attractive or repulsive? Explain.

3. The two different strips of tape have different charges. The top strips have a positive electric charge. They have lost some of their electrons. Since the number of protons has remained the same, the strips are positive. The bottom strips have a negative charge. The bottom strips have gained some electrons. Since the number of protons has remained the same, the strips are negative. The force between the strips is called the electric force.

a) Is the force between two positive strips repulsive or attractive? Use evidence to justify your answer.

b) Is the force between two negative strips repulsive or attractive? Use evidence to justify your answer.

c) When a positive and a negative strip come near each other, is the force attractive or repulsive? Justify your answer.

4. The nucleus has a positive charge due to all of the protons there. The electrons surrounding the nucleus have negative charges.

a) What kind of electric force (attractive or repulsive) exists between the nucleus of an atom and any one of the atom's electrons?

b) What kind of electric force (attraction or repulsion) exists between pairs of protons in the nucleus?

5. The nucleus is a very crowded place. The protons in the nucleus are very close to one another. If these protons are repelling each other by an electrostatic force (and they are!), there must be another force, an attractive force, that keeps them there. The attractive force is the nuclear force, also called the strong force. This force is much stronger than the electric force. It acts between pairs of protons, pairs of neutrons, and protons and neutrons. The electron is not affected by the nuclear force.

   a) Copy and complete the table below in your *Active Chemistry* log. The first row has been completed for you.

6. If the nucleus were too large, the protons on one side of the nucleus are too far away to attract the protons on the other side of the nucleus. The protons can still repel one another since the coulomb electrostatic force is long-range. The repulsive electrostatic force wins and the nucleus won't form.

   A large nucleus will break apart when the electrostatic repulsion between the protons is too great. The repulsion pushes the fragments of the nucleus apart, releasing a great amount of energy. This process of splitting an atom into smaller atoms is called fission. It occurs in uranium when an additional neutron is added and causes instability.

   One example of the fission process can be represented as follows:

   $${}^{235}_{92}\text{U} + {}^{1}_{0}\text{n} \rightarrow {}^{94}_{36}\text{Kr} + {}^{139}_{56}\text{Ba} + 3\ {}^{1}_{0}\text{n} + \text{energy}$$

   a) Is the mass number conserved on both sides of the reaction? What is the total mass number on each side?

   b) Is the atomic number conserved on both sides of the reaction? What is the total atomic number on each side?

   c) Why does the neutron have a mass number of 1?

   d) Why is the atomic number of a neutron equal to 0?

   Small nuclei can also combine to form a larger nucleus and release energy. This process is called fusion.

Particles	Coulomb electrostatic force	Strong, nuclear force
electron-proton	*attractive*	*none*
electron-neutron		
proton-proton		
proton-neutron		
neutron-neutron		

Chem Talk

THE NUCLEUS OF AN ATOM

Discovery of the Neutron

The average atomic masses of some elements were known in Mendeleev's time, even though scientists didn't know much about the actual structure of an atom. In **Part A** of this activity you explored the idea of how the atomic mass relates to the atomic number. Mendeleev began organizing his periodic table by listing all the known elements in order of atomic mass. However, he found that organizing the elements in this way did not always make sense in terms of the behavior of the elements. He concluded that his measurements of atomic mass were incorrect and in those situations used the properties of the elements to place them in the table.

As it turned out, Mendeleev's measurements were not necessarily flawed. Although early models of the nucleus included the **proton**, the proton alone could not account for the fact that the mass of a helium atom is four times the mass of a hydrogen atom while the electric charge on the helium nucleus is only twice that of the hydrogen atom. Lord Rutherford (after discovering that atoms had a nucleus) addressed this problem when he suggested that another particle was present in the nucleus, with about the same mass as the proton but no electric charge. He named this particle the **neutron**.

The neutron was actually discovered in 1932 (by Chadwick, a British physicist), adding a great deal to the understanding of the nucleus of the atom. This discovery did not solve all of the mysteries concerning the atomic masses of some elements. Scientists today refer to protons and neutrons as nucleons since they reside in the nucleus and are almost identical in mass. The mass number tells us the number of nucleons.

Chem Words

proton: a positively charged subatomic particle contained in the nucleus of an atom. The mass of a proton is 1.673×10^{-24}g and has a charge of +1.

neutron: neutral subatomic particle with a mass of 1.675×10^{-24}g located in the nuclei of the atom.

Isotopes

In **Part A** of this activity you also investigated why the atomic mass of an element is not a whole number. Not all atoms of a given element have the same number of neutrons in the nucleus. Only the number of protons, the atomic number, is the same in all atoms of a given element. Atoms of the same element with different number of neutrons in the nucleus are known as **isotopes** (meaning "same number of protons"). Isotopes are identified by their mass number, the sum of the number of neutrons plus protons.

You can refer to an element by its name (chlorine), by its atomic symbol (Cl), or by its atomic number (17). All three identifications are equivalent and used interchangeably in chemistry. The same element can have a different number of neutrons in the nucleus. Chlorine, which must have 17 protons in the nucleus, can have 18 or 20 neutrons. Chlorine with 20 neutrons and chlorine with 18 neutrons are the isotopes of chlorine (^{35}Cl and ^{37}Cl).

Chem Words

isotope: atoms of the same element but different atomic masses due to different number of neutrons.

Electrostatic and Nuclear Forces

In **Part B** of this activity, when you brought the two positive strips near each other, they experienced a repulsive force. This was true for two negative strips as well. When a positive and a negative strip were brought close together, the force was attractive. As you have heard, "opposites attract!"

Inside the nucleus, the protons are repelling one another. Every pair of protons has a repulsive force between them. The force is very large because the distances within the nucleus are very small. The nucleus is between 10,000 and 100,000 times smaller than the atom. The electrical force can be described mathematically.

$$F = \frac{kq_1q_2}{d^2}$$

where F is the force,

k is Coulomb's constant (a number $= 9 \times 10^9$ N m^2/C^2),

q_1 and q_2 are the charges, and
d is the distance between the charges.

As the distance between the charges increases the force weakens. Since the distance in the denominator is squared, if the distance triples the electrical force is 9 times (3^2) weaker or one-ninth as strong.

The question then becomes, what holds the protons together in the nucleus? The protons do have an electrical force pushing them apart but they have the larger nuclear force holding them together. The nuclear force is strong at short range. Anywhere beyond a distance of approximately 10^{-14} m (that's less than one 10-millionth of one 10-millionth of a meter), the nuclear force is zero. Neutrons in the nucleus are also attracted to each other and to protons with the nuclear force. Electrons are not affected by the nuclear force. Electrons belong to a different class of particles than protons and neutrons and do not interact with the strong nuclear force.

The nucleus is held together by a new force—the strong nuclear force. The nuclear force:

- is very, very strong at small distances;
- acts only between nucleons (proton-proton, proton-neutron, neutron-neutron);
- is always attractive;
- is very short range (if nucleons are more than 10^{-14} m apart, the nuclear force is zero).

The atom is held together by the electrostatic coulomb force. The electrostatic force:

- is strong at small distances, weak at large distances;
- acts only between charged particles (proton-proton, electron-electron, proton-electron);
- is attractive or repulsive;
- is long range (the force gets weaker at large distances).

All the nucleons are attracted by the nuclear force. The electrostatic force repelling protons in the nucleus is overwhelmed by the attractive nuclear force between these protons.

Unstable Atoms

You might expect to find nuclei of atoms with all sorts of combinations of neutrons and protons. Yet the quantity of isotopes for each element is rather small, and the number of elements is also limited. Moreover, elements do not occur in nature with atomic number greater than 92, and the highest atomic number for an atom created in the laboratory is 117.

There are two stable masses of chlorine, chlorine-35 and chlorine-37. The key word in this statement is "stable." There are other isotopes of chlorine, both heavier and lighter than chlorine-35 and chlorine-37, but they are not stable. The unstable isotopes can convert to a more stable combination of neutrons and protons, and they do so according to a systematic pattern in time. These other isotopes of chlorine are said to be **radioactive**. Understanding why certain elements are radioactive requires a deeper understanding of the structure of the nucleus. Scientists are still trying to fully understand stability of the elements.

If the nucleus of an atom is too large, the protons on one side of the nucleus are too far away to attract the protons on the other side of the nucleus. The protons can still repel one another since the coulomb electrostatic force is long-range. The interaction between the repulsive electrostatic force and the attractive nuclear force is one determining factor on the maximum size of a nucleus.

The stability of an atom varies with the elements. Light elements become more stable as the atomic mass (the number of nucleons) increases. The most stable element is iron (atomic number 26) with an atomic mass of 56. Elements with larger atomic masses become less stable.

In general, elements with nuclear mass much, much less than 56 can combine to gain mass, become more stable, and give off energy. This process is called **fusion**. Elements with nuclear mass much, much greater than 56 can break apart to lose mass, become more stable, and give off energy. This process is called **fission**.

Fusion is the process of small nuclei combining to increase their mass. The best example of fusion processes is what occurs in the Sun and other stars. The fusion process is ideal for supplying safe energy because it releases very large amounts of energy without leaving much dangerous radioactive residue. However, it is very difficult to accomplish this on an industrial level at the present time. In the future we hope scientists will figure out how to harness the energy of nuclear fusion, because it would be an excellent source of energy for society.

The process of splitting an atom into smaller atoms is called fission. This is the process that is used to produce nuclear energy. It is used to power nuclear submarines and to produce electrical energy in nuclear power plants all over the world.

Chem Words

radioactive: an atom that has an unstable nuclei and will emit alpha, positron, or beta particles in order to achieve more stable nuclei.

fusion: nuclei of lighter atoms combining to form nuclei with greater mass and release of a large amount of energy.

fission: the process of breaking apart nuclei into smaller nuclei and with the release of a large amount of energy.

The use of nuclear energy for the production of electricity is quite apparent as you look at the numerous states that depend on nuclear energy. For example, over 40% of Illinois' electricity is produced by nuclear energy. Nuclear fission does create some major problems: (1) Security, (2) Radiation, (3) Removal of spent rods, and (4) Disposal of waste. With these problems, there is a need for continued research. Numerous universities and government facilities are trying to improve the efficiency of nuclear fission and at the same time, trying to develop nuclear fusion for commercial use.

This ongoing research is expensive and depends on the government, industry, and other organizations to continue supporting this research. If we can learn how to harness nuclear fusion we can alleviate our nation's electrical problems while decreasing pollution. The field of nuclear science is going to continue to grow and the future will provide great opportunities for a young scientist like you to get involved.

Checking Up

1. Explain the difference between atomic mass and atomic number.
2. What two forces are at work in the nucleus of an atom? Explain how each works.
3. What is an isotope?
4. Why are some isotopes unstable?
5. Construct a table or diagram to compare and contrast the nuclear processes of fission and fusion.

Reflecting on the Activity and the Challenge

In **Part A** of this activity you learned that the mass of an atom, concentrated in the nucleus, is due to two types of particles, the proton and the neutron. Elements are identified by their atomic number, the number of protons in the nucleus. The atomic mass, the average mass of an atom of a given element, listed on the periodic table is a reflection of the variety of isotopes of a given element that exist. How will you incorporate your expanded understanding of the contents of an atom's nucleus, average atomic masses and isotopes into your game about the periodic table?

In **Part B** of this activity you also learned that only some combinations of neutrons and protons in a nucleus are stable, depending on the balance between the strong force holding the nuclear particles together and the electric force pushing them apart. The nuclear force is a short-range force. Beyond a distance of approximately 10^{-14} m, the nuclear force has no strength. Within that distance, this force between protons and protons, protons and neutrons, and neutrons and neutrons is quite strong. Recognizing the interplay between the electric force in the nucleus and the strong, attractive nuclear force provides an insight into the size of nuclei and the maximum size of a nucleus. These insights can be incorporated into your periodic table game in a creative way.

Chemistry to Go

1. If lithium loses an electron to become Li+, what is the average atomic mass of the lithium ion? Explain how you arrived at your answer.
2. Hydrogen has 3 isotopes with mass numbers of 1, 2, and 3. Write the complete chemical symbol for each isotope.
3. Give the complete chemical symbol for the element that contains 16 protons, 16 electrons, and 17 neutrons.
4. Complete the table below: (Use the periodic table.)

Chemical symbol	${}^{39}_{19}K$			
Atomic number		9		
Number of protons			15	
Number of electrons				53
Number of neutrons		10	16	
Atomic mass				127

5. Neutrons can be used to bombard the nucleus of an atom like uranium. Why would it be more difficult to inject the nucleus of uranium with a proton?

6. Complete the following reaction: $^{215}_{92}\text{U} + {}^{1}_{0}\text{n} \rightarrow {}^{94}_{38}\text{Sr} +$ _____ $+ 2\,{}^{1}_{0}\text{n}$

7. Radon is a threat to the well-being of people in their homes because it emits radioactive particles at a significant rate. Complete the following radioactive decay equation:

   $^{222}_{86}\text{Rn} \rightarrow {}^{218}_{84}\text{Po} +$ _____

8. Explain why a helium atom is able to exist. What keeps the 2 electrons, 2 protons, and 2 neutrons together?

Inquiring Further

Calculating average atomic mass

If you know the percentages of abundance for the isotopes of a chemical element and the known masses of those isotopes, you can calculate the average atomic mass of that element. The process is similar to calculating the average age of students in your class — add up each person's age and divide by the number of students in your class. However, if you had to average the age of all of the students in your high school, you might choose another route. It would be easier to find out how many students are fourteen, how many are fifteen, and so on. Then you could multiply the number of students in each age group by that age. Then you would add these subsets together and divide by the total number of students.

A similar process is used to average the masses of different isotopes of an element. Consider the element chlorine. There are two stable isotopes of chlorine, chlorine-35 and chlorine-37. Of all the chlorine atoms on Earth, 75.77% of them are the isotope chlorine-35, each having a mass of 34.96885. The other 24.23% of stable chlorine atoms are the isotope chlorine-37, each having a mass of 36.96590. This means that 75.77 out of 100 chlorine atoms have a mass of 34.96885 and 24.23 have a mass of 36.96590. To find the average mass, the number of each isotope is multiplied by that isotope's mass. Then the products are added together. The sum is divided by 100, since the information pertained to 100 chlorine atoms. The result is an average atomic mass of 35.45 for chlorine, the same value stated in the periodic table. The math is shown below:

Chlorine-35 $\quad 34.96885 \times 75.77 = 2649.6$

Chlorine-37 $\quad 36.96590 \times 24.23 = 895.7$

$3545.3 \div 100 = 35.453$

Magnesium, another isotope you investigated, has three stable isotopes as follows:

mass number	isotopic mass	% abundance
24	23.98504	78.99
25	24.98594	10.00
26	25.98259	11.01

Calculate the average atomic mass for magnesium. Describe how you arrived at your answer. You may use the process described above or challenge yourself to develop your own process.

The Periodic Table Assessment

Your Chapter Challenge is to develop a game related to Mendeleev's Periodic Table of the Elements. The game should be interesting, entertaining and informative. It should demonstrate your understanding of the periodic table while it helps other students learn about the periodic table. When people begin your game, they may have no knowledge of chemistry or the periodic table. After they complete the game, they should be able to report on chemistry principles which are reflected in the periodic table.

You can begin work on the **Chapter Challenge** by reviewing the important features of the periodic table that you want to include in your game. Choose a small section of the periodic table and list what you know about each element including its atomic number, atomic mass, chemical properties, electron configuration, nuclear structure and where and why it is placed where it is in the periodic table. Next to each of these items, you may wish to describe how we know some of these details (e.g., describe the Rutherford scattering experiment for evidence of the nucleus; Law of Definite Proportions for evidence of atoms). You should review the nine activities in the chapter to add to your list. You should pay particular attention to the section **Reflecting on the Activity and the Challenge**. You should also compare your list with that given in the **Chemistry You Learned** summary.

You now have to create the game which will highlight the chemistry principles that you have investigated in the past few weeks. There are a wide variety of games. There are traditional board games like Monopoly, Parcheesi, Clue and Risk. There are card games like gin rummy, hearts, solitaire, poker and bridge. There are TV games like *Jeopardy, Wheel of Fortune* and *The Price is Right*. There are video games like SimCity, PacMan and Space Invaders. There are also word games like crossword puzzles and anagrams. You have lots of experience with many different kinds of games. The first hurdle for your team will be to decide on the type of game that you will use as the format for communicating your knowledge of the periodic table. Your game should not merely be a periodic table test. It should be an entertaining approach to learning about the periodic table as well as displaying knowledge.

Once you decide and agree upon a game format, you will have to generate a set of rules. These rules will include whether people work individually or in groups, how one earns points or wins, how to get help during the game and suggested strategies for the game.

After the rules of the game are set, you must integrate your expert knowledge of the periodic table into the game. How will people playing the game learn that the atomic number corresponds to the number of protons and electrons in the neutral atom? How will players learn that isotopes of chlorine affect the stated mass? How will players get credit for remembering that the electron configurations are responsible for chemical behavior. The list of chemistry principles that you first completed can now be expanded so that each principle becomes the basis for a part of the game.

After the game structure is completed, you may have to design a board or some computer application so that the game can be played. Some careful preparation here can make a big difference in how well your game is accepted by others.

You should now practice the game as if you and your team members are playing it for the first time. Carefully critique your own game and edit the rules, approaches and chemistry principles so that the game is truly entertaining and educational. Review all of the criteria listed at the beginning of the chapter and the grading criteria agreed to by your teacher and class. Create a checklist and perform a final review of your checklist. Grade yourself before the presentation of your game to ensure that your team is heading for a great grade.

Chemistry You Learned

- Elements' chemical and physical properties
- Atoms
- Single-displacement reactions
- Double-displacement reactions
- Atomic mass
- Compound
- Law of Definite Proportions
- Avogadro's number and a mole
- Electrons
- Nucleus
- Protons
- Line spectrums produced by different elements
- Energy, frequency, and wavelength produced by the excitation of electrons
- Electromagnetic spectrum
- Bohr's Atomic Theory
- Orbits of electrons
- The spectral lines classified as Balmer series, Lyman series, and Paschen series
- Ionization potential of atom's electron
- Periodicity of the elements
- Atom's energy level and the sublevel energy of the outermost electron
- Electron configuration
- Family or chemical groups on the periodic table
- Group chemical behavior
- Noble gas electron configuration and stability
- Valence electrons
- Ionic compounds and ionic bonding
- Covalent compounds and covalent bonds
- Chemical formula
- Neutrons
- Nuclear stability
- Isotopes
- Fission
- Fusion
- Electrostatic force
- Strong, nuclear force

Chapter 12

12

Cool Chemistry Show

Scenario

The fourth- and fifth-grade students at a local elementary school have been studying chemistry in their classes. Because of the students' overwhelming interest, their teachers have asked your class to present a chemistry science show to their students. The elementary teachers have requested that the show be both interesting and informative.

For the chemistry science show, the fourth-grade teachers are asking your class to include demonstrations and explanations about chemical and physical properties and changes. The fifth-grade teacher wants the students to learn more about acids and bases, and about chemical reactions that involve color changes

Chapter Challenge

You and your classmates are being challenged to present an entertaining and informative chemistry science show to fourth- and fifth-grade students.

- The content of the show should meet the needs and interest of your audience. Remember that the fourth-grade teacher has specifically requested that the show address chemical and physical properties and changes. The fifth-grade teacher wants the students to learn more about acids and bases, and about chemical reactions that involve color changes. Your class may choose to add other presentations to enhance the show.

- All presentations must include a demonstration and an audience-appropriate explanation of the chemistry concepts involved.

- You must provide the teachers with a written summary including directions for your chemistry show with explanations of the chemistry.

- As always, safety is a top priority. You and your classmates will wear safety gear, including safety goggles, appropriate to the presentation being conducted. Presentations including flammable or explosive reactions are not appropriate for the elementary audience and may not be included in the show.

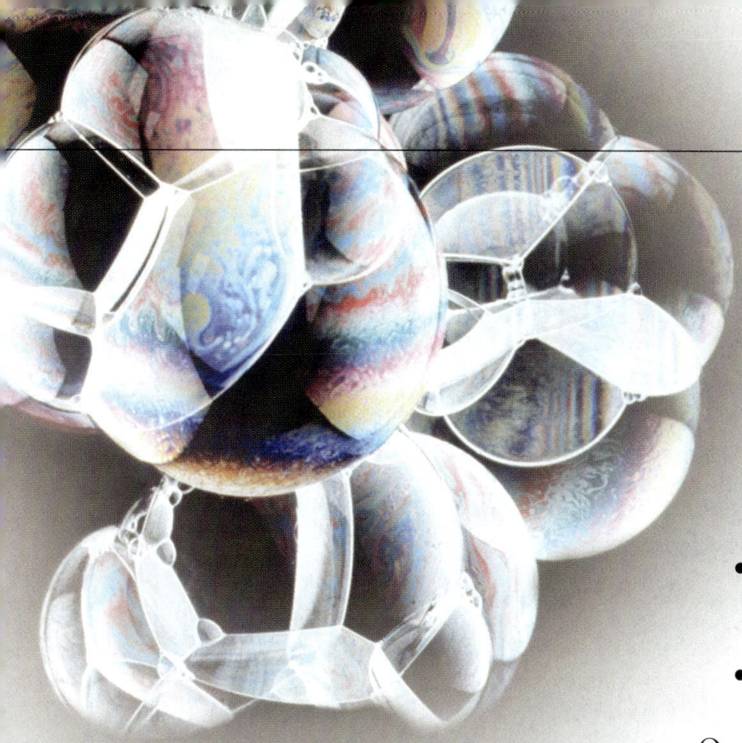

The class, as a whole, is responsible for putting on this Cool Chemistry Show. You will need to coordinate your selection of presentations to provide a show that addresses a variety of chemistry concepts in an entertaining and informative manner.

Criteria

How will your involvement in the Cool Chemistry Show be graded? What qualities should a good presentation have? Discuss these issues in small groups and with your class. You may decide that some or all of the following qualities of your presentation should be graded:

- Chemistry content
  - accuracy
  - meets teacher needs (fourth or fifth grade)
  - number of concepts addressed
- Demonstration
  - carefully planned
  - safety
  - explanation (age-appropriate)
  - adherence to assigned time limits
  - showmanship
  - creativity
  - clarity
  - organization
  - appeal
  - written summary
  - directions for experiment
  - explanation of chemistry
  - appropriateness for an elementary school teacher with limited chemistry background
  - statements concerning safety needs

Once your class has determined the list of criteria for judging the presentations, you should also decide how many points should be given for each criterion. Determining grading criteria in advance will help you focus your time and effort on the various aspects of the presentation. How many points should be assigned to the content and how many should be assigned to the actual presentation? Will each high school student be involved in only one presentation, or more? For each criterion, you should decide on how excellence is defined and how it compares to a satisfactory effort.

Since you will be working with other students in small groups, you will need to determine grading criteria that reward each individual in the group for his or her contribution and also reward the group for the final presentation. You should discuss different strategies and choose the one that is best suited to your situation. Your teacher may provide you with a sample rubric to help you get started.

Activity 1 Chemical and Physical Changes

GOALS

In this activity you will:

- Learn to differentiate between chemical and physical changes.
- Make observations and cite evidence to identify changes as chemical or physical.
- Explore the new properties exhibited when new materials are made from combinations of two or more original materials.
- Design an experiment to test properties of different combinations of materials.

What Do You Think?

There are two basic types of changes that matter can undergo: chemical change and physical change. Consider two wooden matches. One is broken in half and the other is ignited by striking it along the side of the matchbox. In both of these instances matter has changed.

- Which match has undergone a chemical change and which has undergone a physical change? Give specific reasons to support your answer.

Record your ideas about this question in your *Active Chemistry* log. Be prepared to discuss your responses with your small group and with the class.

Investigate

1. Listed on the following page are 15 opportunities for you to observe changes in matter. Your teacher may choose to do some or all of these as a demonstration or set up stations for you to visit. Notice that the directions call for small amounts of substances.

Safety goggles and a lab apron must be worn during this activity.

Make a data table to organize your observations of the matter before and after any change(s) that may occur. Be detailed in your observations.

- Heat an ice cube in a beaker.
- Boil a small amount of water.
- Melt a small amount of candle wax. Then allow the melted wax to cool.
- Break a wooden splint into several pieces.
- Hold a wooden splint in a flame.
- Add a few drops of lemon juice to a small amount of milk.
- Add a few drops of vinegar to a small amount of baking soda ($NaHCO_3$).
- Add a small amount of table salt to water; stir; boil the solution to dryness; cool and record the result.
- Add several drops of iodine solution to a small amount of starch.
- Add a small piece of zinc to a small amount of hydrochloric acid (0.1 M HCl).
- Add a drop of phenolphthalein indicator solution to a solution of sodium hydroxide (0.1 M NaOH).
- Add two drops of sodium carbonate (0.1 M Na_2CO_3) to two drops of sodium hydrogen sulfate (0.1 M $NaHSO_4$).
- Add a few drops of household ammonia to a small amount of a copper (II) sulfate (0.1 M $CuSO_4$) solution.
- Add a few drops of vinegar to a small piece of chalk.
- Sharpen a pencil and collect the shavings.

Dispose of the materials as directed by your teacher. Clean up your work-station and then wash your hands.

2. Organize the information in your data table.

    a) Prepare and complete a chart that organizes your observations into separate columns — one that includes the situations where color changes occurred, one that notes the formation of precipitates (sometimes visible as a cloudy solution), one that includes gas formation (fizz), one to note other changes, and one where no visible change occurred.

3. A physical change involves changes in the appearance of the material but does not involve creation of new materials. A chemical change involves the formation of new materials. Chemical reactions are characterized by a number of changes, including color changes and the formation of a precipitate or gas.

    a) Which of the interactions you observed were chemical changes? Explain your answer.

    b) Which of the interactions you observed were physical changes? Explain your answer.

    c) When you placed the wooden splint into a flame, what other evidence (besides the color change) indicated that a chemical change took place?

    d) Imagine a situation where two colorless solutions are mixed together. There is no color change, no precipitate is formed, and no gas is released. However, heat is released as the solutions are mixed. Is this an example of a chemical or physical change? Explain your choice.

4. Each group will be given a piece of disposable diaper. Place the piece in a beaker.

   a) Predict how much liquid the diaper will be able to hold. Record your prediction in your log.

   b) Design an investigation to measure the amount of liquid that the diaper can absorb. Record your procedure in your log.

   c) With the approval of your teacher, carry out your investigation. Record your results.

   d) Explain how your prediction compared with your observations.

   e) The diaper is made of a material called sodium polyacrylate. When it absorbs water, is this a physical or chemical change? Explain your answer.

5. Your teacher will show you a solution of sodium acetate in a 250-mL flask. Observe the solution carefully.

   a) Record your observations in your *Active Chemistry* log.

   Your teacher will then add one crystal of sodium acetate to the flask.

   b) What happens? Record your observations in your log.

   c) Was this a chemical or physical change?

6. In a large throwaway glass jar, mix 100 mL of sodium silicate (sometimes called water-glass solution) and 400 mL of water.

   Carefully drop solid-colored crystal compounds of cobalt, copper, nickel, iron, and/or manganese in different locations inside the jar.

   a) Is there evidence of a change immediately? In several minutes? In several hours? In several days? In your *Active Chemistry* log, describe the results.

   b) Is the phenomenon you see the result of a physical or a chemical change? Explain your answer.

ChemTalk

CHANGES IN MATTER

Physical and Chemical Changes

In this activity you observed a number of situations that involved changes in matter, both physical and chemical. A **physical change** involves changes in the appearance of the material but does not involve creation of new materials. A change of a solid to a liquid is a physical change. When the candle wax melted it may have appeared different, but it was still wax. After it solidified, it had a similar appearance to the initial product. Dissolving is also a physical change. When you added the salt to the water, the salt crystals seemed to disappear as they dissolved in the water. However, they had only spread out into a solution. A **solution** is a homogeneous mixture of at least two different materials. The material being dissolved is called the **solute**, and the material present in the largest amount is called the **solvent**. When you boiled away the solvent, water, the solute, the salt crystals, remained the same as they were originally.

A **chemical change** involves the formation of new materials. The new materials are called **products** and the starting materials are

Chem Words

physical change: a change that involves changes in the state or form of a substance but does not cause any change in chemical composition.

solution: a homogeneous mixture of two or more substances.

solute: the substance that dissolves in a solvent to form a solution.

solvent: the substance in which a solute dissolves to form a solution.

chemical change: a change that converts the chemical composition of a substance into different substance(s) with different chemical composition.

product: the substance(s) produced in a chemical reaction.

Chem Words

reactants: the starting materials in a chemical reaction.

chemical reaction: a process in which new substance(s) are formed from starting substance(s).

precipitate: an insoluble solid formed in a liquid solution as a result of some chemical reactions.

concentration: a measure of the composition of a solution, often given in terms of moles of solute per liter of solution.

saturated solution: a maximum amount of solute that can be dissolved at a given temperature and pressure.

polymer: a substance that is a macromolecule consisting of many similar small molecules (monomers) linked together in long chains.

called **reactants**. The process that brings about a chemical change is called a **chemical reaction**. Chemical reactions are characterized by a number of changes, including color changes, the formation of a **precipitate** or gas, a release of heat or light. Chemical changes are usually not easy to reverse. When you burned the wooden splint you could not put the charcoal together to form the original splint as you could when you simply broke the splint into pieces.

Saturated and Supersaturated Solutions

The solution your teacher used in the demonstration was a supersaturated solution of sodium acetate. Solutions are commonly described in terms of **concentration**. The concentration of a solution is the ratio of the quantity of solute to the quantity of solution. A dilute solution has fewer solute molecules per volume than a concentrated solution. However, in both the dilute and concentrated solutions more solute can dissolve in the solvent. These solutions are unsaturated.

You probably recognize the term "saturated." When something is saturated, it is full. A saturated sponge is full of water; it can't hold any more. A **saturated solution** is one in which no more solute will dissolve under the given conditions. To say that the sodium acetate solution is supersaturated means that it is "over full." A supersaturated solution contains more solute

particles than it normally would under the given conditions. A supersaturated solution can be made using some solutes. If a saturated solution at a high temperature is allowed to cool undisturbed, all the solute may remain dissolved at the lower temperature. The solution is then supersaturated. As you observed in the activity, such solutions are unstable. By introducing a "seed" crystal the extra solute particles "joined" the crystal and came out of the solution.

Polymers

The chemical material that you were working with when you investigated the absorbency of the diaper was sodium polyacrylate. It is a chemical compound called a **polymer**. It is made up of many (poly) repeating units of a smaller group of elements (the monomer called acrylate). This particular polymer has a unique property. It will absorb more than 800 times its own mass in distilled water. The fascinating ability of this polymer (sodium polyacrylate) to absorb large amounts of water has led to its use in a number of commercial endeavors.

Checking Up

1. What is a physical change? Provide two examples.
2. Explain the meaning of a solution, a solute, and a solvent.
3. What is a chemical change? Provide two examples.
4. What "clues" can you look for to determine if a chemical change has occurred?
5. How do you describe the concentration of a solution?
6. Explain the difference between a saturated and a supersaturated solution.

Reflecting on the Activity and the Challenge

Recall that the fourth-grade teacher has specifically requested that your chemistry show addresses chemical and physical properties and changes. You are right on track for the fourth graders. The fifth-grade teacher wants the students to learn more about chemical reactions that involve color changes. You have seen a few of those, too. If you had to conduct the show based on your experiences so far, which activity would you use? What additional information would you need to be able to explain the chemistry to fourth- and fifth-grade students?

Chemistry to Go

1. Which of the following are chemical changes and why?
   a) Toast turns black after being in the toaster too long.
   b) Water condenses on the outside of a glass of iced tea.

c) Green leaves turn orange, yellow, and red in the fall.

d) Green bananas become yellow.

e) Butter melts on a hot summer day.

2. Think back to a recent lunch or dinner. Describe two physical and two chemical changes that were involved in the meal and explain why you think each was a physical or chemical change.

3. Write a paragraph describing a common activity (such as making a cake or driving a car). Underline the physical changes (use one line) and chemical changes (use two lines) taking place within the activity. Select and describe an activity that is sure to have at least two physical changes and two chemical changes.

4. The following information is obtained for the element aluminum. Identify which are physical (use one line) and which are chemical (use two lines) properties.

   Aluminum is a shiny silver metal and melts at 660°C. When a strip of aluminum is placed in hydrochloric acid, hydrogen gas is released. The density of aluminum is 2.70 g/cm^3. When polished aluminum is exposed to oxygen over a period of time it forms aluminum oxide (Al_2O_3) on the surface of the metal.

5. How would you determine whether a clear solution in a beaker is saturated sugar water or just water? Remember, you do not taste samples in the laboratory.

Preparing for the Chapter Challenge

Describe how you would demonstrate the difference between a physical and chemical change in a "cool" way.

Inquiring Further

Factors affecting solubility and the rate of dissolving

Understanding the factors that affect how quickly a solute dissolves in a solvent is important in many practical applications in manufacturing. Design an investigation to determine the factors that affect solubility. Consider the following:

- nature of the solute and solvent
- temperature
- agitation (stirring or shaking)
- surface area (for example, try using a sugar cube, granulated sugar, and icing sugar)
- pressure of gases

Remember that your investigation must be controlled, if your results are to be reliable. What will be your independent and what will be your dependent variables?

Activity 2 More Chemical Changes

GOALS

In this activity you will:

- Observe several typical examples of evidence that a chemical change is occurring.
- Make generalizations about the combinations of materials that result in the same evidence.
- Make generalizations about materials that tend to react with everything and materials that tend not to react with anything.
- Practice careful laboratory techniques, such as avoiding contamination of reactants, to ensure that results observed are repeatable and unambiguous.

What Do You Think?

Mix 1 cup flour, 1/3 cup sugar, 1 teaspoon of baking powder with a cup of milk and 1 egg, well-beaten. Place the mixture in an oven for 30 minutes.

Add two drops of sodium carbonate (0.1 M Na_2CO_3) to two drops of sodium hydrogen sulfate (0.1 M $NaHSO_4$).

- Which of the instructions above will result in a chemical reaction? Why?
- Describe one similarity and one difference in the above instructions.

Record your ideas about these questions in your *Active Chemistry* log. Be prepared to discuss your responses with your small group and with the class.

Investigate

1. Eight solid materials listed on the next page have been dissolved in distilled water to make solutions. You will combine the solutions (one to one) with each other in an organized manner in order to observe their interactions.

Safety goggles and a lab apron must be worn during this activity.

- barium nitrate $Ba(NO_3)_2$
- sodium hydroxide ($NaOH$)
- sodium hydrogen carbonate ($NaHCO_3$)
- copper (II) sulfate ($CuSO_4$)
- potassium iodide (KI)
- silver nitrate ($AgNO_3$)
- iron (III) nitrate $Fe(NO_3)_3$
- hydrochloric acid (HCl)

After mixing two solutions, make notes on your chart of any changes you observe. Don't overlook any color changes, the formation of a precipitate (sometimes observed as a cloudy solution), the formation of a gas (fizzing or bubbles), or a change in temperature.

Using another dropper, continue by adding three drops of the sodium hydrogen carbonate to the second well.

	$Ba(NO_3)_2$	NaOH	$NaHCO_3$	4	5	6	7	8
$Ba(NO_3)_2$								
NaOH								
$NaHCO_3$								
4								
5								
6								
7								
8								

Silver nitrate will stain skin and clothing. Handle with care.

a) Begin by making a chart to record your data. Your chart will require an entire page of your notebook. Allow plenty of room to record your observations. A sample chart has been provided. Notice that some of the blocks in this chart are shaded, indicating there is no need to mix those particular chemicals. Why do you suppose those particular blocks are shaded?

2. Now it is time to mix the solutions.

Begin with barium nitrate. Add three drops of the barium nitrate solution to each of seven wells of a well plate. Add three drops of sodium hydroxide solution to the first well.

It is important that you do not allow the tip of the dropper of one solution to come in contact with another solution. Your attention to this detail will prevent contamination of solutions.

Continue by adding copper (II) sulfate to the third well, and so on.

a) After mixing the pairs of solutions, make note on your chart of any changes you observe.

Continue by putting three drops of sodium hydroxide into each of seven wells and adding the other solutions.

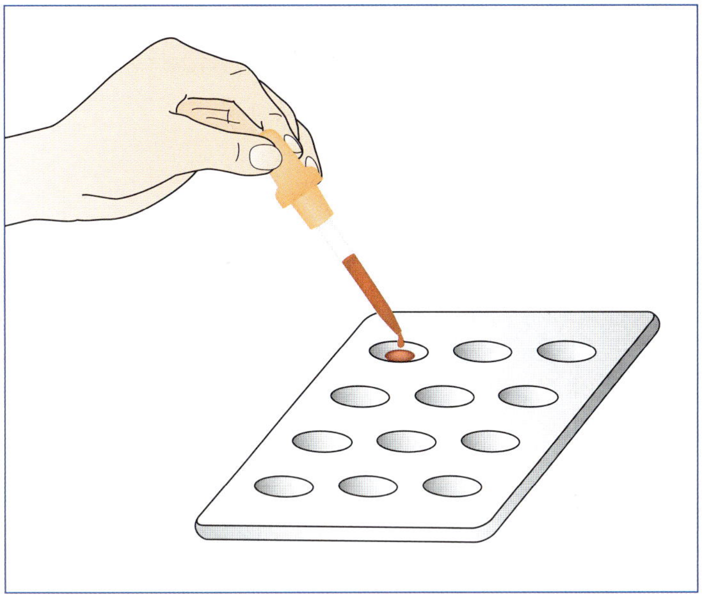

After completing your *entire* chart in this fashion and mixing all possible one-to-one combinations of solutions, clean up your workstation. Your teacher will provide disposal information. Wash your hands.

3. Use your chart to answer the following questions:

   a) Which combination of reactants seems to produce no reaction when mixed together?

   b) Which combination of reactants forms a gas? Can you guess which gas is formed? Try to deduce this from the reactants' names and chemical formulas.

   c) Which combination of reactants produces a color change when mixed together?

   d) Which combination of reactants forms precipitates quickly? Slowly?

   e) Which combination of reactants forms a yellow precipitate? A muddy brown precipitate? A white precipitate? A blue precipitate?

   f) Which combination of reactants produces heat? How could you tell?

   g) What evidence indicates that a chemical change is occurring?

4. Place the following chemicals in a quart-size resealable plastic bag with a zipper seal:

   One teaspoon (scoop)(~28 g) of calcium chloride ($CaCl_2$)

   One teaspoon (scoop)(~28 g) baking soda ($NaHCO_3$)

   Seal the bag and mix the powders.

   a) Record your observations in your *Active Chemistry* log. Did a chemical reaction occur?

   Pour 10 mL of phenol red indicator solution into the bag and seal quickly. Make sure the solids come in contact with the indicator solution.

   b) Observe the reaction and, in your *Active Chemistry* log, describe what you see.

   c) Did a chemical reaction occur in the plastic bag? If so, identify all of the evidence of the chemical change.

   d) For this particular reaction, calcium chloride and sodium hydrogen carbonate combined to produce an aqueous solution of sodium chloride and calcium oxide in addition to the carbon dioxide and water.

What are the names of the reactants? What are the names of the products?

5. Your teacher will provide you with a small amount (~25 mL) of limewater, a solution of calcium hydroxide ($Ca(OH)_2$), in a beaker or flask.

   Gently blow through a straw into the solution for a minute or so. One end of the straw should be submerged in the solution. You are actually bubbling carbon dioxide through the solution.

   a) Did a chemical reaction occur? What is the evidence?

Chem Talk

TESTS FOR CHEMICALS

Chemical Tests for Gases

In this activity you focused on chemical reactions, those processes that result in the formation of new products. You also tested for the presence of some of the new materials. You used **chemical tests** to identify the unknown substances. A chemical test is a form of a diagnostic test. To test for the presence of oxygen, you introduce a glowing splint into a test tube with a small amount of gas. If the splint bursts into a flame, you then know that the gas is oxygen. When you introduce a burning splint into a test tube and heard a loud pop, you assume the gas present to be hydrogen. In this activity you tested for the presence of carbon dioxide. Since carbon dioxide does not burn or support burning, by using a glowing or burning splint, you could not tell if a gas was carbon dioxide.

Chem Words

chemical test: a physical procedure or chemical reaction used to identify a substance.

(If the splint is extinguished you can say that the gas is neither oxygen nor hydrogen and therefore could be carbon dioxide.) The test for carbon dioxide uses limewater, a clear, colorless solution of calcium hydroxide in water. When you blew bubbles into the test tube you were actually blowing carbon dioxide from your lungs into the limewater. The carbon dioxide reacted with the calcium hydroxide forming a precipitate. The precipitate caused the limewater to turn cloudy in appearance.

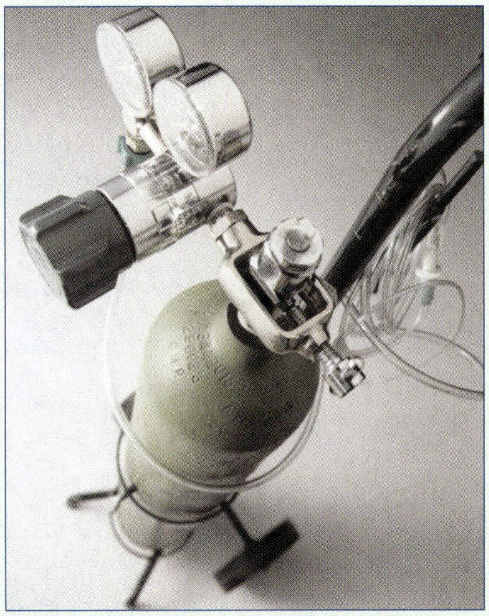

Indicators for Acids and Bases

When acids and bases are involved in a chemical reaction the appearance of the products is often very similar to the appearance of the reactants. (You will learn more about acids and bases in a later activity.) Therefore, indicators are used to determine the presence of an acid or base. Substances that change color when they react with an acid or a base are called **acid-base indicators**. In this activity you used phenol red, an acid-base indicator that turns yellow in the presence of an acid. Chemists use a great variety of acid-base indicators. You may also have used litmus in previous science classes. It is a very common indicator used in school laboratories.

Checking Up

1. What is a chemical test?
2. Describe how you can use a burning or glowing splint to test for hydrogen or oxygen.
3. Why does a glowing splint test not work with carbon dioxide?
4. What test is used to identify the presence of carbon dioxide?
5. What is a precipitate?
6. What are acid-base indicators and how are they useful?

Chem Words

acid-base indicator: a dye that has a certain color in an acid solution and a different color in a base solution.

Reflecting on the Activity and the Challenge

In this activity you saw evidence of chemical changes taking place when you observed a color change, a change in temperature, a gas being emitted, a precipitate being formed, or light being produced. Which of these chemical reactions would be an exciting or informative addition to your class's **Cool Chemistry Show**? Does your class want to be sure to include a variety of reactions that provide different types of evidence of a chemical reaction, or does your class just want to highlight a few of them? These are decisions your class will need to make as you build your **Cool Chemistry Show**.

Chemistry to Go

1. In both **Activity 1** and **Activity 2** you gathered evidence for chemical changes. However, this evidence does not always indicate a chemical change. For instance, a change in color can be evidence of a chemical change. However, when you add water to a powdered drink mix, the color often changes, but a chemical change has not taken place.

   In each of the following situations indicate whether the evidence suggests a chemical change or not:

   a) An acid is dissolved in water and heat is released.

   b) A burning match produces light.

   c) A "seed" crystal is placed in a supersaturated solution and the extra solute particles "join" the crystal and come out of the solution.

   d) A bottle of a carbonated beverage is opened and carbon dioxide is released.

   e) The glowing filament of a light bulb produces light.

   f) A small piece of metal is placed into an acid and hydrogen is released.

   g) Solutions of sodium hydroxide and copper (II) sulfate are mixed and a blue precipitate appears.

2. Anhydrous copper (II) sulfate ($CuSO_4$) is a white solid. When it is dissolved in water, the solution becomes blue. Is this a chemical change? Explain how you would defend your answer.

3. If a glass of carbonated soda drink is allowed to sit out for a period of time, you find that the drink seems to be flat. Discuss this observation in terms of whether this is a physical or chemical change.

Preparing for the Chapter Challenge

Select one of the reactions you observed in this activity that you thought was pretty cool. Describe how you might incorporate it into a possible event in the **Cool Chemistry Show** you are designing. Would it meet the needs of the fourth-grade teacher, the fifth-grade teacher, or both? What additional information would you need to be able to explain the chemistry to the audience?

Activity 3 Chemical Names and Formulas

GOALS

In this activity you will:

- Predict the charges of ions of some elements.
- Determine the formulas of ionic compounds.
- Write the conventional names of ionic compounds.
- Make observations to determine whether there is evidence that chemical changes occur on combining two ionic compounds.

What Do You Think?

Chemistry is the study of matter and its interactions. A great contribution to the study of chemistry was the discovery that all of the world is composed of elements and that there are approximately 100 elements in nature. The periodic table provides valuable information about each element.

- What information is provided for the element shown? What significance does this information have?

20	1.0
2	
Ca	
40.078	
$[Ar]4s^2$	
Calcium	

Record your ideas about these questions in your *Active Chemistry* log. Be prepared to discuss your responses with your small group and the class.

Investigate

1. The periodic table lists the elements in order of their atomic number. The atomic number is the number of protons (positively charged particles) in the nucleus of one atom of that element. For a neutral atom, the number of protons also equals the number of electrons (negatively charged particles).

Electrons are found outside the nucleus. A helium atom, with an atomic number of 2, has 2 protons in the nucleus and 2 electrons surrounding its nucleus.

For each of the following elements, write the symbol for the element and indicate the number of protons an atom of that element would have. (Refer to the periodic table.)

a) copper b) sulfur
c) zinc d) gold
e) oxygen f) carbon
g) silver h) chlorine
i) nitrogen j) hydrogen
k) magnesium l) iodine
m) iron n) calcium
o) aluminum p) sodium
q) potassium r) lead

2. Elements can combine to form compounds. A compound results when two or more different elements bond. Some compounds are comprised of positive and negative ions that are bound by their mutual attraction. An ion is an atom that has lost or gained electrons, and therefore is charged because its protons and electrons no longer balance and cancel each other. For example, when a chlorine atom gains one electron, it becomes a chloride ion with a charge of –1 (remember electrons have negative charge). When a sodium atom loses one electron, it becomes a sodium ion with a charge of +1 (because now there is one more proton than the number of electrons). The resulting compound is sodium chloride (NaCl), which you know as table salt and it is an ionic compound.

a) The chemical formula for the compound of potassium and bromine is KBr. Look at where potassium is located on the periodic table (Group 1) and also where bromine is located (Group 17). Each of these has an ionic charge of 1. Potassium is +1, bromine is –1. List four other compounds that are created from elements in Group 1 combining with elements in Group 17.

b) Magnesium forms an ion with a charge of +2 and oxygen forms an oxide ion with a –2 charge. The chemical formula for magnesium oxide is MgO. List four other compounds that are created from elements in Group 2 combining with elements in Group 16.

3. If the values of the charge on a positive ion and a negative ion are the same, the formula of the resulting compound is simply the chemical symbols of each element. If the values of the charge on a positive ion and a negative ion are not the same, subscripts can be used to balance them. For example, aluminum loses 3 electrons to become an ion with a charge of +3. An iodine atom gains only 1 electron to form an ion with a charge of –1. It takes 3 iodine atoms to accept the 3 electrons given up by aluminum. This is reflected in the formula AlI_3. (Note where the 3 is placed for the 3 iodine atoms.) Another example is $CaCl_2$, where 2 chloride ions (each gaining one electron) and one calcium ion (having lost 2 electrons) combine.

Write the chemical formula and name for the compound formed when the following pairs of elements are combined:

a) calcium and oxygen

b) aluminum and fluorine

c) boron and oxygen

d) strontium and nitrogen

e) barium and selenium

4. Some compounds, like baking soda, sodium hydrogen carbonate ($NaHCO_3$), incorporate polyatomic ions. Polyatomic ions are made up of several elements joined together. In the case of baking soda, the sodium (Na) ion has a charge of +1 and the hydrogen carbonate ion (the polyatomic ion HCO_3) has a charge of –1. (Note: hydrogen carbonate ion is also called bicarbonate ion.)

   Write the chemical formula for each compound below.

   a) potassium nitrate (nitrate: NO_3^{-1})

   b) barium sulfate (sulfate: SO_4^{-2})

   c) potassium sulfate

   d) sodium acetate (acetate: $C_2H_3O_2^{-1}$)

   Write the name for each compound below.

   e) $(NH_4)_2SO_4$ (ammonium: NH_4^{+1})

   f) $Al_2(CO_3)_3$

   g) $LiHCO_3$

   h) HNO_3

5. You have learned about ionic compounds that are made from positive and negative ions. In another class of compounds, called molecules, the atoms are bound by electrons being mutually attracted to the protons in adjacent atoms. These bonds are called covalent bonds, because atoms are sharing electrons. It is often useful to imagine, however, that the atoms inside of molecules are charged. These "imagined charges" are called oxidation numbers.

   a) The formula for carbon dioxide is CO_2. If you pretend this is an ionic compound, what is the charge (oxidation number) of carbon?

   b) Carbon monoxide is CO. What is the oxidation number of carbon now?

   c) Explain how you arrived at your answers.

You must wear safety goggles and a lab apron.

6. Do chemical changes occur every time reactants are mixed? Let's find out. Read the directions for this step so you can prepare a data table to record and describe all that you observe.

   Put equal amounts of baking soda, crushed Alka-Seltzer™ tablet, and baking powder into three separate test tubes respectively. Be sure to label the test tubes!

   Add equal amounts of water to each.

   a) Record your observations. (You should now know the chemical formula for baking soda $NaHCO_3$.)

   b) Place a glowing splint into the top of each test tube. Make note of what happens. A glowing splint bursts into flames in the presence of oxygen, and a glowing splint is extinguished in the presence of carbon dioxide. Which gases were most likely given off for each reaction?

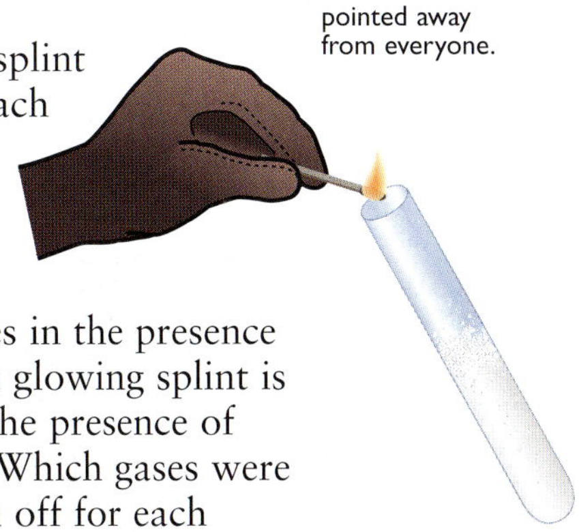

Hold the splint with tongs or wear a heatproof glove. Be sure the mouth of the test tube is pointed away from everyone.

7. When the reactions have stopped completely, your teacher will put three of the test tubes in a beaker of boiling water. Observe what happens.

   a) Make a note of the results in your *Active Chemistry* log.

8. Repeat **Steps 6** and 7 using clean test tubes, fresh reagents, and instead of water add

   • vinegar • ammonia.

9. Your teacher will give you a small amount of a white powdered substance that is either baking soda, Alka-Seltzer, or baking powder.

   a) Write down the number of your unknown and determine which of the three substances it is. Provide evidence to support your conclusion.

10. Clean all apparatus and the laboratory bench when you are finished. Dispose of all chemicals as directed by your teacher. Wash your hands.

Chem Talk

FORMING COMPOUNDS

Ionic Compounds

There are certainly more than 100 physically different materials in this world. With approximately 100 elements, how is it possible to have such a variety of materials? How is it possible to invent new materials for clothing, building, and food? Elements can combine to form compounds. A **compound** results when two or more different elements bond.

Some compounds are comprised of positive and negative **ions** that are bound by their mutual attraction. An ion is an atom that has lost or gained electrons, and therefore is charged because its protons and electrons no longer balance and cancel each other. For example, when an iodine atom gains one electron, it becomes a iodide ion with a charge of −1 (remember electrons have negative charge). A negatively charged ion is called an **anion**. When a potassium atom loses one electron, it becomes a potassium ion with a charge of +1 (because now there is one more proton than the number of electrons). A positively charged ion is called a **cation**. The resulting **ionic compound** is potassium iodide (KI). Potassium iodide is added to most of the table salt you use. Table salt (NaCl) is another example of an ionic compound.

Chem Words

compound: a material that consists of two or more elements united together in definite proportion.

ion: an electrically charged atom or group of atoms that has acquired a net charge, either negative or positive.

anion: a negatively charged ion.

cation: a positively charged ion.

ionic compound: a compound consisting of positive or negative ions.

If you refer to a periodic table you will notice that elements that form positive ions are on the left side of the table and elements that form negative ions are on the right side. Metals combine with nonmetals to form ionic compounds. Also, when two elements combine the name given the negative ion will end with -ide, and the compound is named with the metal or positive ion first. This is true for binary compounds, for example: sodium chloride, potassium bromide, and magnesium oxide. When dissolved in water, you would find that ionic compounds conduct electricity. In this activity you also investigated some compounds formed with **polyatomic ions**. Polyatomic ions are made up of several elements joined together. For example, Milk of Magnesia™ incorporates the hydroxide ion (OH^{-1}) with the magnesium ion to form magnesium hydroxide, $Mg(OH)_2$. The following table lists some polyatomic ions and their charges.

Polyatomic Ions		
nitrate	NO_3^{-1}	negative one, –1
sulfate	SO_4^{-2}	negative two, –2
hydroxide	OH^{-1}	negative one, –1
carbonate	CO_3^{-2}	negative two, –2
hydrogen carbonate	HCO_3^{-1}	negative one, –1
acetate	$C_2H_3O_2^{-1}$	negative one, –1
ammonium	NH_4^{+1}	positive one, +1

Molecular Compounds

You also learned about **molecular compounds** in this activity. When two atoms of molecular elements come together, neither atom gains nor loses an electron. Instead, the bonding electrons are shared between the two atoms. The mutual attraction of two nuclei for a shared pair of bonding electrons is called a **covalent bond**. Molecular compounds are usually formed by nonmetal-nonmetal combinations. You would find that when dissolved in water, molecular compounds do not conduct electricity.

With covalent bonds you also found that it is often useful to imagine that the atoms inside of molecules are charged. These "imagined

Chem Words

polyatomic ion: an ion that consists of two or more atoms that are covalently bonded and have either a positive or negative charge.

molecular compound: two or more atoms bond together by sharing electrons (covalent bond).

covalent bond: a bond formed when two atoms combine and share their paired electrons with each other.

Checking Up

1. If there are only about 100 elements in this world, why are there so many different materials?
2. What is an ion?
3. How are ionic compounds formed?
4. What is a polyatomic ion? Provide an example of a compound formed with a polyatomic ion.
5. How are molecular compounds formed?
6. Distinguish between an ionic and a covalent bond.

Chem Words

oxidation number: a number assigned to an element in a compound designating the number of electrons the element has lost, gained, or shared in forming that compound.

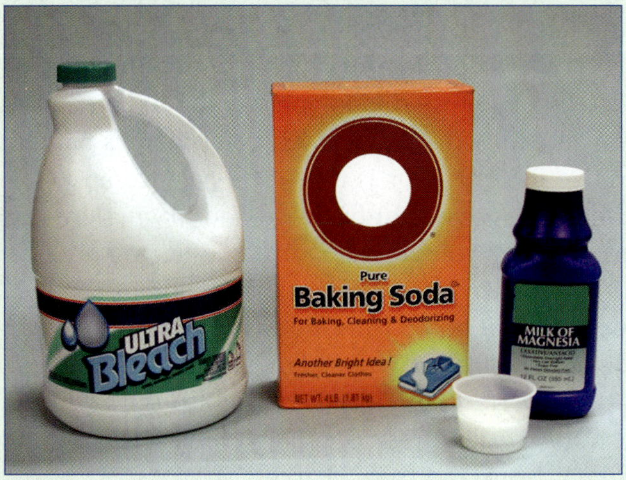

charges" are used as a type of bookkeeping and are called **oxidation numbers**. In both ionic compounds and molecular compounds the atoms achieve a stable state, similar to the noble gases.

Reflecting on the Activity and the Challenge

In this activity you have learned how to write the formulas for many compounds and how to name some compounds. You have also investigated both ionic and molecular compounds. As you prepare your presentation for your **Cool Chemistry Show**, you will want to include your knowledge of formulas, the names of compounds, and the different kinds of compounds. Remember that you will be providing the teacher with an explanation of why you included certain demonstrations, and you will also want to include explanations that are grade-appropriate. Think about how much information you will need to provide for each demonstration.

Chemistry to Go

1. Write the chemical formula and name for the compound formed when the following pairs of elements are combined:
   a) sodium and bromine
   b) potassium and sulfur
   c) magnesium and chlorine
   d) cesium and iodine
   e) aluminum and oxygen

2. Write the chemical formula for each of the following:
   a) hydrogen nitrate (nitric acid)
   b) ammonium hydroxide
   c) calcium carbonate
   d) hydrogen acetate (acetic acid)
3. a) Write the chemical formula for copper (II) sulfate. The (II) indicates that this copper ion has a +2 charge.
   b) Oxygen ions usually have a negative 2 charge (–2). How would formulas for iron (II) oxide differ from iron (III) oxide?
4. You may have noticed that all the elements in the first column of the periodic table (the alkali metals: lithium, sodium, potassium, rubidium, and cesium) have a +1 charge when they combine with negative ions. Another group of positive ions are the alkaline earth metals (beryllium, magnesium, calcium, strontium, and barium), located in the second column of the periodic table. What charge is typical for ions of the alkaline earth metals?
5. The formula for sodium phosphate is Na_3PO_4. What is the charge on the polyatomic phosphate ion? What information did you use to arrive at your answer?
6. When you write the formula for sodium hydroxide, you do not have to put parentheses around the hydroxide polyatomic ion. However, when writing the formula for aluminum hydroxide, you must put parentheses around the hydroxide polyatomic ion.
   a) Write each formula.
   b) Explain why the parentheses are necessary for aluminum hydroxide.
7. a) If the chemical formula for iron (III) chloride is $FeCl_3$, what is the chemical formula for iron (III) nitrate?
   b) If the chemical formula for lead (II) oxide is PbO, what is the chemical formula for lead (II) sulfate?
   c) If the chemical formula for silver chloride is AgCl, what is the chemical formula for silver nitrate?
8. In **Activity 2**, you tested various compounds for chemical changes. (Barium nitrate, sodium hydroxide, sodium hydrogen carbonate, copper (II) sulfate, potassium iodide, silver nitrate, iron (III) nitrate, and hydrochloric acid.) Write the chemical formulas for each of the reactants.

Preparing for the Chapter Challenge

Review any chemical reactions you are considering including in your **Cool Chemistry Show**. Write the formulas of any compounds that you plan to use.

Activity 4 Chemical Equations

GOALS

In this activity you will:

- Represent chemical changes using word equations and chemical equations.
- Distinguish between different classes of chemical reactions.
- Predict the possible products of single-displacement and double-displacement reactions.
- Determine whether a reaction has occurred based on evidence observed.
- Use the principle of conservation of matter to balance chemical reactions.

What Do You Think?

In **Activity 1** you mixed zinc metal and hydrochloric acid. You noticed that a gas was produced from the fizzing that occurred.

- How could you communicate (tell about) the reactants and products of this reaction in a compact way?

Record your ideas about this question in your *Active Chemistry* log. Be prepared to discuss your responses with your small group and with the class.

Investigate

1. Watch closely as your teacher shows you some cool chemistry.

   a) Record your observations in your *Active Chemistry* log.

2. Here's how the cool chemistry was done. Into each of three beakers that appeared empty, your teacher added about 45 mL of 2.0 M ammonium hydroxide solution.

   Before beginning the demonstration, your teacher had also added the following to each beaker:

Beaker One—20 drops of the indicator phenolphthalein solution;

Beaker Two—15 drops of 1 M magnesium sulfate solution;

Beaker Three—15 drops of 1 M copper (II) sulfate solution.

a) Did a chemical reaction take place in each beaker? What evidence do you have to justify your answer?

3. There are many chemical reactions that can occur. You have already observed some of them. Chemists group most of these reactions into four main categories. They are:

- synthesis reactions
- decomposition reactions
- single-displacement reactions
- double-displacement reactions

a) What do the words synthesis and decomposition mean?

4. In a synthesis reaction two or more chemicals combine to form a compound.

$A + B \rightarrow AB$

Here is an example of a synthesis reaction. When magnesium and oxygen react, a white solid, magnesium oxide is formed.

This can be written as a word equation:

Magnesium (solid) and oxygen (gas) produce magnesium oxide (solid).

This can also be written using a chemical equation:

$?Mg + ?O_2 \rightarrow ?MgO$

(The subscript communicates the number of atoms in one molecule. Oxygen is diatomic, that means that it exists as a molecule made up of two atoms.)

a) What do you think are the advantages of writing a reaction using chemical symbols?

Any equation in chemistry must follow scientific laws or principles. The number of atoms of each element must be equal before and after the reaction.

b) In the equation above, how many atoms of oxygen are in the reactants (before the reaction)?

c) How many oxygen atoms are in the product (after the reaction)?

d) What appears to be the problem with the equation? How could you correct this problem?

Write a 2 in front of the MgO.

$?Mg + O_2 \rightarrow 2MgO$

(The number in front of a chemical formula, called a coefficient, communicates the number of molecules or formula units that are involved in the reaction. In this equation there are two molecules of magnesium oxide represented. That is, there is a total of two magnesium atoms and two oxygen atoms.)

e) How many magnesium atoms are now represented in the product?

f) How many reactant atoms of magnesium are shown?

g) What now appears to be the problem with the equation? How could you correct this problem?

Write a two in front of the Mg.

$2Mg + O_2 \rightarrow 2MgO$

h) The chemical equation above is now balanced. (The number of magnesium and oxygen atoms in the product is equal to the number in the reactant.) In your own words, explain the meaning of a balanced equation. How does the chemical equation communicate what happened in the reaction, and how does it follow the Law of Conservation of Matter?

When writing a chemical equation the states of the reactants and products are also given. The following symbols are used:

- (s) for solid
- (l) for liquid
- (g) for gas
- (aq) for aqueous, meaning in a water solution

The complete balanced chemical equation for the reaction of magnesium and oxygen is:

$2\ Mg_{(s)} + O_{2(g)} \rightarrow 2\ MgO_{(s)}$

5. Write a word equation and a balanced chemical equation for each of the following synthesis reactions.

Note that there are eight elements that are diatomic, that means that they exist as a molecule comprised of two atoms. They are hydrogen (H_2), nitrogen (N_2), oxygen (O_2), fluorine (F_2), chlorine (Cl_2), bromine (Br_2), iodine (I_2) and astatine (At_2). If you need to include any of these elements in an uncombined state in a chemical equation, don't forget the 2 as a subscript.

a) Solid carbon (C) burns in air (oxygen gas) to form carbon dioxide gas (CO_2).

b) Hydrogen gas reacts with oxygen gas to form liquid water (H_2O).

c) A piece of solid iron (Fe) over time will react with oxygen to form iron (III) oxide (Fe_2O_3).

d) A piece of solid sodium (Na) is dropped into a container of chlorine gas to produce solid sodium chloride (NaCl).

6. Water can be separated into its elements with an input of energy. The equation for this reaction is:

Water (liquid) and energy produces hydrogen (gas) and oxygen (gas).

$2H_2O_{(l)} + \text{energy} \rightarrow 2\ H_{2(g)} + O_{2(g)}$

a) Is the equation properly balanced? How did you check?

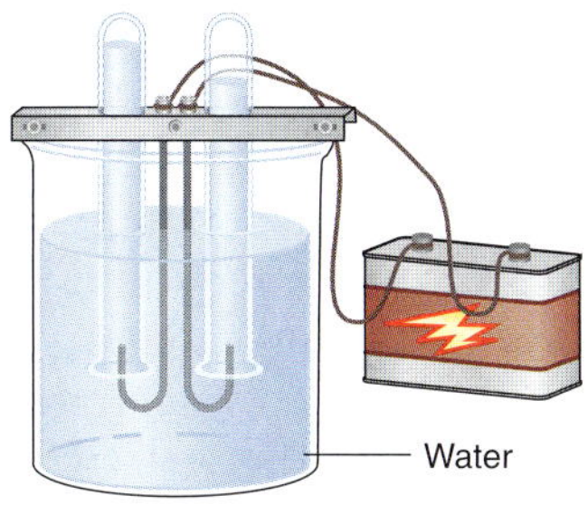

Decomposition of water into $H_{2(g)}$ and $O_{2(g)}$ by electrolysis

When a substance breaks down into its component parts, the process is called a **decomposition** reaction.

form products. Such was the case when solid zinc was dropped into hydrochloric acid in **Activity 1**, forming hydrogen gas and aqueous zinc chloride solution.

The equation for this reaction is:

Zinc (solid) and hydrochloric acid (aqueous) produces hydrogen (gas) and zinc chloride (solution).

$Zn_{(s)} + 2HCl_{(aq)} \rightarrow H_{2(g)} + ZnCl_{2(aq)}$

a) Check to ensure that the chemical equation is properly balanced. Complete the following table in your log.

Number of Atoms

	Before	After	Balanced
Zn	1	1	yes
H			
Cl			

$AB \rightarrow A + B$

Write word and balanced chemical equations for the following decomposition reactions. (Remember that some elements are diatomic –H_2, N_2, etc.):

b) sodium chloride solid ($NaCl_{(s)}$)

c) potassium iodide solid ($KI_{(s)}$)

d) magnesium bromide solid ($MgBr_{2(s)}$)

7. The reactions mentioned above involve elements combining to form compounds or compounds breaking up to form elements.

There are other reactions that involve elements reacting with compounds to

The reaction with zinc and hydrochloric acid is called a single-displacement reaction because zinc replaces the hydrogen in the acid.

$A + BC \rightarrow B + AC$

Write word and chemical equations for the following:

b) A piece of iron (Fe) is added to an aqueous solution of copper (II) sulfate ($CuSO_4$) and produces iron sulfate ($FeSO_4$) and copper.

c) Solid lead (Pb) is added to an aqueous solution of silver nitrate ($AgNO_3$) and produces lead nitrate $Pb(NO_3)_2$ and silver.

d) Aluminum foil (Al) is placed in a beaker of aqueous copper (II) hydroxide ($Cu(OH)_2$) and produces aluminum hydroxide ($Al(OH)_3$) and copper.

Balance each of the equations, if you have not done so.

8. Another type of reaction is a double-displacement reaction.

$AB + CD \rightarrow CB + AD$

Safety goggles and a lab apron must be worn during this activity.

Try some double-displacement reactions on your own in your group. Use the chart below to guide your work. The compounds are in water solution.

a) Record your observations of the reactants before you mix them. For example, record your observations of potassium carbonate and silver nitrate before you mix them.

b) Create a chart in your log to record your observations after you mix the reactants. You may wish to use a chart similar to the one below.

9. Mix three drops of one solution (i.e., potassium carbonate) with three drops of another solution (i.e., silver nitrate), as indicated by the first box on the chart. You can mix these solutions in the well of a spot plate or on a plastic surface. Do not allow the tip of the dropper of one solution to come in contact with another solution. This is important to prevent contamination of solutions.

a) In the chart in your log, record your observations after mixing the reactants.

Continue with the other reactants (i.e., potassium carbonate with copper (II) sulfate; then potassium carbonate with magnesium sulfate; and so on).

b) Record all your observations in your *Active Chemistry* log.

	Silver Nitrate ($AgNO_3$)	Copper (II) Sulfate ($CuSO_4$)	Magnesium Sulfate ($MgSO_4$)	Sodium Hydroxide (NaOH)
Potassium Carbonate (K_2CO_3)	1.	2.	3.	4.
Sodium Hydroxide (NaOH)	5.	6.	7.	8.
Potassium Iodide (KI)	9.	10.	11.	12.
Iron (III) Chloride ($FeCl_3$)	13.	14.	15.	16.

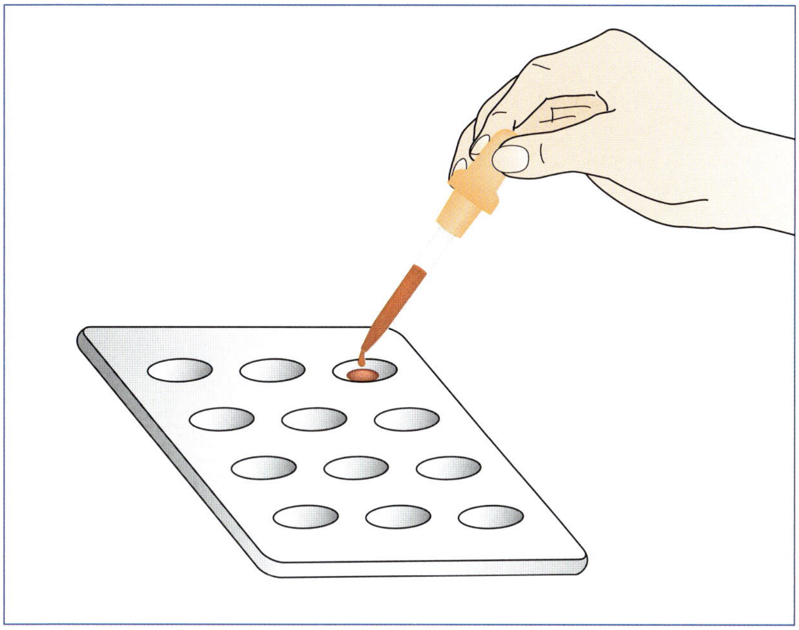

10. Clean all apparatus and the laboratory bench when you are finished. Dispose of all chemicals as directed by your teacher. Wash your hands.
11. Look at the data from the double-displacement reactions that you observed.
    a) Do you think a chemical reaction took place in each case? Explain your answer.
    b) Are you able to predict or identify any of the products that were formed? If so, which ones?
    c) Write word equations and balanced chemical equations for each reaction that you observed. In all cases, assume only two reactants are used and only two products are formed. Use the following formulas to help you write your equations:

copper (II) carbonate ($CuCO_{3(s)}$);	copper (II) hydroxide ($Cu(OH)_{2(s)}$);
copper (II) iodide ($CuI_{2(s)}$);	iron (III) nitrate ($Fe(NO_3)_{3(s)}$);
iron (III) hydroxide ($Fe(OH)_{3(s)}$);	magnesium carbonate ($MgCO_{3(s)}$);
silver carbonate ($Ag_2CO_{3(s)}$);	silver chloride ($AgCl_{(s)}$);
silver hydroxide ($AgOH_{(s)}$);	potassium nitrate ($KNO_{3(aq)}$);
potassium sulfate ($K_2SO_{4(aq)}$);	magnesium hydroxide ($Mg(OH)_{2(s)}$);
silver iodide ($AgI_{(s)}$);	sodium chloride ($NaCl_{(aq)}$);
sodium nitrate ($NaNO_{3(aq)}$);	sodium sulfate ($Na_2SO_{4(aq)}$);

Chem Talk

CHEMICAL REACTIONS

Kinds of Chemical Reactions

A chemical reaction takes place when starting materials (reactants) change to new materials (products). Synthesis, decomposition, single-displacement, and double-displacement reactions are some common kinds of chemical reactions.

Synthesis and Decomposition Reactions

Synthesis means "putting together." In a **synthesis reaction** two or more elements combine to form one or more compounds. In this activity you investigated the reaction of magnesium in oxygen to form magnesium oxide. The opposite kind of reaction is a **decomposition reaction.** In chemical decomposition a compound is separated into its elements.

Chem Words

synthesis reaction: a chemical reaction in which two or more substances combine to form a compound.

decomposition: a chemical reaction in which a single compound reacts to give two or more products.

single-displacement reaction: a reaction in which an element displaces or replaces another element in a compound.

Single-displacement Reactions

A **single-displacement reaction** is one in which an element reacts with a compound to produce a new element and an ionic compound. For example, a single-displacement reaction occurs when you put a strip of zinc in a copper (II) sulfate solution. The zinc metal exchanges places with the copper cations. You also observed that the free copper metal atoms now plate on the zinc strip.

Activity Series

If you put copper metal in a zinc sulfate solution you would find that no reaction would take place, as expected. Zinc exchanges places with copper, but copper will not exchange places with zinc. You have learned something about a property of copper and zinc. Zinc is more active than copper. If you were to experiment with different metals and metallic solutions, you should be able to create an activity series of metals. The activity series of metals can be put into a table that you can use to predict if a reaction will take place. The table looks like the one on the following page. The table permits you to determine how a metal will react in a metal solution. A metal that is more active than another will dissolve into the metal solution and plate out the less

Activity Series of Metals (Most Active to Least Active)	
1. Lithium ($Li \rightarrow Li^+ + e^-$)	8. Iron ($Fe \rightarrow Fe^{2+} + 2e^-$)
2. Potassium ($K \rightarrow K^+ + e^-$)	9. Lead ($Pb \rightarrow Pb^{2+} + 2e^-$)
3. Calcium ($Ca \rightarrow Ca^{2+} + 2e^-$)	10. Hydrogen ($H_2(g) \rightarrow 2H^+ + 2e^-$)
4. Sodium ($Na \rightarrow Na^+ + 1e^-$)	11. Copper ($Cu \rightarrow Cu^{2+} + 2e^-$)
5. Magnesium ($Mg \rightarrow Mg^{2+} + 2e^-$)	12. Mercury ($Hg \rightarrow Hg^{2+} + 2e^-$)
6. Aluminum ($Al \rightarrow Al^{3+} + 3e^-$)	13. Silver ($Ag \rightarrow Ag^+ + e^-$)
7. Zinc ($Zn \rightarrow Zn^{2+} + 2e^-$)	14. Gold ($Au \rightarrow Au^{3+} + 3e^-$)

active metal. Zinc replaced the copper and the copper plated the zinc. For example, let's say that you place a strip of copper in a silver nitrate solution. According to the table, the copper will dissolve into copper ions (Cu^{2+}) and the silver ions will plate out on the copper as silver.

In addition to metals, you will notice that hydrogen gas is also listed in the activity series. In your investigation you found that different metals produced hydrogen gas and metal ions when they reacted with hydrochloric acid. You read above that metals can replace less active metals in metal salt solutions. Metals that are more active than hydrogen can replace the hydrogen from water to form metal hydroxides. As an example, if you were to react potassium metal with water, you would get hydrogen gas and potassium hydroxide solution. The equation is:

$$2K_{(s)} + 2HOH_{(l)} \rightarrow H_{2(g)} + 2\,KOH_{(aq)}$$

Double-displacement Reactions

Double-displacement reactions are different from single-displacement reactions, in that you start with two aqueous phase solutions and when they react they "switch partners." An example of this type of reaction is:

$$Ba(NO_3)_{2(aq)} + Na_2SO_{4(aq)} \rightarrow 2\,NaNO_{3(aq)} + BaSO_{4(s)}$$

Note that the cation of the one compound (Ba^{2+}) exchanged places with the cation (Na+) of the other compound. The solid $BaSO_4$ is a

Chem Words

double-displacement reaction: a chemical reaction in which two ionic compounds "exchange" cations to produce two new compounds.

Chem Words

salts: ionic compounds in which the anion is not a hydroxide ion (OH–) and the cation is not a hydrogen proton (H+).

precipitate. When examining double-displacement reactions, you know that a reaction has taken place if you see:

- a precipitate
- a gas
- water

If none of these are present, then no reaction had occurred.

Solubility Rules

A precipitate will form if the compound is not soluble in water. In the example above, barium sulfate was not soluble in water. This was noted in the equation by referring to it as a solid (s). Chemists have created a set of solubility rules for **salts**. Salts are classified as ionic compounds in which the anion is not a hydroxide ion (OH^-) and the cation is not a hydrogen proton (H^+).

Solubility Rules
1. All salts (defined as ionic compounds) of the alkali (Group I on the periodic table) metals and the ammonium ion are soluble in water.
2. All chlorides, bromides, and iodides are soluble with the exception of silver, lead, and mercury halides.
3. All nitrate, chlorate, perchlorate, and acetate salts are soluble.
4. All sulfates are soluble with the exception of calcium, barium, strontium, and lead.
5. All carbonates, phosphates, chromates, hydroxides, and sulfides are insoluble except when they are combined with alkali metals or the ammonium ion.

In the example above, barium sulfate formed a precipitate. Since barium sulfate is insoluble, this agrees with the solubility rules. If you mixed silver nitrate with sodium chloride, would you expect to get a precipitate? The two products that would form are silver chloride and sodium nitrate. The solubility rule #2 tells you that silver chloride is insoluble and solubility rule #3 tells you that sodium nitrate is soluble. Using these rules, you can now predict whether a mixture will produce a precipitate or not.

Checking Up

1. What is a synthesis reaction? Provide an example.
2. What is a decomposition reaction? Provide an example.
3. Distinguish between a single and a double-displacement reaction.
4. What evidence would you look for to determine if a double-displacement reaction has occurred?
5. Will hydrochloric acid react with a clean strip of copper? Explain your answer.
6. Is calcium sulfate soluble in water? Justify your answer.

Reflecting on the Activity and the Challenge

In this activity you have learned about a number of different types of reactions: synthesis, decomposition, single-displacement and double-displacement reactions. Knowing these types of reactions can help you predict the products of some chemical reactions. You'll need to decide if you want the audience for the **Cool Chemistry Show** to learn about these reactions types. You also learned how to write balanced equations for some of the reactions you observed. Think about a creative way of showing how you can explain balancing chemical equation to your elementary school audience. In this activity you have learned how to write the formulas for many compounds and how to name some compounds. You have also investigated both ionic and molecular compounds. As you prepare your presentation for your **Cool Chemistry Show**, you will want to include your knowledge of formulas, the names of compounds, and the different kinds of compounds. Remember that you will be providing the teacher with an explanation of why you included certain demonstrations, and you will also want to include explanations that are grade-appropriate. Think about how much information you will need to provide each.

Chemistry to Go

1. Baking soda ($NaHCO_3$) has been used in several reactions in previous activities. When heat is applied to baking soda, three compounds are produced. Two of the compounds are gases and the other is a solid. If the two gases are water and carbon dioxide, what is the third product? Explain how you arrived at your answer.
2. When solutions of sodium hydroxide and potassium carbonate are mixed together, no apparent reaction takes place. The same is true when you mix sodium hydroxide and potassium iodide together. Explain this observation.
3. If you mix sodium sulfate and barium nitrate solutions together, you get a white precipitate. What is the precipitate that formed? What information did you use to arrive at your answer?
4. There were five combinations that you mixed together that did not form a precipitate. Complete and balance the following equations:

   a) $K_2CO_{3(aq)} + NaOH_{(aq)} \rightarrow$

   b) $KI_{(aq)} + MgSO_{4(aq)} \rightarrow$

   c) $KI_{(aq)} + NaOH_{(aq)} \rightarrow$

   d) $FeCl_{3(aq)} + CuSO_{4(aq)} \rightarrow$

   e) $FeCl_{3(aq)} + MgSO_{4(aq)} \rightarrow$
5. Use the solubility rules to explain why the products in **Question 4** do not form precipitates.

Activity 5 Chemical Energy

GOALS

In this activity you will:

- Make hot packs and cold packs.
- Observe energy changes when matter changes.
- Determine whether energy changes are endothermic or exothermic from a particular point of reference.

What Do You Think?

Two chemicals at room temperature are mixed together and the temperature cools drastically. You may have used cold packs that work like this when you were injured.

- How do the chemicals in the cold pack lower the temperature?

Record your ideas about this question in your *Active Chemistry* log. Be prepared to discuss your responses with your small group and with the class.

Investigate

1. To make a cold pack, place 10 g of ammonium nitrate in a quart-size, resealable plastic bag. Add 20 mL water to the bag and seal.

   a) Record your observations.

   In an endothermic chemical reaction, energy in the form of heat is absorbed in the process. In an exothermic chemical reaction, energy in the form of heat is given off in the process.

Safety goggles and a lab apron must be worn during this activity.

b) Was the cold pack an example of an exothermic or endothermic chemical reaction?

2. Make a hot pack by placing 20 g of sodium carbonate (or calcium chloride) in a quart-size, resealable plastic bag. Add 20 mL of water to the bag and seal.

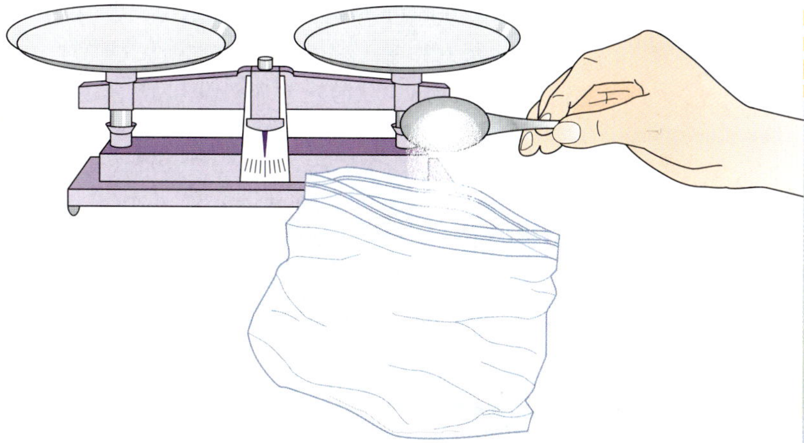

a) Record your observations.

b) Was the reaction exothermic (heat generating) or endothermic (heat absorbing)?

3. To a flask containing 16 g of ammonium thiocyanate, add 32 g of barium hydroxide.

Place a rubber stopper in the mouth of the flask. Shake it vigorously.

Put the stoppered flask on a wood board that has been wet down with puddles of water.

a) Record your observations. Cool chemistry!

b) Was the reaction exothermic (heat generating) or endothermic (heat absorbing)?

4. Using a chemical scoop, transfer a few pellets of sodium hydroxide to a test tube half full of water. Carefully feel the side of the test tube.

a) Record your observations.

b) Was the reaction exothermic (heat generating) or endothermic (heat absorbing)?

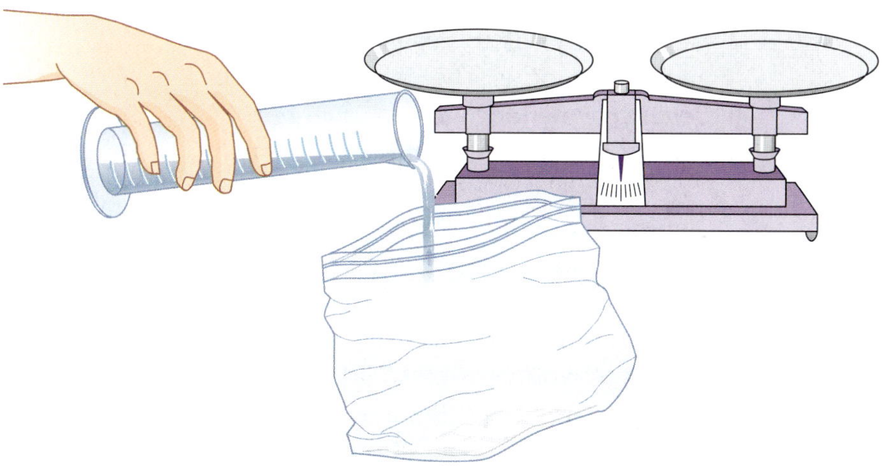

Be careful when working with the sodium hydroxide pellets. Wear rubber gloves and eye protection. If you should accidentally drop a pellet, do not try to pick it up with your bare hands as it may burn them. Use gloved hands to retrieve the pellets.

Chem Talk

ENDOTHERMIC AND EXOTHERMIC REACTIONS

A process is described as **endothermic** when heat energy is absorbed, increasing the internal energy of the system. One example of an endothermic reaction is the cold pack you made with ammonium nitrate. Another example is the decomposition of potassium chlorate. In this reaction, energy must be added to the system in order to cause the decomposition of the potassium chlorate to form the products of oxygen gas and potassium chloride. If you touch a container that holds an endothermic process, it will feel cool to the touch.

An **exothermic** process results when heat energy is released, decreasing the internal energy of the system. If you touch a container that holds an exothermic process, it will feel warm or hot to the touch. One example of an exothermic reaction is the hot pack you made with sodium carbonate. Another example was the combining of sodium hydroxide solution with hydrochloric acid. This reaction produces sodium chloride and water and releases energy to the environment. The terms endothermic and exothermic can be used when describing both physical and chemical changes.

Why is energy so important? In order for a chemical reaction to take place, the particles (reactants) involved in the reaction must interact. Not all collisions result in a chemical reaction. The particles involved must have enough energy to enable them to react with each other. The colliding particles must have enough kinetic energy to break the existing bonds in order for new bonds to be formed. The minimum energy required for a chemical reaction is activation energy. Bond breaking is an endothermic process and requires an addition of energy and bond formation is an exothermic process and requires a release of energy. (Physics reminder: Kinetic energy is the energy of motion $KE = \frac{1}{2}mv^2$.)

Chem Words

endothermic change or reaction: a change in which energy in the form of heat is absorbed from the surrounding environment resulting in an increase in the internal energy of the system.

exothermic change or reaction: a change in which energy in the form of heat is released from a system resulting in a decrease in the internal energy of the system.

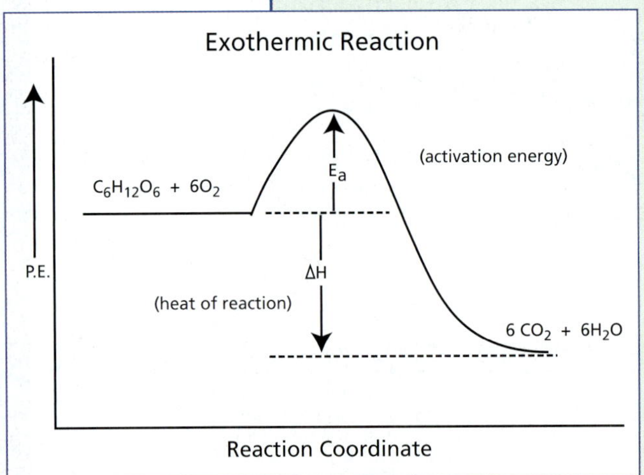

When more energy is released as products form than is absorbed to break the bonds in the reactants, the reaction is said to be exothermic, as shown in the graph. A chemical reaction in which energy is released is called exothermic. The prefix *exo* means exit and *thermo* means heat or energy.

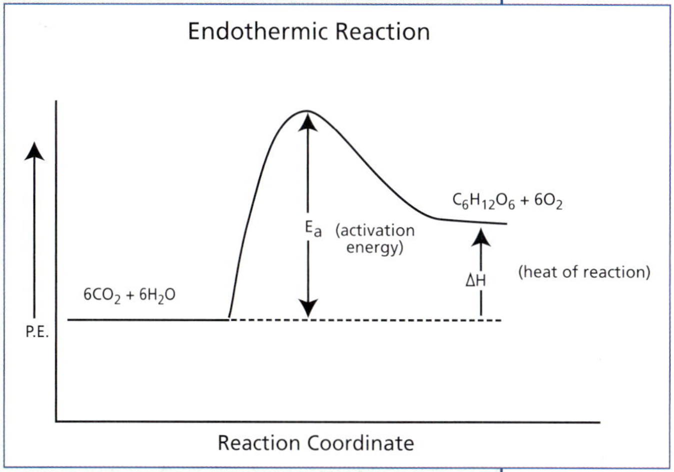

The cells in your body use glucose to get energy for cellular respiration. Glucose plus oxygen (from breathing) provides you with energy.

$C_6H_{12}O_6 + 6O_2 \rightarrow 6CO_2 + 6H_2O + \text{energy}$

When more energy is needed to break the bonds in the reactants than is given off by forming the bonds in the products, the reaction is said to be endothermic—a chemical reaction in which energy is absorbed, as shown in the second graph. In some endothermic reactions, energy must be added in order for the reaction to occur.

In the process of photosynthesis, plants use energy from light to break the bonds of carbon dioxide and water in order to form the products glucose and oxygen. The glucose produced makes the plant grow and the reaction also provides animals with oxygen:

$6CO_2 + 6H_2O + \text{light energy} \rightarrow$
$C_6H_{12}O_6 + 6O_2$

Conservation of Energy

Energy transfer is an important feature in all chemical changes. Energy is transferred whether the chemical reaction takes place in the human body, like the metabolism of carbohydrates, or in a car, like the combustion of gasoline. In both these cases energy is released into the surroundings.

Checking Up

1. Describe an endothermic reaction.
2. Describe an exothermic reaction.
3. Explain whether each of the following is endothermic or exothermic:
   a) bond breaking
   b) bond forming.
4. Distinguish between heat and temperature.

Chem Words

temperature: the measure of the average kinetic energy of all the particles of the material.

The photosynthesis that occurs in living plants and the decomposition of water into hydrogen and oxygen both remove energy from the surroundings. According to the Law of Conservation of Energy, the energy absorbed from the surroundings or released to the surroundings is equal to the change in the energy of the system.

Energy is the great organizing principle of all science. The conservation of energy allows you to better understand the world around you. Energy exists as light energy, heat energy, sound energy, nuclear energy, kinetic energy, chemical energy as well as other forms. The total energy in any closed system remains the same. In the simplest physical systems, it is quite easy to describe the energy changes. When a bowling ball falls, its original gravitational potential energy becomes kinetic energy as the ball increases its speed. After the ball hits the ground, this kinetic energy is converted to sound energy (you hear the crash) and heat energy (you can measure a temperature rise of the ball and ground) and the compression and vibrations of the ground. Each of these can be measured and it is always found that the total energy before an event is equal to the total energy after the event. When a human being is involved, you notice other energy interactions. A person is able to eat food and digest the food. The energy released from this slow-burn of the food (metabolism) is able to keep the body at about 37°C. As all non-living things in the room cool down to room temperature, humans are able to stay warm in the 22°C room environment. People also use the food energy that they ingest for moving muscles, keeping the heart pumping and operating all human functions. Living organisms are superb energy conversion systems.

Heat and Temperature

Both heat and **temperature** have been mentioned in this activity. It is important to note that heat and temperature are not the same, although they are related. Heat is one form of energy. When two materials of different temperatures interact, they exchange heat energy until they arrive at the same temperature. Temperature is a number associated with how hot or cold something is. On a molecular level, temperature is related to the average kinetic energy of the atoms in the material. All particles in a material are in a constant state of motion. The temperature of the material is a measurement of this molecular motion. If the kinetic energy of the particles increases, the temperature increases. Temperature can be defined as the measure of the average kinetic energy of all the particles of a substance. A thermometer is an instrument that measures temperature. Heat is the transfer of energy, which often results in a change in the kinetic energy of particles—a change in the temperature of the system.

Reflecting on the Activity and the Challenge

Energy is involved in any process that requires the breaking or making of bonds. The process may be a chemical or physical change. Sometimes energy changes are not noticed or measured, but other times an energy change is significant enough to be detected. In this activity you explored both endothermic and exothermic processes. As you select activities to include in the **Cool Chemistry Show**, you must be aware of any heat energy released or absorbed. The audience may be interested to learn about the energy changes that accompany chemical processes. In addition, your awareness will ensure that the presentations are safe for both the presenters and the audience.

Chemistry to Go

1. Identify the following changes as endothermic or exothermic. (Ask yourself whether the reaction requires the addition of heat energy to occur or does it release energy in the form of heat.)
   a) Melting ice.
   b) Lighting a match.
   c) Dry ice subliming into carbon dioxide gas.
   d) Frying an egg.
   e) Burning gasoline.
   f) Explosion of hydrogen gas.
2. The water in a teapot is heated on a stovetop. The temperature of the water increases. Is this an endothermic or exothermic process?
3. If a red-hot piece of iron is dropped into a bucket of water, what type of heat change takes place in reference to the water? What type of heat change takes place in reference to the iron?
4. Explain in terms of energy flow how a cold pack works on a sprained ankle.
5. If ice is at –20°C and you apply some heat by sitting a beaker of the ice on a hot plate, explain why the ice does not appear to be melting.

Preparing for the Chapter Challenge

Review the chemical reactions that you have so far planned to feature in your **Cool Chemistry Show**. Have you included both endothermic and exothermic reactions? In a paragraph describe the difference between the two types of reactions. Be sure to mention why energy transfers are important.

Inquiring Further

1. Commercial cold and hot packs

Research several cold and hot packs. What materials are used in the packs, and how is the chemical reaction activated? You should be aware that there are a variety of commercially available hot and cold packs. In general some are reusable and some are not. Those that are not reusable are typically chemical reactions. Those that are reusable are typically physical changes.

Design a process to make a cold pack, using your research. Have your teacher approve your design before you actually try it out.

2. The colligative property of a solvent

A salt solution will depress the freezing point of water. This is commonly known as the colligative property of the solvent. When you add anti-freeze (ethylene glycol) to water, you find that the freezing point of the water is lowered, which prevents the water in the car radiator from freezing at 0°C. It also elevates the boiling point of the water, which will prevent the water solution from boiling. Design an experiment to demonstrate the colligative property of a solvent. With the approval of your teacher carry out your experiment.

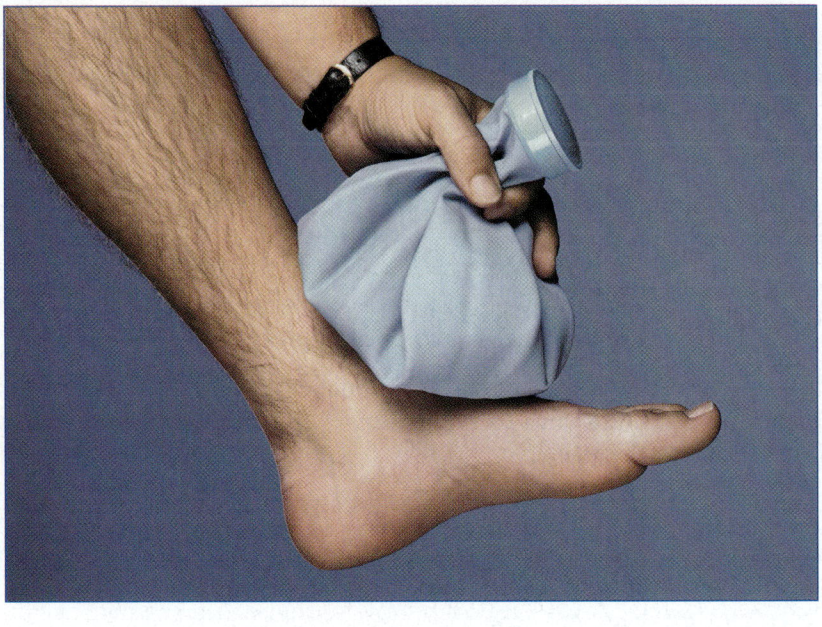

Activity 6 Reaction Rates

GOALS

In this activity you will:

- Discover conditions that make a reaction proceed faster or slower.
- Discuss explanations for why this happens at the molecular level.

What Do You Think?

Have you ever wondered why some chemical reactions, like the burning of a match, take place at a fast rate, while others, like the spoiling of milk, take place slowly? The rate of a chemical reaction is the speed at which the reactants are converted to products.

- What are some factors that influence the rate of a reaction?
- How could you make a reaction take place at a faster rate?
- How could you slow a reaction down?

Record your ideas about these questions in your *Active Chemistry* log. Be prepared to discuss your responses with your small group and with the class.

Investigate

1. One way to study the rate of a reaction is to time the reaction with a stopwatch. Try this!

   Place 20 mL of vinegar into a large test tube.

   To a second test tube, add 10 mL of vinegar and 10 mL of water. Mix well using a stirring rod.

Safety goggles and a lab apron must be worn during this activity.

In a third test tube, add 5 mL of vinegar to 15 mL of water. Mix.

Prepare three equal-sized pieces of polished magnesium ribbon. Set your stopwatch so that it is ready to start immediately.

Add a piece of magnesium ribbon to the first test tube, keeping track of the time the reactions take.

a) Record your observations and time on a data table. Repeat for the other two test tubes.

b) In this step, you changed the concentration of one of the reactants (the vinegar). The vinegar was less concentrated (more dilute) in each successive test tube. Did this impact the reaction rate? If so, describe the relationship between the concentration of the reactant and the resulting reaction rate.

2. Repeat the same reaction above using a well plate and smaller amounts of vinegar, water, and magnesium.

a) Record your design. Include the equipment you will need, the amount of reactants, and any safety procedures.

b) With the approval of your teacher, carry out the procedure. Record your data and results.

c) How do the results compare with the reaction in **Step 1**?

3. Place 10 drops of 0.1 M HCl (weak concentration) into one clean well in your well plate and 10 drops of 1.0 M HCl (strong concentration) into a second clean well. Drop a small piece of zinc (equal size) into each well containing HCl.

a) Record your observations.

b) How do these results compare to your earlier results? Do these results support or refute the relationship you stated in **Step 2 (c)** above?

4. Hydrogen peroxide is sold over the counter in pharmacies to be used as a disinfectant for minor injuries. Because hydrogen peroxide decomposes slowly to form oxygen and water, it is also a source of oxygen gas.

$2H_2O_2(l)$ + light energy $\rightarrow O_2(g) + 2H_2O(l)$

Pour a small amount of hydrogen peroxide (about 15 mL) into each of two test tubes. Add a small amount of manganese dioxide to one of the test tubes.

a) Record your observations.

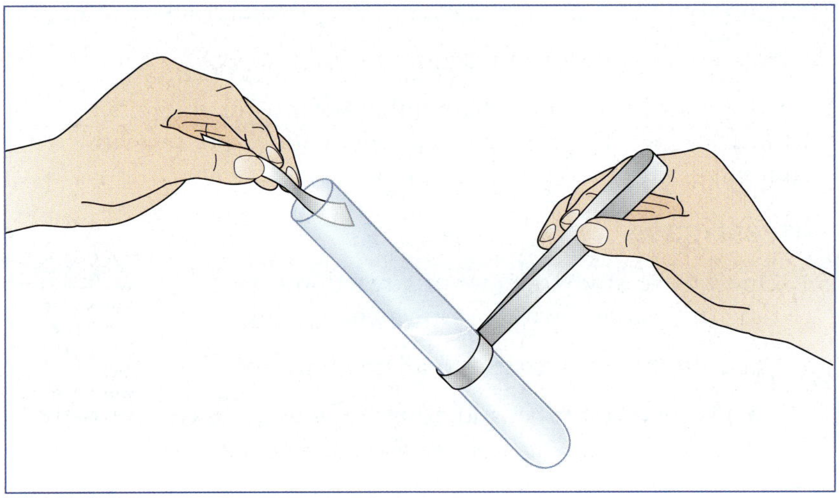

The manganese dioxide did not actually react with the hydrogen peroxide; it simply acted as a catalyst for the decomposition of the hydrogen peroxide. A catalyst is a material that speeds up a reaction without being permanently changed itself. The chemical equation for this reaction is:

$$2H_2O_{2(l)} \xrightarrow{MnO_2} O_{2(g)} + 2\ H_2O_{(l)}$$

5. Design an investigation to prove that the manganese dioxide did not get used up in the reaction.

   Record your design. Include the equipment you will need, the amount of reagents, and any safety procedures.

   a) Record your procedure in your *Active Chemistry* log.

   b) With the approval of your teacher, carry out the experiment.

   c) Record your data and results.

   d) Describe the relationship between the use of a catalyst and the rate of reaction.

6. Pour 200 mL hot water into one beaker and 200 mL ice water into another beaker. Add a tea bag to each.

   a) Record your results.

   b) Repeat the procedure using an Alka-Seltzer tablet in place of the tea bag. Record your observations.

   c) Describe the relationship between temperature and the rate of reaction, based on the two situations you studied.

7. Prepare two beakers, each containing equal amounts (about 200 mL) of room temperature water.

   Obtain two Alka-Seltzer tablets. Crush one and leave the other whole.

   Simultaneously add the crushed table to one beaker and the whole tablet to the other beaker.

   a) Record your observations.

      What factor was being studied in **Step 4 (a)**? Describe the relationship between that factor and the rate of reaction.

8. Use the results of this activity to answer the following:

   a) Describe how you could increase the rate of reaction by altering:

   • concentration • catalyst

   • temperature • surface area

9. Dispose of all chemicals as directed by your teacher. Clean and put away any

equipment as instructed. Clean up your work station. Wash your hands.

10. Chemical systems that highlight reaction rates can be very interesting. Because time is a factor, these systems are often called clock reactions. Your teacher will do the following reaction as a demonstration.

Your teacher will use the following solutions:

Solution A = 0.1 M potassium iodate (KIO_3)

Solution B = 1% starch solution

Solution C = 0.25 M sodium hydrogen sulfite ($NaHSO_3$)

Two rows of five beakers (all the same size!) will be arranged in front of you.

Each of the beakers in the back row contains 100 mL of Solution A, 50 mL of Solution B, and 100 mL of distilled water.

Each beaker in the front row contains 20 mL of Solution C and 130 mL of water.

Your teacher will add the contents of one beaker from the back row to the contents of one of the beakers in the front row.

Use a stopwatch and stop when a color change occurs.

Your teacher will then combine the next set of beakers as you will again use the stopwatch. She or he will continue down the row.

a) Record the time and observations about the change in each case.

Chem Talk

Chem Words

surface area: changing the nature of the reactants into smaller particles increases the surface exposed to react. Successful reaction depends on collision and increasing the area of the reactant increases the chance of a successful collision taking place. Lighting a log is more difficult than lighting wood shavings. The shavings have a greater surface area and speed up the reaction.

FACTORS AFFECTING THE RATE OF A REACTION

In this activity you investigated several common factors that influence reaction rate. They included surface area, concentration of reactants, temperature, and catalysts. On a molecular level, these factors increase the collision frequency of the particles of the materials involved in the reaction.

Consider **surface area**. In water, a sugar cube dissolves in water at a much slower rate than if the same cube is first crushed. The crushed cube has a greater surface area — more parts of the sugar are in contact with the water. In a fireplace, wood chips burn faster than a pile of logs. In both of these cases, the smaller pieces, with their increased surface area, allow the particles that are reacting to come in

contact with each other more often. This increases the collision frequency.

Another factor that influences reaction rate is the concentration of the reactants. An increase in concentration means an increase in the number of particles in the reaction. This results in an increase in the collision frequency. If a chemist wants to increase the rate of a reaction, an increase in the concentration of one or more of the reactants will do the trick.

Altering collision frequency and efficiency can also be accomplished through temperature changes. According to the kinetic theory, particles move faster at higher temperatures and slower at lower temperatures. The faster motion of the particles increases the energy of the particles and increases the probability that particles will collide. As a result, the reaction rate increases.

Catalysts play an important role in many chemical reactions. A catalyst is a substance that speeds up a reaction without being permanently changed itself. The catalyst lowers the action energy of the reaction. Many commercial reactions make use of catalysts, because the catalysts can be recovered, regenerated, and reused. You are probably familiar with the term catalytic converter, a device used in automobiles to improve the efficiency of unleaded gasoline engine's combustion exhaust. The catalyst in this converter is platinum.

Checking Up

1. List four factors that influence the rate of a reaction.
2. For each of the factors you listed above, describe how the factor increases the collision frequency of the particles of the materials involved in the reaction.
3. How is a catalyst different from the reactants and products of a chemical reaction?

Chem Words

catalyst: a substance that changes the speed of a chemical reaction without being permanently changed itself.

Reflecting on the Activity and the Challenge

In this activity you have observed how different factors affect the rate of a reaction. This knowledge can be applied to the presentations you will make in the **Cool Chemistry Show** demonstrations. If you need to cause a reaction to happen faster or slower, you now know what changes you can make. Of course, it is important for you to check with your teacher before making adjustments to any procedure you might be considering.

Chemistry to Go

1. Explain each of the following in terms of the factors that influence reaction rate:
   a) Which will cook faster: Cookies at 50°C or at 150°C?
   b) A bear (and many other animals) hibernates during the winter months. Scientists claim that the low body temperature slows down the animal's metabolism. Explain.
   c) A sugar cube dissolves slower than the same amount of sugar in granulated form.
   d) Antacid tablets are used to neutralize acid in the stomach. Explain why two tablets are faster than one tablet in neutralizing the acid.
   e) If you tried to burn a sugar cube with a match, you would find it very difficult to get the sugar to burn. However, if you put some cigarette ash on the cube, the cube would then burn when you put a flame to it. Explain the purpose of the cigarette ash in changing the burning of the sugar cube.
   f) Why does powdered aspirin dissolve faster than an aspirin tablet in water?
   g) Sugar dissolves more readily in hot tea than in iced tea. Explain.
2. In most cases, if you increase the temperature, the reaction rate increases. Explain this in terms of the collision theory.
3. Imagine that you purchase a lightstick necklace or wristband at a social event and want to make it last as long as possible. What would you do? Why would it help?
4. Explain why the effervescent antacid tablet did not seem to react as fast when it was put in a more dilute solution of vinegar.
5. Explain in terms of the reaction rate factors that you have studied why it is possible for a person who has been submerged in very cold ice water in some

cases to survive, but individuals who have been submerged in warmer water for the same length of time do not survive.

6. Grain elevators have been known to have explosions because of the production of fine grain powders. Explain in terms of the reaction rate factors that you have studied as to why this could happen.

Preparing for the Chapter Challenge

The factors affecting a reaction can be varied to achieve different reaction rates. How could you use this information in developing a presentation for the **Cool Chemistry Show**? Describe one possible scenario.

Inquiring Further

Quantifying the relationship between temperature and reaction rate

You have seen that temperature is a factor that influences the rate of reaction. In general, if the temperature increases, the reaction rate increases. When the temperature decreases, the reaction rate decreases. Can this relationship be quantified? Is there a mathematical relationship between temperature and reaction time? To answer these questions, explore the reaction between magnesium ($Mg_{(s)}$) and vinegar ($CH_3COOH_{(aq)}$). Design and conduct an investigation that will use this reaction to show the relationship between temperature and reaction time in a quantitative way. Have your teacher approve your design before you begin. Remember — the point of the investigation is to see if the relationship between temperature and reaction time is quantifiable. You'll need to monitor both the temperature and the time carefully. Plot your data on a graph to make the relationship explicit. In your notes, include the chemical equation for this reaction.

Activity 7

Acids, Bases, and Indicators—Colorful Chemistry

GOALS

In this activity you will:

- Identify common household acids and bases.
- Identify characteristic properties of acids and bases, and learn to tell the difference between acids and bases.
- See how strong acids and bases behave differently from weak acids and bases.
- Make neutral solution by combining an acid and a base by titration.
- Determine the pH of various solutions using indicators.
- Categorize solutions based on the pH scale.
- Use the mathematical definition of pH.

What Do You Think?

When red cabbage is chopped up and added to boiling water and the resulting mixture is allowed to cool, a special bluish colored solution is made. After the solution is separated from the cabbage, it can be used to indicate if other substances are acids or bases. When household vinegar, a common acid, is added to the cabbage-juice water, the solution turns red. When household ammonia, a common base, is added to the cabbage-juice water, the solution turns green.

- What are some other properties of acids and bases you know about?
- How can you tell the difference between an acid and a base?

Record your ideas about these questions in your *Active Chemistry* log. Be prepared to discuss your responses with your small group and with the class.

Investigate

1. Your teacher will provide you with samples of some of the materials listed below. Place a small amount of each solution in a separate well of a well plate.

   Add a small piece of polished zinc (or magnesium) to each of the solutions.

   a) Make a data table to record your observations.

   hydrochloric acid ($HCl_{(aq)}$)

   lemon or orange juice (citric acid)

   vinegar (acetic acid, $CH_3COOH_{(aq)}$)

   sulfuric acid ($H_2SO_{4(aq)}$)

   mineral water

   carbonated beverage such as Sprite® or Seven Up® (contains $H_2CO_{3(aq)}$)

   milk

   dishwashing solution (Ivory®, Palmolive®, Joy®, etc.)

   sodium hydroxide ($NaOH_{(aq)}$)

   Milk of Magnesia® (contains $Mg(OH)_{2(aq)}$)

   apple juice (malic acid)

   potassium hydroxide ($KOH_{(aq)}$)

   calcium hydroxide ($Ca(OH)_{2(aq)}$)

   household ammonia (NH_3 or $NH_4OH_{(aq)}$)

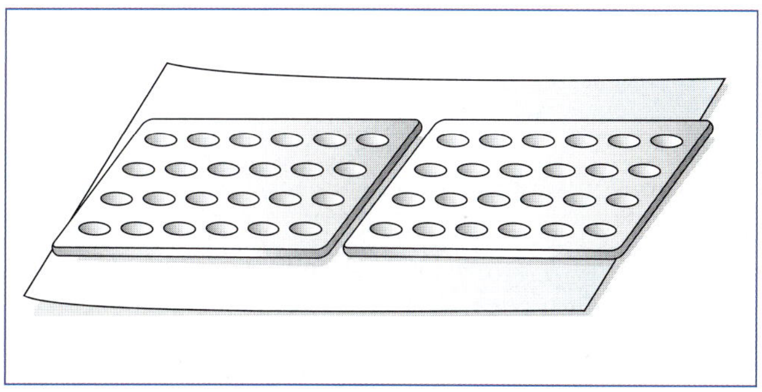

   b) Which substances reacted with the metal? How could you tell? What do these substances have in common? (Consider the chemical formulas listed for some of the substances.)

Safety goggles and a lab apron must be worn during this activity.

   c) Which substances did not react with the metal? What do these substances have in common? (Consider the chemical formulas listed for some of the substances.)

2. Place small amounts of each solution you used in **Step 1** in a separate well of a well plate.

   Test the solutions with one or more common laboratory indicators. (Your teacher will provide acid-base indicators like blue litmus paper, red litmus paper, phenolphthalein, bromothymol blue, methyl red.) Indicator papers are activated simply by dipping a small piece of the paper into the solution and noting any color change. If the indicator is a solution, add a drop or two to the substance being tested and note any color change. You will need to use fresh test solutions if you want to test with more than one indicator solution.

   a) Make up a chart and record your observations.

3. Use your observations as well as previous experiences to answer the following.

   a) Make a list of some of the observable properties for acids and bases.

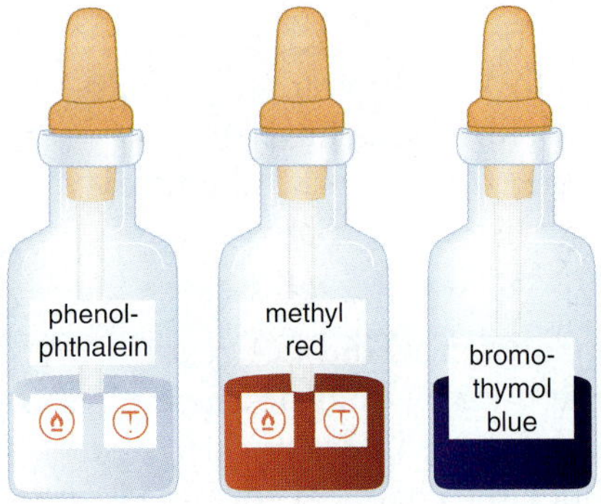

For example:

- How do substances containing acids or bases taste? You should never taste substances in a lab, but you have probably had the opportunity to taste vinegar or lemon juice at home, or you may have accidentally got soap in your mouth.
- How do acids and bases feel. You must be very cautious when handling chemicals both at home and in the lab. However, you've probably had the experience of touching cleaning materials, such as soaps or floor cleaners. Think about vinegar or citrus fruits. How do they feel on a cut on your skin or a canker sore in your mouth?

4. The pH scale can also be used to describe acids and bases. This number scale ranges from 0 to 14. Acid solutions have a pH less than 7. The more acidic a solution is, the lower the pH. Base solutions have a pH greater than 7. The more basic a solution is, the higher the pH. Neutral solutions have a pH of 7.

There are a number of ways you can use to measure pH. You will use pH paper and/or a universal indicator solution. Determine the pH of some of the substances you used in **Step 1.**

a) Make a data table that includes the name of the substance, the pH test, and whether the substance is an acid, a base, or a neutral substance.

b) You may have used both pH paper and universal indicator solution. Both of these are made from a combination of indicators, in order to produce a continuous range of colors throughout the pH scale. Methyl red is a chemical that changes from red to yellow when the pH is between 4.2 and 6.3. Thymolphthalein is a chemical that changes from colorless to blue when the pH is between 9.4 and 10.6. Thymol blue is a chemical that changes from red to yellow when the pH is between 1.2 and 2.8. How could these three chemicals be used to create an indicator scale? What are the limitations of being limited to these three chemicals?

5. Use the pH paper to test additional common household substances in order to determine which are acids and which are bases. (Hint: Try carbonated beverages, tea, coffee, baking powder, mayonnaise, power drinks, pickle juice, window cleaner, stain removers.) Your teacher may give you some pH paper to take home with you.

a) Make a list of common acids and bases found in your school or home. When possible, include both the name and formula for each substance you test.

6. Dispose of all chemicals as directed by your teacher. Clean and put away any equipment as instructed. Clean up your workstation. Wash your hands.

7. Here are two activities that display the characteristics of acids and bases in a colorful way. Your teacher will show you these as demonstrations.

   - Paint a message on a large sheet of paper or poster board using phenolphthalein indicator solution. (How about painting a message announcing your **Cool Chemistry Show**?) Allow the message to dry completely and hang the paper/poster board where everyone can see it. Use a window glass cleaner that contains ammonia water and when you are ready to reveal the message, lightly spray the design with the basic solution. (The secret message can also be revealed with a dilute ammonia solution. As the ammonia evaporates, the secret message that has been revealed will disappear again.)

   - Rinse a small beaker with a strong acid and label it "A." Rinse another small beaker with a strong base and label it "B." Let both beakers air dry. In another beaker (label it "I") add 20 drops of phenolphthalein indicator solution to about 50 mL of distilled water. When you are ready, pour some of the solution from beaker "I" into beaker "A." Then pour the solution from beaker "A" into beaker "B."

   a) Record your observations.

   b) Account for the observations in each case.

Caution must be used with these sprays, because they can cause eye damage if they get into the eyes.

Chem Talk

ACIDS AND BASES

Arrhenius' Definition of Acids and Bases

Acids and **bases** were first classified according to their characteristic properties. As you've experienced, acids and bases have different, distinct interactions with indicators (substances that change color with changes in the acidic or basic nature of another material). Some acids react with metals, while bases do not. Bases have a characteristic bitter taste and slippery feel, while acids have a characteristic sour taste. In fact, the term, acid, comes from the Latin word, *acidus*, which means sour. Acids and bases are also good conductors of electricity.

Chem Words

acid: a substance that produces hydrogen ions in water, or is a proton donor.

base: a substance that releases hydroxide ions (OH^-) in water, or is a proton acceptor.

In the 19th century a chemist named Svante Arrhenius attributed the characteristic properties of acids to their ability to produce hydrogen ions when dissolved in water. If you look at the formulas for many common acids (HCl, H_2CO_3, H_2SO_4), you'll notice that they all have H as a common element. When these acids are added to water, a hydrogen atom can be drawn off into the water solution. The hydrogen atom leaves an electron behind, forming a positive hydrogen ion (H+) and a negative ion. Consider the action of hydrochloric acid in solution:

$$HCl_{(g)} \xrightarrow{\text{water}} H^+_{(aq)} + Cl^-_{(aq)}$$

The chemical equation shown above is valuable because of its simplicity. However, in reality, the hydrogen ion (H^+) is simply a proton and readily attaches itself to a water molecule. The result is called a hydronium ion (H_3O^+).

H^+	+	H_2O	$\rightarrow$	$H_3O^+_{(aq)}$
hydrogen ion		water		hydronium ion

To be more complete, the chemical equation above could be written as shown below. (Your teacher may allow you to use the simpler form of the equation—using the hydrogen ion as opposed to the hydronium ion).

$$HCl_{(g)} + H_2O \rightarrow H_3O^+_{(aq)} + Cl^-_{(aq)}$$

Arrhenius also addressed bases and their characteristic properties. He defined a base as a substance that produces hydroxide ions (OH^-) when dissolved in water.

Let's look at a base using Arrhenius' definition. When solid sodium hydroxide is dissolved in water, both sodium ions and hydroxide ions are produced, as shown in the chemical equation below:

$$NaOH_{(s)} \xrightarrow{H_2O} Na^+_{(aq)} + OH^-_{(aq)}$$

Over time, scientists have extended their definition of acids and bases beyond Arrhenius' definition to be more inclusive. You will learn more about the contributions of scientists like Johannes Bronsted of Denmark, Thomas Lowry of England, and Gilbert Lewis of the United States in further chemistry courses.

Neutralizing Acids and Bases

When acids and bases react together in solution, the hydrogen ions and hydroxide ions react in a one-to-one ratio to produce water. The remaining ions can join to form a salt. The process of an acid and base reacting to form water and a salt is called **neutralization**. Because the hydrogen ions and hydroxide ions have formed water, the solution is said to be neutral. The process of neutralization is shown in the chemical equations below. The chemical formula for water is actually H_2O. In the equations below the formula is written as HOH, so that you can see where the hydrogen and hydroxide ions end up.

$H^+_{(aq)}$	+	$OH^-_{(aq)}$	→	$HOH_{(aq)}$		
hydrogen ion		hydroxide ion		water		
$HCl_{(aq)}$	+	$NaOH_{(aq)}$	→	$HOH_{(aq)}$	+	$NaCl_{(aq)}$
acid		base		water		salt

If a suitable indicator is added to the reaction system, it will change colors when neutralization occurs. The point at which the indicator changes color is called the endpoint.

Consider the reaction of a strong acid (HCl) and a strong base (NaOH), as shown in the equation above. These substances are described as "strong" because they ionize completely in solution. For every HCl molecule, one hydrogen ion is released. For every NaOH, one hydroxide ion is released. These two ions then combine in a one-to-one ratio to form a neutral water molecule. Chemists take advantage of the neutralization process to help determine the concentration of solutions of acids or bases.

The pH Scale

In this activity you observed that one way of describing acids and bases is by examining their effects on indicators. Scientists also use the pH scale to express how acidic or basic a solution is. This number scale ranges from 0 to 14. Acid solutions have a pH less than 7. The more acidic a solution is, the lower the pH. Base solutions have a pH greater than 7. The more basic a solution is, the higher the pH. Neutral solutions have a pH of 7. The pH of a substance can be measured using methods like a pH meter or probe, pH paper, or universal indicator solution.

Chem Words

neutralization: the process of an acid and base reacting to form water and salt.

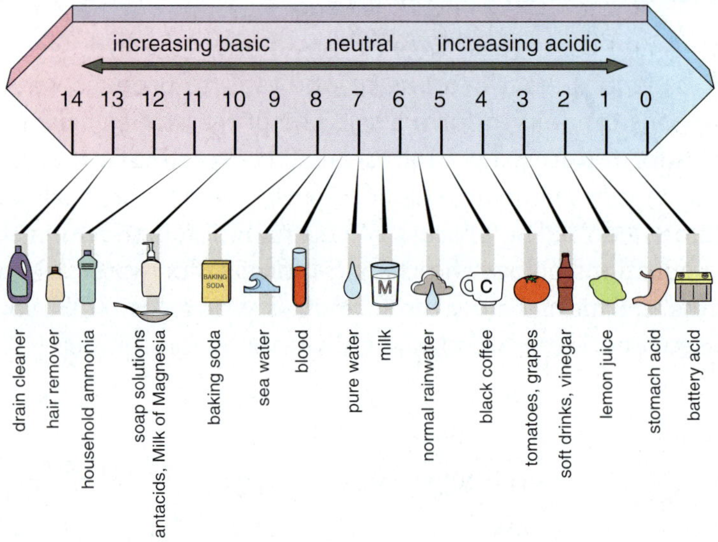

Acid and base indicators are compounds that are sensitive to pH. The color of the indicator changes as the pH of the solution changes. Most indicators are weak acids or weak bases that typically exhibit two different colors under varying pH conditions. The table below shows some common laboratory indicators and the colors they display under different pH conditions.

Common Laboratory Acid-Base Indicators

Indicator	Color Change	pH Range
methyl violet	yellow to blue	0.0 to 1.6
thymol blue	red to yellow	1.2 to 2.8
methyl orange	red to yellow	3.2 to 4.4
bromocresol green	orange to violet	3.8 to 6.4
methyl red	red to yellow	4.2 to 6.3
litmus paper	red to blue	5.5 to 8.0
bromothymol blue	yellow to blue	6.0 to 7.6
phenolphthalein	colorless to red	8.2 to 10
thymolphthalein	colorless to blue	9.4 to 10.6
cabbage juice	red to green	2 to 12 (universal)

The pH scale ranges from 0 to 14 and is used to express the concentration of the hydrogen (H+) or hydronium ion (H_3O^+) of a solution at 25°C. Mathematically, it is defined as the negative logarithm of the hydrogen ion concentration. The term **pH** stands for **p**ower of **H**ydrogen ion. It can be written as:

$$pH = -\log_{10}[H^+]$$

where the brackets [ ] stand for "concentration of" (hydrogen ions in solution). Because pH is a logarithmic scale, the concentration of the hydrogen ion $[H^+]$ actually increases or decreases tenfold for each unit on the scale. An acid with a pH of 2 has a $[H^+]$ that is 10 times greater than an acid with a pH of 3 and 100 times the concentration of an acid with pH 4. A base with a pH of 10 has a $[H^+]$ that is 10 times less than a base with a pH of 9.

Chem Words

pH: a quantity used to represent the acidity of a solution based on the concentration of hydrogen ions ($pH = -\log[H^+]$).

Checking Up

1. Use a chart to compare the properties of acids and bases. Be sure to include headings like taste, feel, pH, and reaction with metals.
2. What characteristic property did Arrhenius attribute to acids and bases?
3. Describe the process that occurs when an acid reacts with a base.
4. Why are litmus paper and phenolphthalein particularly useful indicators for distinguishing between acids and bases?
5. What does pH stand for?
6. How much more acidic is a solution of pH 3 than pH 5?

Reflecting on the Activity and the Challenge

In this activity you expanded your knowledge about acids and bases by becoming familiar with many of their characteristics. You learned about Arrhenius' definition of acids and bases. You also learned a bit about pH, another way of expressing the acid or base nature of substances. This information will all come in handy as you plan your presentation for the **Cool Chemistry Show**. Remember that the fifth-grade teacher has specifically asked that your class includes presentations and information about acids and bases.

Chemistry to Go

1. Identify which of the following characteristics relate to acids and which relate to bases:

   a) taste sour

   b) release hydroxide ions ($OH^{-}_{(aq)}$) when dissolved in water

   c) feel slippery

   d) release hydrogen ions ($H^{+}_{(aq)}$) when dissolved in water

   e) turn pink in the presence of phenolphthalein

   f) react with metals to produce hydrogen gas

   g) taste bitter

   h) turn red cabbage juice indicator green

2. Use Arrhenius' definition of an acid to help you write a chemical equation that shows the acidic nature of the following:

   a) sulfuric acid (H_2SO_4)

   b) carbonic acid (H_2CO_3)

3. Use Arrhenius' definition of a base to help you write a chemical equation that shows the basic nature of the following:

   a) potassium hydroxide (KOH)

   b) calcium hydroxide ($Ca(OH)_2$)

4. If you prepared the same concentration of two strong acids, sulfuric and hydrochloric, why would the pH of sulfuric be smaller than the hydrochloric acid?

5. Distilled water should have a neutral pH of 7, but water often has a pH less than 7. Suggest a reason for this lowering of the pH.

6. If you bubbled carbon dioxide through water, what would the new pH of the solution be?

7. Lemon juice, curdled milk, vinegar, all taste sour. What other properties would you expect them to have in common?

Preparing for the Chapter Challenge

You have seen a number of interesting color changes using acids, bases, and indicators. Choose one or two different cool activities to demonstrate in your show. Describe the procedure you will use and explain the chemistry involved. You may also wish to include an interesting scenario to accompany your "presto-change-o" demonstrations.

Inquiring Further

1. Titration

Titration is the process whereby a measured amount of solution of known concentration of acid (or base) is added to a known amount of a solution of unknown concentration. Research how chemists perform a titration, and the importance of indicators.

With your teacher's permission, demonstrate titration to your class.

2. The changing definition of an acid and base

The definition of acids and bases has changed through time. You are familiar with the earliest definitions that defined acids and bases in terms of their characteristic properties. The traditional definition has been expanded a number of times to include other substances that behave like acids and bases, but don't fit the traditional definition. Research the expansion through time of the definition of acids and bases. Identify the scientists involved and the changes that were made. Consider researching chemists such as Johannes Bronsted of Denmark, Thomas Lowry of England, or the American chemist Gilbert Lewis.

3. Is it pH balanced?

You may have heard the term "pH balanced" used to describe a shampoo or a deodorant. What does this term mean? What is the pH of most shampoos? Deodorants? Is it important for a shampoo or deodorant to be "pH balanced?" Conduct some research, both in and out of lab, to get answers to these questions. Focus just on shampoos or just on deodorants.

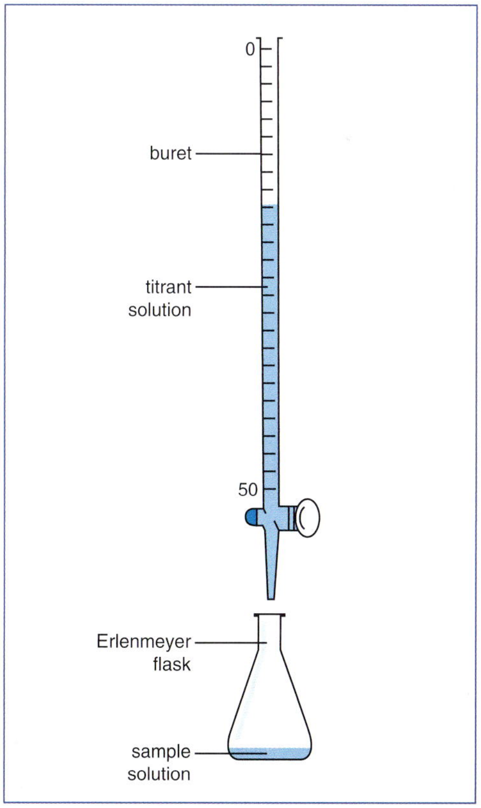

Activity 8 Color Reactions that Involve the Transfer of Electrons

GOALS

In this activity you will:

- Cause different metals to rust by oxidation-reduction (redox) reactions.
- Determine what materials can react with metals, causing the metals to rust.
- Write the word equations and chemical equations for redox reactions.
- Identify the materials that react, and the materials that are simply spectators, in a redox reaction.
- Learn how to impede rusting.

What Do You Think?

What happens to a scratch on a car that is not repaired? What happens to metal barbecue tools that get left out in the rain for a few weeks?

- What is rust and what causes it?

Record your ideas about this question in your *Active Chemistry* log. Be prepared to discuss your responses with your small group and with the class

Investigate

1. Half-fill a test tube with copper (II) sulfate pentahydrate solution ($CuSO_4 \cdot 5H_2O_{(aq)}$). Add a small amount of zinc powder to the test tube. Stopper the test tube, and shake carefully.

   a) Record your observations.

   Dispose of the products as directed by your teacher.

Safety goggles and a lab apron must be worn during this activity.

b) The reaction you just observed was a single-displacement reaction. The zinc displaced the copper. Use this information and your observations to complete the following equations:

zinc + copper (II) sulfate →
______ + ______

c) Write the word equation as a sentence. Also, explain why this would be classified as a single-displacement reaction. (Refer back to **Activity 4.**)

d) Write the equation using the chemical formulas of the products.

$Zn_{(s)} + CuSO_{4(aq)} \rightarrow$
______ + ______

Because the sulfate ion shows on both sides of the equation, it is considered a spectator ion and the equation can be written as shown below:

$Zn_{(s)} + Cu^{2+}{}_{(aq)} \rightarrow Cu_{(s)} + Zn^{2+}{}_{(aq)}$

e) Write a sentence or two describing what happened in the test tube in terms of the zinc and copper.

2. Cut a design or a strip of aluminum from a pie plate or tray. If you cut a strip of aluminum, twist it into an interesting shape.

Place the aluminum in a solution that contains copper (II) ions ($Cu^{2+}{}_{(aq)}$). Possible solutions include: copper (II) nitrate ($Cu(NO_3)_{2(aq)}$) or copper (II) chloride ($CuCl_{2(aq)}$).

a) Observe and record your results. Dispose of the products as directed by your teacher.

b) What evidence do you have that a chemical reaction occurs? What changes have taken place with the aluminum? With the copper ions?

c) Complete this equation

$Al_{(s)} + Cu^{2+}{}_{(aq)} \rightarrow$
______ + ______

3. Repeat **Step 2** using a different metal such as zinc.

a) Record your observations.

b) Write a chemical equation for the reaction that takes place.

4. Repeat **Step 2** again, this time using a strip of copper in a solution of aluminum nitrate ($Al(NO_3)_{3(aq)}$).

a) Record your observations.

b) Write a chemical equation for the reaction that takes place.

Chem Talk

REDOX REACTIONS

Chem Words

oxidation: the process of a substance losing one or more electrons.

reduction: a process in which the substance under consideration gains electron(s).

When zinc solid reacts with copper ions in solution, a change occurs. Atoms of zinc lose electrons to form zinc ions (Zn^{2+}) that dissolve into the solution. Copper ions (Cu^{2+}) gain the electrons from the zinc atoms to form copper atoms that plate out as a solid. Whenever an atom or ion becomes more positively charged in a chemical reaction, as in the case of zinc atoms forming positive zinc ions, the process is called **oxidation**. Oxidation is the process of losing electrons. Whenever an atom or ion becomes less positively charged in a chemical reaction, as in the case of the copper ions forming copper atoms, the process is called **reduction**. Reduction involves a gain of electrons. The processes of oxidation and reduction happen together and as such are commonly referred to as "redox" reactions. An easy way to remember which is oxidation and which is reduction is by remembering "LiOn GRrr,"— Lose electrons Oxidation; Gain electrons Reduction.

The formation of rust is a redox process. Water and oxygen are necessary for the iron metal to corrode (rust). Iron atoms lose electrons to form mostly Fe^{3+} ions with the help of the moisture in the air and the heat of the Sun. Because the atoms have given up electrons to become more positively charged, oxidation of iron has taken place. Molecules of oxygen gain electrons to form O^{2-} ions. The oxygen has accepted electrons and is said to have been reduced. Corrosion can be prevented by painting a surface of iron to prevent moisture and air from coming in contact with the metal. Let's summarize what you have learned about atoms and ions:

The term atom means that the element is neutral; this means that it has the same number of protons and electrons.

Ions mean that the atom (or ion) has gained or lost electron(s). Examples are:

$Na \rightarrow Na^+ + e^-$ (Sodium atom loses 1 electron and now has a net charge of +1.)

$Cl_2 + 2\ e^- \rightarrow 2\ Cl^-$ (The two chlorine atoms gain 1 electron each and the net charge is −1 for each chloride ion.)

Polyatomic ions like the sulfate ion (SO_4^{2-}) imply that there are 2 more electrons than protons in the entire structure.

In some cases you will find that an ion can gain or lose an electron and form a new ion. An example of this type is: $Fe^{2+} \rightarrow Fe^{3+} + e^-$ (The iron in the 2+ state loses 1 more electron and now will be in a 3+ state.)

Reflecting on the Activity and the Challenge

Although many colorful chemical reactions involve the use of acids and bases with indicators, there is an entire group of chemical reactions that produce colorful results through the transfer of electrons. In this activity you became familiar with some of the simple concepts behind redox reactions, and you saw several examples of the color changes they can produce. You and/or your classmates may decide to include some redox reactions in the **Cool Chemistry Show**.

Chemistry to Go

1. Aluminum metal can react to form an ion with a charge of +3. Does the aluminum atom gain or lose electrons to form the Al^{+3} ion?
2. A copper ion with a charge of +2 can react to form an atom of copper. Does the copper ion have to gain or lose electrons in this reaction?
3. The element iron can form two different ions. The iron (II) ion (Fe^{+2}) is commonly called a ferrous ion while the iron (III) ion (Fe^{+3}) is called a ferric ion. When ferrous ions undergo a chemical change to become ferric ions, what process has taken place, oxidation or reduction? Explain your answer.
4. In the reaction you did with zinc metal reacting with copper ions, which substance gains electrons? Which loses electrons?
5. What must take place for copper metal to be oxidized?
6. Galvanized iron nails are used to fasten materials that will be exposed to the outdoors. A galvanized nail is a regular iron nail that is coated with zinc.
   a) Why would a zinc coating be an advantage here? What do you think is the purpose of the zinc?
   b) What two reactants could you use to test this in the laboratory? What results would you expect if you were right about the purpose of the zinc?

Inquiring Further

The Statue of Liberty

In the 1980s the Statue of Liberty in New York harbor underwent extensive renovation. Research the involvement of oxidation-reduction reactions in this renovation. Identify what the problem was and its solution.

Cool Chemistry Show Assessment

Your Chapter Challenge is to present an entertaining and informative chemistry science show to fourth- and fifth-grade students. The show for fourth graders should address chemical and physical properties and changes. The show for fifth graders should help students learn about acids and bases, and about chemical reactions that involve color changes. All presentations must include a demonstration and an audience appropriate explanation of the chemistry concepts involved. Finally, the teachers of the fourth- and fifth-grade students need written directions for your chemistry show with explanations of the chemistry concepts included in the show.

Your team should review the activities that you completed. A useful way of completing this review would be to list the activities, the cool chemistry demonstrations that you can perform, whether these demonstrations would be most appropriate for the fourth- or fifth-grade show and the chemistry concepts of the demonstration. You should pay particular attention to the section **Reflecting on the Activity and the Challenge.** You should also compare your list with that given in the **Chemistry You Learned** summary.

Your team's next step would then be to decide on which demonstrations are best suited to your interests and can be the most interesting, informative, creative and entertaining for the elementary students. Your teacher may wish you to share your decision with the class as a whole so that the full cool chemistry program is not repetitious.

Each demonstration that you plan on producing will be judged along specific criteria. Each demonstration should be carefully planned, should adhere to all safety considerations, should be completed within the assigned time limits and should exhibit showmanship, creativity, clarity and appeal. As you plan and practice your demonstration, you should use these criteria as a checklist to ensure high quality.

Each demonstration must also be accompanied with an explanation of chemistry concepts. You should rehearse your explanation so that it complements the demonstration itself. You may choose to have one team member producing the demonstration, another team member explaining the manipulations and focusing the audience attention on specific observations and a third team member explaining the chemistry concepts. You should decide if your presentation would be strengthened by having some of the chemistry concepts described prior to the demonstration of if the explanation should follow the demonstration.

Your final responsibility is to provide detailed instructions to the fourth- and fifth-grade teachers so that the teacher could repeat the demonstration. This will require a list of materials, step-by-step directions, notes on what should be happening and safety considerations. You must also provide a written explanation of the chemistry concepts illustrated in the **Cool Chemistry Show**. Your chemistry concepts should be more thorough than the oral presentation during the show. Your explanation of chemistry concepts is now targeted to teachers rather than elementary students. You can include more formulas, more background and provide additional experiments which clarify the chemistry concepts.

Your class should generate a grading rubric. The rubric will assign points to the demonstration, the chemistry content and the written summary. It will also assign points to specific aspects of each major category. For instance, how many additional points will be afforded for additional concepts introduced in the demonstration? How many points will be assigned for creativity, showmanship and adherence to time limits of the cool chemistry demonstration? This rubric can be a useful tool for your team to check the quality of your work and to ensure that you have included all necessary parts of the project.

Chemistry You Learned

Chemical and physical changes of matter
Solute, solvent and solution
Saturated and supersaturated solution
Precipitate
Solubility of compounds in water
pH indicators for acids and bases

Cations and anions
Polyatomic ions
Ionic compounds and formulas
Molecular compounds
Chemical equations
Synthesis reactions
Single- and double-displacement reactions

Decomposition reactions
Metal activity series
Endothermic and exothermic reactions
Conservation energy
Kinetic and potential energy
Heat energy
Colligative properties

Reaction rates
Factors that affect the reaction rate:
- **Concentration**
- **Catalyst**
- **Temperature**
- **Surface area or nature of the material**

Acids and bases
Titration
Oxidation and reduction

Chemistry at Work

Marc Pollack

President, Flix FX

"Everyone talks about 'Movie Magic,'" says Marc Pollack, president of the prestigious Hollywood special effects company Flix FX. "So I guess that makes me a magician." But Pollack is clearly more comedian than magician. The 'magic' he creates for movies like *Blackhawk Down*, *Men In Black* and *Cast Away*, in addition to scores of television commercials, museum installments and Las Vegas casinos, is the product not of mysterious hocus-pocus but rather fundamental principles of science. "One of the most important aspects of our work," he continues, "is to push the limits of how chemicals are designed to be used." Among other things, Pollack and his crew at Flix FX use vacuum-forming thermoplastics to make tin-based silicon molds for everything from prehistoric creatures to futuristic robots. Through a combination of trial and error experimentation and traditional research science, they've perfected the process. "Silicon is what we call an R.T.V.," Pollack explains. "That stands for room temperature vulcanization. So depending on the type and amount of catalyst we use, the mold will cure at different rates and with slightly differing properties." By manipulating the ratio of silicon to catalysts they can make strong, realistic molds in the most efficient way possible. "Increasing the amount of catalyst will speed up the curing process but too much catalyst will shorten the life of the mold," he says. "Every job is different so determining that balance is one of our many challenges."

Pollack, who is now a master in the art of using chemicals like silicon, polypropylene, urethane and urethane elastimers, is not a chemist by trade. He actually graduated from film school at SUNY Purchase in the hopes of becoming "the next Steven Spielberg." Then, through a twist of fate, he became a special effects nut and eventually founded Flix FX in 1990. "Now," Pollack says, "Spielberg may one day come to me."

Special effects — Pollack creates both mechanical and physical — is an industry in a constant state of transformation. "The industry is always trying new stuff and that's exciting," Pollack says. "For instance, someone just developed a great water-based breakaway glass for stunts called Smash Glass. It's similar to fiberglass without the dangerous elements associated with that material and can be made to break into either large chunks or tiny little pieces. I can't wait to get my hands on it and break it over someone's head. That's part of my job these days and I love it."

Glossary: Unit 1

Chapters 1–9 Active Physics

acceleration: the change in velocity per unit time.

$$a = \frac{\Delta v}{\Delta t}$$

air resistance: a force by the air on a moving object; the force is dependent on the speed, volume, and mass of the object as well as on the properties of the air, like density.

alternating current: an electric current that reverses in direction.

amplitude: the maximum displacement of a particle as a wave passes; the height of a wave crest; it is related to a wave's energy.

ampere: the SI unit for electric current; one ampere (1 A) is the flow of one coulomb of charge every second.

analog: a description of a continuously variable signal or a circuit; an analog signal can be represented as a series of waves.

angle of incidence: the angle a ray of light makes with the normal to the surface at the point of incidence.

angle of reflection: the angle a reflected ray makes with the normal to the surface at the point of reflection.

antinode: a point on a standing wave where the displacement of the medium is at its maximum.

atom: the smallest particle of an element that has all the element's properties; it consists of a nucleus surrounded by electrons.

atomic mass unit: a standard unit of atomic mass based on the mass of the carbon atom, which is assigned the value of 12.

baryon: a group of elementary particles that are affected by the nuclear force; neutrons and protons belong to this group.

center of mass: the point at which all the mass of an object is considered to be concentrated for calculations concerning motion of the object.

centripetal acceleration: the inward radial acceleration of an object moving at a constant speed in a circle.

$$a = \frac{v^2}{R}$$

centripetal force: a force directed towards the center that causes an object to follow a circular path.

$$F = \frac{mv^2}{R}$$

closed system: a physical system on which no outside influences act; closed so that nothing gets in or out of the system and nothing from outside can influence the system's observable behavior or properties.

concave lens: a lens that causes parallel light rays to diverge; a lens that is thicker at its edges than in the center.

conduction: (of heat) the energy transfer from one material or particle to another when the materials or particles are in direct contact.

constructive interference: the result of superimposing different waves so that two or more waves overlap to produce a wave with a greater amplitude.

convection: the heat transfer resulting from the movement of the heated substance, such as air or water currents.

converging lens: parallel beams of light passing through the lens are brought to a real point or focus (convex lens) (if the outside index of refraction is less than that of the lens material.); also called a convex lens.

convex lens: a lens that causes parallel light rays to converge (if the outside index of refraction is less than that of the lens material); a lens that is thinner at its edges than in the center.

coulomb: the SI unit for electric charge; one coulomb (1 C) is equal to the charge of 6.25 x 10^{18} electrons.

Coulomb's Law: the relationship among electrical force, charges, and the distance between the charges.

$$F = k\frac{q_1 q_2}{d^2}$$

crest: the highest point of displacement of a wave.

critical angle: the angle of incidence for which a light ray passing from one medium to another has an angle of refraction of 90° degrees.

destructive interference: the result of superimposing different waves so that two or more waves overlap to produce a wave with a decreased amplitude.

diffraction: the ability of a wave to spread out as it emerges from an opening or moves beyond an obstruction.

digital: a description of data that is stored or transmitted as a sequence of discrete symbols; usually this means binary data (1s and 0s) represented using electronic or electromagnetic signals.

displacement: the difference in position between a final position and an initial position; it depends only on the endpoints, not on the path; displacement is a vector; it has magnitude and direction.

Doppler Effect: change in frequency of a wave of light or sound due to the motion of the source or the receiver.

electric charge: a fundamental property of matter; charge is either positive or negative.

electric circuit: an electrical device that provides a conductive path for electrical current to move continuously.

electric current: the flow of electric charges through a conductor; electric current is measured in amperes.

electric field: the region of electric influence defined as the force per unit charge.

electrical resistance: opposition of a material to the flow of electrical charge through it: it is measured in ohms (Ω); the ratio of the potential difference to the current.

$$R = \frac{V}{I}$$

electromagnet: a device that uses an electric current to produce a concentrated magnetic field.

electromagnetic waves: transverse waves that are composed of perpendicular electric and magnetic fields that travel at 3 x 10^8 m/s in a vacuum; examples of electromagnetic waves in increasing wavelength are gamma rays, x-rays, ultraviolet radiation, visible light, infrared radiation, microwaves, and radio waves.

electron: a negatively charged particle with a charge of 1.6 x 10^{-19} coulombs and a mass of 9.1 x 10^{-31} kg.

entropy: a measure of the degree of disorder in a system or a substance.

focal length: the distance between the center of a lens and either focal point.

focus: the place at which light rays converge or from which they appear to diverge after refraction or reflection; also called focal point.

force: a push or a pull that is able to accelerate an object; force is measured in newtons; force is a vector quantity.

frame of reference: a vantage point with respect to which position and motion may be described.

free fall: a fall under the influence of only gravity.

frequency: the number of waves produced per unit time; the frequency is the reciprocal of the amount of time it takes for a single wavelength to pass a point.

$$f = \frac{v}{\lambda}$$

friction: a force that acts to resist the relative motion or attempted motion of objects that are in contact with each other.

galvanometer: an instrument used to detect, measure, and determine the direction of small electric currents.

gravitational potential energy: the energy a body possesses as a result of its position in a gravitational field.

$GPE = mgh$

gravity: the force of attraction between two bodies due to their masses.

heat energy: a form of energy associated with the motion of atoms or molecules.

Hooke's Law: the distance of stretch or compression of a spring is directly proportional to the force applied to it.

$F = k\Delta x$

impulse: the product of force and the interval of time during which the force acts; impulse results in a change in momentum.

$\text{impulse} = Ft = \Delta(mv)$

index of refraction: a property of a medium that is related to the speed of light through it; it is calculated by dividing the speed of light in vacuum by the speed of light in the medium.

inertia: the natural tendency of an object to remain at rest or to remain moving with constant speed in a straight line.

inertial frame of reference: unaccelerated point of view in which Newton's Laws hold true.

inverse square relation: the relationship of a force to the inverse square of the distance from the mass (for gravitational forces) or the charge (for electrostatic forces).

ionization energy: the energy required to free an electron from an atom.

joule: the SI unit for work and all other forms of energy; one joule (1J) of work is done when a force of one newton moves an object one meter in the direction of the force.

kinetic energy: the energy an object possesses because of its motion.

$KE = \frac{1}{2}mv^2$

lepton: a group of elementary particles that are not affected by the nuclear force; electrons belong to this group.

light-year: the distance that light travels in one year (9.46×10^{12} km).

longitudinal pulse or wave: a pulse or wave in which the motion of the medium is parallel to the direction of the motion of the wave.

magnetic field: the region of magnetic influence around a magnetic pole or a moving charged particle.

meson: a virtual particle that mediates the strong, nuclear force of an atom; the protons and neutrons exchange mesons; the protons and protons exchange mesons; the neutrons and neutrons exchange mesons.

model: a representation of a process, system, or object.

momentum: the product of the mass and the velocity of an object; momentum is a vector quantity.

$p = mv$

neutron: a subatomic particle that is part of the structure of the atomic nucleus; a neutron is electrically neutral.

Newton's Law of Gravitational Attraction: the relationship among gravitational force, masses, and the distance between the masses.

$$F = \frac{Gm_1 m_2}{d^2}$$

Newton's Laws of Motion:

Newton's First Law of Motion: an object at rest stays at rest and an object in motion stays in motion unless acted upon by an unbalanced, external force.

Newton's Second Law of Motion: if a body is acted upon by an external force, it will accelerate in the direction of the unbalanced force with an acceleration proportional to the force and inversely proportional to the mass.

$F = ma$

Newton's Third Law of Motion: forces come in pairs; the force of object A on object B is equal and opposite to the force of object B on object A.

node: a point on a standing wave where the medium is motionless.

normal: at right angles or perpendicular to.

nuclear (strong) force: a strong force that holds neutrons and protons together in the nucleus of an atom; the force operates only over very short distances.

nuclear fission: a nuclear reaction in which a massive, unstable nucleus splits into two or more smaller nuclei with a release of a large amount of energy.

nuclear fusion: a nuclear reaction in which nuclei combine to form more massive nuclei with the release of a large amount of energy.

nucleon: the building block of the nucleus of an atom; either a neutron or a proton.

nucleus: (of an atom): the positively charged dense center of an atom containing neutrons and protons.

ohm: the SI unit of electrical resistance; the symbol for ohm is Ω.

Ohm's Law: the relationship among resistance, voltage, and current.

$I = \frac{V}{R}$

open system: a physical system on which outside influences are able to act; open so that energy can be added and/or lost from the system.

parallax: the apparent difference of position of an object as seen from two different places, or points of view.

period: the time required to complete one cycle of a wave.

periodic wave: a repetitive series of pulses; a wave train in which the particles of the medium undergo periodic motion (after a set amount of time the medium returns to its starting point and begins to repeat its motion).

photoelectric effect: the emission of electrons from certain metals when light (electromagnetic radiation) of certain frequencies shines on the metals.

photon: a particle of electromagnetic radiation; a quantum of light energy.

pitch: the quality of a sound dependent primarily on the frequency of the sound waves produced by its source

polarized waves: disturbances where the medium vibrates in only one plane.

potential energy: energy that is dependent on the position of the object.

power: the time rate at which work is done and energy is transformed.

$P = \frac{W}{t}$

probability: a measure of the likelihood of a given event occurring.

projectile: an object traveling through the air.

proton: a subatomic particle that is part of the structure of the atomic nucleus; a proton is positively charged.

radiation: (heat transfer): electromagnetic radiation strikes a material that can absorb it, causing the particles in the material to have more energy often resulting in a higher temperature.

radioactive: a term applied to an atom that has an unstable nucleus and can spontaneously emit a particle and become the nucleus of another atom.

ray: the path followed by a very thin beam of light.

real image: an image that will project on a screen or on the film of a camera; the rays of light actually pass through the image.

refraction: the change in direction (bending) of a light beam as it passes obliquely from one medium to a different one.

relativity: the study of the way in which observations from moving frames of reference affect your perceptions of the world.

scalar: a quantity that has magnitude, but no direction.

Snell's Law: describes the relationship between the index of refraction and the ratio of the sine of the angle of incidence and the sine of the angle of refraction.

$$n = \frac{\sin \angle i}{\sin \angle R}$$

solenoid: a coil of wire.

Special Theory of Relativity: the theory of space and time.

specific heat: the amount of energy required to raise the temperature of 1 kg of a material by 1°C.

$$E = mc\Delta t$$

speed: the change in distance per unit time; speed is a scalar, it has no direction.

spring potential energy: the internal energy of a spring due to its compression or stretch.

temperature: a measure of the average kinetic energy of the molecules of a material.

trajectory: the path followed by an object that is launched into the air.

transverse pulse or wave: a pulse or wave in which the motion of the medium is perpendicular to the motion of the wave.

trough: the lowest point on a wave.

vector: a quantity that has both magnitude and direction.

velocity: speed in a given direction; displacement divided by the time interval; velocity is a vector quantity, it has magnitude and direction.

virtual image: an image from which rays of reflected or refracted light appear to diverge, as from an image seen in a plane mirror; no light comes directly from or passes through the image.

volt: the SI unit of electric potential; one volt (1 V) is equal to one joule per coulomb (J/C).

wavelength: the distance between two identical points in consecutive cycles of a wave.

weight: the vertical, downward force exerted on a mass as a result of gravity.

$$F_g = mg$$

work: the product of the displacement and the force in the direction of the displacement; work is a scalar quantity.

$$W = F \cdot d$$

Glossary: Unit II

Chapters 10–12 Active Chemistry

accuracy: how close the measured value is to the standard or accepted value of that measurement.

acid: a substance that produces hydrogen ions in water, or is a proton donor.

acid-base indicator: a dye that has a certain color in an acid solution and a different color in a base solution.

alloy: a substance that has metal characteristics and consists of two or more different elements.

anion: a negatively charged ion.

atom: the smallest representative part of an element.

atomic mass: atomic mass is determined by the mass of the protons and neutrons of the atom.

atomic mass unit (amu): a unit of mass defined as one-twelfth of the mass of a carbon-12 atom.

Avogadro's number: the number of atoms contained in 12 grams of carbon-12. The number is 6.022×10^{23}.

base: a substance that releases hydroxide ions (OH–) in water, or is a proton acceptor.

binary compound: a compound formed from the combining of two different elements.

bounce: the ability of an object to rebound to its original position when dropped from a given height.

catalyst: a substance that changes the speed of a chemical reaction without being permanently changed itself.

cation: a positively charged ion.

chemical change: a change that converts the chemical composition of a substance into different substance(s) with different chemical composition.

chemical formula: the combination of the symbols of the elements in a definite numerical proportion used to represent molecules, compounds, radicals, ions, etc.

chemical group: a family of elements in the periodic table that have similar electron configurations.

chemical property: a characteristic that a substance undergoes in a chemical reaction that produces new substance(s).

chemical reaction: a process in which new substance(s) are formed from starting substance(s).

chemical test: a physical procedure or chemical reaction used to identify a substance.

colligative property: a property such as freezing-point depression or boiling-point elevation whose physical value depends on only the ratio of the particles of solute and solvent particles and not on their chemical identities.

colloid: a mixture containing particles larger than the solute but small enough to remain suspended in the continuous phase of another component. This is also called a colloidal dispersion.

combustion: the rapid reaction of a material with oxygen accompanied by rapid evolution of flame and heat.

composite: a solid heterogeneous mixture of two or more substances that make use of the properties of each component.

compound: a material that consists of two or more elements united together in definite proportion.

concentration: a measure of the composition of a solution, often given in terms of moles of solute per liter of solution.

conductivity: the property of transmitting heat and electricity.

covalent bond: a bond formed when two atoms combine and share their paired electrons with each other.

decomposition: a chemical reaction in which a single compound reacts to give two or more products.

density: the mass per unit volume of a material.

double-displacement reaction: a chemical reaction in which two ionic compounds "exchange" cations to produce two new compounds.

ductility: a property that describes how easy it is to pull a substance into a new permanent shape, such as, pulling into wires.

elasticity: the property of a material to resist deformation and return to its normal size or shape after a force has been applied to it.

electrolysis: the conduction of electricity through a solution that contains ions or through a molten ionic compound that will induce chemical change.

electromagnetic spectrum: the complete spectrum of electromagnetic radiation, such as radio waves, microwaves, infrared, visible, ultraviolet, x-rays, and gamma rays.

electron: a negatively charged particle with a charge of 1.6×10^{-19} coulombs and a mass of 9.1×10^{-31} kg.

electron configuration: the distribution of electrons in an atom's energy levels.

element: a substance in which all of the atoms have the same atomic number.

emulsion: a colloid or colloidal dispersion of one liquid suspended in another.

endothermic change or reaction: a change in which energy in the form of heat is absorbed from the surrounding environment resulting in an increase in the internal energy of the system.

exothermic change or reaction: a change in which energy in the form of heat is released from a system resulting in a decrease in the internal energy of the system.

fission: the process of breaking apart nuclei into smaller nuclei and with the release of a large amount of energy.

flame test: an experimental technique or process in identifying a metal from its characteristic flame color.

frequency: the number of waves per second or cycles per second or hertz (Hz).

fusion: nuclei of lighter atoms combining to form nuclei with greater mass and release of a large amount of energy.

halogens: Group VIIA (17) on the periodic table consisting of fluorine, chlorine, bromine, iodine, and astatine.

hydrocarbon: a molecular compound containing only hydrogen and carbon.

inorganic compound: a compound not based on molecular compounds of carbon.

ion: an electrically charged atom or group of atoms that has acquired a net charge, either negative or positive.

ionic bond: the attraction between oppositely charged ions.

ionic compound: a compound consisting of positive or negative ions.

ionization energy: the energy required to remove an electron from a gaseous atom at ground state.

isotope: atoms of the same element but different atomic masses due to different number of neutrons.

kinetic energy: a form of energy related to the motion of a particle

$(KE = \frac{1}{2}mv^2)$.

Law of Definite Proportions: the composition of a pure substance is always the same or the elements of the compound always combine in the same proportion by mass.

luster: the reflection of light from the surface of a material described by its quality and intensity.

malleability: the property of a material to be able to be hammered into various shapes without breaking.

metal: classes of materials that exhibit the properties of conductivity, malleability, reactivity, and ductility. Metal elements readily lose electrons to form positive ions.

mole: a collection of objects that contains Avogadro's number (6.022×10^{23}).

molecular compound: two or more atoms bond together by sharing electrons (covalent bond).

neutralization: the process of an acid and base reacting to form water and salt.

neutron: neutral subatomic particle with a mass of 1.675×10^{-24} g located in the nuclei of the atom.

noble gas: a family of elements (Group 18 or VIIIA) of the periodic table.

nonmetal: elements that do not exhibit the properties of conductivity, malleability, reactivity, and ductility. These elements tend to form negative ions. The oxides of the elements are acidic.

normal boiling point: the temperature at which the vapor pressure of the pure liquid equals 1 atm.

normal freezing point: the characteristic temperature, at 1 atm, at which a material changes from a liquid state to its solid state.

normal melting point: the characteristic temperature, at 1 atm, at which a material changes from a solid state to its liquid state.

nucleus: the very dense core of the atom that contains the neutrons and protons.

orbit: the path of the electron in its motion around the nucleus of Bohr's hydrogen atom.

orbital: in the quantum mechanical model of an atom, it is the region surrounding the atomic nucleus in which the electron distribution is given by a wave function.

organic compound: a molecular compound of carbon.

oxidation: the process of a substance losing one or more electrons.

oxidation number: a number assigned to an element in a compound designating the number of electrons the element has lost, gained, or shared in forming that compound.

period: a horizontal row of elements in the periodic table.

pH: a quantity used to represent the acidity of a solution based on the concentration of hydrogen ions ($pH = -\log[H+]$).

phase change: the conversion of a substance from one state to another state at a specific temperature and pressure. Example: solid to liquid, liquid to gas, or solid to gas (sublimation).

physical change: a change that involves changes in the state or form of a substance but does not cause any change in chemical composition.

physical property: a property that can be measured without causing a change in the substance's chemical composition.

Planck's constant: a proportionality constant of the energy of a photon to its frequency, derived by Max Planck in 1900. His equation was: $E = hf$ and Planck's constant is $(h) = 6.626 \times 10^{-34}$ J•s.

polyatomic ion: an ion that consists of two or more atoms that are covalently bonded and have either a positive or negative charge.

polymer: a substance that is a macromolecule consisting of many similar small molecules (monomers) linked together in long chains.

polymerization: a chemical reaction that converts small molecules (monomers) into large molecules (polymers).

potential energy: stored energy of the material as a result of its position in an electric, magnetic, or gravitational field.

precipitate: an insoluble solid formed in a liquid solution as a result of some chemical reactions.

precision: the closeness of agreement of several measurements of the same quantity.

product: the substance(s) produced in a chemical reaction.

proton: a positively charged subatomic particle contained in the nucleus of an atom. The mass of a proton is 1.673×10^{-24}g and it has a charge of +1.

pure material: an element or compound that has a defined composition and properties.

pure substance: a substance that contains only one kind of particle.

radioactive: an atom that has an unstable nuclei and will emit alpha, positron, or beta particles in order to achieve more stable nuclei.

reactants: the starting materials in a chemical reaction.

reactivity: a property that describes how readily a material will react with other materials.

reduction: a process in which the substance under consideration gains electron(s).

saturated solution: the maximum amount of solute that can be dissolved at a given temperature and pressure.

single-displacement reaction: a reaction in which an element displaces or replaces another element in a compound.

solute: the substance that dissolves in a solvent to form a solution.

solution: a homogeneous mixture of two or more substances.

solvent: the substance in which a solute dissolves to form a solution.

strength: the property of how well a material withstands the application of a force.

sublimation: The change of state of a solid material to a gas without going through the liquid state.

supersaturated solution: a solution containing more solute than a saturated solution and therefore not at equilibrium. This solution is not stable and cannot be maintained indefinitely.

surface area: changing the nature of the reactants into smaller particles increases the surface exposed to react. Successful reaction depends on collision and increasing the area of the reactant increases the chance of a successful collision taking place. Lighting a log is more difficult than lighting wood shavings. The shavings have a greater surface area and speed up the reaction.

suspension: heterogeneous mixture that contains fine solid or liquid particles in a fluid that will settle out spontaneously. By shaking the container they will again be dispersed throughout the fluid.

synthesis reaction: a chemical reaction in which two or more substances combine to form a compound.

temperature: the measure of the average kinetic energy of all the particles of the material.

texture: the characteristics of the surface of a material, like how smooth, rough, or coarse it is.

titration: an analytical procedure in which the concentration of an unknown solution is added to a standard solution until a color change of some indicator indicates that equivalent quantities have reacted.

Tyndall Effect: the scattering of a light beam as it passes through a colloid.

uniformity: the property of how consistent a material is throughout.

valence electrons: the outermost electrons of an atom. These electrons are used for chemical bonding of atoms.

vaporization: the change of state from a liquid to a gas.

viscosity: a property related to the resistance of a fluid to flow.

wavelength: the distance measured from crest to crest of one complete wave or cycle.

Index: Unit I

Chapters 1–9 Active Physics

D

E

F

Index: Unit II

Chapters 10–12 Active Chemistry

A

B

C

D

E

Illustrations

Abbreviation Code
Tomas Bunk=TB, Dennis Falcon=DF, Burmar Technical Corporation=BTC, Dartmouth Publishing Inc.=DP, Jennifer von Holstein=JV, Murray Fuchs=MF, Kadi Sarv=KS

Front Matter: viii TB, xii BTC, xiii BTC, xiv BTC, xvi TB

Chapter 1: 4 TB, 5 DP, 9 TB, 12-13 TB, 15 TB, BTC, 18 TB, 26 TB, 27 DP, 28 DP, 31 TB, 33 DP, 45 TB, 46 DP, 47 DP, 50 TB, 51 DP, 56 TB, BTC, 57 BTC, 61 TB, 62 DP, 65 DP, 69 TB, DP, 71 DP

Chapter 2: 80 TB, 81 DP, 82 BTC, 83 BTC, 86 TB, 87 BTC, 94 TB, 97 DP, 99 TB, 103 DP, 104 DP, 105 TB, 106 DP, 107 DP, 108 TB, 111 TB, 112 DP, 117 TB, 122 TB, 123 DP, 128 DP, 129 TB, 130 DP, 134 DP

Chapter 3: 142 DP, 151TB, 152-153 DP, 157 DP, 160-161 DP, 167 TB, 171 TB, 172 DP, 177 TB, 182 BTC, 184 TB, 186 DP, 192-193 DP, 194-195 DP, 202 DP

Chapter 4: 208 TB, 209 MF, 213 BTC, 214 MF, 218 TB, 219 BTC, 220-221 BTC, 223 JV, 226-227 JV, 228-229 JV, 231 TB, 223 BTC, 237 BTC, 238 MF, 239 TB, 242 MF, 243 BTC, 245 BTC, 248 JV, 249 BTC, 253 TB, 254 BTC, 259-260 JV, 263 JV, 265 BTC, 266 TB, 267 BTC, 269 BTC, 270-271 BTC, 278 JV, 280-281 JV, 283 JV, 285 MF, 286 TB, 287 JV, 288 BTC, 291 BTC, 294-295 BTC, 296 MF, 297 TB, 299-301 JV, 303-304 JV, 309 TB, 313 JV, 316-317 JV

Chapter 5: 326 TB, 327 BTC, 327-332 BTC, 334 BTC, 339 BTC, 340 TB, 341-342 BTC, 346 TB, 348-349 BTC, 350 TB, 354 TB, 355-358 BTC, 360-361 BTC, 362 TB, 363-364 BTC, 366 BTC, 368-370 BTC, 371 TB, 376 BTC, 377 TB, 378-381 BTC, 383-384, 386 TB, 387 BTC

Chapter 6: 400 TB, 402-404 BTC, 406-409 BTC, 412 BTC, 416 TB, 417 BTC, 421 TB, 425 TB, 429 TB, 433 TB

Chapter 7: 446 TB, 450 TB, 451 BTC, 452 BTC, 455 TB, 456-457 JV, 463 TB, 472 TB, 473 BTC, 476-480 JV, 483-485 JV, 486 TB, 487 BTC, 498 TB, 503 TB

Chapter 8: 512 TB, 515-516 BTC, 517 TB, 518-519 BTC, 521-522 BTC, 523 TB, 526 TB, 527-530 BTC, 533 TB, 534-536 BTC, 539 TB, 540-541 BTC

Chapter 9: 548 TB, 549-550 BTC, 553-554 BTC, 556 TB, 557-558 BTC, 561 TB, 562-564 BTC, 567-568 TB, 576 TB, 577-578, 581 BTC, 587 TB, 595 TB, 600 TB, 605 TB, 612 TB, 613-614 BTC

Chapter 10: 622 TB, 626 TB, 627-628 BTC, 634 TB, 635-635 BTC, 639 BTC, 641 BTC, 644 TB, 645-646 BTC, 650 TB, 656 TB, 657-659 BTC, 666 TB, 667 BTC, 673 TB, 678 TB, 679 BTC, 683 TB

Chapter 11: 694 TB, 696 TB, 697-698 BTC, 702 TB, 704-705 BTC, 713 TB, 714-716 BTC, 718 BTC, 721 TB, 722-723 BTC, 726 BTC, 728 BTC, 731 TB, 734 KS, 740 TB, 748 TB, 755 TB, 757 BTC,

Chapter 12: 772 TB, 774 BTC, 779 TB, 781-782 BTC, 785 TB, 792 TB, 795 BTC, 797 BTC, 802 TB, 803-805 BTC, 809 TB, 810-811, 816 TB, 817-818 BTC, 822 BTC, 825 BTC, 826 TB

Photography

Abbreviation Code
Corbis=CO, Getty Images=GI, The Nobel Foundation= NF, The National Broadcasting Company=NBC, Digital Stock=DS, John Nordland=JN

Front Matter: iii JN, v GI, vii GI, x GI, xi GI, xii GI, xv GI

Chapter 1: 1 GI, 2 GI, 3 NBC, 8 GI, 10 GI, 11 GI, 13 GI, 14 NBC, 17 GI, 19 GI, 25 NBC, 30 NBC, 34 GI, 37 GI, 38 GI, 44 NBC, 49 NBC, 55 GI(top), NBC(bottom), 59 GI, 60 GI(top), 63 GI, 64 GI, 66 GI, 72 GI, 74 NBC, 75 Dean Bell

Chapter 2: 77-78 Nicholas Pinturas/Tony Stone; 79 GI, 82 GI, 84-85 GI, 90 GI, 93 GI, 101 Aldo Torelli/Tony Stone, 116 Donald Johnston/Tony Stone, 119 GI, 125 GI, 135 JN, 137 Donald Johnston/Tony Stone

Chapter 3: 139-140 David Madison/Tony Stone Images , 145 Ed Mahan Photos, 149 GI, 156 Reuters/Gary Hershorn/Archive Photos, 164 GI, 176 GI, 183 GI, 191 GI, 197 GI, 203 Erv Hunt

Chapter 4: 206 GI, 207 GI, 211 GI, 230 GI, 234 GI, 235 GI, 236 GI, 240 GI, 246 GI, 248 GI, 261 GI, 268 GI, 269 GI, 281 NASA, 305 CO, 308 GI, 314 USAF(left) GI(right), 321 Mekea Hurwitz

Chapter 5: 323-325 GI, 331 GI, 339 GI, 344 GI, 345 Courtesy of Steinway & Sons, 347 GI, 352 GI, 361 GI, 365 GI, 370 GI, 375 GI, 380 GI, 385 GI

Chapter 6: 393-394 GI, 396 GI, Amla Sanghvi(lower right); 397 Micheal Simpson/FPG International LLC, 403 GI, 405 GI, 409 GI, 413 GI, 415 GI, 418 GI, 419 Home Depot, 420 GI, 422 GI, 426 GI, 438-439 GI, 442 GI.

Chapter 7: 443-445 GI, 447 GI, 453 GI, 460 GI, 464 GI, 469 KRUPS, Inc. 474 GI, 488 GI, 495 GI

Chapter 8: 509-511 GI, 515 GI, 518 GI, 535 GI

Chapter 9: 545-547 GI, 552 GI, 558 NF, 569 NF, 582-583 GI, 584 NF 591 NF, 592 GI, 601 GI

Chapter 10: 623-625 GI, 629-631 GI, 640-641 GI, 649 GI, 652-653 GI, 655 GI, 661 GI, 665 GI, 669 GI, 672 GI, 675 JN, 676 GI, 680 GI, 682 GI, 684-688 GI

Chapter 11: 691-693 GI, 693(top) The Nature of Physical Science (www.hcc.hawaii.edu) 706 Liberty on Line (libertyonline.hypermall.com/Aristotle), 707 CHF Chemistry Web Quest, 709 GI, 711 GI, 717 NF, 724 NF, 725 DS, 727 GI, 730 GI, 751 GI, 761 photo by Doug Brown of Robert Hobbs, Physics Instructor, Bellevue Community College, Bellevue, WA, 764 DS

Chapter 12: 770-772 GI, 775 GI, 776 JN, 783 GI, 790 JN, 793 photo by A.R.Smith, Department of Chemistry, Appalachian State University, 805 GI, 808 DS, 813 GI, 805 GI, 823 JN, 824 GI

Periodic Table of the Elements

KEY

1 1.0 1 H 1.00794 $1s^1$ Hydrogen	Atomic Number Electronegativity Oxidation Number Symbol Average Atomic Mass Electron Configuration Name

- Gases at room temperature
- Liquids at room temperature
- Solids at room temperature
- Metals
- Metalloids
- Nonmetals

GROUP

PERIOD	1	2	3	4	5	6	7	8	9
	IA/1A Alkali Metals	IIA/2A Alkaline Earth Metals	IIIB/3B (Transition Metals)	IVB/4B	VB/5B	VIB/6B	VIIB/7B	VIIIB/8B	
1	1 2.1 1 H 1.00794 $1s^1$ Hydrogen								
2	3 1.0 1 Li 6.941 $1s^2 2s^1$ Lithium	4 1.5 2 Be 9.012182 $1s^2 2s^1$ Beryllium							
3	11 0.9 1 Na 22.98977 $[Ne]3s^1$ Sodium	12 1.2 2 Mg 24.3050 $[Ne]3s^2$ Magnesium							
4	19 0.8 1 K 39.0983 $[Ar]4s^1$ Potassium	20 1.0 2 Ca 40.078 $[Ar]4s^2$ Calcium	21 1.3 3 Sc 44.95591 $[Ar]4s^2 3d^1$ Scandium	22 1.5 3,4 Ti 47.867 $[Ar]4s^2 3d^2$ Titanium	23 1.6 2,3,4,5 V 50.9415 $[Ar]4s^2 3d^3$ Vanadium	24 1.6 2,3,6 Cr 51.9961 $[Ar]4s^2 3d^4$ Chromium	25 1.5 2,3,4,6,7 Mn 54.93805 $[Ar]4s^2 3d^5$ Manganese	26 1.8 2,3 Fe 55.847 $[Ar]4s^2 3d^6$ Iron	27 1.9 2,3 Co 58.93320 $[Ar]4s^2 3d^7$ Cobalt
5	37 0.8 1 Rb 85.4678 $[Kr]5s^1$ Rubidium	38 1.0 2 Sr 87.62 $[Kr]5s^2$ Strontium	39 1.2 3 Y 88.90585 $[Kr]5s^2 4d^1$ Yttrium	40 1.4 4 Zr 91.224 $[Kr]5s^2 4d^2$ Zirconium	41 1.6 3,5 Nb 92.90638 $[Kr]5s^2 4d^3$ Niobium	42 1.8 2,3,4,5,6 Mo 95.94 $[Kr]5s^2 4d^4$ Molybdenum	43 1.9 7 Tc 98.9072 $[Kr]5s^2 4d^5$ Technetium	44 2.2 2,3,4,6,8 Ru 101.07 $[Kr]5s^2 4d^6$ Ruthenium	45 2.2 2,3,4 Rh 102.90550 $[Kr]5s^2 4d^7$ Rhodium
6	55 0.7 1 Cs 132.90543 $[Xe]6s^1$ Cesium	56 0.9 2 Ba 137.327 $[Xe]6s^2$ Barium	57 1.1 3 *La 138.9055 $[Xe]6s^2 4f^1$ Lanthanum	72 1.3 4 Hf 178.49 $[Xe]6s^2 4f^{14} 5d^2$ Hafnium	73 1.5 5 Ta 180.9479 $[Xe]6s^2 4f^{14} 5d^3$ Tantalum	74 1.7 2,3,4,6,8 W 183.85 $[Xe]6s^2 4f^{14} 5d^4$ Tungsten	75 1.9 2,4,6,7 Re 186.207 $[Xe]6s^2 4f^{14} 5d^5$ Rhenium	76 2.2 2,3,4,6,8 Os 190.23 $[Xe]6s^2 4f^{14} 5d^6$ Osmium	77 2.2 2,3,4,6 Ir 192.22 $[Xe]6s^2 4f^{14} 5d^7$ Iridium
7	87 0.7 1 Fr 223.0197 $[Rn]7s^1$ Francium	88 0.9 2 Ra 226.0254 $[Rn]7s^2$ Radium	89 1.1 3 **Ac 227.0278 $[Rn]7s^2 5f^1$ Actinium	104 Rf (261) $[Rn]7s^2 5f^{14} 6d^2$ Rutherfordium	105 Db (262) $[Rn]7s^2 5f^{14} 6d^3$ Dubnium	106 Sg (263) $[Rn]7s^2 5f^{14} 6d^4$ Seaborgium	107 Bh (262) $[Rn]7s^2 5f^{14} 6d^5$ Bohrium	108 Hs (265) $[Rn]7s^2 5f^{14} 6d^6$ Hassium	109 Mt (265) $[Rn]7s^2 5f^{14} 6d^7$ Meitnerium

*Lanthanide Series	58 1.1 3,4 Ce 140.115 $[Xe]6s^2 4f^2$ Cerium	59 1.1 3,4 Pr 140.90765 $[Xe]6s^2 4f^3$ Praseodymium	60 1.1 3 Nd 144.24 $[Xe]6s^2 4f^4$ Neodymium	61 1.1 3 Pm 144.9125 $[Xe]6s^2 4f^5$ Promethium	62 1.2 2,3 Sm 150.36 $[Xe]6s^2 4f^6$ Samarium
**Actinide Series	90 1.3 4 Th 232.0381 $[Rn]7s^2 5f^2$ Thorium	91 1.5 4,5 Pa 231.03588 $[Rn]7s^2 5f^3$ Protactinium	92 1.4 3,4,5,6 U 238.0289 $[Rn]7s^2 5f^4$ Uranium	93 1.4 3,4,5,6 Np 237.0482 $[Rn]7s^2 5f^5$ Neptunium	94 1.3 3,4,5,6 Pu 244.0642 $[Rn]7s^2 5f^6$ Plutonium